Set Theoretical
Aspects of
Real Analysis

MONOGRAPHS AND RESEARCH NOTES IN MATHEMATICS

Series Editors

Published Titles

Iterative Optimization in Inverse Problems, Charles L. Byrne

Modeling and Inverse Problems in the Presence of Uncertainty, H. T. Banks, Shuhua Hu, and W. Clayton Thompson

Sinusoids: Theory and Technological Applications, Prem K. Kythe

Blow-up Patterns for Higher-Order: Nonlinear Parabolic, Hyperbolic Dispersion and Schrödinger Equations, Victor A. Galaktionov, Enzo L. Mitidieri, and Stanislav Pohozaev

Set Theoretical Aspects of Real Analysis, Alexander B. Kharazishvili

Forthcoming Titles

Stochastic Cauchy Problems in Infinite Dimensions: Generalized and Regularized Solutions, Irina V. Melnikova and Alexei Filinkov

Signal Processing: A Mathematical Approach, Charles L. Byrne

Monomial Algebra, Second Edition, Rafael Villarreal

Groups, Designs, and Linear Algebra, Donald L. Kreher

Geometric Modeling and Mesh Generation from Scanned Images, Yongjie Zhang

Difference Equations: Theory, Applications and Advanced Topics, Third Edition, Ronald E. Mickens

Method of Moments in Electromagnetics, Second Edition, Walton C. Gibson

The Separable Galois Theory of Commutative Rings, Second Edition, Andy R. Magid

Dictionary of Inequalities, Second Edition, Peter Bullen

Actions and Invariants of Algebraic Groups, Second Edition, Walter Ferrer Santos and Alvaro Rittatore

Practical Guide to Geometric Regulation for Distributed Parameter Systems, Eugenio Aulisa and David S. Gilliam

Analytical Methods for Kolmogorov Equations, Second Edition, Luca Lorenzi

Handbook of the Tutte Polynomial, Joanna Anthony Ellis-Monaghan and Iain Moffat

Application of Fuzzy Logic to Social Choice Theory, John N. Mordeson, Davendar Malik and Terry D. Clark

Microlocal Analysis on R^n and on NonCompact Manifolds, Sandro Coriasco

Cremona Groups and Icosahedron, Ivan Cheltsov and Constantin Shramov

Forthcoming Titles (continued)

Special Integrals of Gradshetyn and Ryzhik: the Proofs – Volume I, Victor H. Moll

Special Integrals of Gradshetyn and Ryzhik: the Proofs – Volume II, Victor H. Moll

Symmetry and Quantum Mechanics, Scott Corry

Lineability and Spaceability in Mathematics, Juan B. Seoane Sepulveda, Richard W. Aron, Luis Bernal-Gonzalez, and Daniel M. Pellegrinao

Line Integral Methods and Their Applications, Luigi Brugnano and Felice Iaverno

MONOGRAPHS AND RESEARCH NOTES IN MATHEMATICS

Set Theoretical Aspects of Real Analysis

Alexander B. Kharazishvili

Tbilisi State University

Georgia

CRC Press
Taylor & Francis Group
Boca Raton London New York

CRC Press is an imprint of the
Taylor & Francis Group, an **informa** business

A CHAPMAN & HALL BOOK

CRC Press
Taylor & Francis Group
6000 Broken Sound Parkway NW, Suite 300
Boca Raton, FL 33487-2742

First issued in paperback 2020

ISBN-13: 978-1-4822-4201-0 (hbk)
ISBN-13: 978-0-367-65907-3 (pbk)

Library of Congress Cataloging-in-Publication Data

Kharazishvili, A. B.
    Set theoretical aspects of real analysis / Alexander B. Kharazishvili, Tbilisi State University, Tbilisi, Georgia.
        pages cm. -- (Monographs and research notes in mathematics)
    "A Chapman & Hall book."
    Includes bibliographical references and index.
    ISBN 978-1-4822-4201-0 (alk. paper)
    1. Mathematical analysis. 2. Set theory. I. Title.

QA300.K43 2014
514'.7--dc23   2014013375

Visit the Taylor & Francis Web site at
http://www.taylorandfrancis.com

and the CRC Press Web site at
http://www.crcpress.com

Table of Contents

Preface ix

1. ZF theory and some point sets on the real line 1

2. Countable versions of AC and real analysis 21

3. Uncountable versions of AC and
   Lebesgue nonmeasurable sets 35

4. The Continuum Hypothesis and
   Lebesgue nonmeasurable sets 53

5. Measurability properties of sets and functions 67

6. Radon measures and nonmeasurable sets 87

7. Real-valued step functions with strange
   measurability properties 107

8. A partition of the real line into
   continuum many thick subsets 123

9. Measurability properties of Vitali sets 137

10. A relationship between the measurability
    and continuity of real-valued functions 151

11. A relationship between
    absolutely nonmeasurable functions and
    Sierpiński–Zygmund type functions 167

12. Sums of absolutely nonmeasurable
    injective functions 181

13. A large group of absolutely
    nonmeasurable additive functions 195

14. Additive properties of certain classes of
    pathological functions 209

15. Absolutely nonmeasurable homomorphisms
    of commutative groups 225

16. Measurable and nonmeasurable sets
    with homogeneous sections 239

17. A combinatorial problem on translation
    invariant extensions of the Lebesgue measure 253

18. Countable almost invariant
    partitions of G-spaces 269

19. Nonmeasurable unions of measure zero
    sections of plane sets 287

20. Measurability properties of well-orderings 299

    Appendix 1: The axioms of set theory 317

    Appendix 2: The Axiom of Choice and
    Generalized Continuum Hypothesis 341

    Appendix 3: Martin's Axiom and
    its consequences in real analysis 355

    Appendix 4: ω_1-dense subsets of the real line 371

    Appendix 5: The beginnings
    of descriptive set theory 381

    Bibliography 411

    Subject Index 429

Preface

Modern mathematical analysis is an enormously broad scientific discipline having numerous, more or less independent branches carrying their own research ideologies. Among those branches one may observe:

Non-smooth and Convex Analysis (including connections with optimization theory and various problems of finding extremum values of functionals);

Infinite-dimensional Analysis (containing the study and research of smooth infinite-dimensional manifolds);

Abstract Harmonic Analysis (including the general theory of topological groups and Pontryagin's duality theory which is a far-going generalization of the classical Fourier transform);

The Theory of Banach Spaces (especially, beautiful geometric topics for such spaces);

Applied Analysis (among diverse topics here, applications to game theory and economical models should be especially mentioned).

Obviously, besides those indicated above, there are a lot of other domains of modern mathematical analysis. At present, all contemporary areas of analysis are intensively developed and the spectrum of their applications becomes substantially wider.

However, the classical and traditional domains of analysis still remain important and attractive for researchers. It is interesting to see how in those domains old approaches and methods meet new ones and how they successfully interact with each other. The phrase "there is always something new to be found in something old" is relevant here. First of all we mean fairly standard themes of real analysis and classical measure theory. Undoubtedly, basic ideas and concepts which appeared and then were extensively elaborated in real analysis and Lebesgue measure theory, after many years have found their ways to modern mathematical disciplines. Being substantially transformed, those ideas and concepts have changed their form into more delicate and sophisticated paradigms. For example, the well-known and rather elementary Borel–Lebesgue lemma on open coverings of a closed bounded subinterval of the real line $\mathbf{R}$ has been transformed into the fundamental notion of quasi-compactness of a topological space, which plays a prominent role in various questions of contemporary mathematics (cf. [19], [49], [153], [156], [241]).

The techniques of abstract set theory and general topology inspired further development of real analysis and classical measure theory. Discovery of a lot of new deep interrelations between these areas of mathematics stimulated their further progress and produced many new interesting (sometimes unexpected and even paradoxical) results. We refer the reader, e.g., to the books [17], [20], [25], [38], [62]-[68], [93], [115], [147], [190], [203], [206], [268], and especially to a comprehensive treatise:

E. Schechter, Handbook of Analysis and its Foundations, Academic Press, New York, 1997.

In this context, we also wish to mention numerous issues of the journal "Real Analysis Exchange", where different aspects of the subject are systematically discussed. Moreover, exciting and influential expository surveys appear in this journal periodically.

The present book is devoted to a circle of questions in real analysis and classical measure theory, which are of a somewhat set-theoretic flavor. We are focused on certain logical and set-theoretical aspects of real analysis, but not only on these aspects or similar problems of foundational nature. In some respects, the topics considered in the book are connected with those which were partially considered in our previous monographs [127], [133], [137]. However, these three monographs are independent from the point of view of the presentation of material, so they can be studied separately. The same can be definitely said on this book – our attempt was to give the material in a maximally readable and liquid form, accessible for graduate and under-graduate students.

For the reader's convenience, we would like to give a short review of the questions which are under further consideration.

In our opinion, it is natural to start with the beginnings of set-theoretic real analysis. For this reason, in Chapter 1 we say a few words about the basic role of Zermelo–Fraenkel set theory (the commonly used abbreviation: **ZF**) for the foundations of real analysis. Since real analysis heavily relies on deep structural properties of various subsets (briefly, point sets) of the real line **R** or of a finite-dimensional Euclidean space, we envisage more or less thoroughly the following two classical statements which are provable by effective methods, i.e., are provable within **ZF** theory:

(i) the Cantor–Bendixson theorem concerning the perfect subset property for all uncountable closed point sets;

(ii) the Lebesgue theorem on the existence of a partition of **R** into ω_1 many subsets, where ω_1 denotes, as usual, the least uncountable ordinal number.

In the same chapter, we also indicate one important result of Feferman and Levy [57] (see also [102]) stating that there is a model of **ZF** theory, in which the real line **R** can be represented as the union of a countable family of countable sets. In such a model many unexpected and strange facts may be detected; several of them are mentioned in exercises to Chapter 1. This circumstance shows, by the way, that elementary mathematical analysis is rather limited in its scope, so needs some extension of **ZF** theory.

Chapter 2 is devoted to those forms (versions, variants) of the Axiom of Choice (briefly, **AC**), which are connected with the concept of countability. Such forms turn out to be necessary in the most fundamental concepts of analysis. For example, there are two classical definitions of the continuity of real-valued functions at a fixed point of **R**. They are due to Cauchy and Heine, respectively. For establishing the equivalence of these two definitions, a certain version of **AC** is needed. The above-mentioned fact was first discovered by Sierpiński in his very important article [232]. This article was the first work in which the author demonstrated the necessity of certain variants of the Axiom of Choice in concrete questions of mathematical analysis. Following Sierpiński's article, we consider in Chapter 2 several relevant examples and statements from real analysis, which are in close connection with the Axiom of Choice. The reader will see from the text of Chapter 2 (cf. also Chapter 1, [87], [93], [102]) that the absolute rejection of this axiom may lead to a total collapse, so is not acceptable for most of researchers working in advanced modern areas of mathematics such as abstract algebra, general topology, functional analysis, etc. Let us stress once more that even elementary mathematical analysis substantially exploits **AC** or, at least, some of weak versions of **AC**.

In Chapter 3 we are focused on uncountable forms of the Axiom of Choice, which yield the existence of subsets of **R** with paradoxical and, in some sense, pathological features. First of all, we mean here Lebesgue nonmeasurable subsets of the real line **R** and subsets of **R** not possessing the Baire property. It should be underlined that the (more or less philosophical) question on the existence of such pathological subsets of **R** was under permanent attention of several great mathematicians of the twentieth century. In this respect, we may recall the names of Lebesgue, Luzin, Sierpiński, Hausdorff, Gödel, Kolmogorov, Novikov, and others. In their works and publications they repeatedly declared that it is no hope to give any precise construction of a Lebesgue nonmeasurable point set. Many years have passed and, as finally turned out, the intuition of the above-mentioned famous mathematicians was absolutely right. Namely, as was demonstrated by Solovay in his remarkable work [253], under the assumption of the existence of a certain large uncountable cardinal number, there are models of **ZF** & **DC** set theory in which all subsets of **R** become measurable in the Lebesgue sense (here **DC** denotes a special restricted form of **AC**; see Chapter 2 for more details). Consequently, it is impossible to establish, even within **ZF** & **DC** theory, the existence of Lebesgue nonmeasurable subsets of **R**. After this result, it became clear that all known classical constructions of Lebesgue nonmeasurable sets in **R** (or in a finite-dimensional Euclidean space) need to exploit uncountable forms of the Axiom of Choice. Thus, Chapter 3 contains a brief review of classical and standard constructions of Lebesgue nonmeasurable point sets and of Lebesgue nonmeasurable real-valued functions. We especially underline those places in the above-mentioned constructions, where uncountable forms of the Axiom of Choice are essentially utilized. Exercises for

Chapter 3 contain some additional material about Lebesgue nonmeasurable sets and functions.

Chapter 4 deals with special kinds of paradoxical point sets, namely, with the so-called Sierpiński subsets of **R**. A set $S \subset \mathbf{R}$ is called a Sierpiński set if S is uncountable but its intersection with each first category subset of **R** is at most countable (see, e.g., [147], [152], [188], [190], [203]). Sierpiński sets are extremely bad from the point of view of measurability in the Lebesgue sense. More precisely, any uncountable subset of a Sierpiński set S is nonmeasurable with respect to the standard Lebesgue measure λ on **R** (at the same time, S is of first category in **R**, so is small in the topological sense). Actually, Sierpiński sets are so paradoxical that their existence cannot be established within **ZF** & **AC** set theory (commonly used abbreviation: **ZFC**). Consequently, if one wants to have such point sets, he or she must enrich **ZFC** theory by adding to it new axioms. The famous Continuum Hypothesis (**CH**) of Cantor completely suffices for this purpose. By using **CH**, we present in Chapter 4 the classical transfinite construction of a Sierpiński set and then indicate some typical properties of Sierpiński sets. However, it should be mentioned that there is a model of **ZFC** theory in which the Continuum Hypothesis fails to be true and in which there exists a Sierpiński set of cardinality continuum (abbreviation: **c**). In the same chapter, assuming again **CH**, we consider one remarkable partition of the Euclidean plane $\mathbf{R}^2$ consisting of two sets, first of which meets every straight line parallel to the axis of abscissae in at most countably many points and the second set meets every straight line parallel to the axis of ordinates in at most countably many points. This partition was introduced by Sierpiński in his short note [233] and possesses very strange properties from the measure-theoretical viewpoint. Moreover, Sierpiński demonstrated in [233] that such a partition exists if and only if **CH** is valid.

Chapter 5 occupies one of the central places in the book. In this chapter we envisage measurability properties of sets and functions from a quite general position. Namely, we fix an abstract set E and a class $\mathcal{M}$ of σ-finite measures on E. Further, for this class $\mathcal{M}$ and for an arbitrary function $f : E \to \mathbf{R}$, we introduce and discuss the following three notions:

(i) the absolute measurability of f with respect to $\mathcal{M}$;

(ii) the relative measurability of f with respect to $\mathcal{M}$;

(iii) the absolute non-measurability of f with respect to $\mathcal{M}$.

If f is at most two-valued, i.e., coincides with the characteristic function (indicator) of a subset X of E, then we naturally come to the corresponding notions of absolute measurability, relative measurability and absolute nonmeasurability of X with respect to the same class $\mathcal{M}$.

In our opinion, these notions are much deeper than the usual concept of the measurability (nonmeasurability) of a function $f : E \to \mathbf{R}$ (or of a set $X \subset E$) with respect to a fixed measure μ on E. In fact, many results of measure theory can be formulated in terms of the above-mentioned notions and quite often it turns out that such an approach is convenient, fruitful and leads to a better understanding of the subject. So, as we have said above, conceptual move is made in Chapter 5 to the studying of absolutely measurable, relatively measurable and absolutely nonmeasurable real-valued functions on E. At first sight, absolutely nonmeasurable functions and absolutely nonmeasurable sets seem to be so pathological and paradoxical that their role in mathematical analysis and measure theory is expected to be minimal or very limited. But we show in Chapter 5 and also in subsequent sections of the book that absolutely nonmeasurable sets and functions naturally occur in numerous questions and topics of real analysis and measure theory. In the same Chapter 5, several typical examples illustrating the introduced notions are indicated and corresponding comments are given. Some of those examples are envisaged more thoroughly in further sections of the book.

The Lebesgue measure on a finite-dimensional Euclidean space is a very special member of a large class of measures which turn out to be extremely important in different topics of functional analysis, probability theory and general topology. We mean the so-called Radon measures which are characterized by the property that every measurable set can be inner approximated by its compact subsets (of course, the approximation is meant here in the measure-theoretical sense). The theory of Radon measures on Hausdorff topological spaces constitutes a substantial part of modern topological measure theory (cf. [17], [20], [26], [76], [89], [178], [213], [272]). It suffices to recall, in this connection, that the theory of Haar measures on locally compact topological groups is a particular case of the theory of Radon measures. Earlier, representatives of famous Bourbaki's school were primarily interested in study of Radon measures given on locally compact topological spaces. However, it became clear later that many useful facts concerning Radon measures remain valid without assuming that a ground space is locally compact. In Chapter 6 we present a few fundamental statements about σ-finite Radon measures on Hausdorff topological spaces: the τ-smoothness of such measures, connections of Radon probability measures with perfect probability spaces, extensions of Radon measures, some canonical operations over Radon measures, etc. But our main attention is again concentrated on the question of the existence of a nonmeasurable subset of a Hausdorff topological space E equipped with a nonzero σ-finite Radon measure μ which vanishes at all singletons of E. In Chapter 6, following Mauldin [184], we show the existence of such bad subsets of E with respect to μ. Therefore, Chapter 6 may be regarded as a natural continuation of Chapter 3 where various approaches leading to Lebesgue nonmeasurable point sets are briefly discussed.

In the family of all real-valued functions defined on a ground set E, we may distinguish those members which have rather simple descriptive structure, for instance, we may consider the so-called step functions on E whose ranges are at most countable subsets of $\mathbf{R}$. In Chapter 7 we deal with those Lebesgue measurable real-valued step functions on $\mathbf{R}$ which possess rather strange descriptive properties. The material presented in that chapter is motivated by the following circumstance. In his remarkable work [232], Sierpiński gave an example of a Lebesgue measurable function $f : \mathbf{R} \to \mathbf{R}$ such that there exists no Borel function $\phi : \mathbf{R} \to \mathbf{R}$ satisfying the relation

$$f(x) \leq \phi(x) \qquad (x \in \mathbf{R}).$$

In other words, Sierpiński's function f is not bounded from above by any Borel real-valued function on $\mathbf{R}$. The construction of f given in [232] is based on the existence of a Lebesgue nonmeasurable subset of $\mathbf{R}$, so essentially exploits an uncountable form of the Axiom of Choice. We present a different construction which enables us to strengthen Sierpiński's result. Moreover, assuming Martin's Axiom (**MA**) and using so-called generalized Luzin subsets of $\mathbf{R}$, we prove that there exists a real-valued step function on $\mathbf{R}$ which is not bounded from above and from below by any real-valued Borel function on $\mathbf{R}$ and which is absolutely measurable with respect to the class of completions of all σ-finite Borel measures on $\mathbf{R}$.

Chapter 8 may be treated as a further development of the topic presented in Chapter 3 where some classical constructions of Lebesgue nonmeasurable point sets on the real line $\mathbf{R}$ are touched upon. As is well known, among such constructions those which belong to Vitali [266], Hamel [90], and Bernstein [14] respectively, play an outstanding role. In Chapter 8 we analyze in more details these three constructions of Lebesgue nonmeasurable sets and also envisage the theorem of Luzin and Sierpiński [168] stating that there exists a partition of $\mathbf{R}$ into continuum many subsets of $\mathbf{R}$, each of which is thick with respect to the Lebesgue measure λ on $\mathbf{R}$. In particular, all members of this partition turn out to be nonmeasurable in the Lebesgue sense. The main result of Chapter 8 shows that such a partition can also be obtained by starting with the existence of a Lebesgue nonmeasurable subset of $\mathbf{R}$ and exploiting countably infinite products of probability measures. In addition to this fact, we indicate the role of **DC** axiom in the process of obtaining partitions of $\mathbf{R}$ into continuum many thick sets with respect to the Lebesgue measure.

Chapter 9 is entirely devoted to Vitali's classical construction of a Lebesgue nonmeasurable set on the real line $\mathbf{R}$. In 1905 Vitali [266] carried out his construction and so he positively answered the question about the existence of such pathological sets. Recall that the question was originally raised by Lebesgue in the same year (see his fundamental manuscript [163]); however, Lebesgue was not satisfied by Vitali's result, because Vitali's argument substantially utilized uncountable forms of the Axiom of Choice. In Chapter 9, our goal is to highlight some aspects of the measurability properties of Vitali's sets from the viewpoint stated in Chapter 5. Namely, we discuss the relative measurability and absolute nonmeasurability of certain Vitali sets. We also strengthen Vitali's classical theorem and demonstrate that the union of an arbitrary nonempty finite family

of Vitali sets is absolutely nonmeasurable with respect to the class of all translation invariant extensions of the Lebesgue measure λ on $\mathbf{R}$. Unlike the case
of a single Vitali set, this more general situation needs an essentially different
method based on the celebrated theorem of Banach stating that there exists an
extension of any positive finitely additive translation invariant functional from
a given ring of bounded subsets of $\mathbf{R}$ to the family of all bounded subsets of $\mathbf{R}$
(see, e.g., [6] and [268]). The behavior of Vitali sets with respect to translation
quasi-invariant extensions of the Lebesgue measure λ is also envisaged in Chapter 9. In exercises for the same chapter we provide the reader with additional
information about various extraordinary measurability properties of Vitali sets.

Undoubtedly, the most fundamental role in mathematical analysis is played
by the following two notions: continuity and measurability. At first sight, the
second notion is of a more general character. However, it frequently turns out
that the measurability of a real-valued function f is closely connected with the
continuity of an appropriate restriction of f to a relatively big subset of the domain of f. There are many important examples of connections of this type, for
instance, Luzin's classical theorem concerning the C-property of Lebesgue measurable real-valued functions (see, e.g., [17], [22], [23], [167], [197], [199], [203],
[224]). Such connections are primarily caused by various regularity properties
of those measure spaces on which measurable real-valued functions are given.
Remind, for example, perfect probability spaces, Radon spaces, and Prokhorov
spaces (see [64], [76], [78], [89], [178], [213], [265], and Chapter 3). The main
goal of Chapters 10 and 11 is to express more or less adequately some explicit
or implicit relationships between the measurability and continuity in terms of
absolutely nonmeasurable functions and so-called universal measure zero sets.
Notice that, from the measure-theoretical point of view, absolutely nonmeasurable functions are very bad while universal measure zero sets are very small.
In the same two chapters, we also deal with totally discontinuous real-valued
functions or, equivalently, with functions of Sierpiński–Zygmund type. Recall
that Sierpiński–Zygmund type functions on $\mathbf{R}$ are extremely bad from the topological stand-point, i.e., their restrictions to all subsets of $\mathbf{R}$ of cardinality $\mathbf{c}$ are
discontinuous. In this context, structural properties of absolutely nonmeasur-
able functions and of Sierpiński–Zygmund type functions are compared to each
other.

We would like to remark that in the process of writing the book, our strong
intention was to maximally present Chapters 1–11 in a form accessible for graduate and under-graduate students. So, we think that these eleven chapters collectively may be regarded as a small lecture course of concrete topics in classical
Lebesgue measure theory and real analysis. The rest of this book, consisting of
subsequent chapters, deals with more special material about set-theoretic real
analysis.

In Chapter 12 we continue our discussion of various types of absolutely
nonmeasurable functions. Here we primarily are interested in representation

theorems of a certain kind. It is an old and well-known fact that any function acting from $\mathbf{R}$ into itself is representable as a sum of two injective functions (see [165] or Sierpiński's excellent and comprehensive monograph [243] on classical set theory). The main purpose of Chapter 12 is to show the validity of several analogues of this fact. Namely, we establish that, under Martin's Axiom, every function acting from $\mathbf{R}$ into itself can be expressed as a sum of two very bad (from the measure-theoretical viewpoint) injective functions. An appropriate algebraic version of this result is valid, too: assuming again Martin's Axiom, it is demonstrated that every endomorphism of the additive group $(\mathbf{R}, +)$ can be written as a sum of two very bad (from the measure-theoretical viewpoint) injective endomorphisms of $(\mathbf{R}, +)$. In the same chapter, we also deal with those functions whose graphs are thick subsets of the plane $\mathbf{R}^2$ with respect to the standard two-dimensional Lebesgue measure λ_2 on $\mathbf{R}^2$, i.e., the graphs meet every λ_2-measurable set of strictly positive measure. It turns out that such functions are not absolutely nonmeasurable with respect to the class of all measures on $\mathbf{R}$ extending the standard one-dimensional Lebesgue measure $\lambda \ (= \lambda_1)$ on $\mathbf{R}$. On the other hand, under Martin's Axiom, the sum of two such functions can be absolutely nonmeasurable with respect to the class of all nonzero σ-finite diffused measures on $\mathbf{R}$ (the diffuseness of a measure μ on $\mathbf{R}$ means that $\mu(\{x\}) = 0$ for each point $x \in \mathbf{R}$).

In Chapters 13 and 14, we are interested in some additive properties of the following three typical families of pathological functions:

(1) continuous nowhere differentiable functions on the closed unit interval $[0, 1]$;

(2) Sierpiński–Zygmund functions, i.e., those functions whose restrictions to all subsets of $\mathbf{R}$ of cardinality continuum are discontinuous;

(3) absolutely nonmeasurable functions, i.e., those functions which are non-measurable with respect to all nonzero σ-finite diffused measures on $\mathbf{R}$.

Obviously, the functions belonging to the first class are very bad from the differential point of view, and we would like to repeat once more that the functions belonging to the second class are very bad from the topological point of view, while the functions belonging to the third class can be regarded as very bad from the measure-theoretical point of view.

Notice that there are many works devoted to additive properties of various families of pathological functions (cf. [9], [13], [37], [73], [74], [75], [85], [129], [133], [140], [198], [218]). Our main goal in Chapters 13 and 14 is to consider some direct analogues between such properties. Of course, our presentation is far from being comprehensive. We only wish to indicate certain parallels and interrelations between additive properties for the above-mentioned three families of pathological functions. In particular, concerning the family (1) (respectively, the family (2)) we point out that there are large vector spaces, all nonzero members of which are continuous nowhere differentiable functions (respectively, are additive Sierpiński–Zygmund functions). In the case (3), an analogous fact is established, namely, there exists a large group (not a vector space over $\mathbf{R}$) of additive functions such that all nonzero members of the group are absolutely nonmeasurable.

Chapter 15 is devoted to absolutely nonmeasurable homomorphisms acting from uncountable commutative groups either into the additive group $(\mathbf{R}, +)$ or into the additive group $(\mathbf{T}, +)$, where $\mathbf{T}$ denotes the one-dimensional unit torus (i.e., the circle group). It is well known that, for any commutative locally compact topological group $(G, +)$, there are many continuous homomorphisms (the so-called characters) of $(G, +)$ into $(\mathbf{T}, +)$. Moreover, such homomorphisms separate points in G. A similar result holds true for real characters, namely, if $(G, +)$ belongs to a certain, sufficiently large class of commutative locally compact groups, then the family of all continuous homomorphisms acting from $(G, +)$ into $(\mathbf{R}, +)$ separates points of G (see, for instance, [95]). At the same time, there are several constructions of everywhere discontinuous homomorphisms from commutative locally compact groups into $(\mathbf{T}, +)$ (into $(\mathbf{R}, +)$). Such homomorphisms can be constructed by the method of transfinite recursion for concrete groups $(G, +)$. Furthermore, if $(G, +)$ is assumed to be endowed with the standard Haar measure ν, then everywhere discontinuous homomorphisms from $(G, +)$ into $(\mathbf{T}, +)$ turn out to be nonmeasurable with respect to the completion of ν. Nevertheless, these bad group homomorphisms are sometimes useful for constructing nonseparable translation invariant extensions of ν and they become measurable with respect to such extensions (cf. [41], [95], [137], [144]). So the natural question arises whether there exist ultimately bad homomorphisms ϕ acting from an uncountable commutative group $(G, +)$ into $(\mathbf{R}, +)$ (or into $(\mathbf{T}, +)$). Here the bad status of ϕ means that ϕ must be absolutely nonmeasurable with respect to the class of all nonzero σ-finite G-quasi-invariant measures on G. In Chapter 15 we discuss this question and describe all those commutative groups $(G, +)$ for which such homomorphisms ϕ do exist.

In Chapter 16 we deal with those sets in the Euclidean plane $\mathbf{R}^2$, which have homogeneous (again from the measure-theoretical viewpoint) horizontal and vertical linear sections. We also deal with those sets in the Euclidean space $\mathbf{R}^3$, which have homogeneous (from the same viewpoint) sections produced by planes parallel to one of the three coordinate planes of $\mathbf{R}^3$. To be more precise, here the homogeneity of linear (respectively, planar) sections means that all of them are of equal one-dimensional (respectively, two-dimensional) Lebesgue measure. In particular, we show that some of the sets with homogeneous sections can be measurable in the sense of the standard two-dimensional Lebesgue measure λ_2 on $\mathbf{R}^2$ (in the sense of the standard three-dimensional Lebesgue measure λ_3 on $\mathbf{R}^3$) and some of them can be nonmeasurable with respect to λ_2 (with respect to λ_3). It needless to say here that the topic we touch upon in Chapter 16 is mainly motivated by the classical theorem of Fubini concerning double and iterated integrals of Lebesgue integrable real-valued functions of two variables. Also, it is well known that Fubini's theorem plays an important role in many topics of mathematical analysis and probability theory (see, for instance, [17], [20], [197], [199], [203]). This theorem has a very long history. Its prototype is Cavalieri's principle which many times, during years, decades and centuries, was applied (mostly at intuitive level) for calculating the areas and volumes of various geometric figures. In this context, it should be recalled that much earlier than Cavalieri, the great ancient Greek mathematician Archimedes successfully utilized the above-mentioned principle. As a result, he was able to obtain a beautiful formula for the volume of a three-dimensional Euclidean ball. Namely, he expressed its volume as a simple function of its radius:

$$v = \frac{4}{3}\pi r^3.$$

Of course, after introducing the rigorous concepts of general measure theory, the spectrum of applications of Cavalieri's principle became substantially wider and the principle itself was fully replaced by Fubini's theorem, which sometimes is applicable even to nonmeasurable sets (a more detailed explanation is given in Chapter 16; see also [70], [135], [209]).

Chapter 17 is completely dedicated to a concrete question concerning translation invariant extensions of the classical Lebesgue measure λ_n on the Euclidean space $\mathbf{R}^n$, where $n \geq 1$. This question is of somewhat combinatorial flavor. Namely, we prove that, for every natural number $k \geq 2$, there exist subsets

$$A_1, \ A_2, \ \ldots, \ A_k$$

of $\mathbf{R}^n$ such that any $k - 1$ of them can be made measurable (simultaneously) with respect to a certain translation invariant measure on $\mathbf{R}^n$ extending λ_n, but there is no nonzero σ-finite translation quasi-invariant measure on $\mathbf{R}^n$, for which all sets $A_1, A_2, \ldots, A_k$ become measurable. Notice that, in general, the above-mentioned extension of λ_n depends on a choice of $k - 1$ members of $\{A_1, A_2, ..., A_k\}$. Here we wish to stress that the technique developed for solving this question is based on an appropriate analogue of the classical Sierpiński partition of $\mathbf{R}^3$ into three subsets, each of which is finite with respect to the corresponding coordinate axis (see, e.g., Sierpiński's widely known monograph [243]). In connection with the main result obtained in Chapter 17, a related open problem of a more complicated combinatorial nature is formulated, which seems to be interesting and, in our opinion, deserves to be investigated.

Chapter 18 is concerned with some kinds of countable partitions of a base (ground) set E, which are almost invariant with respect to a given group G of transformations of E. In particular, for a nonzero σ-finite G-quasi-invariant measure μ which is G-ergodic and has the so-called Steinhaus property, it is shown that every nontrivial countable μ-almost invariant partition of E has a μ-nonmeasurable member. In this connection, it should be mentioned that several interesting countable partitions of the real line $\mathbf{R}$ into pairwise congruent subsets were known many years ago. Historically, the first example of such a partition was presented by Vitali [266] in 1905 (cf. Chapters 3, 9). Recall that, with the aid of an uncountable form of the Axiom of Choice, Vitali constructed a set $V \subset \mathbf{R}$ having the following properties:

(a) $(V + p) \cap (V + q) = \emptyset$ for any two distinct rational numbers p and q;

(b) the union of the sets $V + q$, where q ranges over the field $\mathbf{Q}$ of all rational numbers, coincides with $\mathbf{R}$.

The set V is, in fact, a selector of the quotient group $\mathbf{R}/\mathbf{Q}$ and simultaneously is the first example of a subset of $\mathbf{R}$ which is Lebesgue nonmeasurable and does not possess the Baire property (this V is usually called a Vitali subset of $\mathbf{R}$).

Some time later, Sierpiński in [232] and [238] gave another example of a countable partition of $\mathbf{R}$ into pairwise congruent sets. Namely, he constructed a countable disjoint family $\{X_i : i \in I\}$ of subsets of $\mathbf{R}$ such that:

(*) all sets X_i $(i \in I)$ are translates of each other and together form a covering of $\mathbf{R}$;

(**) all sets X_i $(i \in I)$ are thick with respect to the Lebesgue measure λ on $\mathbf{R}$; in particular, each X_i is nonmeasurable in the Lebesgue sense.

The main purpose of Chapter 18 is to present some related results concerning countable almost invariant partitions of the real line (of a finite-dimensional Euclidean space equipped with an appropriate transformation group). Notice that, in the process of constructing nontrivial countable almost invariant partitions of the space $\mathbf{R}^n$, where $n \geq 1$, an essential role is played by certain invariant extensions of the Lebesgue measure λ_n on $\mathbf{R}^n$.

In Chapter 19 we consider subsets Z of the Euclidean plane $\mathbf{R}^2$ having the property that all horizontal and vertical sections of Z are of λ-measure zero. As was underlined earlier, such Z do not need to be of λ_2-measure zero (recall that Sierpiński provided various examples of λ_2-nonmeasurable subsets of $\mathbf{R}^2$, all horizontal and vertical sections of which either are empty or contain at most one point). Supposing that $\mathrm{pr}_1(Z)$ is not of λ-measure zero, we are interested in the question whether the union of some horizontal sections of Z turns out to be a λ-nonmeasurable subset of $\mathbf{R}$. A positive answer to this question is obtained under Martin's Axiom and assuming certain regularity properties of the original set Z. The argument is substantially based on the classical Luzin–Jankov–von Neumann theorem about the existence of measurable selectors (see, e.g., [115], [137], [191], [264]). Furthermore, it is demonstrated that regularity properties of Z are rather essential and several applications of the above-mentioned result are given to vector sums of measure zero sets and to the continuity of measurable homomorphisms acting from a Polish group into an arbitrary topological group (see [160]).

Chapter 20 is devoted to interrelations between the following two fundamental mathematical structures: measures and well-orderings. Recall that, as was demonstrated by Zermelo [275] in 1904 (with the aid of the Axiom of Choice), any infinite set E admits a well-ordering of all of its elements, and there are many such well-orderings for E. Also, it is clear that if E is uncountable, then there are many nonzero σ-finite measures defined on various σ-algebras of sub sets of E and vanishing at all singletons in E. So the natural question arises whether these well-orderings and measures can behave coherently or, in other words, whether these two structures can be compatible in a reasonable sense As an answer to this question, we note in Chapter 20 that well-orderings are either negligible or nonmeasurable with respect to the completions of σ-finite product measures. In addition, several consequences of this fact are discussed in the light of some classical problems posed by Hilbert, Lebesgue and Luzin in the first decades of the twentieth century. To be more precise, consider an arbitrary base (ground) set E and suppose that some subset of E is well-ordered by a certain binary relation $G \subset E \times E$. Thus, we assume that G is the graph of a well-ordering of the set $\mathrm{pr}_1(G)$ and $\mathrm{pr}_1(G) \subset E$. We also suppose that a nonzero σ-finite measure μ is given on E and, in addition, this μ is continuous (i.e., diffused), which means that the equality $\mu(\{x\}) = 0$ holds true for each element $x \in E$. Actually, in Chapter 20 we analyze some relationships between G and μ, from the point of view of the compatibility of these two important structures, and demonstrate that the compatibility of G and μ is an atypical phenomenon.

For the reader's convenience, the main text of the book is provided with five appendices. In fact, the reader may begin (or prefer) to read appendices independently of the main text and then to study the material of concrete chapters.

Appendix 1 contains a list of the axioms of set theory and some direct logical consequences of these axioms. Actually, we give brief information on the Zermelo–Fraenkel formal system of set theory, underline once more the role of the Axiom of Choice in this system, consider the notion of transitive classes of sets, introduce ordinals in the sense of von Neumann and compare them with the standard definition of ordinal numbers as order types of well-ordered sets. We also touch upon the so-called von Neumann's Universe which may be regarded as a canonical model of **ZFC** theory. For more detailed information about various axiomatic systems of modern set theory, see, e.g., [10], [18], [40] [42], [47], [60], [87], [103], [109], [111], [148], [154], [164], [172], [173], [187], [231] [251].

In Appendix 2 we are focused on close connections between the Axiom of Choice and Generalized Continuum Hypothesis. Namely, we present a proof of Sierpiński's delicate theorem stating that a certain form of the Generalized Continuum Hypothesis implies the Axiom of Choice. Some other profound relationships between these two fundamental statements of classical set theory are pointed out. In addition, it is shown how the Generalized Continuum Hypothesis essentially simplifies the arithmetic of infinite cardinal numbers.

In Appendix 3 we present Martin's Axiom (**MA**) in its standard formulation in terms of partially ordered sets. As is well known, this axiom serves as a useful alternative to the Continuum Hypothesis **CH**. Indeed, a lot of statements which hold true under **CH** have straightforward valid analogues under **MA**. At the same time, **MA** does not bound from above the size of **c** and this circumstance demonstrates the advantage of **MA** over **CH**. Further, in real analysis **MA** allows one to prove the **c**-completeness (**c**-additivity) of the two classical σ-ideals on **R**; of course, we mean here the σ-ideal of all Lebesgue measure zero subsets of **R** and the σ-ideal of all first category subsets of **R**. It directly follows from the **c**-completeness of these two σ-ideals that all those subsets of **R** whose cardinalities are strictly less than **c** are small in the sense of measure and category, i.e., they are of λ-measure zero and of first category in **R**. Also, it is possible to show under **MA** that there exist a generalized Luzin set and a generalized Sierpiński set on **R**. Other important facts in real analysis connected with Martin's Axiom are also mentioned, for instance, the statement that, assuming **MA** with the negation of the Continuum Hypothesis, there exist no Luzin sets and no Sierpiński sets on **R**, and there is no Suslin line. At present, a rich literature is dedicated to Martin's Axiom and to its numerous consequences. We especially refer the reader to [10], [62], [87], [103], [106], [148], [223].

The material of Appendix 4 is focused on certain set-theoretical questions which naturally arise during the study of the structure of the real line **R**. Recall that before Dedekind and Cantor, the notion of a real number was absolutely unclear and misty. These two great mathematicians made first steps in the direction of giving a precise definition of real numbers. Their original definitions were radically different and were based, respectively, on Dedekind cuts and Cauchy's sequences. Both of these approaches were then extensively developed, namely, Dedekind's theory was organically included in the abstract theory of Galois correspondences, while Cantor's theory was extended to metric and more general uniform topological structures introduced by Bourbaki's school. Besides, both approaches directly lead to the axiomatic description of **R** as a unique complete linearly ordered field. The uniqueness property of **R** is closely connected with the universality of $(\mathbf{Q}, \leq)$ in the class of all countable linearly ordered sets (this result is also due to Cantor). The latter fact initiated further research work of Hausdorff and Sierpiński on universal linearly ordered sets for a given cardinality and, much later, the research work of model theorists on homogeneous, universal and saturated models (see [29]). In Appendix 4, starting with Cantor's theorem stating that any countably infinite dense linearly ordered set without the least and greatest elements is isomorphic to $(\mathbf{Q}, \leq)$, we discuss a natural analogue of this theorem for the so-called ω_1-dense sets. Such an analogue was thoroughly analyzed by Baumgartner [11] who formulated the corresponding axiom and showed its consistency with **ZFC** set theory. Furthermore, we demonstrate negative effects of Baumgartner's axiom on the existence of some classical subsets of **R**, e.g., Luzin sets and Sierpiński sets.

Appendix 5 presents the beginnings of classical descriptive set theory. We

touch upon elementary properties of Borel and analytic (Suslin) subsets of uncountable Polish topological spaces and apply those properties to certain questions of measurability of sets and functions. Luzin's separation principle for analytic sets and Luzin–Jankov–von Neumann theorem on measurable uniformization of analytic sets receive special attention. Difficulties of logical nature, which
arise even in connection with the structure of co-analytic sets in Polish spaces,
are also indicated. Difficulties of the same type essentially increase when so-
called projective point sets enter the scene. These sets were first introduced
by Luzin and Sierpiński, in 1925, and then were extensively studied by many
authors. Moreover, deep connections of projective sets with foundations of
mathematics (especially with the theory of infinite games, recursion theory, and
theory of large cardinals) were recognized during further progress and development. Since universal measure zero point sets play a substantial role in most
chapters of this book, Luzin's specific construction of an uncountable universal
measure zero subset of $\mathbf{R}$ is given, which heavily relies on delicate structural
properties of analytic and co-analytic sets in $\mathbf{R}$. As far as we know, his construction was historically first; afterwards, a number of other constructions of
uncountable universal measure zero subsets of $\mathbf{R}$ were carried out. Of course,
we have to restrict our consideration only to several facts from classical descriptive set theory. The standard monographs, textbooks or surveys devoted to this
beautiful and extremely important mathematical area are [108], [115], [152],
[154], [191], [264]; see also [17], [25], [103], [167], and Martin's article in [10].
Besides, quite a lot of works are concerned with numerous nontrivial applications of descriptive set theory in diverse domains of modern mathematics such
as general topology, functional analysis, optimization, stochastic processes, and
so on.

We would like to finish this Preface with a few words about exercises which
are scattered throughout the whole text of the book (even including appendices).
The difficulty of exercises varies very substantially from simple observations to
quite advanced results. Almost all of them are more or less connected with
the material of the corresponding chapters or appendices, and many of them
serve to supply the reader with additional useful information on set-theoretical
aspects of real analysis and measure theory. Relatively difficult exercises are
marked by asterisks and are provided with hints or necessary explanations.

A. B. Kharazishvili

1. ZF theory and some point sets on the real line

When dealing with various problems of real analysis and classical Lebesgue measure theory, working mathematicians often need to use rather delicate set-theoretical arguments, facts or statements.

In this context, it can definitely be said that one of the first profound investigations concerning the role of purely set-theoretical and logical approaches to advanced topics of mathematical analysis was done by Sierpiński in his remarkable work [232] (unfortunately, we do not know whether this work is translated into English, while a Russian version of [232], with some inessential changes in it, does exist [238]).

During the first two decades of the twentieth century, numerous extravagant and paradoxical applications of the Axiom of Choice (the commonly used abbreviation: **AC**) have appeared and it was quite natural that in [232] Sierpiński primarily concentrated his attention on this axiom and, also, on some of its important logical consequences. On the other hand, he vividly demonstrated that even the simplest statements of classical analysis require the help of concrete forms of **AC**. Thus, in certain respects, the work [232] may be regarded as a starting point for further development of set-theoretic real analysis and, moreover, as a prototype of the so-called reverse mathematics (see, e.g., [91], [248] or [249] about the latter term).

Actually, a broad ideological program was outlined by Sierpiński in [232] and, more thoroughly, in his fundamental monograph [243]. The program was partially carried out by him, and the work in this direction was intensively continued during subsequent decades by many other mathematicians.

According to Sierpiński's program, it is most desirable to distinguish between theorems which can be proved without the aid of **AC** and those which are not provable without the help of this axiom. Furthermore, specifying proofs based on **AC**, one can:

(1) ascertain that the proof in question makes use of some particular case of **AC**;

(2) determine a special case of **AC** which is sufficient for the proof of the theorem in question;

(3) determine a particular case of **AC** which is both sufficient and necessary for the proof of the theorem in question.

Notice that various examples have been given by Sierpiński, in which all the above-mentioned cases (1) through (3) were realized.

At present, the Axiom of Choice is necessarily added to the quite concise list of other axioms of set theory, which themselves constitute the so-called **ZF** (Zermelo–Fraenkel) theory.

The much stronger theory **ZF** & **AC** is briefly denoted by **ZFC** and reflects the conjunction: Zermelo & Fraenkel & Choice.

In fact, **ZFC** set theory serves as a basis of all contemporary mathematics.

For more details about **ZF** and **ZFC**, see widely known textbooks and monographs, e.g., [10], [38], [60], [87], [103], [148], [154], [164] (cf. also Appendix 1 of this book).

To begin our presentation of material devoted to certain set-theoretical aspects of real analysis, let us first recall one of the possible formulations of **AC**. Actually, the following version is equivalent (of course, within **ZF** theory) to the classical formulation of **AC** given by Zermelo [275] in 1904.

If $\{X_i : i \in I\}$ is any family of nonempty sets, then there exists a family $\{x_i : i \in I\}$ of elements such that $x_i \in X_i$ for each index $i \in I$.

In other words, **AC** states that if the sets X_i are nonempty for all indices $i \in I$, then the Cartesian product $\prod\{X_i : i \in I\}$ is nonempty, too.

The above-mentioned family $\{x_i : i \in I\}$ is often called a selector of the family $\{X_i : i \in I\}$.

It should be noticed here that much earlier than Zermelo, another great mathematician Peano encountered some version of **AC** during his studies in the theory of ordinary differential equations (see [207]). However, Peano radically rejected **AC** in his above-mentioned work [207]. So he was forced to argue in that manner which completely avoids **AC**. As is known, he finally succeeded in establishing within **ZF** set theory his famous theorem on the existence of a local solution of any first-order ordinary differential equation with a continuous right-hand side.

Certainly, the Axiom of Choice may be regarded as one of the most intriguing statements among all reasonable and admissible set-theoretical axioms or hypotheses. Because of an extreme importance of **AC**, it makes sense to distinguish between several weaker forms of **AC** (see some analysis of such forms in Chapters 2 and 3; for more details, we refer the reader to [58], [60], [87], [93], [102], [189], [222], and [243]).

If a set of indices I is finite (i.e, if card(I) is a natural number), then **AC** says nothing new. Indeed, in this case it can easily be proved by induction on card(I) that a required selector $\{x_i : i \in I\}$ does always exist (see Exercise 2 in the present chapter).

However, some profound problems of logical and philosophical nature arise even in this almost trivial case (cf. [60], [243]). For instance, consider the situation where I is a one-element set, i.e., $I = \{i\}$. In other words, suppose that we are given a nonempty set $X_i = Y$ and, for this Y, we are required to find an element $y \in Y$. At first sight, from the purely logical point of view, there is no difficulty in this task because of the evident equivalence

$$Y \neq \emptyset \Leftrightarrow (\exists y)(y \in Y).$$

But if we are looking for a concrete, in somewhat reasonable sense, element $y \in Y$, then quite interesting questions may be posed and studied. At an intuitive level, one of such questions can be specified in the following manner.

Suppose that a completely concrete set Y is given and we know that Y is not empty. Are we able to indicate a completely concrete element $y \in Y$?

Unfortunately, in most cases an answer to the question is negative, and the main reason for this phenomenon is a famous result of Gödel, namely, his first incompleteness theorem concerning sufficiently rich formal systems (see, e.g., [10], [148], [187], [231]).

To say more precisely, according to the above-mentioned theorem of Gödel, any extension $\mathcal{T}$ of **ZF** set theory (or of the much weaker formal arithmetic) turns out to be incomplete whenever the list of axioms of $\mathcal{T}$ is recursively decidable. The latter means that one has an algorithm which, for every formula of $\mathcal{T}$, decides whether the formula is an axiom of $\mathcal{T}$. So, in this situation, there always exists a statement S of $\mathcal{T}$ (without free variables) such that neither S nor $\neg S$ can be deduced within $\mathcal{T}$. Now, in the same $\mathcal{T}$ we may consider the relation $R(y)$ of one free variable y, defined as follows:

$$(S \Rightarrow y = 0) \ \& \ (\neg S \Rightarrow y = 1).$$

From the definition of $R(y)$ we immediately get that

$$R(y) \Rightarrow y \in \{0, 1\}.$$

Therefore, by virtue of the Separation Scheme (see Appendix 1), we get the completely concrete set
$$E = \{y : R(y)\}.$$

This E has a unique element e but it is impossible to prove within $\mathcal{T}$ theory that $e = 0$ (or, similarly, that $e = 1$).

Of course, the reader may declare that this argument does not lead to any, more or less valuable, purely mathematical fact. Nevertheless, the obtained conclusion points out to some unexpected logical (perhaps, philosophical) possibilities and, consequently, inspires working mathematicians to be rather careful in reasonings of this sort and to keep such possibilities in view, when dealing with various set-theoretical constructions.

The first substantial case of the Axiom of Choice is where a set I of indices is countably infinite (denumerably infinite). We will devote the next chapter to a discussion of this case and of further variants of **AC** closely connected with the countability phenomenon. In particular, we will see that under those variants traditional areas of mathematics such as classical point set theory, Lebesgue measure theory and real analysis become more or less adequate to common mathematical intuition.

However, a few deep and important facts of classical theory of point sets in $\mathbf{R}$ (or in the Euclidean n-dimensional space $\mathbf{R}^n$, where $n \geq 1$) are known, which can be proved within $\mathbf{ZF}$ theory. Therefore, in some lucky cases $\mathbf{AC}$ does not look as an indispensable axiom for studying nontrivial properties of point sets in $\mathbf{R}$ (respectively, in $\mathbf{R}^n$).

Here we wish to review two of such remarkable facts.

Let X be a subset of $\mathbf{R}$.

A point $x \in \mathbf{R}$ is a condensation point for (of) X if each neighborhood of x contains uncountably many points from X. As usual, we denote by X^c the set of all condensation points for X. Our goal is to show effectively (i.e., within $\mathbf{ZF}$ theory) that the set $X \setminus X^c$ is denumerable whenever X is closed in $\mathbf{R}$. Actually this is an effective version of the famous Cantor–Bendixson theorem (cf. [49], [152], [197]).

We argue in the following manner (see [232], [238]). Let the symbol $\mathbf{N}$ denote the set of all natural numbers. Actually, $\mathbf{N}$ may be identified with the least infinite ordinal (cardinal) number ω (see Appendix 1).

First, we fix a sequence $\{\triangle_n : n < \omega\}$ of all open intervals in $\mathbf{R}$ with rational endpoints.

If Y is an arbitrary subset of $\mathbf{R}$, then we denote by the symbol Y' the set of all accumulation points for Y (recall that a point $y \in \mathbf{R}$ is an accumulation point for (of) Y if each neighborhood of y contains infinitely many points from Y).

Now, for every ordinal number α which is strictly less than the first uncountable ordinal ω_1, we are going to define a concrete set P_α.

If $\alpha = 0$, then we put $P_\alpha = X$.

If $\alpha = \beta + 1$, then we put

$$P_\alpha = P_\beta \cap P'_\beta.$$

If α is a limit ordinal, then we put

$$P_\alpha = \cap \{P_\beta : \beta < \alpha\}.$$

Proceeding in this manner, we come to the effectively determined and decreasing (by the inclusion relation) ω_1-sequence of sets $\{P_\alpha : \alpha < \omega_1\}$. Finally, we denote

$$P = \cap \{P_\alpha : \alpha < \omega_1\}.$$

Now, pick an arbitrary point $x \in X \setminus P$. Then x does not belong to some P_β, where $\beta < \omega_1$, and we may suppose that β is the least ordinal number having this property. From the definition of the sets P_α ($\alpha < \omega_1$) it follows that β differs from zero and is not a limit ordinal, so we must have

$$\beta = \alpha + 1, \quad x \in P_\alpha, \quad x \notin P_{\alpha+1}.$$

It can easily be seen that there exists an interval $\triangle_n$ such that

$$P_\alpha \cap \triangle_n = \{x\},$$

and we may assume that n takes the minimum value, so the interval $\triangle_n$ is uniquely determined.

Thus, we can effectively associate to each point $x \in X \setminus P$ the natural index $n = n(x)$ such that the set $P_\alpha \cap \triangle_n$ coincides with the singleton $\{x\}$. Clearly, the following relation holds:

$$(x \in X \setminus P \ \& \ y \in X \setminus P \ \& \ x \neq y) \Rightarrow (n(x) \neq n(y)).$$

This circumstance directly implies that the set $X \setminus P$ is effectively denumerable:

$$X \setminus P = \{x_0, \ x_1, \ \ldots, \ x_i, \ \ldots\},$$

where we have

$$i < j \Leftrightarrow n(x_i) < n(x_j) \qquad (i < \omega, \ j < \omega).$$

Further, as was already stated above, for any x_i there is a uniquely determined countable ordinal $\alpha(x_i) = \alpha_i$ such that

$$x_i \in P_{\alpha_i}, \quad x_i \notin P_{\alpha_i+1}.$$

Let us denote

$$\gamma = \inf\{\beta : \beta \text{ is an ordinal } \& \ (\forall i < \omega)(\alpha_i < \beta)\}$$

and let us check that if $\alpha < \gamma$, then $\alpha = \alpha_i$ for some $i < \omega$. Indeed, take $\alpha < \gamma$ and observe that there exists $j < \omega$ such that $\alpha \leq \alpha_j$. The point x_j belongs to P_{α_j} but does not belong to P_{α_j+1}. Notice now that $P_\alpha \neq P_{\alpha+1}$ (otherwise, we directly come to the equality $P_{\alpha_j} = P_{\alpha_j+1}$ which yields a contradiction). Therefore, there exists a point

$$x \in P_\alpha \setminus P_{\alpha+1} \subset X \setminus P$$

and we have

$$x = x_i, \quad \alpha = \alpha(x_i) = \alpha_i$$

for some natural index i. The argument just presented shows also that γ is a countable ordinal number, i.e., $\gamma < \omega_1$, and $P_\gamma = P$.

One more consequence of the above argument is that P_γ is dense in itself, i.e., P_γ does not contain isolated points. Indeed, supposing to the contrary, we readily get

$$P_\gamma \setminus P_{\gamma+1} \neq \emptyset, \quad P \subset P_{\gamma+1}$$

and so we come to a contradiction with the equality $P = P_\gamma$.

Summarizing all these considerations, we obtain the following classical statement of Cantor and Bendixson.

Theorem 1. *Within* **ZF** *set theory, every set $X \subset \mathbf{R}$ admits a decomposition in the form*

$$X = X_1 \cup X_2 \quad (X_1 \cap X_2 = \emptyset),$$

where X_1 is a denumerable set and X_2 $(= P = P_\gamma)$ is dense in itself. Moreover, in this decomposition X_2 is largest in the sense that no nonempty subset of X_1 is dense in itself.

If X is closed in $\mathbf{R}$, then all sets P_α $(\alpha < \omega_1)$ are closed, too, the set X_2 $(= P = P_\gamma)$ is also closed and, consequently, X_2 is perfect. So all points of X_2 are condensation points of X and $X_2 = X^c$.

Remark 1. Theorem 1 and its proof remain valid for any topological space with a countable base (see Exercise 11 for this chapter).

Another important fact which also belongs to **ZF** set theory and which we are going to recall here is one old result of Lebesgue [163]. Actually, it indicates (within **ZF** theory) a deep interrelation between the two standard and ultimately important objects in mathematics: the cardinality of the continuum $\mathbf{c}$ and the first uncountable cardinal number ω_1.

Theorem 2. *Within* **ZF** *set theory, there exists a partition*

$$\{X_\xi : \omega \leq \xi \leq \omega_1\}$$

of the half-open unit interval $]0, 1]$. Similarly, there exists an analogous partition

$$\{Y_\xi : \omega \leq \xi \leq \omega_1\}$$

of the whole real line $\mathbf{R}$.

Proof. It suffices to show the validity of this statement for $]0, 1]$. Fix a bijective enumeration $\{r_n : 1 \leq n < \omega\}$ of all rational numbers. Now, take any real number $t \in \,]0, 1]$. As is well known, t can be uniquely represented in the following form:

$$t = 2^{-n_1} + 2^{-n_2} + \ldots + 2^{-n_k} + \ldots,$$

where $(n_1, n_2, ..., n_k, ...)$ is a strictly increasing sequence of nonzero natural indices. Consider the corresponding injective sequence

$$R(t) = (r_{n_1}, r_{n_2}, \ldots, r_{n_k}, \ldots)$$

of rational numbers. Only two cases are possible.

(1) The set $R(t)$ is well-ordered with respect to the order induced by the standard order $\leq$ of $\mathbf{R}$.

In this case, we denote by $\xi = \xi(t)$ the ordinal type of $R(t)$ and put $t \in X_\xi$.

(2) The set $R(t)$ is not well-ordered with respect to the order induced by the standard order $\leq$ of $\mathbf{R}$.

In this case, we put $t \in X_{\omega_1}$.

Now, in view of the Cantor classical theorem (see Theorem 1 from Appendix 4), for each ordinal ξ satisfying $\omega \leq \xi < \omega_1$, there exists an injective sequence $(r_{n_1}, r_{n_2}, \ldots, r_{n_k}, \ldots)$ of rational numbers, which forms a set of order type ξ and for which the corresponding sequence of indices

$$(n_1, n_2, \ldots, n_k, \ldots)$$

is strictly increasing. So the real number t defined by

$$t = 2^{-n_1} + 2^{-n_1} + \ldots + 2^{-n_k} + \ldots$$

belongs to the set X_ξ and we thus have $X_\xi \neq \emptyset$.

Also, it can readily be seen that there is an injective sequence

$$(r_{m_1}, r_{m_2}, \ldots, r_{m_k}, \ldots)$$

of rational numbers which is not well-ordered with respect to the standard linear order $\leq$ on $\mathbf{R}$ and for which

$$m_1 < m_2 < \ldots < m_k < \ldots.$$

Therefore, the real number t' determined by the analogous equality

$$t' = 2^{-m_1} + 2^{-m_1} + \ldots + 2^{-m_k} + \ldots$$

belongs to the set X_{ω_1} and we thus have $X_{\omega_1} \neq \emptyset$.

This yields the required partition $\{X_\xi : \omega \leq \xi \leq \omega_1\}$ of $]0, 1]$ and finishes the proof of Theorem 2.

Remark 2. The beautiful and deep result presented above implies several important consequences in **ZF** theory. For example, the reader can easily observe that:

(a) there exists a surjection of $\mathbf{R}$ (of $[0, 1]$) onto ω_1;

(b) the inequality $2^{\omega_1} \leq 2^{\mathbf{c}}$ holds true.

We shall see in the sequel that the stronger inequality $\omega_1 \leq \mathbf{c}$ (or, equivalently, the existence of an injection acting from ω_1 into $\mathbf{R}$) cannot be established within **ZF** and even within the enriched **ZF** & **DC** theory. The precise formulation of **DC** axiom will be given in Chapter 2.

Remark 3. Let us briefly touch upon one more important fact of real analysis, which is provable within **ZF** set theory. For this purpose, recall that a function $f : \mathbf{R} \to \mathbf{R}$ belongs to Baire class 0 (the commonly used abbreviation: $f \in \mathcal{B}_0(\mathbf{R}, \mathbf{R}) = \mathcal{C}(\mathbf{R}, \mathbf{R})$) if f is continuous on the whole **R**. Further, a function

$$f : \mathbf{R} \to \mathbf{R}$$

belongs to Baire class 1 (the abbreviation: $f \in \mathcal{B}_1(\mathbf{R}, \mathbf{R})$) if there exists a sequence of continuous functions

$$f_n : \mathbf{R} \to \mathbf{R} \quad (n < \omega)$$

which is pointwise convergent to f. One can pose the question about the cardinality of the class $\mathcal{B}_1(\mathbf{R}, \mathbf{R})$. Since all continuous real-valued functions on **R** belong to $\mathcal{B}_1(\mathbf{R}, \mathbf{R})$, we have the trivial inequality

$$\mathrm{card}(\mathcal{B}_1(\mathbf{R}, \mathbf{R})) \geq \mathbf{c}$$

and this inequality is effective, i.e., its validity does not need the aid of **AC**. It turns out that the opposite inequality

$$\mathrm{card}(\mathcal{B}_1(\mathbf{R}, \mathbf{R})) \leq \mathbf{c}$$

can also be proved effectively, but the corresponding argument is much more difficult (see, for instance, [151], [152]). Actually, the opposite inequality may be reduced to choosing effectively, for any given function

$$f \in \mathcal{B}_1(\mathbf{R}, \mathbf{R}),$$

a certain sequence

$$\{f_n : n < \omega\} \subset \mathcal{B}_0(\mathbf{R}, \mathbf{R})$$

pointwise convergent to f. This effective choice is indeed possible and is closely connected with a beautiful characterization of f in terms of the continuity points of the restrictions of f to all nonempty closed subsets of **R** (this characterization was obtained by Baire in his classical manuscript [4]; see also [152], [197]).

It should also be noticed, in connection with Remark 3, that the analogous question on the cardinality of the next Baire class $\mathcal{B}_2(\mathbf{R}, \mathbf{R})$ cannot be resolved within **ZF** theory and even within **ZF** & **DC** set theory (see Exercise 4 from Chapter 3; as was mentioned above, **DC** axiom is introduced and envisaged in Chapter 2).

Of course, it is reasonable to recall here that the class $\mathcal{B}_2(\mathbf{R}, \mathbf{R})$ is defined quite similarly to the class $\mathcal{B}_1(\mathbf{R}, \mathbf{R})$. Namely, a function

$$f : \mathbf{R} \to \mathbf{R}$$

belongs to $\mathcal{B}_2(\mathbf{R}, \mathbf{R})$ if and only if there exists a sequence of functions

$$\{f_n : n < \omega\} \subset \mathcal{B}_1(\mathbf{R}, \mathbf{R})$$

which is pointwise convergent to f.

Analogously, by using transfinite recursion, the Baire classes $\mathcal{B}_\alpha(\mathbf{R}, \mathbf{R})$ are defined for all ordinal numbers $\alpha < \omega_1$. Various deep descriptive properties of these classes were extensively studied by many mathematicians (see, e.g., [115], [152], [154], [167], [191] and references therein). In particular, denoting

$$\mathcal{B}(\mathbf{R}, \mathbf{R}) = \cup\{\mathcal{B}_\alpha(\mathbf{R}, \mathbf{R}) : \alpha < \omega_1\},$$

we obtain the class $\mathcal{B}(\mathbf{R}, \mathbf{R})$ of all real-valued Baire functions on $\mathbf{R}$, and it was shown by Lebesgue (assuming a certain weak form of $\mathbf{AC}$) that:

(i) $\mathcal{B}(\mathbf{R}, \mathbf{R})$ coincides with the class of all real-valued Borel functions on $\mathbf{R}$;

(ii) for each ordinal $\alpha < \omega_1$, the set

$$\mathcal{B}_\alpha(\mathbf{R}, \mathbf{R}) \setminus \cup\{\mathcal{B}_\beta(\mathbf{R}, \mathbf{R}) : \beta < \alpha\}$$

is not empty.

A more detailed presentation about Lebesgue's fundamental results (i) and (ii) which turned out to be a starting point for the emergence of descriptive set theory and have inspired further development of this theory, may be found in [103], [108], [115], [152], [167], [191] (see also Appendix 5).

EXERCISES

1. In view of one of the axioms of $\mathbf{ZF}$ theory (see Appendix 1), any family of sets $\{E_i : i \in I\}$ can be treated as a family of subsets of some set, namely, as a family of subsets of $E = \cup\{E_i : i \in I\}$. We thus have

$$\{E_i : i \in I\} \subset \mathcal{P}(E),$$

where $\mathcal{P}(E)$ denotes the power set of E. Consequently, the Axiom of Choice admits the following formulation:

For every set E, there exists a mapping

$$f : \mathcal{P}(E) \setminus \{\emptyset\} \to E$$

such that $f(X) \in X$ whenever X is a nonempty subset of E.

This mapping f is usually called a choice function for $\mathcal{P}(E) \setminus \{\emptyset\}$.

Suppose we are given such a choice function $f : \mathcal{P}(E) \setminus \{\emptyset\} \to E$.

Let $\mathbf{b}$ be a cardinal number satisfying the inequality $\mathbf{b} \leq \mathrm{card}(E)$. Consider the set

$$A = \{x \in E : \mathrm{card}(f^{-1}(x)) \leq \mathbf{b}\}$$

and let $\mathbf{a}$ denote the cardinality of A.

Verify the validity of the relation $2^{\mathbf{a}} \leq 1 + \mathbf{ab}$.

For this purpose, observe that if Y is an arbitrary nonempty subset of A then $f(Y) \in Y \subset A$.

Further, introduce the set B of all those elements $x \in E$, which have the following property: for an arbitrary nonempty set $X \in f^{-1}(x)$, the inequality $\mathrm{card}(X) \leq \mathbf{b}$ holds true.

Verify the validity of the relation $\mathrm{card}(B) \leq \mathbf{b}$.

2. Prove the finite version of **AC** within **ZF** set theory. In other words, for any finite sequence $\{X_1, X_2, ..., X_n\}$ of nonempty sets, show effectively the existence of a sequence $\{x_1, x_2, ..., x_n\}$ such that

$$x_1 \in X_1, \quad x_2 \in X_2, \quad \ldots, \quad x_n \in X_n.$$

For this purpose, use induction on n.

Also, prove within the same **ZF** set theory that the union of an arbitrary finite family $\{Y_1, Y_2, ..., Y_n\}$ of countable sets is countable, too.

For this purpose, use again induction on n.

3*. Work in **ZF** set theory and demonstrate that, for every uncountable subset X of the unit segment $[0, 1]$, there exists at least one condensation point of X (clearly, belonging to $[0, 1]$). In other words, if $X \subset [0, 1]$ is uncountable, then $X^c \neq \emptyset$.

For this purpose, utilize the standard method of constructing an appropriate decreasing (by inclusion) sequence of closed subintervals of $[0, 1]$ whose lengths tend to zero.

Show also, within the same theory, that the following three assertions are equivalent:

(a) the union of any countable family of countable subsets of $\mathbf{R}$ is a countable set;

(b) for any uncountable set $X \subset [0, 1]$, there exist at least two condensation points of X;

(c) for any uncountable set $X \subset [0, 1]$, the set $X^c \cap X$ is uncountable, too.

Argue as follows. First, observe that the implication (a) $\Rightarrow$ (b) is almost trivial.

To demonstrate the validity of the converse implication (b) $\Rightarrow$ (a), assume (b) and suppose to the contrary that there exists a disjoint family $\{X_n : n < \omega\}$ of countable subsets of $\mathbf{R}$ such that the set

$$X = \cup\{X_n : n < \omega\}$$

is uncountable. Then construct another disjoint family $\{Y_n : n < \omega\}$ satisfying these two conditions:

(i) for each $n < \omega$, the set Y_n is effectively equinumerous with the set X_n;

(ii) Y_n is contained in the open interval $]1/(n+2), 1/(n+1)[$ for any $n < w$. Deduce from (i) and (ii) that the set

$$Y = \cup\{Y_n : n < w\}$$

is effectively equinumerous with X, so Y is uncountable as well, but only 0 can be a condensation point for Y. The obtained contradiction establishes the implication (b) $\Rightarrow$ (a).

Finally, observe that the implications (a) $\Rightarrow$ (c) and (c) $\Rightarrow$ (b) are trivially valid within **ZF** theory.

4. Verify that the following two assertions are equivalent within **ZF** set theory:

(a) there exists a countable covering of the real line $\mathbf{R}$, consisting of countable subsets of $\mathbf{R}$;

(b) there exists a function $g : \mathbf{R} \to \mathbf{R}$ having the property that, for each uncountable set $X \subset \mathbf{R}$, the restriction $g|X$ is unbounded.

Remark 4. As is known, there exists a model of **ZF** set theory, in which $\mathbf{R}$ can be expressed as the union of countably many countable sets (see [57] or [102]). This important fact indicates once again that, in order to have an adequate imagination of the continuum, there is no getting rid of the Axiom of Choice.

5*. Add to **ZF** set theory the axiom stating that the real line $\mathbf{R}$ can be represented as the union of countably many countable sets.

Show in this enriched theory that the cardinal w_1 is not regular, i.e., there exists a countable family $\{\eta_n : n < w\}$ of ordinals, all of which are strictly less than w_1 and for which we have

$$\sup\{\eta_n : n < w\} = w_1.$$

Argue as follows. Fix two partitions of $\mathbf{R}$:

$$\{X_n : n < w\}, \quad \{Y_\xi : \xi < w_1\},$$

where all X_n are at most countable (notice that the existence of the second partition $\{Y_\xi : \xi < w_1\}$ is guaranteed by Theorem 2 of this chapter). Further, take any $n < w$ and introduce the set

$$\Xi_n = \{\xi < w_1 : X_n \cap Y_\xi \neq \emptyset\}.$$

Verify the validity of the equality

$$w_1 = \cup\{\Xi_n : n < w\}.$$

Then consider the following two possible cases.

(a) At least one set Ξ_n is unbounded from above in ω_1.

In this case, let Ξ_n be such a set and let $\{x_0, x_1, ..., x_k, ...\}$ be an enumeration of all points of the set X_n. For each element x_k, there exists a unique ordinal η_k such that $x_k \in Y_{\eta_k}$. Verify that

$$\sup\{\eta_k : k < \omega\} = \omega_1.$$

(b) All sets Ξ_n are bounded from above in ω_1.

In this case, for each natural number n, denote $\eta_n = \sup(\Xi_n)$ and verify that

$$\sup\{\eta_n : n < \omega\} = \omega_1.$$

Remark 5. The previous exercise shows that the regularity of ω_1 is not provable within the framework of **ZF** set theory, i.e., the statement that ω_1 is representable as the union of countably many countable sets does not contradict **ZF**. On the other hand, the next exercise shows that, for the second uncountable cardinal ω_2, the situation radically differs from the case of ω_1.

6. Demonstrate, within **ZF** set theory, that ω_2 cannot be represented as the union of a countable family of countable sets.

For this purpose, utilize the existence of an effective bijection between the set ω_1 and the product set $\omega_1 \times \omega$.

Generalize this result and prove, within the same **ZF** set theory, that for any ordinal number α the cardinal $\omega_{\alpha+2}$ cannot be represented as the union of an ω_α-sequence of sets, each of which is equinumerous with ω_α.

7*. As in Exercise 5, add to **ZF** theory the axiom stating that **R** can be represented as the union of countably many countable sets.

Show in this enriched theory that every infinite subset of **R** is infinite in the Dedekind sense (for the definition of infinite sets in Dedekind's sense, see Appendix 1).

Argue as follows. Fix a partition $\{X_n : n < \omega\}$ of **R**, where all X_n are at most countable. Let X be an infinite subset of **R** (this means that, for any natural number k, the cardinality of X is greater than or equal to k). Consider two alternatives.

(a) For each $n < \omega$, the set $X \cap X_n$ is finite.

In this case, construct effectively a bijection of the set X onto **N**, which trivially implies that X is infinite in the Dedekind sense.

(b) There exists $n < \omega$ such that the set $X \cap X_n$ is infinite.

In this case, utilize the fact that X_n is countable, so $X \cap X_n$ is countably infinite and there is a bijection of $X \cap X_n$ onto its proper subset. Obviously, this bijection admits an extension to a bijection of the set X onto a proper subset of X.

8*. Again, add to **ZF** theory the axiom stating that **R** admits a representation in the form of the union of countably many countable sets.

Show in this enriched theory that ω_1 and $\mathbf{c}$ are incomparable as cardinal numbers.

For this purpose, suppose otherwise, i.e., $\mathbf{c} < \omega_1$ or $\omega_1 \leq \mathbf{c}$, and consider the following two cases.

(a) $\mathbf{c} < \omega_1$.

This case directly leads to a contradiction by virtue of the definition of ω_1. Indeed, any proper initial subinterval of ω_1 is at most countable, while the cardinal $\mathbf{c}$ is uncountable in view of the classical Cantor theorem (see, e.g., Appendix 1).

(b) $\omega_1 \leq \mathbf{c}$.

In this case, there exists a subset of $\mathbf{R}$ whose cardinality is equal to ω_1. According to our assumption, $\mathbf{R}$ is representable as the union of countably many countable sets. Since $\mathbf{R}$ and $\mathbf{R}^\omega$ are effectively equinumerous, the same is true for $\mathbf{R}^\omega$; in other words, the equality

$$\mathbf{R}^\omega = \cup\{Y_n : n < \omega\}$$

holds true, where all sets Y_n $(n < \omega)$ are at most countable. Let X be a subset of $\mathbf{R}^\omega$ whose cardinality is ω_1. This X can be equipped with a well-ordering $\preceq$ isomorphic to the canonical well-ordering of ω_1. Further, for each $n < \omega$, the set $\mathrm{pr}_n(Y_n)$ is at most countable (without the aid of $\mathbf{AC}$). So one may put

$$y_n = \min_{\preceq}(X \setminus \mathrm{pr}_n(Y_n)) \quad (n < \omega).$$

It remains to check that the obtained sequence

$$\{y_n : n < \omega\} \in \mathbf{R}^\omega$$

does not belong to the set $\cup\{Y_n : n < \omega\}$, which again leads to a contradiction.

Taking into account the above result and Remark 2, construct a partition $\mathcal{P}$ of $\mathbf{R}$ whose cardinality is strictly greater than $\mathbf{c}$.

For this purpose, take any set Z such that

$$Z \cap \mathbf{R} = \emptyset, \quad \mathrm{card}(Z) = \omega_1.$$

Then define a surjection

$$f : \mathbf{R} \to Z \cup \mathbf{R}$$

so that

$$f|[0,1] = Z, \quad f|(\mathbf{R} \setminus [0,1]) = \mathbf{R}.$$

Finally, check that the partition

$$\mathcal{P} = \{f^{-1}(t) : t \in Z \cup \mathbf{R}\}$$

of $\mathbf{R}$ is as required.

9. Prove that there exists an effective choice function for the family of all nonempty open subsets of the Euclidean space $E = \mathbf{R}^n$. Moreover, show that there exists an effective choice function for the family of all those nonempty sets which are effectively of type G_δ in E, i.e., which can be effectively represented as the intersection of a sequence of open sets in E.

Starting with this fact, demonstrate that there exists an effective choice function for the family of all nonempty closed subsets of the Euclidean space $E = \mathbf{R}^n$. Give a direct argument for this case by using induction on n.

Extend the above results to the case of an arbitrary complete separable metric space and, more generally, to the case of a complete metric space which contains a well-orderable everywhere dense subset.

For the family of all nonempty closed bounded convex subsets of $\mathbf{R}^n$, obtain the existence of a continuous choice function (here the above-mentioned family is assumed to be endowed with the standard Hausdorff metric; see, e.g., [49], [152], [153] about this metric).

10. Two persons are playing the following infinite game on $\mathbf{R}$. The first player takes a countable partition $\{X_n : n < \omega\}$ of $\mathbf{R}$ such that $\operatorname{card}(X_n) = \mathbf{c}$ for all $n < \omega$. The second player picks some member X_{n_0} of this partition. Then the first player takes a countable partition $\{X_{n_0,n} : n < \omega\}$ of X_{n_0} such that $\operatorname{card}(X_{n_0,n}) = \mathbf{c}$ for all $n < \omega$. The second player picks some member X_{n_0,n_1} of this partition, and so on. Proceeding in this manner, they obtain a decreasing sequence of sets

$$X_{n_0} \supset X_{n_0,n_1} \supset \cdots \supset X_{n_0,n_1,\ldots,n_k} \supset \cdots .$$

If the set $\cap\{X_{n_0,n_1,\ldots,n_k} : k < \omega\}$ is not empty, then the first player wins, otherwise the second player wins.

Demonstrate, within $\mathbf{ZF}$ theory, that the first player has a winning strategy and describe it.

For this purpose, consider an appropriate effective bijection of $\mathbf{R}$ onto the canonical Baire space $\mathbf{N}^{\mathbf{N}}$.

Remark 6. It is useful to compare the result of the previous exercise with the result of Exercise 11 from Chapter 2.

11. Check that the direct analogue of the Cantor–Bendixson theorem holds true, within $\mathbf{ZF}$ theory, for an arbitrary set X in a topological space E with a countable base.

Also, for this more general case, give another proof by considering in E the family of all those subsets of X which are dense in themselves.

12. Show within $\mathbf{ZF}$ set theory that:
(a) any well-ordered (by inclusion) family of open (respectively, closed) subsets of $\mathbf{R}$ is at most countable;

(b) any well-ordered (by the standard linear ordering) family of real numbers is at most countable.

Moreover, prove the following strengthened version of (b).

Let X be a subset of $\mathbf{R}$ endowed with the order induced by the standard linear order $\leq$ on $\mathbf{R}$. Suppose that, for any $x \in X$, the implication

$$\{y \in X : x < y\} \neq \emptyset \;\Rightarrow\; (\exists z \in X)(z = \min\{y \in X : x < y\})$$

holds true. Then the set X is at most countable.

13*. Let $f : \mathbf{R} \to \mathbf{R}$ be a function such that there are at most countably many discontinuity points of it.

Demonstrate that:

(a) the set $D(f)$ of all discontinuity points of f is effectively countable;

(b) $f \in \mathcal{B}_1(\mathbf{R}, \mathbf{R})$.

For this purpose, utilize the fact that the set $D(f)$ is of type F_σ in $\mathbf{R}$, i.e., $D(f)$ is representable as the union of a countable family of closed subsets of $\mathbf{R}$, and keep in mind the Cantor–Bendixson theorem (more precisely, its effective version).

Check that the Dirichlet function, i.e., the characteristic function $\chi_{\mathbf{Q}}$ of the set $\mathbf{Q}$ of all rational numbers, belongs to the class $\mathcal{B}_2(\mathbf{R}, \mathbf{R})$ but does not belong to the class $\mathcal{B}_1(\mathbf{R}, \mathbf{R})$.

14. Let E be a topological space such that:

(a) E satisfies the first countability axiom, i.e., any point of E possesses a countable fundamental system of its neighborhoods (or, by another terminology, any point of E possesses a countable local base);

(b) there exists an everywhere dense subset X of E which is well-ordered by some relation $\preceq$.

Demonstrate that the following two assertions are equivalent within **ZF** set theory;

(i) a function $f : E \to \mathbf{R}$ is continuous in the Cauchy sense, i.e., for each point $x \in E$ and for any neighborhood V of $f(x)$, there exists a neighborhood U of x such that $f(U) \subset V$;

(ii) a function $f : E \to \mathbf{R}$ is continuous in the Heine sense, i.e., for each point $x \in E$ and for any sequence $\{x_n : n < \omega\} \subset E$ converging to x, the sequence $\{f(x_n) : n < \omega\}$ converges to $f(x)$.

15*. Let λ denote the standard Lebesgue measure on the real line $\mathbf{R}$.

Show, within **ZF** theory, that there exists a homeomorphism

$$f : [0, 1] \to [0, 1]$$

such that, for some λ-measurable set $X \subset \mathbf{R}$ with $\lambda(X) > 0$, the set $f(X)$ is of λ-measure zero.

For this purpose, consider the classical Cantor set $C \subset [0,1]$ and another set $C' \subset [0,1]$ which is homeomorphic to C and has strictly positive λ-measure. Let $\phi : C' \to C$ be a homeomorphism. Check that ϕ can be extended to the desired homeomorphism f.

Another way to obtain the required result is to construct a surjective strictly increasing continuous function $g : [0,1] \to [0,1]$ whose derivative is equal to zero λ-almost everywhere on $[0,1]$ (cf. [23], [133], [197], [200] and Exercise 13 from Chapter 2).

16. Suppose that there exists (in **ZF** theory) a well-ordering of a base (ground) set E.

Verify within the same theory that there exists a linear ordering of the power set $\mathcal{P}(E)$.

For this purpose, equip E with some well-ordering relation $\preceq$, consider the family of two-valued functions

$$\mathcal{F} = \{0,1\}^E,$$

endowed with the lexicographical order with respect to $\preceq$, and check that $\mathcal{F}$ is linearly ordered.

Working again in **ZF** theory, infer from the fact stated above that if there exists a well-ordering of the real line **R**, then there exists a linear ordering of the power set $\mathcal{P}(\mathbf{R})$.

Remark 7. It is useful to compare the presented result with Exercise 4 from Chapter 3.

17*. Show, within **ZF** set theory, that there exists a continuous surjection

$$f : [0,1] \to [0,1]^2.$$

Any such f is called a Peano type curve, because Peano was the first mathematician who gave a concrete recursive construction of a continuous surjection of $[0,1]$ onto $[0,1]^2$.

One way to obtain the required f is to define by an explicit formula a continuous function acting from the Cantor set $C \subset [0,1]$ onto $[0,1]$, and then to utilize the fact that C and $C \times C$ are homeomorphic to each other.

Show that the required f can additionally satisfy the equalities $f(0) = (0,0)$ and $f(1) = (1,0)$.

Deduce from the stated above that, for any natural number n, there exists a continuous surjection

$$g : \mathbf{R} \to \mathbf{R}^n$$

such that the following condition is fulfilled: for every bounded subset Y of $\mathbf{R}^n$, the set $g^{-1}(Y)$ is also bounded.

18. Verify that it is impossible to prove, within **ZF** set theory, the existence of a subset of **R** which is not Borel.

For this purpose, keep in mind a model of **ZF** where the real line **R** is representable in the form of a union of countably many countable sets.

Remark 8. Consider the set X_{ω_1} described in the proof of Theorem 2 of this chapter. Obviously, this set is constructed effectively, i.e., within **ZF** theory. By virtue of Exercise 18, it is impossible to prove in the same theory that X_{ω_1} is not Borel. On the other hand, by using some weak form of **AC**, it can be demonstrated that X_{ω_1} is a non-Borel subset of **R** (for more details, see [115], [152], [167]).

19. A topological space X is called Lindelöf if every open covering of X contains some (at most) countable subcovering.

A topological space Y is called hereditarily Lindelöf if all subspaces of Y are Lindelöf.

According to the terminology of Bourbaki [19], a topological space Z is said to be quasi-compact if every open covering of Z contains some finite subcovering.

A topological space T is said to be compact if T is quasi-compact and Hausdorff simultaneously.

Verify the validity of the following assertions:

(a) a closed subspace of a Lindelöf space is also Lindelöf;

(b) a continuous image of a Lindelöf space is also Lindelöf;

(c) the topological sum of a countable family of Lindelöf spaces is a Lindelöf space;

(d) the union of a countable family of Lindelöf subspaces of a topological space E is also a Lindelöf subspace of E (in particular, any countable topological space is Lindelöf);

(e) a topological space E is hereditarily Lindelöf if and only if each open subspace of E is Lindelöf;

(f) every quasi-compact topological space is Lindelöf.

20. Give an example of a compact topological space containing an open subspace which is not Lindelöf.

For this purpose, take the closed interval $[0, \omega_1]$ of ordinal numbers, equipped with its order topology, and consider its proper initial subinterval $[0, \omega_1[$.

Another example of this sort can be obtained in the following manner. Take any uncountable set X endowed with the discrete topology and denote by X^* Alexandrov's one-point compactification of X. Check that X^* is as required.

21. Recall that a linearly ordered set $(E, \preceq)$ is conditionally complete (or complete in the Dedekind sense, or Dedekind complete) if, for every nonempty bounded from above subset Z of E, there exists $\sup(Z)$ in E. For example, $(\mathbf{R}, \leq)$ is a Dedekind complete linearly ordered set.

Also, recall that a linearly ordered set $(E, \preceq)$ satisfies the Suslin condition (or the countable chain condition, briefly, ccc) if every disjoint family of nonempty open intervals in E is at most countable. For example, the same $(\mathbf{R}, \leq)$ contains a countable everywhere dense subset, so satisfies ccc.

Let $(E, \preceq)$ be a conditionally complete linearly ordered set satisfying ccc, and let this E be endowed with its order topology.

Demonstrate that E is a hereditarily Lindelöf regular topological space.

For this purpose, take into account the following two easily verified facts:

(a) any bounded closed subinterval of E is compact, so E is a locally compact space;

(b) E is representable in the form of a countable union of closed bounded subintervals of E.

22*. Let E be an arbitrary regular Lindelöf topological space.

Demonstrate that E is a normal space.

Argue as follows. Take any two disjoint closed subsets A and B in E. For A, there exists a sequence $U_0, U_1, \dots , U_n, \dots$ of open sets in E such that

$$A \subset \cup\{U_n : n < \omega\}, \quad B \cap \mathrm{cl}(U_n) = \emptyset \quad (n < \omega).$$

Similarly, for B, there exists a sequence $V_0, V_1, \dots , V_n, \dots$ of open sets in E such that

$$B \subset \cup\{V_n : n < \omega\}, \quad A \cap \mathrm{cl}(V_n) = \emptyset \quad (n < \omega).$$

Further, for any natural number n, put

$$G_n = U_n \setminus \cup\{\mathrm{cl}(V_m) : m \leq n\}, \quad H_n = V_n \setminus \cup\{\mathrm{cl}(U_m) : m \leq n\}$$

and define two sets

$$G = \cup\{G_n : n < \omega\}, \quad H = \cup\{H_n : n < \omega\}.$$

Finally, check that:

(a) G and H are disjoint open sets in E;

(b) $A \subset G$ and $B \subset H$.

Deduce from the statement above that any regular hereditarily Lindelöf space E is perfectly normal, i.e., E is normal and every closed set in E is of type G_δ.

23*. Let E be an arbitrary topological space.

Show that these two assertions are equivalent:

(a) E is hereditarily Lindelöf;

(b) for any uncountable subset X of E, there exists a condensation point of X belonging to X (i.e., $X \cap X^c \neq \emptyset$).

The implication (a) $\Rightarrow$ (b) is quite easy, so is left aside.

To establish the converse implication (b) $\Rightarrow$ (a), argue as follows. Suppose that (b) is satisfied but E is not hereditarily Lindelöf. This means that there exist an uncountable ordinal number ξ and a family $\{U_\zeta : \zeta < \xi\}$ of open sets in E such that the open set

$$U = \cup\{U_\zeta : \zeta < \xi\}$$

cannot be covered by any countable subfamily of $\{U_\zeta : \zeta < \xi\}$. One may assume, without loss of generality, that $\mathrm{card}(\xi)$ takes the minimum value, $\xi = \omega_\alpha$ for some ordinal number $\alpha > 0$, and that

$$U_\zeta \setminus \cup\{U_\eta : \eta < \zeta\} \neq \emptyset \quad (\zeta < \xi).$$

For each ordinal $\zeta < \xi$, pick a point x_ζ from the set $U_\zeta \setminus \cup\{U_\eta : \eta < \zeta\}$ and denote

$$X = \{x_\zeta : \zeta < \xi\}.$$

Clearly, the set X is uncountable. So, by virtue of (b), X contains in itself at least one condensation point. Let η be the smallest ordinal for which x_η is a condensation point of X. Check that η is uncountable as well and, consequently, the set

$$Y = \{x_\zeta : \zeta < \eta\} \subset X$$

must contain a condensation point of Y, which yields a contradiction with the definition of η. The obtained contradiction proves (b) $\Rightarrow$ (a).

24. Let E be a hereditarily Lindelöf topological space, all singletons in which are closed, and let A be any subset of E.

Verify that the set A^c is always perfect and the difference $A \setminus A^c$ is at most countable.

25. Show that the following two assertions on a Lindelöf topological space E are equivalent:
(a) E is quasi-compact;
(b) every infinite subset of E has an accumulation point in E.

26. Recall that a topological space E is Baire if no nonempty open subset of E is of first category in E.

Let X be a Baire space without isolated points and such that all singletons in X are closed.

Check that all points in X are its condensation points.

27. Following Sierpiński, consider the family $\mathcal{T}$ of all those subsets of $\mathbf{R}$ which admit a representation in the form $U \setminus D$, where a set U is open in $\mathbf{R}$ and D is a finite or countably infinite set in $\mathbf{R}$.

Verify that:

(a) $\mathcal{T}$ is a topology on $\mathbf{R}$ strictly extending the standard Euclidean topology of $\mathbf{R}$ (in particular, $(\mathbf{R}, \mathcal{T})$ is a Hausdorff space);

(b) $(\mathbf{R}, \mathcal{T})$ is not separable (moreover, every countable set in this space is closed and nowhere dense);

(c) $(\mathbf{R}, \mathcal{T})$ is hereditarily Lindelöf;

(d) $(\mathbf{R}, \mathcal{T})$ is not regular.

Remark 9. We shall return to the space $(\mathbf{R}, \mathcal{T})$ in our further considerations, especially, in connection with property (d) of this space. It is easy to give examples of hereditarily Lindelöf topological spaces with arbitrarily large cardinalities, in which all singletons are closed. Indeed, motivated by Exercise 27, for any uncountable set E consider the family $\mathcal{T}$ of subsets of E defined as follows:

$$\mathcal{T} = \{E \setminus D : D \subset E, \ \mathrm{card}(D) \leq \omega\}.$$

Then $(E, \mathcal{T})$ is a hereditarily Lindelöf space, all singletons in which are closed. On the other hand, if a hereditarily Lindelöf topological space X is simultaneously Hausdorff, then we necessarily have the inequality $\mathrm{card}(X) \leq \mathbf{c}$. This inequality can readily be deduced from one important combinatorial theorem of Erdös and Rado [54] (see also Kunen's article in [10] or [103], [106], [223], and Appendix 1).

28. Recall that a topological space E is Polish if E is homeomorphic to some complete separable metric space.

Work in $\mathbf{ZF}$ set theory and prove that no nonempty Polish topological space is of first category.

29. Check, within $\mathbf{ZF}$ set theory, that any closed bounded subset of the real line $\mathbf{R}$ is compact.

2. Countable versions of AC and real analysis

As we have already mentioned in Chapter 1, the number of significant statements of real analysis which are provable within **ZF** set theory is rather limited. So, for further development of this classical area of mathematics, certain variants of the Axiom of Choice are absolutely necessary.

One of the nontrivial weak versions of **AC** is when a set I of indices is countably infinite (or, according to another terminology, denumerably infinite). In this situation **AC** takes the following form:

If $\{X_i : i \in I\}$ is an arbitrary countable family of nonempty sets, then there exists a family $\{x_i : i \in I\}$ of elements such that $x_i \in X_i$ for each index $i \in I$.

For the above-mentioned form, the abbreviation **CC** is frequently utilized (reflecting the countable choice for nonempty sets). Denoting, as usual, by **N** the set of all natural numbers, we may present **CC** in the following equivalent form within **ZF** theory:

If $\{X_n : n \in \mathbf{N}\}$ is a sequence of nonempty sets, then there exists a sequence $\{x_n : n \in \mathbf{N}\}$ of elements such that $x_n \in X_n$ for each $n \in \mathbf{N}$.

This form is permanently exploited in classical mathematical analysis. As a typical example of its application, we may indicate the equivalence of the continuity (at a point) in Cauchy's sense and Heine's sense for any function $f : \mathbf{R} \to \mathbf{R}$. A more detailed explanation concerning this equivalence will be given later in the present chapter.

Actually, in many topics of standard lecture courses of mathematical analysis and of the much more advanced theory of real functions it suffices to use the following particular case of **CC** which is denoted by **CC(R)** and reflects the countable choice for nonempty subsets of the real line **R**:

If $\{X_n : n \in \mathbf{N}\}$ is a sequence of nonempty subsets of **R**, then there exists a sequence $\{x_n : n \in \mathbf{N}\}$ of points of **R** such that $x_n \in X_n$ for each $n \in \mathbf{N}$.

However, there are also numerous facts in classical mathematical disciplines which need versions of **AC** much stronger than **CC(R)** (cf. Chapter 3). To confirm this circumstance, let us consider one typical example concerning certain restrictions of functions acting from the real line **R** into itself.

Example 1. Suppose that a function $f : \mathbf{R} \to \mathbf{R}$ is given. It is natural to ask whether this function is bounded on a sufficiently large subset of $\mathbf{R}$. In particular, one may pose the question:

Does there exist a subset $X \subset \mathbf{R}$ with $\operatorname{card}(X) = \operatorname{card}(\mathbf{R})$ such that the restriction $f|X$ is bounded?

It turns out that in order to give a positive answer to this question, we must be sure that the cardinality of the continuum

$$\mathbf{c} = \operatorname{card}(\mathbf{R}) = 2^{\operatorname{card}(\mathbf{N})}$$

is not cofinal with the least infinite ordinal (cardinal) number ω which is usually identified with $\mathbf{N}$. In other words, we must have a guarantee that $\mathbf{R}$ cannot be represented as the union of countably many sets, all of which have cardinalities strictly less than $\mathbf{c}$. For this purpose, we need some substantial version of $\mathbf{AC}$. Actually, an argument based on $\mathbf{AC}$ and leading to the non-cofinality of $\mathbf{c}$ with ω is not difficult and is left to the reader as a useful exercise (see Exercises 3 and 4 for this chapter).

In most parts of this book we assume appropriate forms of the Axiom of Choice and they, naturally, depend on concrete questions which are under discussion.

Among those forms, the Axiom of Dependent Choice (or Principle of Dependent Choices denoted usually by $\mathbf{DC}$) should be especially mentioned, because it is sufficient for development of classical mathematical disciplines such as real analysis, Lebesgue measure theory, and elementary topology of Euclidean spaces and their subsets.

The $\mathbf{DC}$ axiom first introduced by Bernays and Tarski is as follows:

Let X be a nonempty set and let $S(x, y)$ be a binary relation on X satisfying the condition

$$(\forall x \in X)(\exists y \in X)S(x, y).$$

Then there is a sequence $\{x_n : n < \omega\}$ of elements of X such that $S(x_n, x_{n+1})$ for all natural numbers n.

Also, it makes sense to give here one restricted version of the $\mathbf{DC}$ axiom, usually denoted by $\mathbf{DC}(\mathbf{R})$:

Let $S(x, y)$ be a binary relation on $\mathbf{R}$ satisfying the condition

$$(\forall x \in \mathbf{R})(\exists y \in \mathbf{R})S(x, y).$$

Then there is a sequence $\{x_n : n < \omega\}$ of points in $\mathbf{R}$ such that $S(x_n, x_{n+1})$ for all natural numbers n.

In connection with the Axiom of Dependent Choice, an interesting and somewhat unexpected circumstance should be indicated. Namely, it turns out that, within **ZF** set theory, **DC** is equivalent to the classical Baire theorem stating that any nonempty complete metric space is of second category. The validity of this equivalence was first shown by Blair [15].

Theorem 1. *The following two assertions are equivalent within* **ZF** *theory:*
(1) **DC** *axiom;*
(2) *every nonempty complete metric space is of second category.*

Proof. Indeed, the reader can check that the standard proofs of the Baire theorem belong to **ZF** & **DC** theory (see, for instance, [49] or [152]). Suppose now that the assertion formulated in (2) is valid in **ZF** theory and consider an arbitrary nonempty set X with a binary relation $S(x, y)$ on it satisfying the condition

$$(\forall x \in X)(\exists y \in X)S(x, y).$$

Equip X with the discrete metric and introduce the product metric space

$$E = X^\omega = X \times X \times \ldots \times X \times \ldots .$$

A straightforward verification (within **ZF**) allows us to assert that E is a complete metric space. The elements of E will be denoted by $e = (e_n)_{n<\omega}$. Further, for any two natural numbers n and m such that $n < m$, define the set

$$U_{n,m} = \{e \in E \ : \ S(e_n, e_m)\}$$

and, for each $n < \omega$, put

$$V_n = \cup\{U_{n,m} : n < m < \omega\}.$$

It is not difficult to check that all the sets V_n $(n < \omega)$ are open and everywhere dense in E. Consequently, according to the Baire theorem, we have

$$\cap\{V_n \ : \ n < \omega\} \neq \emptyset.$$

Let $v = (v_n)_{n<\omega}$ be an arbitrary element from $\cap\{V_n \ : \ n < \omega\}$. Then we can easily construct (by recursion) a strictly increasing sequence

$$\{n(k) \ : \ k < \omega\} \subset \omega = \mathbf{N}$$

such that $S(v_{n(k)}, v_{n(k+1)})$ for all $k < \omega$. Thus, denoting $x_k = v_{n(k)}$, we get

$$(\forall k < \omega)S(x_k, x_{k+1}).$$

In this way, we have demonstrated within **ZF** set theory the equivalence between the **DC** axiom and Baire theorem.
Theorem 1 has thus been proved.

As was already mentioned, the theory **ZF** & **DC** may be regarded as a basis of all classical mathematical disciplines. The weaker **ZF** set theory does not suffice, because, for example, there are some models of this theory, in which the real line **R** is representable as the union of a countable family of countable sets (see [57], [102]). In those models a number of other surprising phenomena may be detected (cf. Exercises 5 and 8 from Chapter 1).

The formulation of the axiom **CC(R)** was given at the beginning of this chapter. Now, we would like to consider one version of **CC(R)** which is denoted by **PCC(R)** (the partial countable choice for nonempty subsets of **R**) and is formulated as follows:

If $\{X_n : n \in \mathbf{N}\}$ is a sequence of nonempty subsets of **R**, then there exists an infinite set $M \subset \mathbf{N}$ such that $\prod\{X_n : n \in M\} \neq \emptyset$.

Clearly, we have the implication

$$\mathbf{CC(R)} \Rightarrow \mathbf{PCC(R)}.$$

At first sight, **CC(R)** appears a much stronger assertion than **PCC(R)**. However, it turns out that these two axioms are equivalent within **ZF** set theory (see Exercises 8 and 9 for this chapter).

In his remarkable article [232] Sierpiński linked the axiom **PCC(R)** with two standard definitions of continuity of real-valued functions at a fixed point of **R**.

Let $f : \mathbf{R} \to \mathbf{R}$ be a function and let $x \in \mathbf{R}$. Recall that:

(C) f is continuous at x in the sense of Cauchy if, for any real $\varepsilon > 0$, there exists a real $\delta > 0$ such that

$$(\forall y \in \mathbf{R})(|y - x| < \delta \Rightarrow |f(y) - f(x)| < \varepsilon);$$

(H) f is continuous at x in the sense of Heine (or, according to modern terminology, f is sequentially continuous at x) if, for any sequence of points $\{y_n : n \in \mathbf{N}\} \subset \mathbf{R}$ tending to x, the sequence $\{f(y_n) : n \in \mathbf{N}\}$ tends to $f(x)$.

The next statement is an almost trivial fact of **ZF** set theory.

Theorem 2. *Within* **ZF** *theory, the continuity of f at x in the sense of Cauchy implies the continuity of f at x in the sense of Heine.*

The proof of Theorem 2 is very easy and is left to the reader.

However, the converse implication, which says that the continuity of f at x in Heine's sense implies the continuity of f at x in Cauchy's sense, is not provable within **ZF** theory, so needs some form of the Axiom of Choice.

Actually, as a simple and fairly standard argument shows, the axiom **CC(R)** is sufficient for obtaining the converse implication (see Exercise 6). Since, as was mentioned above, $\mathbf{CC}(\mathbf{R})$ and $\mathbf{PCC}(\mathbf{R})$ are equivalent within $\mathbf{ZF}$ theory, the axiom $\mathbf{PCC}(\mathbf{R})$ is also sufficient for deducing the continuity in the sense of Cauchy from the continuity in the sense of Heine.

Moreover, Sierpiński was able to establish in [232] the necessity of $\mathbf{PCC}(\mathbf{R})$ for the equivalence of these two classical definitions of continuity of real-valued functions at a point of $\mathbf{R}$. To say more precisely, Sierpiński obtained the following much stronger result.

Theorem 3. *The equivalence of Cauchy's and Heine's definitions of the continuity at 0 of any function $f : [0,1] \to \mathbf{R}$ implies, within $\mathbf{ZF}$ set theory, the axiom* $\mathbf{PCC}(\mathbf{R})$.

Proof. Assume that these two definitions are equivalent and let us try to show the validity of $\mathbf{PCC}(\mathbf{R})$ axiom.

Take an arbitrary family $\{Y_n : 1 \le n < \omega\}$ of nonempty subsets of $\mathbf{R}$. We must find an injective infinite sequence

$$\{m(1), \ m(2), \ \ldots, \ m(i), \ \ldots\} \subset \mathbf{N}$$

and a sequence $\{y_{m(i)} : 1 \le i < \omega\}$ of points in $\mathbf{R}$ such that $y_{m(i)} \in Y_{m(i)}$ for each natural index $i \ge 1$.

For this purpose, define a function $f : [0,1] \to \mathbf{R}$ as follows:

(a) $f(0) = 0$ and $f(1/n) = 0$ for all natural numbers $n \ge 1$;

(b) supposing that $1/(n+1) < x < 1/n$, put $f(x) = 1$ if the number

$$r_n(x) = \frac{2n(n+1)x - 2n - 1}{1 - |2n(n+1)x - 2n - 1|}$$

belongs to Y_n; otherwise put $f(x) = 0$.

By proceeding in this manner, f will be defined on the whole closed interval $[0,1]$.

First, let us check that the function f is not continuous at 0 in the sense of Cauchy. Indeed, choose any real $\delta > 0$ and find the smallest natural number n such that $1/n < \delta$. Let y be a point from the set Y_n which, by virtue of our assumption, is not empty. Consider the real number

$$x = \frac{1}{2n(n+1)}\left(2n + 1 + \frac{y}{1 + |y|}\right).$$

Observe that $r_n(x) = y$ by the above definition. Keeping in mind the elementary inequality

$$-1 < \frac{y}{1 + |y|} < 1,$$

we obtain that

$$1/(n+1) < x < 1/n,$$

so $0 < x < \delta$ and $f(x) = 1$. Since $f(0) = 0$ and $\delta > 0$ can be arbitrarily small, we infer that the function f is not continuous at 0 in the Cauchy sense. Therefore, f is not continuous at 0 in the Heine sense, either, and this means that there exists a sequence of points

$$\{x_n : 1 \leq n < \omega\} \subset [0, 1]$$

tending to 0, for which the corresponding sequence $\{f(x_n) : 1 \leq n < \omega\}$ does not tend to 0. In particular, there exists an infinite subset M of ω such that

$$f(x_m) \neq 0 \quad (m \in M).$$

Taking into account the fact that the range of f is contained in $\{0, 1\}$, we deduce that

$$f(x_m) = 1 \quad (m \in M).$$

The partial sequence $\{x_m : m \in M\}$ also converges to 0, hence there are infinitely many natural numbers k having the property that the open interval $]1/(k+1), 1/k[$ contains at least one member of $\{x_m : m \in M\}$.

Keeping in mind the latter circumstance, we can effectively define two strictly increasing infinite sequences of natural numbers

$$\{k_1, \ k_2, \ \ldots, \ k_i, \ \ldots\}$$

$$\{m(1), \ m(2), \ \ldots, \ m(i), \ \ldots\} \subset M$$

such that, for any $i \in \mathbf{N} \setminus \{0\}$, the point $x_{m(i)}$ belongs to $]1/(k_i + 1), 1/k_i[$.

Now, it is easy to see that

$$y_{m(i)} = r_{m(i)}(x_{m(i)}) \in Y_{m(i)} \quad (i = 1, 2, \ldots),$$

so the infinite subset $\{m(i) : i \in \mathbf{N} \setminus \{0\}\}$ of $\mathbf{N}$ is as desired.

This completes the proof of Theorem 3.

Remark 1. From the theorem just proved it follows that these three assertions are equivalent in **ZF** set theory:

(1) the axiom $\mathbf{CC(R)}$;

(2) the axiom $\mathbf{PCC(R)}$;

(3) the equivalence of both definitions (Cauchy's and Heine's) of the continuity of any function $f : [0, 1] \to \mathbf{R}$ at 0.

In modern real analysis, there are many notions of smallness for subsets of **R**. For example, all first category sets in **R** and all Lebesgue measure zero sets in **R** are usually treated as two different kinds of small subsets in **R**. A radical difference between them is caused by the existence of a partition $\{X, Y\}$ of **R**, where X is a first category set in **R** and Y has Lebesgue measure zero (see [33], [77], [115], [147], [152], [190], [203]). However, there are also a lot of similarities and analogies between these two families of small sets. Such analogies are thoroughly specified and considered in several textbooks, monographs, and surveys (see, e.g., [33], [37], [115], [152], [188], [190], [203]).

Some much more delicate notions of smallness for subsets in **R** will be discussed in subsequent sections of this book. For instance, we will be dealing many times with the so-called universal measure zero sets (see especially Chapter 5 where it is shown that universal measure zero subsets of **R** are in a close connection with so-called absolutely nonmeasurable real-valued functions).

As a rule, a definition of a small subset of **R** is introduced in a manner which guarantees that the union of any countable family of small sets is also a small set.

The ultimate concept of smallness in analysis is, of course, the notion of a countable (denumerable) set. The widely known theorem of classical set theory, due to Cantor, states that the union of an arbitrary countable family of countable sets is countable, too. Of course, Cantor proved this theorem by using the Axiom of Choice which was exploited by him at intuitive level. At present, we have a more precise formulation of this theorem.

Theorem 4. *In the theory* **ZF** & **CC***:*
(1) the union of countably many countable sets is again countable;
(2) the union of countably many first category subsets of a topological space is again of first category in the same space.

We leave an easy proof of Theorem 4 to the reader.

Remark 2. As has already been indicated, there exists a model of **ZF** theory, in which **R** is representable as the union of a countable family of countable sets. This circumstance underlines the necessity of a certain form of **AC** for the validity of the assertions (1) and (2) of Theorem 4.

EXERCISES

1. The following original and somewhat restricted version of **AC** was given by Zermelo:

If $\{X_i : i \in I\}$ is any family of nonempty pairwise disjoint sets, then there exists a family $\{x_i : i \in I\}$ of elements such that $x_i \in X_i$ for each index $i \in I$.

Show, within **ZF** theory, that the above-mentioned restricted version is equivalent to the full version of **AC** which was formulated in this chapter.

In addition, consider the analogous restricted versions of the axioms **CC**, **CC(R)**, **PCC(R)**, respectively, and prove the equivalence (again within **ZF** theory) of each of those versions to the corresponding axiom.

For **CC(R)** and **PCC(R)**, take into account the fact that there is an effectively determined bijection between the set **R** and the product set **N** × **R**.

2. Work in **ZF** theory and check that the **DC** axiom implies the **CC** axiom.
Also, demonstrate, within **ZF** & **DC** theory, that if $(E, \leq)$ is a linearly ordered set, then the following two assertions are equivalent:

(a) $(E, \leq)$ is well-ordered;

(b) there exists no infinite strictly decreasing (with respect to $\leq$) sequence of elements of E.

Remark 3. It was established in [104] that, within **ZF** set theory, the **CC** axiom does not imply the **DC** axiom. Within the same theory, **DC** does not imply **AC** (see [56], [253]).

3. Prove that the cardinal $\mathbf{c}$ is not cofinal with the least infinite cardinal ω, i.e., the real line $\mathbf{R}$ cannot be represented in the form

$$\mathbf{R} = \cup\{X_n : n < \omega\},$$

where $\mathrm{card}(X_n) < \mathbf{c}$ for each $n < \omega$.

Argue in the following manner. In view of the effectively provable equality $\mathrm{card}(\mathbf{R}) = \mathrm{card}(\mathbf{R}^\omega)$, it suffices to show that

$$\mathbf{R}^\omega \neq \cup\{Y_n : n < \omega\},$$

where $\mathrm{card}(Y_n) < \mathbf{c}$ for each $n < \omega$. Suppose otherwise, i.e.,

$$\mathbf{R}^\omega = \cup\{Y_n : n < \omega\}, \quad (\forall n < \omega)(\mathrm{card}(Y_n) < \mathbf{c}),$$

and, for any $n < \omega$, consider the canonical projection

$$\mathrm{pr}_n : \mathbf{R}^\omega \to \mathbf{R}.$$

Since $\mathrm{card}(Y_n) < \mathbf{c}$, one may infer by virtue of **AC** that

$$\mathrm{card}(\mathrm{pr}_n(Y_n)) \leq \mathrm{card}(Y_n) < \mathbf{c},$$

so there exists a point $y_n \in \mathbf{R} \setminus \mathrm{pr}_n(Y_n)$. Check that the obtained sequence $\{y_n : n < \omega\} \in \mathbf{R}^\omega$ cannot belong to the set $\cup\{Y_n : n < \omega\}$, so a contradiction has been obtained.

More precisely, verify that the cardinal $\mathbf{c}$ becomes not cofinal with ω in the case when **ZF** theory is enriched by adding to it these two set-theoretical assumptions:

(*) **CC(R)**;

(**) if Z is a subset of $\mathbf{R}$ with $\mathrm{card}(Z) < \mathbf{c}$ and $Z' \subset \mathbf{R}$ is a surjective image of Z, then $\mathrm{card}(Z') < \mathbf{c}$.

4. Another proof leading to the result of the previous exercise was presented by Luzin (as indicated in Sierpiński's article [232]). Luzin's argument is based on the fact that any uncountable compact set K in $\mathbf{R}$ produces a disjoint family

$\{K_i : i \in I\}$ of uncountable compact subsets of K, where $\text{card}(I) = \mathbf{c}$ (the above-mentioned fact can readily be obtained by using appropriate Peano-type continuous surjections; see Exercise 17 from Chapter 1).

Try to restore all details of Luzin's argument and show again, with the aid of $\mathbf{AC}$, that the cardinal $\mathbf{c}$ is not cofinal with the cardinal ω.

5. Let g be an arbitrary mapping of $\mathbf{R}$ into itself.

Check, within $\mathbf{ZF}$ & $\mathbf{CC}$ theory, that there exists a set $X \subset \mathbf{R}$ of second Baire category (respectively, of strictly positive outer Lebesgue measure) such that the restriction $g|X$ is bounded (cf. Exercise 4 from Chapter 1 and Example 1 of the present chapter).

In addition, demonstrate that if the cardinal $\mathbf{c}$ is not cofinal with the cardinal ω, then, for any function $f : \mathbf{R} \to \mathbf{R}$, there exists a set $Y \subset \mathbf{R}$ such that $\text{card}(Y) = \mathbf{c}$ and the restriction $f|Y$ is bounded.

6*. Recall $\mathbf{CC}(\mathbf{R})$ axiom which states that, for any family $\{X_n : n < \omega\}$ of nonempty subsets of $\mathbf{R}$, there exists at least one selector of $\{X_n : n < \omega\}$.

Suppose that the principle $\mathbf{CC}(\mathbf{R})$ holds true (i.e., work in $\mathbf{ZF}$ & $\mathbf{CC}(\mathbf{R})$ set theory) and show that:

(i) the continuity (at a point) of a function $f : \mathbf{R} \to \mathbf{R}$ in Heine's sense implies the continuity (at the same point) of f in Cauchy's sense;

(ii) ω_1 is a regular cardinal number, i.e., ω_1 cannot be represented in the form $\omega_1 = \sup\{\xi_n : n < \omega\}$, where all ordinal numbers ξ_n $(n < \omega)$ are strictly less than ω_1.

For (ii), argue as follows. Consider the closed half-line

$$\mathbf{R}_+ = \{t \in \mathbf{R} : t \geq 0\}$$

with its partition into the half-open intervals $[n, n+1[$, where n ranges over the set ω of all natural numbers. Let $\{\xi_n : n < \omega\}$ be an arbitrary countable family of countable ordinals. Remembering that, for any $n < \omega$, the set $\mathbf{Q} \cap [n, n+1[$ contains a subset isomorphic to ξ_n (see Appendix 4) and taking into account the fact that the family of all infinite subsets of $\mathbf{Q}$ is effectively equinumerous with $\mathbf{R}$, define a sequence of sets $\{X_n : n < \omega\}$ satisfying the relations:

(a) $X_n \subset \mathbf{Q} \cap [n, n+1[$ for each $n < \omega$;

(b) X_n is isomorphic to ξ_n for each $n < \omega$.

Then observe that:

(c) the set $\cup\{X_n : n < \omega\}$ is effectively countable and well-ordered;

(d) the order type α of $\cup\{X_n : n < \omega\}$ is such that $\xi_n \leq \alpha$ for all ordinal numbers ξ_n $(n < \omega)$.

Finally, conclude from (c) and (d) that ω_1 is a regular cardinal.

7. Verify within $\mathbf{ZF}$ set theory that the product space of a countable family of complete metric spaces is a complete metric space.

8^*. Recall that the abbreviation **PCC** (the partial countable choice) stands for the following assertion:

If $\{Y_n : n \in \mathbf{N}\}$ is a sequence of nonempty sets, then there exists an infinite set $M \subset \mathbf{N}$ such that $\prod\{Y_m : m \in M\} \neq \emptyset$.

Demonstrate, within **ZF** set theory, that $\mathbf{CC} \Leftrightarrow \mathbf{PCC}$.

Argue in the following manner. The implication $\mathbf{CC} \Rightarrow \mathbf{PCC}$ is trivial. To show the validity of the converse implication, assume **PCC**, take an arbitrary countable family $\{X_n : n \in \mathbf{N} \setminus \{0\}\}$ of nonempty sets and denote

$$Y_n = \prod\{X_k : 1 \leq k \leq n\} \qquad (n \in \mathbf{N} \setminus \{0\}).$$

Obviously, $Y_n \neq \emptyset$ for each $n \in \mathbf{N} \setminus \{0\}$ (see Exercise 2 from Chapter 1). So the countable family $\{Y_n : n \in \mathbf{N} \setminus \{0\}\}$ of nonempty sets is obtained and, according to **PCC**, there exists an infinite set $M \subset \mathbf{N} \setminus \{0\}$ such that

$$\prod\{Y_m : m \in M\} \neq \emptyset.$$

Further, fix an element

$$\{y_m : m \in M\} \in \prod\{Y_m : m \in M\}.$$

Clearly, each y_m $(m \in M)$ can be uniquely written in the form

$$y_m = (x_{1,m}, x_{2,m}, ..., x_{m,m}) \in \prod\{X_k : 1 \leq k \leq m\},$$

where $x_{k,m} \in X_k$ for any natural index $k \in [1, m]$. Now, for each $n \in \mathbf{N} \setminus \{0\}$, put

$$m(n) = \min\{m \in M : n \leq m\}$$

and verify that $\{x_{n,m(n)} : n \in \mathbf{N} \setminus \{0\}\}$ is a selector of $\{X_n : n \in \mathbf{N} \setminus \{0\}\}$.

9. Let us return to the $\mathbf{CC(R)}$ axiom and consider its restricted version $\mathbf{PCC(R)}$:

For any sequence $\{X_n : n < \omega\}$ of subsets of $\mathbf{R}$, there exists an infinite set $M \subset \omega$ such that $\prod\{X_m : m \in M\} \neq \emptyset$.

At first sight, $\mathbf{PCC(R)}$ looks like a somewhat weaker version of $\mathbf{CC(R)}$.

However, show that $\mathbf{PCC(R)}$ is equivalent to $\mathbf{CC(R)}$ within **ZF** set theory.

To do this, utilize an argument similar to that outlined in the previous exercise and keep in mind the fact that if $\{X_k : 1 \leq k \leq m\}$ is a given finite family of subsets of $\mathbf{R}$, then the product set

$$\prod\{X_k : 1 \leq k \leq m\} \subset \mathbf{R}^m$$

is effectively equinumerous with a concrete subset of $\mathbf{R}$.

10. Two persons play the infinite game alternatively choosing uncountable subsets X_n of $\mathbf{R}$ so that

$$X_0 \supset X_1 \supset \ldots \supset X_n \supset \ldots .$$

Demonstrate, with the aid of **AC**, that no matter how the first player chooses his sets, the second player can always achieve the validity of the relation

$$\cap\{X_n : n < \omega\} = \emptyset.$$

Generalize this result to the case where an arbitrary infinite set E is taken instead of $\mathbf{R}$ and all the chosen subsets X_n ($n < \omega$) of E should be of cardinality $\mathrm{card}(E)$.

11*. Two persons play the following infinite game on ω_1. The first player takes a countable partition $\{Y_n : n < \omega\}$ of ω_1 such that $\mathrm{card}(Y_n) = \omega_1$ for all $n < \omega$. The second player picks some member Y_{n_0} of this partition. Then the first player takes a countable partition $\{Y_{n_0,n} : n < \omega\}$ of Y_{n_0} such that $\mathrm{card}(Y_{n_0,n}) = \omega_1$ for all $n < \omega$. The second player picks some member Y_{n_0,n_1} of this partition, and so on. Proceeding in this manner, they come to a decreasing sequence of sets

$$Y_{n_0} \supset Y_{n_0,n_1} \supset \ldots \supset Y_{n_0,n_1,\ldots,n_k} \supset \ldots .$$

If the set $\cap\{Y_{n_0,n_1,\ldots,n_k} : k < \omega\}$ is not empty, then the first player wins, otherwise the second player wins.

Consider **ZF** & **DC** & **AD** set theory, where **AD** stands for the Axiom of Determinacy introduced by Steinhaus and Mycielski [196]. Prove, within this theory, that the second player has a winner strategy and describe it.

For this purpose, take into account the deep result of Solovay stating that, in the above-mentioned theory, ω_1 is a two-valued measurable cardinal number, i.e., there exists a nontrivial ω_1-complete ultrafilter of subsets of ω_1 (in this connection, see [103] and Appendix 5).

Remark 4. Comparing the preceding exercise with Exercise 10 from Chapter 1, one can vividly see the difference between $\mathbf{R}$ and ω_1 from the point of view of the theory of infinite games.

12. A topological space E is called scattered if no nonempty subset of E is dense in itself.

By using the method of transfinite induction, demonstrate with the aid of **AC** that any such space E admits a representation in the form

$$E = \{x_\xi : \xi < \alpha\},$$

where α is some ordinal, the family $\{x_\xi : \xi < \alpha\}$ is injective and, for each $\xi < \alpha$, the point x_ξ is isolated in the set $\{x_\zeta : \xi \leq \zeta < \alpha\}$.

Verify that:

(a) in the Cantor–Bendixson theorem (i.e., in Theorem 1 of Chapter 1), the set X_1 is scattered and the set X_2 is a largest (with respect to the inclusion relation) subset of X which is dense in itself;

(b) every scattered set in $\mathbf{R}$ is effectively denumerable;

(c) any countable closed subset of $\mathbf{R}$ is scattered and, consequently, is effectively denumerable.

13*. Let $h : \mathbf{R} \to \mathbf{R}$ be an increasing function and let $F \subset \mathbf{R}$ be a closed set such that the derivative of h exists and vanishes at all points of F, i.e., we have $h'(x) = 0$ for any $x \in F$.

Prove that the set $h(F)$ is of λ-measure zero, where λ denotes the standard Lebesgue measure on $\mathbf{R}$.

Argue as follows. Without loss of generality, it can be supposed that F is bounded, so $\lambda(F) < +\infty$. Let $G \subset \mathbf{R}$ be an open bounded set containing F. Consequently,

$$\lambda(F) \leq \lambda(G) < +\infty.$$

Fix a real number $\varepsilon > 0$ and construct, by using the method of transfinite recursion, a family

$$\mathcal{T} = \{[a_\xi, b_\xi[\; : \xi \in \Xi\}$$

of nonempty half-open intervals, where Ξ is a maximal initial subinterval of ω_1 for which the following conditions are satisfied:

(a) $a_0 = \inf(F)$ and, for each $\xi \in \Xi$, the point a_ξ belongs to F;

(b) if $\xi \in \Xi$, $\zeta \in \Xi$ and $\xi < \zeta$, then $b_\xi \leq a_\zeta$;

(c) for each ordinal $\xi \in \Xi$, the inequality

$$\lambda(h([a_\xi, b_\xi[)) \leq \varepsilon(b_\xi - a_\xi)$$

holds true;

(d) for any ordinal $\xi \in \Xi$, the interval $[a_\xi, b_\xi[$ is contained in G;

(e) for each ordinal $\xi \in \Xi$, the inclusion

$$F \cap [a_0, \sup\{b_\zeta : \zeta < \xi\}[\; \subset \cup\{[a_\zeta, b_\zeta[\; : \; \zeta < \xi\}$$

holds true.

Taking into account the Suslin condition for $\mathbf{R}$ (see Exercise 21 from Chapter 1), conclude that Ξ is a proper initial subinterval of ω_1, so both Ξ and $\mathcal{T}$ are countable. Verify that

$$F \subset \cup\{[a_\xi, b_\xi[\; : \; \xi \in \Xi\}$$

and, consequently,

$$\lambda(h(F)) \leq \lambda(\cup\{h([a_\xi, b_\xi[) : \xi \in \Xi\}) \leq \varepsilon\lambda(G).$$

Keeping in mind the arbitrary smallness of ε, obtain the desired result.

Observe that the above argument only exploits the fact that the right derivative of h exists at all points of F and vanishes at them. Moreover, the same argument is applicable in the more general case when at each point of F there exists at least one right derived number of h equal to zero.

Remark 5. It worth noticing that the method described in the previous exercise is due to Borel and Lebesgue (sometimes, it is called the method of transfinite chains of half-open intervals). By using this method, it becomes possible in many situations to avoid reasonings based on statements about the existence of coverings of certain types. Among such statements a distinguished place is occupied by Vitali's covering theorem (see, e.g., [17], [23], [133], [197]).

14. Work in **ZF** & **DC** theory and prove Alexandrov's classical theorem asserting that, for any nonempty compact metric space K, there exists a continuous surjection $g : C \to K$, where C denotes the Cantor subset of **R**.

Deduce Peano's result (on the existence of continuous surjections of $[0,1]$ onto $[0,1]^2$) from the above-mentioned theorem of Alexandrov.

Moreover, observe that Alexandrov's theorem trivially implies the existence of a continuous surjection of C (of $[0,1]$) onto the Hilbert cube $[0,1]^\omega$.

Finally, check that the existence of a continuous surjective mapping of C (of $[0,1]$) onto $[0,1]^\omega$ can be established within **ZF** set theory.

15*. Let X be a compact topological space such that there exists a continuous surjection

$$f : X \to X \times X.$$

Show that then there exists a continuous surjection of X onto X^ω.

Argue in the following manner. First, denote $f = (f_1, f_2)$ and put

$$g_0 = f_1, \quad g_1 = f_1 \circ f_2, \quad g_2 = f_1 \circ f_2 \circ f_2, \quad g_3 = f_1 \circ f_2 \circ f_2 \circ f_2, \quad \dots .$$

Further, define $g : X \to X^\omega$ by the formula

$$g(x) = (g_0(x), g_1(x), g_2(x), g_3(x), ...) \qquad (x \in X).$$

To demonstrate that g is the required continuous surjection, check by induction that, for every natural number n and for any points

$$y_0 \in X, \quad y_1 \in X, \quad y_2 \in X, \quad \dots, \quad y_n \in X,$$

there exists a point $x \in X$ such that

$$y_0 = g_0(x), \quad y_1 = g_1(x), \quad y_2 = g_2(x), \quad \dots, \quad y_n = g_n(x).$$

This yields that the set $g(X)$ is everywhere dense in X^ω. Finally, utilize the assumption that X is quasi-compact and Hausdorff.

Conclude from the stated above that if a compact space Y contains at least two distinct points and $\text{card}(Y) < \mathbf{c}$, then there exists no continuous surjection of Y onto $Y \times Y$.

Remark 6. The result of Exercise 15 implies, in particular, that if the negation of the Continuum Hypothesis holds and Z is a compact space of cardinality ω_1, then there exists no continuous surjection of Z onto $Z \times Z$. Further, let the set $\mathbf{N}$ of all natural numbers be equipped with the discrete topology. Obviously, the two discrete spaces $\mathbf{N}$ and $\mathbf{N} \times \mathbf{N}$ are homeomorphic to each other. At the same time, it is clear that there exists no surjection of $\mathbf{N}$ onto the product space $\mathbf{N}^{\omega}$. This trivial example shows that in Exercise 15 the assumption of compactness of X cannot be weakened by the assumption of local compactness of X.

16. Give an example of a compact topological space E where the continuity in the Cauchy sense differs from the continuity in the Heine sense.

Observe that such a space E cannot satisfy the first countability axiom.

Remark 7. It is shown in [55] that the following statement does not contradict **ZFC** set theory: there exists a compact topological space X with $\text{card}(X) = 2^{\omega} = \mathbf{c}$, in which there are no non-trivial convergent sequences of points (i.e., every convergent sequence of points in X is eventually constant).

It is not hard to see that in such a space X no closed subset can be countably infinite.

17. Let $f : \mathbf{R} \to \mathbf{R}$ be a function.

Work in **ZF** set theory and demonstrate that the following three assertions are equivalent:

(a) f is locally uniformly continuous;

(b) the restriction of f to any bounded subinterval of $\mathbf{R}$ is uniformly continuous;

(c) for every convergent sequence $\{x_n : n < \omega\}$ of points of $\mathbf{R}$, the corresponding sequence $\{f(x_n) : n < \omega\}$ is convergent, too.

For this purpose, keep in mind Exercise 14 from Chapter 1.

3. Uncountable versions of AC and Lebesgue nonmeasurable sets

The fundamental result of Solovay [253] states that, under the assumption of the existence of an uncountable strongly inaccessible cardinal number, there is a model of **ZF** & **DC** set theory in which all subsets of the real line **R** turn out to be measurable in the Lebesgue sense, i.e., all of them are λ-measurable, where λ stands for the ordinary Lebesgue measure on **R**.

Consequently, it is impossible to establish, within the same **ZF** & **DC** theory, the existence of Lebesgue nonmeasurable subsets of **R**.

After the above-mentioned result of Solovay, it has become clear that the standard constructions of Lebesgue nonmeasurable sets in **R** (or in a finite-dimensional Euclidean space) cannot be carried out without the help of un-countable forms of the Axiom of Choice. Obviously, the analogous phrase may be said about Lebesgue nonmeasurable functions. Indeed, as is well known, a function

$$g : \mathbf{R} \to \mathbf{R}$$

is Lebesgue nonmeasurable if and only if at least one set $g^{-1}(\Delta_n)$ is Lebesgue nonmeasurable, where $\{\Delta_n : n < \omega\}$ denotes the family of all open intervals in **R** with rational endpoints.

Let us recall several classical statements and facts which are concerned with the existence of λ-nonmeasurable subsets of **R**. For additional information about subsets of **R** with strange or bad descriptive properties, see, e.g., [24], [27], [32], [33], [37], [63], [71], [77], [98], [128], [133], [137], [143], [147], [152], [162], [168], [185], [188], [190], [203], [216], [242], [262], [268].

The first of such bad sets in **R** (in the historical context and from the view-point of its importance) was discovered by Vitali [266] in 1905. The idea of Vitali's construction is very simple and we would like to recall it briefly. The set **Q** of all rational numbers is a countably infinite subgroup of the additive group $(\mathbf{R}, +)$, so **Q** produces the canonical equivalence relation $S(x, y)$:

$$x \in \mathbf{R} \ \& \ y \in \mathbf{R} \ \& \ x - y \in \mathbf{Q}.$$

Therefore, one can consider the quotient set (factor set) with respect to $S(x, y)$. Denoting this quotient set by the symbol $\mathbf{R}/\mathbf{Q}$ and applying an uncountable form of **AC**, one can pick a selector V of $\mathbf{R}/\mathbf{Q}$, which is usually called a Vitali set, and then one can prove that V is a λ-nonmeasurable subset of $\mathbf{R}$.

We thus have the following classical result due to Vitali.

Theorem 1. *The set V is nonmeasurable with respect to λ and, in addition, does not possess the Baire property with respect to the ordinary Euclidean topology on* $\mathbf{R}$.

Actually, Vitali's argument substantially relies on the translation invariance of λ and on the fact that all bounded subsets of $\mathbf{R}$ have finite λ-measure. Another method for establishing the nonmeasurability of V with respect to λ is based on the so-called Steinhaus property of all λ-measurable sets of strictly positive λ-measure (see, e.g., Exercise 1 for this chapter).

A more detailed analysis of Vitali's construction is presented in Chapter 9 where some generalizations and extensions of Vitali's theorem are also given.

The second approach to establishing the fact that there are λ-nonmeasurable subsets of $\mathbf{R}$ is closely connected with the existence of a Hamel basis of $\mathbf{R}$ and with the existence of nontrivial solutions of Cauchy's functional equation

$$f(x+y) = f(x) + f(y) \qquad (x \in \mathbf{R}, \ y \in \mathbf{R}),$$

where f is an unknown function acting from $\mathbf{R}$ into itself.

In 1905 Hamel [90] considered the additive group $(\mathbf{R}, +)$ as a vector space over the field $\mathbf{Q}$ of all rational numbers, proved with the aid of an uncountable form of **AC** the existence of a basis for this vector space (called then a Hamel basis) and, as a byproduct, obtained the next statement.

Theorem 2. *There exist functions $g : \mathbf{R} \to \mathbf{R}$ which satisfy Cauchy's functional equation, i.e.,*

$$g(x+y) = g(x) + g(y) \qquad (x \in \mathbf{R}, \ y \in \mathbf{R}),$$

and are discontinuous at each point of $\mathbf{R}$.

At present, such functions g are regarded as nontrivial solutions of Cauchy's equation, because all trivial solutions f of the same equation are of the form

$$f(x) = rx \qquad (x \in \mathbf{R}),$$

where r is a real parameter (clearly, all f of this form are continuous).

A few years later after Hamel's work [90], it was demonstrated that every nontrivial solution of Cauchy's equation is nonmeasurable in the Lebesgue sense and does not possess the Baire property (see, for instance, [33], [133], [137], [147] or Exercise 10 of this chapter). In subsequent sections of the book we will be dealing with various real-valued additive functions on $\mathbf{R}$ which have much stronger nonmeasurability properties, e.g., they turn out to be nonmeasurable with respect to a large class of σ-finite measures on $\mathbf{R}$ (not only with respect to the standard Lebesgue measure λ). Notice that, sometimes, a stronger nonmeasurability property of functions requires the aid of certain extra set-theoretical assumptions: the Continuum Hypothesis, Martin's Axiom, etc.

We would like to remark in this place that a Hamel basis itself is not necessarily nonmeasurable with respect to λ. Indeed, as was shown by Sierpiński [234], there are Hamel bases in $\mathbf{R}$ which have λ-measure zero and, simultaneously, are of first category (see Exercise 17 for this chapter). On the other hand, there are also λ-thick (λ-massive) Hamel bases in $\mathbf{R}$ which, consequently, turn out to be nonmeasurable with respect to λ (cf. Exercise 16 for the present chapter). It should be underlined that the above-mentioned important facts do not need additional set-theoretical hypotheses, i.e., they are valid within $\mathbf{ZFC}$ set theory.

The third approach leading to the existence of Lebesgue nonmeasurable subsets of $\mathbf{R}$ is due to Bernstein (1908). The transfinite construction presented in his work [14] relies on certain topological properties of λ. Actually, Bernstein utilizes the fact that λ is a Radon measure, which means that any λ-measurable set can be inner approximated by compact sets or, more precisely, the equality

$$\lambda(X) = \sup\{\lambda(K) : K \text{ is compact and } K \subset X\}$$

holds true for each set $X \in \mathrm{dom}(\lambda)$. By taking into account this fact, Bernstein applies the method of transfinite recursion and constructs (with the aid of an uncountable form of $\mathbf{AC}$) his remarkable set $B \subset \mathbf{R}$ having the property that

$$B \cap K \neq \emptyset, \quad (\mathbf{R} \setminus B) \cap K \neq \emptyset$$

for any uncountable compact subset K of $\mathbf{R}$.

This B is usually called a Bernstein set in $\mathbf{R}$.

We thus have within $\mathbf{ZFC}$ set theory the following statement.

Theorem 3. *There exist Bernstein subsets of* $\mathbf{R}$.

Actually, the Bernstein construction may be considered as a very particular case of one purely set-theoretical fact (in this connection, see Exercise 1 from Chapter 8).

A simple argument allows one to assert that B cannot be λ-measurable. The same argument works in a more general situation where a nonzero σ-finite diffused Radon measure μ is given on $\mathbf{R}$, instead of λ. Moreover, from the existence of a Bernstein set it easily follows that every μ-measurable set $X \subset \mathbf{R}$ with $\mu(X) > 0$ contains a subset nonmeasurable with respect to μ (cf. Exercise 12 for this chapter).

In 1922 Sierpiński and Zygmund in their joint paper [246] constructed a function $g : \mathbf{R} \rightarrow \mathbf{R}$ possessing the property which may be regarded as extremely bad from the purely topological point of view. This property says that, for every set $X \subset \mathbf{R}$ with $\mathrm{card}(X) = \mathbf{c}$, the restriction $g|X$ is discontinuous.

Such a function g is usually called a Sierpiński–Zygmund function.

We thus have the next statement (again, within **ZFC** set theory).

Theorem 4. *There exist Sierpiński–Zygmund functions.*

In their construction of g, Sierpiński and Zygmund utilized a very particular case of Lavrentiev's theorem on extensions of continuous functions (see, e.g., [49] [152]) and also a well-ordering of the family of all continuous functions acting from uncountable G_δ-subsets of $\mathbf{R}$ into $\mathbf{R}$ (of course, the existence of such a well-ordering needs an uncountable variant of **AC**). By using Luzin's well-known characterization of all Lebesgue measurable functions, it is easy to show that the function g is not measurable in the Lebesgue sense and does not possess the Baire property (see Exercise 20). Moreover, a similar argument establishes that g is not measurable with respect to the completion of any nonzero σ-finite diffused Borel measure on $\mathbf{R}$.

In our further considerations we will be dealing with Sierpiński–Zygmund type functions which have some interesting additional properties (see especially Chapters 11 and 14). Here we leave aside the details of the classical construction of a Sierpiński–Zygmund function, because much more delicate constructions of such functions will be presented in subsequent sections of this book.

At present, it is well known that the concepts of filters and ultrafilters play a very important role in set theory, general topology and abstract algebra (especially, in model theory). In mathematical analysis and topology the notion of a filter is fundamental for introducing the general concept of convergence (cf., for example, [19], [49]). The question of the existence of nontrivial ultrafilters in an infinite set directly leads to the Axiom of Choice. Even in the case of the least infinite cardinal number ω, the existence of a nontrivial ultrafilter of subsets of ω is a remarkable fact and requires an uncountable version of **AC**. For instance, this fact may be regarded as a starting point for developing the so-called nonstandard analysis (see, e.g., [10], [48], [217]). Also, the same fact is closely connected with subsets of $\mathbf{R}$ nonmeasurable in the Lebesgue sense. More precisely, in 1938 Sierpiński [239] investigated ultrafilters in the power set of ω and showed the validity of the following statement.

Theorem 5. *Within* **ZF** & **DC** *set theory, the existence of a nontrivial ultrafilter of subsets of ω implies the existence of a subset of Cantor's space $\{0,1\}^\omega$, which is nonmeasurable with respect to the completion μ' of the Haar probability measure μ on $\{0,1\}^\omega$.*

Since μ' is canonically isomorphic with the restriction of λ to the unit segment $[0,1]$, one immediately obtains from Theorem 5 the existence of a subset of $[0,1]$ nonmeasurable with respect to λ. For more details, see Exercise 18 for this chapter.

Remark 1. As was pointed out, some results concerning the existence of λ-nonmeasurable sets and functions remain true if λ is replaced by the completion of an arbitrary nonzero σ-finite diffused Borel measure on $\mathbf{R}$, and the corresponding arguments do not need any essential changes. In order to establish analogous results for a wider class of measures, a radically different approach is needed. This approach looks as follows. Instead of a concrete measure on $\mathbf{R}$, we fix a class $\mathcal{M}$ of nonzero σ-finite measures on $\mathbf{R}$ and study the measurability properties of various subsets of $\mathbf{R}$ (or of various real-valued functions on $\mathbf{R}$) with respect to $\mathcal{M}$. Those sets or functions may be good (i.e, they may be measurable for all measures from $\mathcal{M}$), may be relatively good (i.e., may be measurable for some measures from $\mathcal{M}$) and, finally, may be very bad (i.e., may be nonmeasurable with respect to all measures from $\mathcal{M}$). Such a classification of real-valued functions on $\mathbf{R}$ and of subsets of $\mathbf{R}$ highlights much deeper properties connected with the concept of measurability and deserves to be envisaged more thoroughly. We will be dealing with this approach in subsequent sections of this book (see, e.g., Chapters 5 and 13).

In the present chapter, we only want to underline close connections of uncountable forms of the Axiom of Choice with the existence of Lebesgue nonmeasurable subsets of the real line $\mathbf{R}$ or of the Euclidean n-dimensional space $\mathbf{R}^n$, where $n \geq 1$.

Let λ_n denote the standard n-dimensional Lebesgue measure on $\mathbf{R}^n$. The reader can easily verify that the following three assertions are equivalent within **ZF** & **DC** set theory:

(i) there exists a λ-nonmeasurable subset of $\mathbf{R}$;

(ii) for some natural number $n \geq 1$, there exists a λ_n-nonmeasurable subset of $\mathbf{R}^n$;

(iii) for any natural number $n \geq 1$, there exists a λ_n-nonmeasurable subset of $\mathbf{R}^n$.

We would like to stress that these three assertions are equivalent, because all measures λ_n ($n \geq 1$) are Borel isomorphic to the one-dimensional Lebesgue measure λ (see Exercise 3 for this chapter).

Now, we want to discuss one more remarkable result about Lebesgue nonmeasurable real-valued functions, which is also due to Sierpiński (cf. [93], [243]). However, we will formulate and prove this result in a manner slightly different from the original approach of Sierpiński.

For any natural number n, let us consider a function

$$f_n : \mathbf{R} \to \{0, 1\}$$

defined by the formula

$$f_n(x) = [2^n x] - 2[2^{n-1} x] \qquad (x \in \mathbf{R}),$$

where $[t]$ stands for the largest natural number which does not exceed $t \in \mathbf{R}$. Notice that if, for some natural number l and for some odd integer k, we have $x = k/2^l$, then

$$f_l(x) = 1, \qquad (\forall p)(l < p < \omega \Rightarrow f_p(x) = 0).$$

In particular, we see that

$$\mathrm{ran}(f_n) = \{0, 1\} \qquad (n \in \mathbf{N}).$$

Thus, we get the sequence $\{f_n : n \in \mathbf{N}\}$ of functions acting from $\mathbf{R}$ into $\{0, 1\}$. If we equip the set $\{0, 1\}$ with the discrete topology, then the product space $\{0, 1\}^{\mathbf{R}}$ may be regarded as a compact topological space by virtue of the classical Tychonoff theorem (see, e.g., [19], [49], [153]). Therefore, for the described sequence $\{f_n : n \in \mathbf{N}\}$, there exists at least one accumulation point in $\{0, 1\}^{\mathbf{R}}$. Take any such point and denote it by f. Sierpiński showed that f cannot be Lebesgue measurable.

Theorem 6. *The function f indicated above is not measurable in the Lebesgue sense.*

Proof. From the properties of functions of our sequence $\{f_n : n \in \mathbf{N}\}$, we easily infer that f satisfies the following two conditions:

(1) $f(x+r) = f(x)$ for any $x \in \mathbf{R}$ and for any rational number r of the form $r = k/2^l$, where k is an odd integer and l is a natural number;

(2) $f(-x) = 1 - f(x)$ for any nonzero $x \in \mathbf{R}$ which cannot be represented in the form $k/2^l$ where k is an odd integer and l is a natural number.

These two properties of the function f completely suffice to demonstrate that f is nonmeasurable in the Lebesgue sense.

Indeed, suppose to the contrary that f is Lebesgue measurable. Then condition (1) and the metrical transitivity (i.e., ergodicity) of the Lebesgue measure λ with respect to any dense subgroup of $\mathbf{R}$ (see Exercise 2) imply that f must be equivalent to a constant function. Since

$$\mathrm{ran}(f) \subset \{0, 1\},$$

we have either $f(x) = 0$ for λ-almost all points $x \in \mathbf{R}$ or $f(x) = 1$ for λ-almost all points $x \in \mathbf{R}$. On the other hand, taking into account the relation

$$f(-x) = 1 - f(x)$$

for λ-almost all $x \in \mathbf{R}$, we deduce that if f is λ-equivalent to 0, then it must be λ-equivalent to 1 as well, and conversely. This is impossible, of course. The contradiction obtained shows that f is nonmeasurable in the Lebesgue sense.

Theorem 6 has thus been proved.

Remark 2. In fact, the preceding argument establishes a much stronger result. Indeed, let us return to our sequence of functions $\{f_n : n \in \mathbf{N}\}$. Since, for every natural number n, the set of discontinuity points of the function f_n is locally finite, we infer that f_n belongs to the first Baire class and, in particular, is a Borel function. Hence, f_n is Lebesgue measurable as well. The proof of Theorem 6 shows that no subsequence of this sequence is pointwise convergent, because the pointwise limit of any sequence of Lebesgue measurable functions must be Lebesgue measurable, too. In this context, let us recall that the problem of the existence of a sequence of Lebesgue measurable functions, all accumulation points of which (in the Tychonoff topology) are nonmeasurable in the Lebesgue sense, was first formulated by Banach and then was positively solved by Sierpiński who presented the construction described above. By the way, we also may conclude that the compactness of the product space $\{0, 1\}^{\mathbf{R}}$ or even the countable compactness of $\{0, 1\}^{\mathbf{R}}$ (see Exercise 21) implies, within **ZF** & **DC** set theory, the existence of a Lebesgue nonmeasurable function belonging to this space.

Remark 3. Sierpiński devoted many of his works to the study of those point sets in $\mathbf{R}$ which are nonmeasurable in the Lebesgue sense (see, for instance, [244] and [245]). He also gave various constructions of such sets and investigated purely logical aspects of the existence of point sets with bad descriptive properties (cf. Chapter 4). As was mentioned earlier in this chapter, one of his ingenious constructions starts with the existence of a nontrivial ultrafilter in the Boolean algebra $\mathcal{P}(\mathbf{N})$ of all subsets of $\mathbf{N}$ (of course, $\mathbf{N}$ is identified with ω). The corresponding argument is outlined in Exercise 18. Notice that this construction turned out to be fruitful for further much deeper investigations of measurability properties of various filters in the Boolean algebra $\mathcal{P}(\mathbf{N})$. In particular, by exploiting such properties Raisonnier and Shelah were able to prove their famous result stating that in the theory **ZF** & **DC** the inequality $\omega_1 \leq \mathbf{c}$ implies the existence of a Lebesgue nonmeasurable subset of $\mathbf{R}$. For more details, see [214] and [229].

EXERCISES

1. Let X be an arbitrary λ-measurable subset of $\mathbf{R}$ such that $\lambda(X) < +\infty$. Show, within **ZF** & **DC** theory, that the equality

$$\lim_{t \to 0} \lambda(X \cap (t + X)) = \lambda(X)$$

holds true.

Infer from this equality that if $\lambda(X) > 0$, then there exists a real $\delta > 0$ such that $X \cap (X + t) \neq \emptyset$ whenever $|t| < \delta$. In other words, if $\lambda(X) > 0$, then the difference set

$$X - X = \{x - y : x \in X, \ y \in X\}$$

contains some neighborhood of the origin of $\mathbf{R}$. This fact is usually called the Steinhaus property of X (see [257]; cf. also [202], [270]).

Give a direct proof of the Steinhaus property of X by using the classical theorem of Lebesgue stating the existence of density points in X.

Let Z be either a λ-measurable subset of $\mathbf{R}$ with $\lambda(Z) > 0$ or a second category subset of $\mathbf{R}$ possessing the Baire property, let n be a natural number, and let ε be a strictly positive real number.

Prove that there exist $z \in \mathbf{R}$ and $h \in \mathbf{R}$ satisfying the following two relations:

$$0 < |h| < \varepsilon, \quad \{z, \ z+h, \ z+2h, \ \ldots, \ z+nh\} \subset Z.$$

Also, show that Z contains an infinite subset such that all distances between its points are rational numbers.

By using the Steinhaus property, show that no Vitali set is λ-measurable.

Formulate and prove an appropriate topological analogue of the Steinhaus property in terms of the Baire category and Baire property. Exploiting this analogue, show that no Vitali set possesses the Baire property.

2. Let X be an arbitrary λ-measurable subset of $\mathbf{R}$ with $\lambda(X) > 0$ and let T be a countable everywhere dense set in $\mathbf{R}$.

Prove, within **ZF** & **DC** theory, that the equality

$$\lambda(\mathbf{R} \setminus \cup\{t + X : t \in T\}) = 0$$

holds true. This fact is usually called the metrical transitivity (or ergodicity) of the Lebesgue measure λ with respect to the everywhere dense set T of translations (shifts) of $\mathbf{R}$.

For establishing the metrical transitivity of λ, apply once again the theorem of Lebesgue on the existence of density points in X.

Further, let $f : \mathbf{R} \to \mathbf{R}$ be a Lebesgue measurable function such that, for any $t \in T$, the equality

$$f(x + t) = f(x)$$

is valid for λ-almost all points x of $\mathbf{R}$.

Check that f is constant λ-almost everywhere on $\mathbf{R}$.

For this purpose, utilize the above-mentioned metrical transitivity of λ with respect to T.

3. Work in **ZF** & **DC** set theory and define a Borel isomorphism between the restriction of λ to $[0, 1]$ and the completion μ' of the product measure

$$\mu = \otimes\{\mu_n : n \in \mathbf{N}\},$$

where for each $n \in \mathbf{N}$ the probability measure μ_n is defined on the power set of $\{0, 1\}$ and

$$\mu_n(\{0\}) = \mu_n(\{1\}) = 1/2.$$

Further, by taking into account a canonical isomorphism between the measure μ and the product measure $\mu \otimes \mu$, show that λ and λ_2 are Borel isomorphic to each other.

Extend this result and obtain the existence of a Borel isomorphism between the measure λ_n, where $n \geq 1$, and the measure λ $(= \lambda_1)$.

By using the latter fact, prove (in the same **ZF** & **DC** theory) that the following three assertions are equivalent:

(i) there exists a λ-nonmeasurable subset of $\mathbf{R}$;

(ii) for some natural number $n \geq 1$, there exists a λ_n-nonmeasurable subset of $\mathbf{R}^n$;

(iii) for any natural number $n \geq 1$, there exists a λ_n-nonmeasurable subset of $\mathbf{R}^n$.

4^*. Let $(\mathbf{Q}, +)$ denote, as usual, the subgroup of all rational numbers in the additive group $(\mathbf{R}, +)$.

Prove, within **ZF** & **DC** set theory, that if there exists a linear ordering of the quotient set $\mathbf{R}/\mathbf{Q}$, then there exists a λ-nonmeasurable subset of $\mathbf{R}$.

For this purpose, argue in the following manner. First of all, denote

$$\{V_i : i \in I\} = (\mathbf{R}/\mathbf{Q}) \setminus \{\mathbf{Q}\}$$

and suppose that $\preceq$ is a linear order on I. For each index $i \in I$, check that $-V_i = V_j$, where $j = j(i)$ is some (uniquely determined by i) index from I and $i \neq j$. Since the disjunction $i \prec j \ \lor \ j \prec i$ holds true, a function

$$f : \mathbf{R} \setminus \mathbf{Q} \to \mathbf{R}$$

can be defined by putting:

$f(x) = 1$ if $x \in V_i$ and $i \prec j$,

$f(x) = 0$ if $x \in V_i$ and $j \prec i$.

Further, verify that:

(a) $f(x + q) = f(x)$ for any $x \in \mathbf{R} \setminus \mathbf{Q}$ and $q \in \mathbf{Q}$;

(b) $f(-x) = 1 - f(x)$ for any $x \in \mathbf{R} \setminus \mathbf{Q}$.

Conclude from the relations (a) and (b) that the function f cannot be λ-measurable (cf. the proof of Theorem 6).

Deduce from the stated above that:

(*) it is impossible to prove, within **ZF** & **DC** set theory, that the family of all countable subsets of $\mathbf{R}$ (even the family $\mathbf{R}/\mathbf{Q}$) admits a linear ordering of all of its members;

(**) it is impossible to prove, within **ZF** & **DC** set theory, that the cardinality of the family of all countable subsets of $\mathbf{R}$ is less than or equal to $\mathbf{c}$;

(***) it is impossible to prove, within **ZF** & **DC** set theory, that the class $\mathcal{B}_2(\mathbf{R}, \mathbf{R})$ has cardinality less than or equal to $\mathbf{c}$; consequently, it is impossible to prove, within the same theory, that the cardinality of the class $\mathcal{B}(\mathbf{R}, \mathbf{R})$ of all Borel functions acting from $\mathbf{R}$ into $\mathbf{R}$ does not exceed $\mathbf{c}$.

5*. Consider again the family of sets

$$\{V_i : i \in I\} = (\mathbf{R}/\mathbf{Q}) \setminus \{\mathbf{Q}\}$$

described in the previous exercise, and introduce one more family of sets

$$\{-V_i \cup V_i \cup \{t\} : i \in I, t \in \{0,1\}\}.$$

Define a correspondence G between these two families by putting

$$G(V_i) = \{-V_i \cup V_i \cup \{t\} : t \in \{0,1\}\} \qquad (i \in I).$$

Verify that G is a $(2-2)$-correspondence, so there exists an injective selector for G (by virtue of Hall's combinatorial theorem; see Appendix 1). In other words, there exists an injective function g such that $\mathrm{dom}(g) = I$ and

$$g(i) = -V_i \cup V_i \cup \{t_i\} \qquad (i \in I),$$

where $t_i \in \{0,1\}$ for all $i \in I$. Further, define a function

$$f : \mathbf{R} \setminus \mathbf{Q} \to \{0,1\}$$

by putting:
$$f(x) = t_i \qquad (x \in V_i, \ i \in I).$$

Show that f is not measurable in the Lebesgue sense.

Conclude from this result that Hall's combinatorial theorem on the existence of an injective selector for any $(2-2)$-correspondence between two sets cannot be proved within **ZF** & **DC** theory.

6*. Let n be a nonzero natural number and let the symbol $\mathbf{AC}(n)$ abbreviate the following assertion:

For any family of n-element sets, there exists a selector of the family.

Obviously, the assertion $\mathbf{AC}(1)$ holds in **ZF** set theory.

On the other hand, demonstrate, within **ZF** & **DC** set theory, that the assertion $\mathbf{AC}(2)$ implies the existence of a λ-nonmeasurable subset of **R**.

For this purpose, take into account the previous exercise.

Conclude that the assertion $\mathbf{AC}(2)$ cannot be proved within **ZF** & **DC** theory.

Remark 4. The results presented in Exercises 4, 5 and 6 are essentially due to Sierpiński.

7. Consider the space $E = \mathbf{R}^{\mathbf{N}}$ of all real-valued sequences. We shall say that two sequences $(x_n)_{n \in \mathbf{N}}$ and $(y_n)_{n \in \mathbf{N}}$ from this space are equivalent via a permutation (in short, P-equivalent) if there exists a permutation ϕ of **N** such that

$$y_n = x_{\phi(n)} \qquad (n \in \mathbf{N}).$$

Obviously, P-equivalence is a particular case of an equivalence relation on E. Let E/P denote the corresponding quotient set (factor set).

Show, within **ZF** & **DC** set theory, that if there exists a selector of E/P, then there exists a λ-nonmeasurable subset of **R** (this old result is due to Lebesgue).

Conclude from this fact that the existence of such a selector cannot be proved within **ZF** & **DC** set theory.

To show the validity of the formulated result, utilize preceding exercises.

8. A graph (V, E), where V is the set of vertices and E is the set of edges, is called bichromatic (2-chromatic, bipartite) if there exists a coloring of all vertices by exactly two colors (red and blue, say) in such a manner that the endpoints of any edge of (V, E) carry distinct colors.

In other words, (V, E) is bichromatic if and only if V admits a partition $\{V_1, V_2\}$ such that each $e \in E$ is simultaneously incident to a vertex from V_1 and to a vertex from V_2.

By utilizing the Axiom of Choice, demonstrate that the following two assertions are equivalent:

(a) (V, E) is bichromatic;

(b) the length of any simple cycle in (V, E) is even.

Remark 5. In connection with Exercise 8, it should be noticed that the implication (a) $\Rightarrow$ (b) trivially holds in **ZF** set theory, but the converse implication (b) $\Rightarrow$ (a) cannot be proved without the aid of uncountable forms of the Axiom of Choice, because the validity of this implication implies the existence of a Lebesgue nonmeasurable subset of the real line **R**. For more details, see the next exercise.

9*. Let $p > 1$ be an odd natural number. We define the structure of a graph over the set **R** of all real numbers. First of all, we put $V = \mathbf{R}$. Secondly, if r and t are two distinct real numbers, then we put $\{r, t\} \in E$ if and only if there exists an integer k such that $|r - t| = p^k$.

Verify that the graph (V, E) contains no simple cycles with odd lengths.

By starting with this fact, infer that (V, E) is a bichromatic graph, so V admits a partition $\{X, Y\}$ such that any edge $e \in E$ has one of its endpoints in X and the other endpoint in Y.

Show that, for each $\varepsilon > 0$, there exists $h \in \mathbf{R}$ satisfying the following two relations:

(i) $|h| < \varepsilon$;

(ii) $h + X = Y$ and, consequently, $(h + X) \cap X = \emptyset$.

Deduce from (i) and (ii) that neither X nor Y are measurable in the Lebesgue sense, i.e., $X \notin \operatorname{dom}(\lambda)$ and $Y \notin \operatorname{dom}(\lambda)$.

For this purpose, utilize the translation (shift) invariance of λ and the Steinhaus property of λ-measurable sets with strictly positive measure (see Exercise 1).

Summarizing all the facts above, conclude that the equivalence between the assertions (a) and (b) of Exercise 8 cannot be proved within the theory **ZF & DC**.

Remark 6. The result presented in Exercise 9 was first obtained in [262]. For some related results, see also [93].

10*. By using the theorem on the existence of a basis for every vector space, prove that there are precisely $2^{\mathbf{c}}$ automorphisms of the additive group $(\mathbf{R}, +)$.

Consequently, there are precisely $2^{\mathbf{c}}$ endomorphisms of the same group.

Let $g : \mathbf{R} \to \mathbf{R}$ be a nontrivial solution of Cauchy's functional equation, i.e., g is an endomorphism of $(\mathbf{R}, +)$ not representable in the form

$$g(x) = ax \qquad (x \in \mathbf{R}),$$

where a is a real parameter.

Show that:

(a) g is discontinuous everywhere on $\mathbf{R}$;

(b) g is not measurable in the Lebesgue sense and does not possess the Baire property;

(c) the graph of g is everywhere dense in $\mathbf{R}^2$.

For establishing (b), apply the Steinhaus property mentioned in Exercise 1.

11. As was demonstrated earlier (see Exercise 4 of this chapter), **ZF & DC** theory allows one to deduce that the existence of a linear order on the set $\mathbf{R}^{\mathbf{R}}$ (even on the power set $\mathcal{P}(\mathbf{R})$) implies the existence of a Lebesgue nonmeasurable subset of $\mathbf{R}$. In this connection, Sierpiński wrote that there is no hope to effectively indicate a linearly ordered set of cardinality strictly greater than $\mathbf{c}$ (see [238], p. 125).

Show that this remark by Sierpiński is not quite correct.

For this purpose, argue within **ZF** theory, take the Hartogs number $h(\mathbf{c})$ (see Appendix 1) and consider the ordinal sum

$$\alpha = \mathbf{c} + h(\mathbf{c})$$

of $\mathbf{c}$ and $h(\mathbf{c})$. Check that:

(a) α is a linearly ordered set;

(b) the cardinality of α is strictly greater than $\mathbf{c}$.

12. Prove that, for any λ-measurable set $X \subset \mathbf{R}$ with $\lambda(X) > 0$, there exists a λ-nonmeasurable set $Y \subset X$.

For this purpose, consider the set $Y = B \cap X$, where B is some Bernstein subset of $\mathbf{R}$, and verify that Y is as required.

Generalize this result to the case of an uncountable Polish topological space E equipped with the completion of a nonzero σ-finite diffused Borel measure on E.

13. Give an example of two functions

$$f : \mathbf{R} \to \mathbf{R}, \quad g : \mathbf{R} \to \mathbf{R}$$

such that:
  (a) f is a homeomorphism of $\mathbf{R}$ onto itself;
  (b) g is Lebesgue measurable;
  (c) the composition $g \circ f$ is not Lebesgue measurable.
  For this purpose, take as f any homeomorphism of $\mathbf{R}$ onto itself, which transforms some closed set $X \subset \mathbf{R}$ with $\lambda(X) > 0$ onto the set of λ-measure zero (cf. Exercise 15 from Chapter 1). Let $Y \subset X$ be a λ-nonmeasurable set (see the previous exercise). Take the characteristic function of $f(Y)$ as g, and check that the composition $g \circ f$ is not measurable in the Lebesgue sense.

14*. A set $X \subset \mathbf{R}$ is called totally imperfect if there exists no nonempty perfect subset of $\mathbf{R}$ entirely contained in X.
  For example, if B is a Bernstein set in $\mathbf{R}$, then both B and $\mathbf{R} \setminus B$ are totally imperfect sets. Also, any set $Y \subset \mathbf{R}$ with $\mathrm{card}(Y) < \mathbf{c}$ is totally imperfect.
  Prove, within **ZF** & **DC** theory, that the existence of a totally imperfect subset Z of $\mathbf{R}$ having cardinality $\mathbf{c}$ implies the existence of a Lebesgue nonmeasurable subset of $\mathbf{R}$.
  For this purpose, consider an injection $f : \mathbf{R} \to \mathbf{R}$ such that $f(\mathbf{R}) = Z$ and verify that f is not a Lebesgue measurable function (by virtue of the C-property of Luzin).
  By using the method of transfinite recursion, construct a Bernstein subset A of $\mathbf{R}$ such that all distances between distinct points of A differ from each other. Consequently, the inequality

$$\mathrm{card}((A + t) \cap A) \leq 1$$

is fulfilled for any $t \in \mathbf{R} \setminus \{0\}$.

15. Let μ be the completion of a nonzero σ-finite diffused Borel measure on $\mathbf{R}$ (in particular, μ may coincide with λ) and let B be a Bernstein subset of $\mathbf{R}$.
  Check that B is μ-thick (according to another terminology, μ-massive) set, i.e., the equality

$$\mu_*(\mathbf{R} \setminus B) = 0$$

holds true, where μ_* stands for the inner measure associated with μ. Taking into account this equality with the analogous equality $\mu_*(B) = 0$, infer that B is nonmeasurable with respect to μ.
  Verify that the restriction of the characteristic function χ_B to any nonempty perfect subset of $\mathbf{R}$ is discontinuous and deduce from this fact that B and χ_B do not possess the Baire property.

16*. By using the method of transfinite recursion, construct a Hamel basis H in $\mathbf{R}$ which simultaneously is a Bernstein subset of $\mathbf{R}$.

Conclude that H is not λ-measurable and, moreover, H is nonmeasurable with respect to the completion of any nonzero σ-finite diffused Borel measure on $\mathbf{R}$.

By using an analogous method, construct a Vitali set in $\mathbf{R}$ which simultaneously is a Bernstein subset of $\mathbf{R}$. Consequently, there are λ-thick Vitali sets in $\mathbf{R}$ (cf. the previous exercise).

On the other hand, show that, for any real number $\varepsilon > 0$, there exists a Vitali set $V \subset \mathbf{R}$ such that $\lambda^*(V) < \varepsilon$, where λ^* denotes, as usual, the outer measure associated with the Lebesgue measure λ on $\mathbf{R}$;

17^*. Let C denote the classical Cantor subset of the unit segment $[0, 1]$. Within $\mathbf{ZF}$ set theory, check the validity of the relation

$$\{x + y : x \in C,\ y \in C\} = C + C = [0, 2].$$

Furthermore, by starting with this relation, show that there exists a set $X \subset \mathbf{R}$ satisfying the following three conditions:

(a) X is of first category in $\mathbf{R}$;
(b) X has λ-measure zero;
(c) $X + X = \{x + x' : x \in X,\ x' \in X\} = \mathbf{R}$.

Then argue within $\mathbf{ZFC}$ theory, use the Kuratowski–Zorn lemma and prove the existence of a Hamel basis H of $\mathbf{R}$ entirely contained in X.

Conclude that H is of first category and $\lambda(H) = 0$, so H turns out to be measurable in the Lebesgue sense.

Deduce from the stated above that there exist two λ-measure zero sets Y_1 and Y_2 such that the algebraic sum

$$Y_1 + Y_2 = \{y_1 + y_2 : y_1 \in Y_1,\ y_2 \in Y_2\}$$

is nonmeasurable with respect to λ.

Remark 7. The result of Exercise 17 is due to Sierpiński [234]. It shows, in particular, that the algebraic sum of two Lebesgue measurable sets can be nonmeasurable in the Lebesgue sense. In other words, the operation of vector sum of two sets is bad from the point of view of Lebesgue measurability. A somewhat similar result is known for Borel subsets of $\mathbf{R}$. Namely, there exist two Borel sets $Z_1 \subset \mathbf{R}$ and $Z_2 \subset \mathbf{R}$ for which the algebraic sum

$$Z_1 + Z_2 = \{z_1 + z_2 : z_1 \in Z_1,\ z_2 \in Z_2\}$$

is not Borel (see, e.g., [219]). Notice, in this context, that if A and B are any two analytic (i.e., Suslin) subsets of $\mathbf{R}$, then $A + B$ is also an analytic set in $\mathbf{R}$, because $A + B$ may be regarded as a continuous image of the analytic product set $A \times B$.

18^*. Consider any nontrivial ultrafilter Φ in the power set of ω. This Φ can be canonically identified with a certain subset F of the Cantor space $C = \{0, 1\}^\omega$

which is a compact topological group with respect to the product topology and with respect to the addition operation modulo 2. Let μ' denote the completion of the Haar measure μ on C. This μ' is Borel isomorphic with the restriction of λ to the unit segment $[0, 1]$.

Work in **ZF** & **DC** theory and verify the validity of the following two relations:

(a) the set F is invariant under some countable everywhere dense subgroup of C;

(b) there exists an element $z \in C$ such that

$$(z + F) \cap F = \emptyset, \quad (z + F) \cup F = C.$$

By using the metrical transitivity of μ' (cf. Exercise 2 of this chapter), deduce from the relations (a) and (b) that F is nonmeasurable with respect to μ', so there is also a λ-nonmeasurable subset of $[0, 1]$.

19*. Prove that there exists a function $f : \mathbf{R} \to \mathbf{R}$ satisfying the following two conditions:

(a) for any λ-measurable set $X \subset \mathbf{R}$ with $\lambda(X) > 0$, the set $f(X)$ coincides with $\mathbf{R}$;

(b) for any second category set $Y \subset \mathbf{R}$ having the Baire property, the set $f(Y)$ coincides with $\mathbf{R}$.

Construct the required f by using the method of transfinite recursion (one may also apply Sierpiński's lemma on disjoint subsets; see Exercise 4 from Chapter 7).

Conclude that f is not measurable with respect to λ and does not possess the Baire property.

Remark 8. It makes sense to compare the result of the previous exercise with Exercise 5 from Chapter 2.

20. Let $g : \mathbf{R} \to \mathbf{R}$ be a Sierpiński–Zygmund function.

Demonstrate that:

(a) g is nonmeasurable with respect to the completion of any nonzero σ-finite diffused Borel measure on $\mathbf{R}$;

(b) g does not have the Baire property.

For (a), utilize the C-property of Luzin. For (b), take into account the fact that if a function $f : \mathbf{R} \to \mathbf{R}$ possesses the Baire property, then there exists a first category set $X \subset \mathbf{R}$ such that the restriction $f|(\mathbf{R} \setminus X)$ is continuous.

21. A topological space E is called countably quasi-compact if any countable open covering of E contains a finite subcovering.

A Hausdorff countably quasi-compact space is called countably compact.

Show that the following two conditions are equivalent within **ZF** & **CC** theory:

(a) E is countably quasi-compact;

(b) every countably infinite subset of E possesses at least one accumulation point in E.

Check that the set $[0, \omega_1[$ equipped with its order topology is a noncompact locally compact, and countably compact space.

22*. Let E be a base (ground) set, $\mathcal{S}$ be a σ-algebra of subsets of E and let μ be a measure whose domain coincides with $\mathcal{S}$.

The triplet $(E, \mathcal{S}, \mu)$ is usually called a measure space (see, e.g., [17], [20], [26], [89], [199], [265]).

Suppose that μ is a σ-finite measure and X is a μ-thick subset of E, i.e., $\mu_*(E \setminus X) = 0$. Denote

$$\mathcal{S}_X = \{X \cap Z : Z \in \mathcal{S}\}.$$

Verify that $\mathcal{S}_X$ is a σ-algebra of subsets of X.

Further, for any set $Z \in \mathcal{S}$, put $\mu_X(X \cap Z) = \mu(Z)$.

Check that the functional μ_X is a well-defined σ-finite measure on the σ-algebra $\mathcal{S}_X$. So the σ-finite measure space $(X, \mathcal{S}_X, \mu_X)$ is determined.

Let $f : X \to \mathbf{R}$ be an arbitrary μ_X-measurable function.

Show that there exists a μ-measurable function $f^* : E \to \mathbf{R}$ which is an extension of f.

For this purpose, consider first the case when the range of f is countable. In the general case, uniformly approximate f by μ_X-measurable functions, all ranges of which are countable.

23. Let $(E, \mathcal{S}, \mu)$ be a measure space satisfying the Suslin condition, i.e., every disjoint family of μ-measurable sets such that all of them are of strictly positive μ-measure is at most countable.

Let $\{X_i : i \in I\}$ be an arbitrary family of μ-measurable sets.

Demonstrate that there exists a countable set $J \subset I$ such that

$$\mu(X_i \setminus \cup\{X_j : j \in J\}) = 0$$

whenever $i \in I \setminus J$.

For this purpose, suppose otherwise and, by using the method of transfinite recursion, construct an ω_1-sequence $\{Y_\xi : \xi < \omega_1\}$ of pairwise disjoint μ-measurable sets such that:

(a) $(\forall \xi < \omega_1)(\mu(Y_\xi) > 0)$;

(b) for each ordinal $\xi < \omega_1$, the set Y_ξ is contained in some member of the family $\{X_i : i \in I\}$.

The relations (a) and (b) directly lead to a contradiction and hence yield the required result.

24*. Verify that the disjunction

$$(\omega_1 \leq \mathbf{c}) \vee (\mathbf{c} \leq \omega_1)$$

cannot be established within **ZF** & **DC** set theory.

In order to check this fact, consider two possible cases.

(1) $\omega_1 < \mathbf{c}$. In this case, any set $X \subset \mathbf{R}$ with $\mathrm{card}(X) = \omega_1$ does not include a nonempty perfect subset.

(2) $\omega_1 = \mathbf{c}$. In this case, construct a Bernstein set $B \subset \mathbf{R}$ within **ZF** & **DC** theory and keep in mind that B does not contain a nonempty perfect subset.

Conclude that the above-mentioned disjunction implies the existence of an uncountable set in $\mathbf{R}$ which does not contain a nonempty perfect subset. Then refer to the model of Solovay [253] which satisfies **ZF** & **DC** and in which every uncountable set in $\mathbf{R}$ contains a nonempty perfect subset.

25*. Give an example of a partition $\{X_i : i \in I\}$ of $[0,1]$ into uncountable closed sets such that, for any selector $\{x_i : i \in I\}$ of this partition, the equality

$$\lambda^*(\{x_i : i \in I\}) = 1$$

is valid. In particular, the set $\{x_i : i \in I\}$ is not measurable with respect to the Lebesgue measure λ.

For this purpose, argue as follows. Identify $\mathbf{c}$ with the initial ordinal number of cardinality continuum and consider the family $\{F_\xi : \xi < \mathbf{c}\}$ of all those closed subsets of the unit square $[0,1]^2$, which satisfy the condition

$$0 < \lambda_2(F_\xi) < 1 \qquad (\xi < \mathbf{c}).$$

Keeping in mind Fubini's theorem, define by transfinite recursion the family of sets $\{\{x_\xi\} \times P_\xi : \xi < \mathbf{c}\}$, where:

(a) all x_ξ ($\xi < \mathbf{c}$) are pairwise distinct points of $[0,1]$;

(b) for each ordinal $\xi < \mathbf{c}$, the set P_ξ is a closed subset of $[0,1]$ with $\lambda(P_\xi) > 0$ and $\{x_\xi\} \times P_\xi \subset F_\xi$.

Further, consider a Borel bijection

$$h : [0,1] \to [0,1]^2$$

which establishes an isomorphism between the measures λ and λ_2 restricted, respectively, to $[0,1]$ and $[0,1]^2$ (recall that the existence of h is a well-known fact of the Lebesgue measure theory; see Exercise 3 of this chapter). Denote

$$Y_\xi = h^{-1}(\{x_\xi\} \times F_\xi) \qquad (\xi < \mathbf{c})$$

and check that the disjoint family $\{Y_\xi : \xi < \mathbf{c}\}$ of uncountable Borel subsets of $[0,1]$ is such that

$$\lambda^*(\{y_\xi : \xi < \mathbf{c}\}) = 1$$

for every selector $\{y_\xi : \xi < \mathbf{c}\}$ of $\{Y_\xi : \xi < \mathbf{c}\}$.

Finally, by starting with this family $\{Y_\xi : \xi < \mathbf{c}\}$, obtain the required partition $\{X_i : i \in I\}$ of $[0,1]$.

Remark 9. The result of the previous exercise is due to Sierpiński. It is useful to compare this result with the following well-known statement on the existence of sufficiently good selectors:

Let E be a Polish topological space, R be an equivalence relation in E and let E/R denote the quotient set (factor set) canonically associated with R (clearly, E/R forms a partition of E). Suppose that the following conditions are satisfied:

(1) all R-equivalence classes are closed subsets of E;

(2) for any closed subset F of E, the set

$$R(F) = \{X : X \in R/E, \ X \cap F \neq \emptyset\}$$

is Borel in E.

Then there exists a Borel selector of the partition E/R of E.

For the proof of this statement, see, e.g., [137]. Many other useful results on the existence of measurable selectors can be found, e.g., in [17], [33], [63], [115], [155], [191] (see also Appendix 5). Those results have a wide range of applications in various branches of mathematics: probability theory and stochastic processes, optimization theory, the theory of differential equations, etc.

26. Verify that Sierpiński's function f in Theorem 6 does not possess the Baire property.

For this purpose, taking into account the topological analogue of metrical transitivity (cf. Chapter 18), argue similarly to the proof of Theorem 6.

4. The Continuum Hypothesis and Lebesgue nonmeasurable sets

In the previous chapter we were concerned with various point sets which are nonmeasurable with respect to the Lebesgue measure λ on the real line $\mathbf{R}$. Here we would like to continue the discussion of this topic by assuming an additional set-theoretical hypothesis.

First, we are going to consider some subsets of $\mathbf{R}$, the so-called Sierpiński sets, which are utterly bad from the point of view of measurability in the Lebesgue sense. Since the existence of such sets cannot be established within $\mathbf{ZFC}$ theory, in order to have them some extra axioms must be added to $\mathbf{ZFC}$. Undoubtedly, among many set-theoretical assumptions which are not provable within $\mathbf{ZFC}$, the most famous is the Continuum Hypothesis (the commonly used abbreviation: $\mathbf{CH}$).

Recall that the Continuum Hypothesis was first formulated by Cantor who also made several attempts to prove it, and then $\mathbf{CH}$ was especially distinguished by Hilbert in his celebrated lecture (Paris, 1900), where he presented a list of important open mathematical problems at the time. He announced Cantor's hypothesis as Problem 1 in this list (see [97]).

There are two standard but essentially different formulations of $\mathbf{CH}$. Here we give them as assertions (*) and (**) stated below.

(*) the equality $2^\omega = \omega_1$ is valid;

(**) there exists no cardinal number $\mathbf{a}$ such that $\omega < \mathbf{a} < 2^\omega$.

Notice that the implication (*) $\Rightarrow$ (**) trivially holds within $\mathbf{ZF}$ set theory, but the converse implication (**) $\Rightarrow$ (*) is fulfilled only by assuming some version of the Axiom of Choice.

There is a rich literature devoted to various aspects of the Continuum Hypothesis and of its natural extension which is called the Generalized Continuum Hypothesis (abbreviation: $\mathbf{GCH}$; for more details, see Appendix 1). Also, it is well known that both $\mathbf{CH}$ and $\mathbf{GCH}$ are independent of $\mathbf{ZFC}$ set theory (see [10], [39], [40], [42], [79], [103], [148], [251]).

Further, according to Sierpiński's classical result, the implication

$$\mathbf{GCH} \Rightarrow \mathbf{AC}$$

holds true even within **ZF** set theory (see, e.g., [243]). Here **GCH** is stated in the form analogous to (**), namely:

If **b** is an arbitrary infinite cardinal number, then there is no cardinal **a** such that $\mathbf{b} < \mathbf{a} < 2^{\mathbf{b}}$.

A more detailed explanation concerning the above-mentioned remarkable implication may be found in Appendix 1 (see also [112], [243], [256]).

At this moment, **CH** is interesting for us from the point of view that it allows one to construct many sets in **R** (or in a finite-dimensional Euclidean space) with diverse paradoxical or pathological properties. One of them is a so-called Sierpiński set in **R** introduced by Sierpiński.

Let us recall the standard definition of a Sierpiński set (see, e.g., [33], [147], [152], [188], [190], [203]).

A set $X \subset \mathbf{R}$ is a Sierpiński set if $\mathrm{card}(X) > \omega$ and $\mathrm{card}(X \cap Y) \leq \omega$ for every λ-measure zero subset Y of **R**.

Denoting by $\mathcal{I}(\lambda)$ the σ-ideal of all λ-measure zero sets, we see, by virtue of the above definition, that a Sierpiński set S almost avoids all members of $\mathcal{I}(\lambda)$.

The following important statement was proved by Sierpiński in 1924.

Theorem 1. *Under the Continuum Hypothesis, there exist Sierpiński sets.*

Proof. According to our assumption, the equality $\mathbf{c} = \omega_1$ holds, so we may denote by $\{X_\xi : \xi < \omega_1\}$ the family of all Borel subsets of **R** having λ–measure zero.

Applying the method of transfinite recursion, let us define an injective family of points

$$\{x_\xi : \xi < \omega_1\} \subset \mathbf{R}.$$

Suppose that, for an ordinal $\zeta < \omega_1$, the partial family of points $\{x_\xi : \xi < \zeta\}$ has already been determined. Consider the set

$$X'_\zeta = (\cup\{X_\xi : \xi < \zeta\}) \cup \{x_\xi : \xi < \zeta\}.$$

Since ζ is at most countable, the set X'_ζ is of Lebesgue measure zero. Consequently, there exists a point $x \in \mathbf{R} \setminus X'_\zeta$. We then put $x_\zeta = x$.

Proceeding in this manner, we are able to construct the family $\{x_\xi : \xi < \omega_1\}$ of points of **R**.

In view of our construction, the following relations are valid:
(i) $\{x_\xi : \xi < \omega_1\}$ is an injective family;
(ii) for any two ordinals $\xi < \omega_1$ and $\zeta < \omega_1$ such that $\xi < \zeta$, we have $x_\zeta \notin X_\xi$.

Let us define

$$X = \{x_\xi : \xi < \omega_1\}$$

and let us demonstrate that X is a Sierpiński subset of **R**. Indeed, relation (i) implies

$$\mathrm{card}(X) = \omega_1 > \omega.$$

Thus, X is an uncountable set. Relation (ii) shows that, for each ordinal $\xi < \omega_1$, we have

$$\text{card}(X \cap X_\xi) \leq \text{card}(\{x_\zeta : \zeta \leq \xi\}) \leq \omega,$$

whence it follows that, for any set $Y \subset \mathbf{R}$ of Lebesgue measure zero, the set $X \cap Y$ is at most countable (since Y is contained in some set X_α, where $\alpha < \omega_1$).

This completes the proof of Theorem 1.

Remark 1. In connection with the above theorem, it should be noticed that the Continuum Hypothesis is not necessary for the existence of Sierpiński sets, because there are models of **ZFC** theory in which **CH** fails to hold and there exist Sierpiński sets on **R**. More precisely, there are models of **ZFC** in which the negation of the Continuum Hypothesis is valid and there exist Sierpiński sets of cardinality $\mathbf{c}$ (cf. [150]).

Theorem 2. *Let X be a Sierpiński subset of $\mathbf{R}$. Then:*
(1) X is of first category in $\mathbf{R}$;
(2) every uncountable subset of X is again a Sierpiński set.

Proof. As is well known (see [33], [77], [152], [190], [203]), there exists a partition $\{A, B\}$ of $\mathbf{R}$ such that the set A has λ-measure zero and the set B is of first category in $\mathbf{R}$. More generally, the reader can check that every σ-finite diffused Borel measure on a separable topological space E satisfying the first countability axiom is concentrated on a first category subset of E. Obviously, we may write the inclusion

$$X \subset (B \cup (X \cap A)).$$

According to the definition of X, we have the inequality $\text{card}(X \cap A) \leq \omega$. So $B \cup (X \cap A)$ is a first category subset of $\mathbf{R}$. Consequently, the set X is of first category, too, i.e., relation (1) holds true.

Relation (2) directly follows from the definition of Sierpiński sets.

Theorem 3. *Let X be an uncountable subset of $\mathbf{R}$. The following two assertions are equivalent within $\mathbf{ZF}$ & $\mathbf{DC}$ theory:*
(1) X is a Sierpiński set in $\mathbf{R}$;
(2) each uncountable subset of X is nonmeasurable in the Lebesgue sense.

Proof. (1) $\Rightarrow$ (2). Let (1) be satisfied. We have to prove that every uncountable set $Z \subset X$ is nonmeasurable with respect to λ. Suppose otherwise, i.e., Z is λ-measurable. If $\lambda(Z) = 0$, then

$$\omega < \text{card}(Z) = \text{card}(Z \cap X) \leq \omega,$$

which yields a contradiction. So the inequality $\lambda(Z) > 0$ must be valid. In this case, there exists an uncountable set $Z' \subset Z$ of λ-measure zero (see Exercise 1 for this chapter). Consequently,

$$\omega < \text{card}(Z') = \text{card}(Z' \cap X) \leq \omega,$$

which again contradicts the definition of X. The contradiction obtained in both possible cases shows us that Z cannot be measurable with respect to λ.

$(2) \Rightarrow (1)$. Let (2) be satisfied. We have to prove that X is a Sierpiński set in $\mathbf{R}$. Since X is uncountable, it suffices to verify that $\mathrm{card}(X \cap Y) \leq \omega$ for an arbitrary λ-measure zero set Y. Indeed, the set $X \cap Y$ being a subset of Y is of λ-measure zero and, in particular, is measurable with respect to λ. So, in view of (2), the set $X \cap Y$ cannot be uncountable, and we get $\mathrm{card}(X \cap Y) \leq \omega$ which immediately yields (1).

Theorem 3 has thus been proved.

Remark 2. The reader can easily see that relation (2) of Theorem 3 in a precise form expresses the ultimate nonmeasurability of Sierpiński's set X with respect to λ. In other words, Sierpiński sets turn out to be extremely bad from the point of view of the Lebesgue measurability. Many other intriguing properties of Sierpiński sets may be found in [33], [152], [188], [190] and in some exercises for this chapter. We will be dealing with various kinds of Sierpiński sets in further sections of this book. We will also introduce and envisage the so-called generalized Sierpiński sets on $\mathbf{R}$. Their properties are very similar to properties of Sierpiński sets.

Remark 3. In connection with relation (2) of Theorem 3, a delicate moment should be especially underlined: if one wants to establish, within **ZFC** theory, the existence of an uncountable set $X \subset \mathbf{R}$ which satisfies (2), i.e., is extremely bad from the point of view of λ-measurability, then he or she necessarily fails, because the existence of Sierpiński sets cannot be proved within this theory (cf. Exercise 18 for the present chapter).

Returning to the Continuum Hypothesis, one may observe that **CH** tries to link in some concrete manner the following two fundamental objects in mathematics: the cardinality $\mathbf{c}$ of the real line $\mathbf{R}$ and the first uncountable cardinal ω_1. Both $\mathbf{c}$ and ω_1 are defined effectively, but none of them looks like a constructive object. Indeed, the nature of the power set $\mathcal{P}(\mathbf{N})$ whose cardinality coincides with $\mathbf{c}$ is absolutely unclear, because our information about this object is reduced to a certain axiom of **ZFC** theory that only guarantees the existence of $\mathcal{P}(\mathbf{N})$ (see Appendix 1). Similar to the above fact, only under some forms of the Axiom of Choice, e.g., under **CC**, are we able to treat ω_1 as a concrete object constructed by transfinite recursion over the countable ordinals (see, e.g., Exercise 5 from Chapter 1 where it is shown that ω_1 may be singular within **ZF** set theory).

Further, both $\mathbf{c}$ and ω_1 are effectively uncountable cardinal numbers and, moreover, ω_1 is the least uncountable cardinal among the cardinalities of all those uncountable sets which can be made well ordered. So the natural question arises whether ω_1 and $\mathbf{c}$ are comparable (of course, as cardinal numbers). In other words, one might wish to know whether the disjunction

$$\omega_1 \leq \mathbf{c} \ \vee \ \mathbf{c} \leq \omega_1$$

holds true (cf. Exercise 8 from Chapter 1). It turns out that this disjunction is provable only under a rather strong version of the Axiom of Choice. Indeed, as we will see below, the above disjunction implies the existence of a Lebesgue nonmeasurable subset of $\mathbf{R}$ (in this connection, cf. also Exercise 24 from Chapter 3).

First, let us prove the following statement due to Sierpiński (cf. [233]).

Theorem 4. *In* **ZF** & **DC** *theory the equality* $\mathbf{c} = \omega_1$ *implies the existence of a Lebesgue nonmeasurable set on* $\mathbf{R}$.

Proof. Suppose that $\mathbf{c} = \omega_1$. In other words, we assume that **CH** is fulfilled in the strong form (*). Consequently, we may represent the real line $\mathbf{R}$ as an injective ω_1-sequence of points:

$$\mathbf{R} = \{x_\xi : \xi < \omega_1\}.$$

In the Euclidean plane $\mathbf{R}^2 = \mathbf{R} \times \mathbf{R}$ consider the following set:

$$Z = \{(x_\xi, x_\zeta) : \xi \leq \zeta < \omega_1\}.$$

This Z has the property that all of its horizontal sections (by lines) are at most countable and all of its vertical sections (by lines) are co-countable, i.e., are complements of at most countable subsets of vertical lines. Now, the assumption that Z is measurable with respect to the standard two-dimensional Lebesgue measure λ_2 on $\mathbf{R}^2$ directly leads to a contradiction, because, by virtue of the classical Fubini theorem, we must simultaneously have

$$\lambda_2(Z) = 0, \quad \lambda_2(\mathbf{R}^2 \setminus Z) = 0,$$

whence it follows that

$$\lambda_2(\mathbf{R}^2) = \lambda_2(Z) + \lambda_2(\mathbf{R}^2 \setminus Z) = 0,$$

which is impossible. Therefore, Z turns out to be λ_2-nonmeasurable.

Recall now the well-known and important fact from classical measure theory, according to which λ_2 is Borel isomorphic to the standard one-dimensional Lebesgue measure $\lambda_1 (= \lambda)$ on $\mathbf{R}$. This means that within the theory **ZF** & **DC** there exists a Borel isomorphism

$$g : \mathbf{R}^2 \to \mathbf{R}$$

which is also an isomorphism between the two measure spaces $(\mathbf{R}^2, \operatorname{dom}(\lambda_2), \lambda_2)$ and $(\mathbf{R}, \operatorname{dom}(\lambda_1), \lambda_1)$ (cf. Exercise 3 from Chapter 3). Since Z is not λ_2-measurable, it can readily be seen that the set $g(Z)$ is not measurable with respect to λ_1.

Remark 4. The proof of Theorem 4 uses the partition $\{Z, \mathbf{R}^2 \setminus Z\}$ of $\mathbf{R}^2$ which was first pointed out by Sierpiński in [233]. This partition plays a remarkable role in various questions of set theory, measure theory and real analysis (see, e.g., [247] and Chapters 16, 17 of the present book).

Now, we would like to formulate without proof the result of Shelah [229] and Raisonnier [214], which is much stronger and deeper than Theorem 4.

Theorem 5. *In* **ZF** *&* **DC** *set theory the relation* $\omega_1 \leq \mathbf{c}$ *implies the existence of a Lebesgue nonmeasurable subset of* **R**.

A detailed proof of Theorem 5 is given, e.g., in [128]. Chapter 20 of the present book contains some further comments concerning this theorem. Moreover, in the same Chapter 20 the general question of compatibility of well-orderings and σ-finite product measures is discussed.

Theorem 5 immediately yields the next statement.

Theorem 6. *In* **ZF** *&* **DC** *set theory the disjunction*

$$\omega_1 \leq \mathbf{c} \ \vee \ \mathbf{c} \leq \omega_1$$

implies the existence of a Lebesgue nonmeasurable subset of **R**. *Consequently, the above disjunction is not provable in this theory.*

Remark 5. It directly follows from Theorem 6 that if in **ZF** & **DC** theory there exists a well-ordering of **R**, then there exists a Lebesgue nonmeasurable subset of **R**. In Exercise 12 for the present chapter a different proof of this fact is outlined, which does not rely on the fundamental Theorem 5 (see also Exercise 24 from Chapter 3).

EXERCISES

1. Let μ be the completion of a σ-finite diffused Borel measure on **R** and let X be a μ-measurable set in **R** with $\mu(X) > 0$.

Work in **ZF** & **DC** theory and show that there exists a subset Y of X such that:

(a) $\mathrm{card}(Y) = \mathbf{c}$;

(b) $\mu(Y) = 0$.

Argue as follows. Since μ is a Radon measure and $\mu(X) > 0$, there exists an uncountable compact set $K \subset X$ with $\mu(K) > 0$. Therefore, $\mathrm{card}(K) = \mathbf{c}$. Let $\{K_i : i \in I\}$ be a partition of K into continuum many compact subsets, all of which are of cardinality continuum (cf. Exercise 4 from Chapter 2). Keeping in mind the σ-finiteness of μ, conclude that there exists an index $i_0 \in I$ such that $\mu(K_{i_0}) = 0$. So one may put $Y = K_{i_0}$.

2. Verify that:

(a) the family of all Sierpiński subsets of $\mathbf{R}$ is countably additive;

(b) the same family is invariant under the group of all those transformations of $\mathbf{R}$ which preserve the σ-ideal $\mathcal{I}(\lambda)$ (in particular, the family is invariant under the group of all translations of $\mathbf{R}$);

(c) the complement of a Sierpiński set is λ-thick in $\mathbf{R}$;

(d) no Sierpiński set can be a Vitali set or a Bernstein set.

3. Under the Continuum Hypothesis, give an example of a nonzero σ-finite diffused Borel measure μ on $\mathbf{R}$, for which there exists a Sierpiński subset of $\mathbf{R}$ which has μ^*-measure zero and hence is μ'-measurable, where μ' denotes the usual completion of μ.

Under the same hypothesis, construct a λ-thick Sierpiński subset of $\mathbf{R}$.

For this purpose. argue similarly to the proof of Theorem 1.

4. Let ε be a strictly positive real number.

Assuming the Continuum Hypothesis, construct a Sierpiński subset of $\mathbf{R}$ whose outer Lebesgue measure is less than ε.

A set $Z \subset \mathbf{R}$ is called perfectly meager (or an always of first category set) if, for every nonempty perfect subset P of $\mathbf{R}$, the set $Z \cap P$ is of first category in P.

Check that any Sierpiński set in $\mathbf{R}$ is perfectly meager (this fact essentially strengthens assertion (1) of Theorem 2).

5. We have already mentioned, in connection with Theorem 1, that the existence of Sierpiński subsets of $\mathbf{R}$ is possible in some models of set theory, where the Continuum Hypothesis does not hold. Moreover, there is a model of **ZFC** in which the negation of the Continuum Hypothesis is valid and there exist Sierpiński sets of cardinality $\mathbf{c}$.

Verify that in such a model there exists a Lebesgue nonmeasurable subset Y of $\mathbf{R}$ with $\operatorname{card}(Y) = \omega_1 < \mathbf{c}$.

6*. As usual, denote by $\operatorname{dom}(\lambda)$ the class of all λ-measurable subsets of $\mathbf{R}$ and introduce the family

$$\mathcal{T}_d = \{Z \in \operatorname{dom}(\lambda) : \text{all points of } Z \text{ are its density points}\}.$$

Verify that:

(a) $\mathcal{T}_d$ is a topology on $\mathbf{R}$ strictly extending the standard Euclidean topology of $\mathbf{R}$ ($\mathcal{T}_d$ is usually called the density topology of $\mathbf{R}$);

(b) $(\mathbf{R}, \mathcal{T}_d)$ is a Baire topological space satisfying the Suslin condition (i.e., the countable chain condition, which means that any disjoint family of nonempty members of $\mathcal{T}_d$ is at most countable);

(c) $(\mathbf{R}, \mathcal{T}_d)$ is a regular topological space;

(d) a set $Y \subset \mathbf{R}$ is of first category in $(\mathbf{R}, \mathcal{T}_d)$ if and only if $\lambda(Y) = 0$;

(e) a set $Z \subset \mathbf{R}$ has the Baire property in $(\mathbf{R}, \mathcal{T}_d)$ if and only if $Z \in \operatorname{dom}(\lambda)$.

Further, let X be a Sierpiński set in $\mathbf{R}$ endowed with the topology induced by $(\mathbf{R}, \mathcal{T}_d)$.

Prove that:

(f) X is a perfectly normal hereditarily Lindelöf topological space;

(g) X is not separable.

For this purpose, keep in mind that X is a regular space and take into account the result of Exercise 22 from Chapter 1.

Remark 6. It can be shown that the space $(\mathbf{R}, \mathcal{T}_d)$ is completely regular (in this connection, see [80], [258]).

7. Let $(X, \leq)$ be a Suslin line (see Appendix 2) equipped with its order topology.

Demonstrate that X is a nonseparable perfectly normal hereditarily Lindelöf topological space.

For this purpose, take into account Exercises 21 and 22 of Chapter 1.

By adding to X the least and greatest elements, deduce that there exists a nonseparable hereditarily Lindelöf compact topological space.

Remark 7. According to the standard terminology of general topology, an L-space is any regular hereditarily Lindelöf topological space which is not hereditarily separable. As has been shown above in Exercises 6 and 7, every Sierpiński set and every Suslin line turn out to be L-spaces. As we know, for the existence of a Sierpiński set on $\mathbf{R}$, the Continuum Hypothesis completely suffices (see Theorem 1 of this chapter), while the existence of a Suslin line needs much more delicate set-theoretical assumptions (cf. Appendix 2). However, under the same **CH**, Kunen was able to construct a compact L-space (see his work [149]). In this context, it should be noticed that the existence of L-spaces cannot be established within **ZFC** set theory. More detailed information on L-spaces (and on dual S-spaces) may be found, e.g., in [49] and [223].

8. Assume that $2^\omega < 2^{\omega_1}$ (notice that this inequality trivially follows from the Continuum Hypothesis). Let X be an uncountable subset of $\mathbf{R}$.

Demonstrate that the following two assertions are equivalent:

(a) X is a Sierpiński set in $\mathbf{R}$;

(b) for any λ-measurable function $f : \mathbf{R} \to \mathbf{R}$, the restriction $f|X$ is a Borel function on X.

9. Let X be a subset of $\mathbf{R}$ such that $\lambda^*(X) > 0$, let E be a λ-measurable hull of X and let λ_X denote the measure on X produced by the restriction of λ to E (see Exercise 22 from Chapter 3). Suppose again that $2^\omega < 2^{\omega_1}$.

Demonstrate that the following two assertions are equivalent:

(a) X is a Sierpiński set in $\mathbf{R}$;

(b) every λ_X-measurable real-valued function on X is Borel.

In order to show this equivalence, use the previous exercise and Exercise 22 from Chapter 3.

10. Let E be an uncountable Polish topological space and let μ be the completion of a nonzero σ-finite diffused Borel measure on E.

A subset X of E is called a Sierpiński set in E (with respect to μ) if X is uncountable and $\operatorname{card}(X \cap Y) \leq \omega$ for every μ-measure zero set Y in E.

Prove, assuming the Continuum Hypothesis, that there exists a Sierpiński set with respect to μ.

Further, take $E = \mathbf{R}^2$ and $\mu = \lambda_2$.

By assuming again **CH**, show that there exists a Sierpiński set $X \subset E$, with respect to μ, such that all points of X are in general position, i.e., no three distinct points of X are collinear.

11*. Prove that these two assertions are equivalent within **ZFC** theory:
(a) the Continuum Hypothesis (**CH**);
(b) there exists a set $Z \subset \mathbf{R}^2$ having the property that all horizontal sections (by lines) of Z are at most countable and all vertical sections (by lines) of Z are co-countable, i.e., are complements of at most countable subsets of vertical lines.

Argue in the following manner. In fact, the implication (a) $\Rightarrow$ (b) was established in the proof of Theorem 4 of this chapter. So it remains to demonstrate the validity of the converse implication (b) $\Rightarrow$ (a). Let Z be a subset of $\mathbf{R}^2$ satisfying (b). Take any set $Y \subset \mathbf{R}$ with $\operatorname{card}(Y) = \omega_1$ and put

$$D = Z \cap (\mathbf{R} \times Y).$$

Check that, for the set D, the following two relations hold:
(i) $\operatorname{card}(D) \leq \omega_1$;
(ii) $\operatorname{pr}_1(D) = \mathbf{R}$.
Finally, deduce from (i) and (ii) that $\mathbf{c} \leq \omega_1$, so $\mathbf{c} = \omega_1$.

12*. Work in **ZF** & **DC** set theory and suppose that the real line $\mathbf{R}$ can be made well-ordered by some relation $\preceq$.

Demonstrate that under this assumption there exists a λ-nonmeasurable subset of $\mathbf{R}$.

Argue in the following manner. First, represent $\mathbf{R}$ in the form of an injective α-sequence $\{x_\xi : \xi \prec \alpha\}$, where α is some ordinal number. Let $\beta \preceq \alpha$ be the least ordinal number such that

$$\lambda^*(\{x_\xi : \xi \prec \beta\}) > 0,$$

where λ^* denotes, as usual, the outer measure associated with λ. If the set $X = \{x_\xi : \xi \prec \beta\}$ is λ-nonmeasurable, then there is nothing to prove. If X is λ-measurable, then in the λ_2-measurable product set $X \times X$ consider the set

$$Z = \{(x_\xi, x_\zeta) : \xi \preceq \zeta \prec \beta\}$$

and apply to Z an argument similar to the proof of Theorem 4 of this chapter.

Conclude that Z is nonmeasurable with respect to λ_2, which implies the existence of a λ-nonmeasurable subset of $\mathbf{R}$.

Generalize the obtained result to the case where λ is replaced by the completion μ of a nonzero σ-finite diffused Borel measure on $\mathbf{R}$.

13. Again, work in **ZF** & **DC** set theory and suppose that the real line $\mathbf{R}$ can be made well-ordered by some relation $\preceq$.

By using an argument different from that given in Exercise 12, show the existence of a λ-nonmeasurable subset of $\mathbf{R}$.

For this purpose, take into account the fact that any well-ordering of $\mathbf{R}$ produces some linear ordering of the power set $\mathcal{P}(\mathbf{R})$, hence produces some linear ordering of the family of all countable subsets of $\mathbf{R}$. Then apply the result of Exercise 4 from Chapter 3.

14. Formulate and prove the result analogous to Exercise 12 in terms of the Baire property. Namely, demonstrate, within **ZF** & **DC** set theory, that if the real line $\mathbf{R}$ admits a well-ordering of all of its points, then there exists a subset of $\mathbf{R}$ which does not have the Baire property.

For this purpose, argue as in Exercise 12, but refer to a certain topological analogue of Fubini's theorem, which is known as the Kuratowski-Ulam theorem (see, e.g., [152], [203]). Another way to obtain the required result is to use the trick indicated in Exercise 13.

15. A subset X of $\mathbf{R}$ is called a Luzin set if $\mathrm{card}(X) > \omega$ and $\mathrm{card}(X \cap Y) \leq \omega$ for every first category set Y in $\mathbf{R}$.

Thus, denoting by $\mathcal{K}(\mathbf{R})$ the σ-ideal of all first category sets in $\mathbf{R}$, one can observe by virtue of the above definition that a Luzin set X almost avoids all members of $\mathcal{K}(\mathbf{R})$.

Demonstrate, assuming the Continuum Hypothesis, that there are Luzin sets on $\mathbf{R}$.

For this purpose, apply the method of transfinite recursion to the σ-ideal $\mathcal{K}(\mathbf{R})$ (similarly to the construction of a Sierpiński set in the proof of Theorem 1).

Remark 8. In connection with the above exercise, it should be noticed that the Continuum Hypothesis is not necessary for the existence of Luzin sets, because there are models of **ZFC** theory in which **CH** fails to hold and there exist Luzin sets on $\mathbf{R}$. More precisely, there are models of **ZFC** theory, in which the negation of the Continuum Hypothesis is valid and there exist Luzin sets of cardinality $\mathbf{c}$ (cf. [150]). It can easily be seen that in such a model there is a subset of $\mathbf{R}$ with cardinality ω_1 ($< \mathbf{c}$), which does not possess the Baire property.

16. Verify that:
(a) the family of all Luzin sets in $\mathbf{R}$ is countably additive;

(b) the same family is invariant under the group of all homeomorphisms of $\mathbf{R}$ and, in particular, is invariant under the group of all translations of $\mathbf{R}$;

(c) any Luzin set has λ-measure zero, i.e., the family of all Luzin sets is contained in the σ-ideal $\mathcal{I}(\lambda)$;

(d) an uncountable set $X \subset \mathbf{R}$ is a Luzin set if and only if every uncountable subset of X does not possess the Baire property;

(e) no Luzin set can be a Vitali set or a Bernstein set.

Remark 9. We will be dealing with various kinds of Luzin sets in further sections of this book. We will also introduce and envisage the so-called generalized Luzin sets on $\mathbf{R}$. Their properties are very similar to corresponding properties of Luzin sets.

17. Assume that $2^{\omega} < 2^{\omega_1}$. Let X be an uncountable subset of $\mathbf{R}$.

Demonstrate that the following two assertions are equivalent:

(a) X is a Luzin set in $\mathbf{R}$;

(b) for any function $f : \mathbf{R} \to \mathbf{R}$ possessing the Baire property, the restriction $f|X$ is a Borel function on X.

18. Assume Martin's Axiom (**MA**) with the negation of the Continuum Hypothesis (see Appendix 3 where some information about **MA** is given).

Show that, under this assumption, there are neither Sierpiński sets nor Luzin sets on the real line $\mathbf{R}$.

19*. Let $[\mathbf{R}]^{\leq \omega}$ denote the family of all (at most) countable subsets of $\mathbf{R}$.

Show the equivalence of the following two assertions:

(a) for every function $f : \mathbf{R} \to [\mathbf{R}]^{\leq \omega}$, there exist two distinct points $x \in \mathbf{R}$ and $y \in \mathbf{R}$ such that $x \notin f(y)$ and $y \notin f(x)$;

(b) the negation of the Continuum Hypothesis.

Argue as follows. Let (a) be satisfied and suppose to the contrary that (b) is not valid, so $\mathbf{c} = \omega_1$. Let $\{x_\xi : \xi < \omega_1\}$ be a bijective enumeration of all points of $\mathbf{R}$. Define

$$f(x_\xi) = \{x_\zeta : \zeta < \xi\} \qquad (\xi < \omega_1)$$

and check that f does not fulfill (a). This contradiction shows that the implication (a) $\Rightarrow$ (b) holds true.

Suppose now that (b) is satisfied and take any mapping $f : \mathbf{R} \to [\mathbf{R}]^{\leq \omega}$. Let $X \subset \mathbf{R}$ be a set of cardinality ω_1. Consider the set

$$Y = X \cup (\cup \{f(x) : x \in X\}).$$

Since $\text{card}(Y) = \omega_1$ and $\text{card}(\mathbf{R}) > \omega_1$, there exists a point $y \in \mathbf{R} \setminus Y$. Also, since X is uncountable and $f(y)$ is at most countable, there exists a point $x \in X \setminus f(y)$. Thus, for the points x and y, the relations

$$x \notin f(y), \quad y \notin f(x)$$

are true, which establishes the validity of the implication (b) $\Rightarrow$ (a).

Remark 10. The equivalence (a) $\Leftrightarrow$ (b) of Exercise 19 is due to Sierpiński. Assertion (a) of the same exercise is sometimes called Freiling's axiom of symmetry (see [61]).

20*. For any set X, denote by $[X]^{<\omega}$ the family of all finite subsets of X. Let X be an infinite set whose cardinality is not cofinal with ω and let $g : X \to [X]^{<\omega}$ be a mapping.

Demonstrate that there exists a set $Y \subset X$ satisfying the following two conditions:

(a) $\mathrm{card}(Y) = \mathrm{card}(X)$;

(b) $y \notin f(y')$ for any two distinct elements $y \in Y$ and $y' \in Y$.

Argue as follows. First of all, reduce the general situation to the case where $\mathrm{card}(f(x)) \leq n$ for some fixed natural number n and for all $x \in X$. Then apply simultaneously the method of induction on n and transfinite induction up to α where α denotes the least ordinal number equinumerous with $\mathrm{card}(X)$.

As a consequence, infer from the stated above that if $h : \mathbf{R} \to [\mathbf{R}]^{<\omega}$ is a mapping, then there exists a set $Z \subset \mathbf{R}$ such that $\mathrm{card}(Z) = \mathbf{c}$ and

$$z \notin f(z'), \quad z' \notin f(z)$$

for any two distinct points $z \in Z$ and $z' \in Z$.

Remark 11. If in the previous exercise $\mathrm{card}(X) = \omega_1$, then the required result can be deduced from the Δ-system lemma (see Exercise 7 of Appendix 3).

21*. Work in **ZF** set theory and show that there exists a Borel function

$$f : \mathbf{R}^\omega \to \mathbf{R}$$

such that $f(x) \notin \mathrm{ran}(x)$ for each sequence $x \in \mathbf{R}^\omega$.

In order to check this fact, analyze Cantor's classical proof of the uncountability of **R**. Actually, for any $x \in \mathbf{R}^\omega$, construct effectively a decreasing (by inclusion) sequence of closed bounded intervals, whose lengths tend to zero and whose endpoints are rational numbers, such that x does not belong to the intersection of all these intervals.

On the other hand, denoting by $[\mathbf{R}]^\omega$ the family of all countably infinite subsets of **R**, demonstrate that it is impossible to prove, within **ZF** & **DC** theory, the existence of a function

$$g : [\mathbf{R}]^\omega \to \mathbf{R}$$

such that $g(x) \notin x$ for each set $x \in [\mathbf{R}]^\omega$.

In order to establish this fact, first verify that if such a g does exist, then there exists a subset of $\mathbf{R}$ whose cardinality is equal to ω_1. Finally, take into account the results of Solovay [253] and of Shelah [229] and Raisonnier [214].

22. A subset X of $\mathbf{R}$ is called a universally Baire set if, for any topological space E and for any continuous mapping $f : E \to \mathbf{R}$, the pre-image $f^{-1}(X)$ has the Baire property in E.

Verify that the following two assertions are valid:

(a) the family of all universally Baire subsets of $\mathbf{R}$ forms a σ-algebra containing the family of all analytic subsets of $\mathbf{R}$ (in particular, every Borel set in $\mathbf{R}$ is universally Baire);

(b) if $X \subset \mathbf{R}$ is universally Baire, then X is measurable in the Lebesgue sense.

For (a), take into account the facts that any analytic subset of $\mathbf{R}$ (of E) can be regarded as the result of A-operation over some A-system of closed subsets of $\mathbf{R}$ (of E) and that the analytic sets in E have the Baire property (see Appendix 5).

For (b), use the density topology on $\mathbf{R}$ which is stronger than the usual Euclidean topology of $\mathbf{R}$ (see Exercise 6).

Remark 12. Universally Baire sets were introduced relatively recently and play a substantial role in some important topics of foundations of mathematics. In particular, these sets highlight new profound aspects of the theory of large cardinals and of their connections with the Continuum Hypothesis (for more details, see [110], [271]).

23*. Let $\{A, B\}$ be a partition of $\mathbf{R}$ such that A is of Lebesgue measure zero and B is of first category in $\mathbf{R}$.

Prove that:

(a) if a set $X \subset \mathbf{R}$ is of second category, then $X + A = \mathbf{R}$;

(b) if a set $Y \subset \mathbf{R}$ is of strictly positive outer Lebesgue measure, then $Y + B = \mathbf{R}$.

Deduce from (a) and (b) the result of Rothberger [221] stating that if there exist a Luzin set L and a Sierpiński set S, then

$$\text{card}(L) = \text{card}(S) = \omega_1.$$

24*. Verify that:

(a) if $L \subset \mathbf{R}$ is a Luzin set and $f : L \to \mathbf{R}$ is a homeomorphism between L and $f(L)$, then $f(L)$ is not, in general, a Luzin set;

(b) if $S \subset \mathbf{R}$ is a Sierpiński set and $g : S \to \mathbf{R}$ is a homeomorphism between S and $g(S)$, then $g(S)$ is not, in general, a Sierpiński set.

For this purpose, equip $\mathbf{R}$ with the standard Lebesgue measure λ, equip the Cantor space $C = \{0, 1\}^\omega$ with the completion of the canonical Borel probability product measure μ, and take into account the following two statements of **ZF** & **DC** theory (see, e.g., [33]):

(i) there exists a Borel isomorphism f between $\mathbf{R}$ and C such that both mappings f and f^{-1} preserve the category of sets;

(ii) there exists a Borel isomorphism g between $\mathbf{R}$ and C such that both mappings g and g^{-1} preserve measure zero sets.

Finally, keep in mind the fact that C is topologically contained in $\mathbf{R}$.

25. Prove the following recursive formula due to Bernstein and Hausdorff:

$$\omega_{\xi+1}^{\mathbf{a}} = \omega_{\xi+1} \cdot \omega_{\xi}^{\mathbf{a}},$$

where $\mathbf{a}$ is an arbitrary nonzero cardinal and ξ is any ordinal number.

For this purpose, keeping in mind the regularity of $\omega_{\xi+1}$, consider separately two possible cases: $\mathbf{a} < \omega_{\xi+1}$ and $\mathbf{a} \geq \omega_{\xi+1}$.

Further, assume the Continuum Hypothesis. By using the above recursive formula and induction on n, show the validity of the equality

$$\omega_n^{\omega} = \omega_n,$$

where n is an arbitrary nonzero natural number.

Remark 13. In **ZFC** theory there are many important formulas (equalities and inequalities) concerning various types of sums and products of cardinal numbers (see, e.g., [18], [103], [154], [243]). Among such formulas of cardinal arithmetic the most intriguing is Shelah's inequality

$$(\omega_{\omega})^{\omega} \leq 2^{\omega} + \omega_{\alpha},$$

where $\alpha = \omega_4$. For the proof, see Shelah's extensive monograph [230].

26. Assuming **CH** and applying the method of transfinite induction, demonstrate that there exist a Luzin set X and a Sierpiński set Y such that

$$X + X = \{x + x' : x \in X,\ x' \in X\} = \mathbf{R},$$

$$Y + Y = \{y + y' : y \in Y,\ y' \in Y\} = \mathbf{R}.$$

5. Measurability properties of sets and functions

Here we are going to consider some properties of those sets and functions which are measurable (or, on the contrary, nonmeasurable) with respect to various classes of measures.

In this context, the concept of an absolutely nonmeasurable set (function) with respect to a given class of measures is introduced and examined.

Further, Sierpiński–Zygmund type functions (which may be treated as utterly bad from the purely topological point of view) are discussed in connection with their absolute non-measurability with respect to the class of completions of all nonzero σ-finite diffused Borel measures on the real line $\mathbf{R}$.

Also, for an uncountable commutative group $(G, +)$, absolutely negligible subsets and absolutely nonmeasurable subsets of G with respect to the class of all nonzero σ-finite translation quasi-invariant measures on G are introduced and some of their properties are envisaged.

We begin our presentation with several general definitions from measure theory.

Let E be a set and let $\mathcal{M}$ be a class of measures on E (of course, it may often happen that measures belonging to $\mathcal{M}$ are defined on different σ-algebras of subsets of E).

Definition 1. We say that a function

$$f : E \to \mathbf{R}$$

is absolutely (or universally) measurable with respect to $\mathcal{M}$ if f is measurable with respect to all measures from $\mathcal{M}$. Accordingly, we say that a set $X \subset E$ is absolutely (or universally) measurable with respect to $\mathcal{M}$ if the characteristic function

$$\chi_X : E \to \{0, 1\}$$

is absolutely measurable with respect to $\mathcal{M}$.

It directly follows from Definition 1 that the family of all absolutely measurable functions with respect to $\mathcal{M}$ forms a linear algebra (over $\mathbf{R}$) closed under pointwise limits of its members. Accordingly, the family of all absolutely measurable sets with respect to $\mathcal{M}$ forms a σ-algebra of subsets of E.

It makes sense to touch upon here a special but extremely important case of E and $\mathcal{M}$.

Let a ground set E be an uncountable Polish topological space and let $\mathcal{M}$ be the class of completions of all σ-finite Borel measures on E. It is well known that in this concrete situation the absolutely measurable subsets of E are precisely the Radon subspaces of E (see, e.g., [17], [20], [33], [265] or Theorems 2 and 3 from the next Chapter 6).

In connection with the notion of absolute measurability, the following definition seems also to be rather natural.

Definition 2. We say that a function

$$g : E \to \mathbf{R}$$

is relatively measurable with respect to $\mathcal{M}$ if there exists at least one measure $\mu \in \mathcal{M}$ such that g is μ-measurable. Accordingly, we say that a set $Y \subset E$ is relatively measurable with respect to $\mathcal{M}$ if the characteristic function

$$\chi_Y : E \to \{0, 1\}$$

is relatively measurable with respect to $\mathcal{M}$.

We now introduce the principal definition of this chapter.

Definition 3. We say that a function

$$h : E \to \mathbf{R}$$

is absolutely nonmeasurable with respect to $\mathcal{M}$ if there exists no measure $\mu \in \mathcal{M}$ such that h is μ-measurable. Accordingly, we say that a set $Z \subset E$ is absolutely nonmeasurable with respect to $\mathcal{M}$ if the characteristic function

$$\chi_Z : E \to \{0, 1\}$$

is absolutely nonmeasurable with respect to $\mathcal{M}$. .

So, for any given class $\mathcal{M}$ of measures on E and for any function

$$f : E \to \mathbf{R},$$

we have the following three notions:
(1) the absolute measurability of f with respect to $\mathcal{M}$;
(2) the relative measurability of f with respect to $\mathcal{M}$;
(3) the absolute non-measurability of f with respect to $\mathcal{M}$.

As was mentioned in the Preface, we incline to think that these notions are much deeper than the usual concept of the measurability (nonmeasurability) of a function $f : E \to \mathbf{R}$ with respect to a fixed measure μ on E. In fact, many results of measure theory can be formulated in terms of the above-mentioned notions and quite often it turns out that the introduced notions and the corresponding terminology are more convenient, more useful and lead to a better understanding of the subject.

Some of the subsequent sections of this book serve to confirm the above concepts.

Of course, if the class $\mathcal{M}$ consists only of a single measure μ, i.e.,

$$\mathcal{M} = \{\mu\},$$

then the concepts (1) through (3) are reduced to the ordinary concepts of measurability (nonmeasurability) with respect to μ. Namely, in this very particular case, (1) and (2) are equivalent and state that a function f is μ-measurable, while (3) simply states that f is μ-nonmeasurable.

The situation essentially changes and becomes much more complicated when $\mathcal{M}$ is a sufficiently rich class of measures. In such a case, the concepts of absolute measurability, relative measurability and absolute nonmeasurability substantially differ from each other.

At first sight, absolutely nonmeasurable functions (absolutely nonmeasurable sets) seem to be so pathological and paradoxical that their role in mathematical analysis and measure theory is expected to be very limited. But we will show in our further considerations that absolutely nonmeasurable sets and functions naturally occur in various questions and topics of real analysis and measure theory. Moreover, quite frequently it may happen that a function (set) absolutely nonmeasurable with respect to a concrete class $\mathcal{M}$ of measures on E is relatively measurable with respect to another class $\mathcal{M}'$ of measures on E.

Let us give several typical examples illustrating the notions just introduced. Some of them will be envisaged and discussed more thoroughly in other chapters of this book.

Example 1. In Chapter 3 we were briefly concerned with Vitali subsets of the real line $\mathbf{R}$. It turns out that any Vitali set is absolutely nonmeasurable with respect to the class of all those measures on $\mathbf{R}$ which are translation invariant extensions of the Lebesgue measure λ (see, e.g., [128], [197], [268]). On the other hand, there exist Vitali's subsets of $\mathbf{R}$ which are relatively measurable with respect to the class of all those measures on $\mathbf{R}$ which are translation quasi-invariant extensions of λ (cf. [139] or Chapter 9 where Vitali sets are considered in more details).

Example 2. Let H be an infinite-dimensional separable Hilbert space (over the field $\mathbf{R}$) and let B denote the unit ball of this space. Then B is absolutely nonmeasurable with respect to the class of all nonzero σ-finite H-quasi-invariant measures on H (see [122] or [137]). As an easy consequence, it follows from the absolute nonmeasurability of B that there exists no nonzero σ-finite translation quasi-invariant Borel measure on H (in connection with this well-known fact and further related results see, for instance, the monograph by Skorokhod [250] and references therein). Also, it is known that the above-mentioned fact remains true for any non-locally compact Polish group, instead of H, and also for any nonseparable metrizable group whose topological weight is not measurable in the Ulam sense (see Exercise 5 for this chapter).

Example 3. The so-called Bernstein subsets of topological spaces play an essential role in many topics of general topology, measure theory and real analysis. Recall that a subset B of a topological space E is a Bernstein set (in E) if neither B nor $E \setminus B$ contains a topological copy of the Cantor space $\{0, 1\}^\omega$ (cf. [33], [147], [152], [188], [190], [203] and Chapter 3). Any Bernstein subset of an uncountable Polish topological space E is absolutely nonmeasurable with respect to the class $\mathcal{CBM}_0(E)$ of completions of all nonzero σ-finite diffused (i.e., vanishing at the singletons) Borel measures on E. In Exercise 1 for the present chapter we suggest the reader prove this fact. More precisely, the family of all Bernstein subsets of E coincides with the family of all absolutely nonmeasurable sets with respect to the same class $\mathcal{CBM}_0(E)$.

Example 4. Let $\mathbf{c}$ denote the cardinality of the continuum. Recall that a function $f : \mathbf{R} \to \mathbf{R}$ is a Sierpiński–Zygmund function if the restriction of f to each subset of $\mathbf{R}$ of cardinality $\mathbf{c}$ is discontinuous (see [152], [246], and Chapter 3). It is not difficult to see that all Sierpiński–Zygmund functions on $\mathbf{R}$ are absolutely nonmeasurable with respect to the class $\mathcal{CBM}_0(\mathbf{R})$ mentioned in Example 3. The characteristic function of a Bernstein subset of $\mathbf{R}$ enables us to conclude that there exist absolutely nonmeasurable functions with respect to the same class of measures, which essentially differ from Sierpiński–Zygmund functions.

It is not hard to describe all those functions

$$h : \mathbf{R} \to \mathbf{R}$$

which are relatively measurable with respect to the class $\mathcal{CBM}_0(\mathbf{R})$. Namely, the following statement is valid.

Theorem 1. *A function $h : \mathbf{R} \to \mathbf{R}$ is relatively measurable with respect to $\mathcal{CBM}_0(\mathbf{R})$ if and only if h admits a representation in the form*

$$h = g \circ f,$$

where $g : \mathbf{R} \to \mathbf{R}$ is a Lebesgue measurable function and f is a Borel isomorphism of $\mathbf{R}$ onto itself.

Proof. Suppose first that a function $h : \mathbf{R} \to \mathbf{R}$ can be represented in the above-mentioned form, i.e., $h = g \circ f$. Denote by λ the standard Lebesgue measure on $\mathbf{R}$ and let λ_0 be the restriction of λ to the Borel σ-algebra $\mathcal{B}(\mathbf{R})$ of $\mathbf{R}$. Further, define a Borel measure μ_0 on $\mathbf{R}$ by the equality

$$\mu_0(B) = \lambda_0(f(B)) \quad (B \in \mathcal{B}(\mathbf{R})).$$

Obviously, the measure μ_0 is nonzero, σ-finite and continuous (i.e., diffused). The Borel isomorphism f^{-1} is also an isomorphism between λ and the completion μ of μ_0. So we readily deduce that h is μ-measurable.

Conversely, suppose that $h : \mathbf{R} \to \mathbf{R}$ is relatively measurable with respect to the class $\mathcal{CBM}_0(\mathbf{R})$. Then there exists a nonzero σ-finite continuous Borel measure μ_0 on $\mathbf{R}$ such that h is measurable with respect to the completion μ of μ_0. We may assume, without loss of generality, that μ_0 (hence μ) is a probability measure. Let λ' denote a probability measure on $\mathbf{R}$ which is equivalent to λ. This means that both λ and λ' have the same domain and the σ-ideal $\mathcal{I}(\lambda)$ of all λ-measure zero sets coincides with the σ-ideal $\mathcal{I}(\lambda')$ of all λ'-measure zero sets. According to a well-known theorem of classical measure theory, there exists a Borel isomorphism $f : \mathbf{R} \to \mathbf{R}$ which transforms μ onto λ' (see, e.g., [17], [33], [115]). This implies that $h \circ f^{-1}$ is a Lebesgue measurable function. Finally, denoting

$$g = h \circ f^{-1},$$

we easily obtain $h = g \circ f$, which shows that h admits the required representation. This finishes the proof of the theorem.

In connection with Theorem 1, let us recall that there are many Lebesgue nonmeasurable functions among the compositions $g \circ f$ of the above-mentioned type (see, e.g., Exercise 13 from Chapter 3).

Example 5. Let μ be a σ-finite measure on E and let the symbol $\mathcal{M}(\mu)$ denote the class of all those measures on E which extend μ. Then, in accordance with the main result of [2], any real-valued step function on E (i.e., any real-valued function on E with at most countable range) is relatively measurable with respect to $\mathcal{M}(\mu)$. In other words, for every disjoint countable family of subsets of E, there exists a measure ν from $\mathcal{M}(\mu)$ such that all members of this family turn out to be measurable with respect to ν (see, for instance, Exercise 7 from Chapter 7).

Let E be an uncountable set and let $\mathcal{M}(E)$ denote the class of all nonzero σ-finite diffused measures on E (their domains may be diverse σ-algebras of subsets of E).

It is easy to verify that there are no absolutely nonmeasurable sets with respect to $\mathcal{M}(E)$. Moreover, for an arbitrary measure $\mu \in \mathcal{M}(E)$ and for each subset X of E, there exists a measure ν on E extending μ and satisfying the relation $X \in \mathrm{dom}(\nu)$ (a proof of this fact is sketched in Exercise 19 of the present chapter). By using the method of induction, it is not difficult to show that the analogous fact is valid for any measure $\mu \in \mathcal{M}(E)$ and for any finite family of subsets of E. On the other hand, the situation may significantly change for some $\mu \in \mathcal{M}(E)$ and for some countable families of subsets of E.

Remark 1. If $\{X_1, X_2, ..., X_n\}$ is an arbitrary finite family of subsets of E, then it produces the disjoint finite covering of E consisting of all those sets which are representable in the form

$$X_1' \cap X_2' \cap ... \cap X_n',$$

where each X_i' is either X_i or $E \backslash X_i$. By virtue of Example 5, for any $\mu \in \mathcal{M}(E)$, there exists a measure ν extending μ and such that all the the above-mentioned sets $X_1' \cap X_2' \cap ... \cap X_n'$ become ν-measurable. It immediately follows from this observation that all initial sets $X_1, X_2, ..., X_n$ are ν-measurable, too. As has already been pointed out, for some sequences

$$\{X_1, \ X_2, \ ... \ , \ X_n, \ ... \ \}$$

of subsets of an uncountable set E, the analogous statement may be false. This circumstance is closely connected with the existence of uncountable universal measure zero subsets of $\mathbf{R}$ which will be discussed below in the present chapter and, more thoroughly, in subsequent sections of the book.

We want to stress once more that if E is an uncountable ground set, X is a subset of E and $\mu \in \mathcal{M}(E)$, then there always exists a measure $\mu' \in \mathcal{M}(E)$ extending μ and such that $X \in \mathrm{dom}(\mu')$. Consequently, there do not exist subsets of E which are absolutely nonmeasurable with respect to the class $\mathcal{M}(E)$. On the other hand, we shall see later that the case of real-valued functions defined on E essentially differs from the case of subsets of E.

Example 6. Let E be a set, $f : E \to \mathbf{R}$ be a function and let, for some point $t_0 \in \mathbf{R}$, the relation

$$\mathrm{card}(f^{-1}(t_0)) > \omega$$

be satisfied. In this case we can assert that f is relatively measurable with respect to the class $\mathcal{M}(E)$. Indeed, it is not difficult to define a complete diffused probability measure μ_0 on E such that

$$f^{-1}(t_0) \in \mathrm{dom}(\mu_0), \quad \mu_0(f^{-1}(t_0)) = 1.$$

Consequently, for any set $T \subset \mathbf{R}$, we have $\mu_0(f^{-1}(T)) = 1$ if $t_0 \in T$, and $\mu_0(f^{-1}(T)) = 0$ if $t_0 \notin T$. This fact immediately implies that f is measurable with respect to the measure μ_0 and hence is relatively measurable with respect to the class $\mathcal{M}(E)$.

Remark 2. From Example 6 it follows, in particular, that if an original ground set E satisfies the relation

$$\mathrm{card}(E) > 2^\omega = \mathbf{c},$$

then every function $f : E \to \mathbf{R}$ is relatively measurable with respect to the class $\mathcal{M}(E)$. Indeed, in this case there always exists a point $t_0 \in \mathbf{R}$ for which we have $\mathrm{card}(f^{-1}(t_0)) > \omega$.

Definition 4. Let E be a topological space all singletons of which belong to the Borel σ-algebra $\mathcal{B}(E)$. We say that E is a universal measure zero space (or an absolute null space) if there exists no nonzero σ-finite continuous (i.e., diffused) Borel measure on E.

It directly follows from this definition that if a topological space E is universal measure zero and a topological space E' is Borel isomorphic to E, then E' is also universal measure zero.

Several nontrivial examples of such spaces E are given in exercises for this chapter. For instance, see Exercise 11 where it is indicated that some non-discrete Hausdorff universal measure zero spaces can be of arbitrarily large cardinality. See also Exercise 23 where it is pointed out that any Luzin subset of $\mathbf{R}$ is universal measure zero.

The natural question arises whether it is possible to prove within **ZFC** set theory that there exist uncountable universal measure zero subspaces of the real line $\mathbf{R}$. This highly nontrivial and important question was investigated from different points of view. By using delicate arguments, several eminent authors (Hausdorff, Luzin, Sierpiński, Marczewski, and others) have established that there are uncountable universal measure zero subspaces of $\mathbf{R}$. One classical construction of such a subspace of $\mathbf{R}$ was done by Luzin and is presented, e.g., in [152] (see also Appendix 5). This construction is based on the canonical decomposition of a non-Borel analytic subset of $\mathbf{R}$ into its Borel components (the so-called constituents). There are also other interesting methods for establishing the existence of uncountable universal measure zero subspaces of $\mathbf{R}$ (see, for instance, [83], [211]). Moreover, some constructions of uncountable universal measure zero sets and of uncountable sets with a much stronger property of "smallness" are presented in [208], [273].

Various ideas and approaches are used in the above-mentioned constructions (e.g., Marczewski's characteristic function, Ulam's transfinite matrix, Fubini type argument, and so on). In Exercise 12 for this chapter it is shown, within **ZF** & **DC** theory, that the existence of a universal measure zero subspace U of $\mathbf{R}$ with $\mathrm{card}(U) = \omega_1$ is equivalent to one purely set-theoretical statement concerning the internal structure of ω_1.

By taking into account Example 6 and using the notion of a universal measure zero space, a characterization of absolutely nonmeasurable functions with respect to the class $\mathcal{M}(E)$ can be obtained for any ground set E.

Theorem 2. *Let E be an arbitrary set and let*

$$f : E \to \mathbf{R}$$

be a function. The following two conditions are equivalent:

(1) f is absolutely nonmeasurable with respect to the class $\mathcal{M}(E)$;

(2) *for each $r \in \mathbf{R}$, the set $f^{-1}(r)$ is at most countable and the set $\mathrm{ran}(f)$ is a universal measure zero subspace of $\mathbf{R}$.*

Proof. Suppose first that f is absolutely nonmeasurable with respect to $\mathcal{M}(E)$. Then the argument given in Example 6 shows that the inequality

$$\mathrm{card}(f^{-1}(r)) \le \omega$$

must be satisfied for any $r \in \mathbf{R}$. Let us check that the set $\mathrm{ran}(f)$ has universal measure zero. Indeed, assuming to the contrary that $\mathrm{ran}(f)$ is not a universal measure zero subset of $\mathbf{R}$, consider some Borel diffused probability measure ν on $\mathrm{ran}(f)$ and denote

$$\mathcal{S} = \{f^{-1}(Z) : Z \in \mathrm{dom}(\nu)\}.$$

Obviously, $\mathcal{S}$ is a σ-algebra of subsets of E and the family of countable sets

$$\{f^{-1}(r) : r \in \mathrm{ran}(f)\}$$

forms a disjoint covering of E. We now put

$$\mu(f^{-1}(Z)) = \nu(Z) \qquad (Z \in \mathrm{dom}(\nu)).$$

In this manner, the probability measure μ on the σ-algebra $\mathcal{S}$ is well defined and, according to the definition of μ, the function f becomes μ-measurable. Clearly, the completion μ' of μ is a diffused measure and f remains measurable with respect to μ'. However, this circumstance contradicts our assumption that f is absolutely nonmeasurable with respect to the class $\mathcal{M}(E)$. The contradiction obtained shows the necessity of (2) for the absolute nonmeasurability of f with respect to $\mathcal{M}(E)$.

Now, assume that condition (2) is fulfilled for a given function f and let us establish that f is absolutely nonmeasurable with respect to the class $\mathcal{M}(E)$. Suppose for a moment that there exists a measure μ belonging to $\mathcal{M}(E)$ such that f is measurable with respect to μ. We may assume, without loss of generality, that μ is a probability measure (because any nonzero σ-finite measure is equivalent to an appropriate probability measure). Denoting by $\mathcal{B}(\mathrm{ran}(f))$ the Borel σ-algebra of the subspace $\mathrm{ran}(f)$ of $\mathbf{R}$, we may put

$$\nu(Z) = \mu(f^{-1}(Z)) \qquad (Z \in \mathcal{B}(\mathrm{ran}(f))).$$

So we get a Borel probability measure ν on the space $\mathrm{ran}(f) \subset \mathbf{R}$ and we see that ν is diffused in view of the inequality $\mathrm{card}(f^{-1}(r)) \le \omega$ for all $r \in \mathbf{R}$. But this circumstance contradicts the fact that $\mathrm{ran}(f)$ has universal measure zero.

The obtained contradiction finishes the proof of Theorem 2.

Remark 3. The equivalence of conditions (1) and (2) of Theorem 2 implies, in particular, that if a function $f : E \to \mathbf{R}$ is absolutely nonmeasurable with respect to the class $\mathcal{M}(E)$, then the restriction of f to any set $X \subset E$ is absolutely nonmeasurable with respect to the class $\mathcal{M}(X)$. In addition to this observation, it should also be pointed out that some restrictions of very good functions to certain uncountable subsets of E can be absolutely nonmeasurable. For instance, if f is the identical transformation of $\mathbf{R}$ and X is an uncountable universal measure zero subset of $\mathbf{R}$, then the restriction $f|X$ is absolutely nonmeasurable with respect to $\mathcal{M}(X)$ (e.g., by virtue of the above-mentioned equivalence of the conditions (1) and (2) of Theorem 2).

Remark 4. The existence of uncountable universal measure zero subsets of $\mathbf{R}$ easily implies that, for every set E with $\mathrm{card}(E) = \omega_1$, there exist real-valued injective functions on E which are absolutely nonmeasurable with respect to the class $\mathcal{M}(E)$. On the other hand, the existence of a function absolutely nonmeasurable with respect to the class $\mathcal{M}(\mathbf{R})$ cannot be established within **ZFC** set theory, because there is a model of this theory in which $\omega_1 < \mathbf{c}$ and all uncountable universal measure zero subsets of $\mathbf{R}$ have cardinality ω_1 (for more details, see [188] and references therein).

Theorem 3. *Let E be a ground set, $g : E \to \mathbf{R}$ be a function such that*

$$\mathrm{card}(g^{-1}(t)) \leq \omega \qquad (t \in \mathbf{R}),$$

and let $f : \mathbf{R} \to \mathbf{R}$ be an absolutely nonmeasurable function with respect to the class $\mathcal{M}(\mathbf{R})$. Then the composition $f \circ g$ is an absolutely nonmeasurable function with respect to the class $\mathcal{M}(E)$.

In particular, the composition of any two absolutely nonmeasurable functions with respect to the class $\mathcal{M}(\mathbf{R})$ is again an absolutely nonmeasurable function with respect to $\mathcal{M}(\mathbf{R})$.

Proof. It suffices to apply a characterization of absolutely nonmeasurable functions formulated in Theorem 2.

Example 7. Under the Continuum Hypothesis (or under Martin's Axiom which is much weaker than **CH**), there exists a Sierpiński–Zygmund function on $\mathbf{R}$ absolutely nonmeasurable with respect to the class $\mathcal{M}(\mathbf{R})$. Moreover, such a function can even be additive, i.e., can be a nontrivial solution of the Cauchy functional equation (see [131]).

Example 8. There exists a Sierpiński–Zygmund function which is relatively measurable with respect to the class of all translation invariant measures on $\mathbf{R}$ extending the standard Lebesgue measure λ (see [132]). However, we do not know whether such a function can be additive. In other words, it is unknown whether there exists a Sierpiński–Zygmund function

$$f : \mathbf{R} \to \mathbf{R}$$

such that:

(a) f is relatively measurable with respect to the class of all those measure
on $\mathbf{R}$ which are translation invariant extensions of λ;

(b) f is a homomorphism of the additive group $(\mathbf{R}, +)$ into itself.

As was shown earlier (see Theorem 2), there is a close connection between
absolutely nonmeasurable real-valued functions and small (in a certain sense
subsets of $\mathbf{R}$. We now wish to introduce the notions of absolutely small (ab
solutely negligible) sets and of absolutely nonmeasurable sets in uncountabl
commutative groups.

Let $(G, +)$ be a commutative group and let μ be a measure defined on som
σ-algebra of subsets of G.

We recall that μ is a translation invariant measure on G (or a G-invariant
measure on G) if $\mathrm{dom}(\mu)$ is a translation invariant σ-algebra of sets and

$$\mu(g + X) = \mu(X + g) = \mu(X)$$

for each element $g \in G$ and for each set $X \in \mathrm{dom}(\mu)$.

Recall also that a measure μ defined on some σ-algebra of subsets of G i
said to be a translation quasi-invariant measure on G (or a G-quasi-invarian
measure on G) if both $\mathrm{dom}(\mu)$ and $\mathcal{I}(\mu)$ are translation invariant classes of sets

Clearly, every translation invariant measure on G is simultaneously transla
tion quasi-invariant. The converse assertion is not true, in general.

Let us formulate the precise definition of sets which are small from the view
point of the theory of invariant (quasi-invariant) measures. Here we restric
ourselves to the case of σ-finite invariant (quasi-invariant) measures given o
uncountable commutative groups.

Definition 5. Let $(G, +)$ be an uncountable commutative group and let X
be a subset of G. We say that X is G-absolutely negligible in G if, for every
σ-finite G-invariant (respectively, G-quasi-invariant) measure μ on G, there ex
ists a G-invariant (respectively, G-quasi-invariant) extension μ' of μ such that
$\mu'(X) = 0$.

It turns out that G-absolutely negligible sets play an essential role in ques
tions concerning extensions of G-invariant (G-quasi-invariant) measures. For
instance, it is known that G can be covered by countably many G-absolutely
negligible sets, which implies that every nonzero σ-finite G-invariant (G-quasi
invariant) measure on G admits a proper G-invariant (G-quasi-invariant) exten
sion (for more details, see [37], [126], [128], [272] and references therein).

One useful characterization of G-absolutely negligible sets is given in Exercise
25 for this chapter

Absolutely nonmeasurable subsets of commutative groups can be defined in
the following manner.

Definition 6. Let $(G, +)$ be again an uncountable commutative group
We say that a set $Y \subset G$ is G-absolutely nonmeasurable in G if Y is absolutely

nonmeasurable with respect to the class of all nonzero σ-finite G-quasi-invariant measures on G.

It is known that every uncountable commutative group $(G, +)$ contains a G-absolutely nonmeasurable subset (see, for instance, Chapter 11 of [128]).

The next example shows that the algebraic sum of two copies of an absolutely negligible set can be an absolutely nonmeasurable set. This fact may be considered as a certain analogue of the classical Sierpiński result on the existence of two Lebesgue measure zero sets whose algebraic sum is not measurable in the Lebesgue sense (see Exercise 17 from Chapter 3).

Example 9. If E is a vector space over the field $\mathbf{Q}$ of all rational numbers and $\mathrm{card}(E)$ is greater than or equal to the cardinality of the continuum, then there exists an E-absolutely negligible set $Z \subset E$ such that the vector sum

$$Z + Z = \{z + z' : z \in Z, \ z' \in Z\}$$

is E-absolutely nonmeasurable in E. For the proof, see [137].

As a straightforward consequence of the above-mentioned fact, we have the following statement:

If the Continuum Hypothesis holds and E is an arbitrary uncountable vector space over $\mathbf{Q}$, then there exists an E-absolutely negligible set $X \subset E$ such that $X + X$ is E-absolutely nonmeasurable in E.

The direct analogue of this statement holds true (under the same hypothesis) for a sufficiently large class of uncountable commutative groups. However, we do not know whether the analogous result remains valid for all uncountable commutative groups. In other words, the following question remains open for an uncountable commutative group $(G, +)$:

Does there exist a G-absolutely negligible set $X \subset G$ such that $X + X$ is a G-absolutely nonmeasurable subset of G?

In this context, let us point out that in order to obtain a positive answer to the question for all uncountable commutative groups, it suffices to get a positive answer only for the commutative groups $(G, +)$ of cardinality ω_1.

Remark 5. One can readily deduce from Example 9 that, under the assumptions imposed therein, there always exists an E-absolutely negligible set $X \subset E$ such that

$$X + X = E.$$

Indeed, the required result follows from the easily provable fact that if one has some E-absolutely nonmeasurable set Y in E, then there is a countable family $\{h_i : i \in I\}$ of elements of E, for which the equality

$$\cup\{h_i + Y \ : \ i \in I\} = E$$

holds true. Thus, the above equality is necessary (but not sufficient) for Y to be an E-absolutely nonmeasurable set in E.

EXERCISES

1. Let E be an uncountable Polish topological space and let X be a subset of E. Show that these three assertions are equivalent:

(a) X is a Bernstein set in E;

(b) X is absolutely nonmeasurable with respect to the class $\mathcal{CBM}_0(E)$ of completions of all nonzero σ-finite continuous Borel measures on E;

(c) X is absolutely nonmeasurable with respect to the class of all nonzero σ-finite continuous Radon measures on E.

2. Let B be a Bernstein subset of $\mathbf{R}$ considered as a topological space with respect to the induced topology.

Check that B is a Baire topological space, i.e., no nonempty open subset of B is of first category in B.

Remark 6. The result stated in Exercise 2 shows, in particular, that among Baire subspaces of $\mathbf{R}$ one can meet those ones which have pathological properties from the point of view of topological measure theory.

3. We say that a σ-finite diffused (continuous) complete measure μ on $\mathbf{R}$ is admissible if, for any μ-measurable set X, there exists a Borel set $Y \subset \mathbf{R}$ satisfying the equality

$$\mu(X \triangle Y) = 0,$$

where the symbol $\triangle$ denotes, as usual, the operation of symmetric difference of two sets.

Let $\mathcal{M}$ be the class of all admissible measures on $\mathbf{R}$.

Check that $\mathcal{M}$ properly contains the class of completions of all nonzero σ-finite diffused Borel measures on $\mathbf{R}$.

Let B be a Bernstein subset of $\mathbf{R}$ and let $\chi_B : \mathbf{R} \to \mathbf{R}$ denote its characteristic function.

Demonstrate that χ_B is relatively measurable with respect to $\mathcal{M}$.

4. Let $\mathcal{M}$ be the class of all admissible measures on $\mathbf{R}$ (see the above exercise) and let

$$f : \mathbf{R} \to \mathbf{R}$$

be a Sierpiński-Zigmund function.

Assuming the Continuum Hypothesis (or much weaker Martin's Axiom), prove that f is absolutely nonmeasurable with respect to $\mathcal{M}$.

Remark 7. The two preceding exercises show the substantial difference (from the point of view of relative measurability) between Sierpiński–Zygmund functions and characteristic functions of Bernstein subsets of $\mathbf{R}$.

5*. Let $(G, \cdot)$ be a non-locally compact Polish topological group.

Demonstrate that there exists no nonzero σ-finite Borel measure on G which is quasi-invariant with respect to a second category subgroup of the group of all left translations of G.

For this purpose, keep in mind the circumstance that any σ-finite Borel measure on G is concentrated on some σ-compact subset of G.

Further, let $(\Gamma, \cdot)$ be any nonseparable metrizable group whose topological weight is not measurable in the Ulam sense (see Appendix 1).

Prove that there exists no nonzero σ-finite Borel measure on Γ which is quasi-invariant with respect to an everywhere dense subgroup of the group of all left translations of Γ.

For this purpose, take into account the fact that any σ-finite Borel measure on Γ is concentrated on some closed separable subset of Γ.

6. Infer from Theorem 2 of this chapter that if a given function

$$f : E \to \mathbf{R}$$

is absolutely nonmeasurable with respect to the class $\mathcal{M}(E)$, then for any set $X \subset E$ the restriction $f|X$ is absolutely nonmeasurable with respect to the class $\mathcal{M}(X)$.

7. Show the validity of the statements formulated in Remarks 4 and 5 of this chapter.

8*. Check that, for obtaining a positive answer to the question formulated before Remark 5, it suffices to get a positive answer in the particular case of any commutative group $(G, +)$ with $\operatorname{card}(G) = \omega_1$.

For this purpose, utilize the following purely group-theoretical assertion.

If $(G, +)$ is an arbitrary uncountable group, then there exists a commutative group $(H, +)$ such that:

(a) $\operatorname{card}(H) = \omega_1$;

(b) H is a homomorphic image of G.

9. Recall that a σ-finite measure μ on a ground set E is separable if the associated Hilbert space $L_2(\mu)$ of all square μ-integrable real-valued functions on E is separable. Otherwise, μ is called a nonseparable measure.

Verify that if a σ-finite measure μ is defined on a countably generated σ-algebra of subsets of E, then μ is necessarily separable (the converse assertion is not true, in general).

Show that if

$$f : E \to \mathbf{R}$$

is an arbitrary function, then there exists a countably generated σ-algebra $\mathcal{S}$ of subsets of E such that f turns out to be $\mathcal{S}$-measurable.

Deduce from this fact that if f is measurable with respect to some nonzero σ-finite measure μ on E, then f is also measurable with respect to some nonzero σ-finite separable measure on E which is the restriction of μ to a certain countably generated σ-algebra of subsets of E.

In other words, every function $f : E \to \mathbf{R}$, which is relatively measurable with respect to the class of all σ-finite nonseparable measures on E, turns out to be relatively measurable with respect to the class of all nonzero σ-finite separable measures on E.

10. Let $(S, \leq)$ be a Suslin line (see Appendix 2).

Taking into account the fact that the cardinality of the family of all closed subsets of S is equal to $\mathbf{c}$, prove that there exists a set $B \subset S$ such that

$$\mathrm{card}(B \cap F) = \mathbf{c}, \quad \mathrm{card}((S \setminus B) \cap F) = \mathbf{c}$$

for any uncountable closed set $F \subset S$. This B may be regarded as an analogue in S of a classical Bernstein set in $\mathbf{R}$.

Check that the above-mentioned set B is absolutely nonmeasurable with respect to the class $\mathcal{CBM}_0(S)$ of completions of all nonzero σ-finite continuous Borel measures on S.

11. Let E be an infinite discrete topological space. Obviously E is locally compact. Denote by E^* Alexandrov's one-point compactification of E.

Show that the following two assertions are equivalent:

(a) $\mathrm{card}(E)$ is not measurable in the Ulam sense;

(b) E^* is a universal measure zero topological space.

Conclude from this equivalence that the existence of compact universal measure zero spaces with arbitrarily large cardinalities does not contradict **ZFC** set theory.

12*. Demonstrate, within **ZF** & **DC** theory, that the following two assertions are equivalent:

(a) there exists a universal measure zero subset U of $\mathbf{R}$ with $\mathrm{card}(U) = \omega_1$;

(b) there exists a countable family $\{\Xi_n : n < \omega\}$ of subsets of ω_1, which separates elements in ω_1 and is such that no nonzero σ-finite diffused measure μ on ω_1 admits an extension μ' for which

$$\{\Xi_n : n < \omega\} \subset \mathrm{dom}(\mu').$$

Argue as follows. First, suppose (a) and starting with U consider the family

$$\{\Xi_n : n < \omega\} = \{U \cap \Delta_n : n < \omega\},$$

where $\{\Delta_n : n < \omega\}$ denotes the countable family of all open intervals in $\mathbf{R}$ with rational endpoints. Identify U with ω_1 by some bijection and check that $\{\Xi_n : n < \omega\}$ satisfies (b), i.e., the implication (a) $\Rightarrow$ (b) holds true.

Now, suppose (b) and starting with the corresponding family $\{\Xi_n : n < \omega\}$ of subsets of ω_1, define a function

$$\chi : \omega_1 \to \{0,1\}^\omega$$

by putting

$$\chi(\xi) = \{r_n(\xi) : n < \omega\} \qquad (\xi < \omega_1),$$

where $r_n(\xi) = 1$ if $\xi \in \Xi_n$, and $r_n(\xi) = 0$ if $\xi \notin \Xi_n$. Check that χ is an injection and the set $\chi(\omega_1)$ is a universal measure zero subspace of the Cantor space $\{0,1\}^\omega$. Finally, applying the existence of a Borel isomorphism between $\{0,1\}^\omega$ and $\mathbf{R}$, conclude that (a) is valid, so the implication (b) $\Rightarrow$ (a) holds true, too.

13*. Let E be an uncountable complete metric space whose topological weight is not measurable in the Ulam sense and let μ be a σ-finite Borel measure on E. Verify that there exists a μ-measure zero set $X \subset E$ with $\mathrm{card}(X) = \mathrm{card}(E)$.
   For this purpose, consider the two possible cases:
   (a) $\mathrm{card}(E) \leq \mathbf{c}$;
   (b) $\mathrm{card}(E) > \mathbf{c}$.
   In both cases utilize the fact that μ is concentrated on some closed separable subspace of E.

14. By using the method of transfinite recursion, construct an injective function

$$f : \mathbf{R} \to \mathbf{R}$$

whose graph is a thick subset of $\mathbf{R}^2$ with respect to the standard two-dimensional Lebesgue measure λ_2.
   For any such function f, verify that:
   (a) f is discontinuous at all points of $\mathbf{R}$;
   (b) f is relatively measurable with respect to the class $\mathcal{M}(\lambda)$ of all those measures on $\mathbf{R}$ which extend λ.

Remark 8. The previous exercise shows, in particular, that there exist real-valued functions on $\mathbf{R}$ which are relatively good from the measure-theoretical point of view, but are very bad from the topological standpoint.

15. Let E be an uncountable Polish topological space and let μ be the completion of a nonzero σ-finite diffused Borel measure on E. Further, let X be a Sierpiński set in E with respect to μ (recall that the existence of X follows from the Continuum Hypothesis; see Exercise 10 of Chapter 4).
   Check that X does not contain an uncountable universal measure zero subspace.

16. Let $(G, +)$ be a commutative group and let X be a subset of G.

This X is called G-negligible (in G) if the following two conditions are sat-isfied:

(a) there exists a nonzero σ-finite translation invariant (translation quasi-invariant) measure μ_0 on G such that $X \in \text{dom}(\mu_0)$;

(b) for any σ-finite translation invariant (translation quasi-invariant) mea-sure μ on G such that $X \in \text{dom}(\mu)$, the equality $\mu(X) = 0$ holds true.

Now, take $(\mathbf{R}^2, +)$ as $(G, +)$ and show that:

(a) the graph of any function $f : \mathbf{R} \to \mathbf{R}$ is $\mathbf{R}^2$-negligible;

(b) there are two $\mathbf{R}^2$-negligible subsets X and Y in $\mathbf{R}^2$ such that the set $X \cup Y$ is not $\mathbf{R}^2$-negligible.

17. Let $(G, +)$ and $(H, +)$ be two commutative groups and let

$$f : G \to H$$

be an epimorphism (i.e., a surjective homomorphism).

Verify that if Y is an H-negligible subset of H, then its pre-image $f^{-1}(Y)$ is a G-negligible subset of G.

18*. In this exercise Marczewski's method is sketched, which describes some extensions of σ-finite measures given on an arbitrary ground set E (cf. [174], [176]).

Let E be a set, μ be a σ-finite measure defined on a σ-algebra $\mathcal{S}$ of subsets of E, and let $\mathcal{I}$ be a σ-ideal of subsets of E such that the inner μ-measure of any member of $\mathcal{I}$ is equal to zero.

Denote by S' the family of all those sets $T \subset E$ which admit a representation in the form

$$T = (X \setminus Y) \cup Z \qquad (X \in \mathcal{S}, Y \in \mathcal{I}, Z \in \mathcal{I}).$$

Verify that:

(a) $\mathcal{S}'$ is the σ-algebra generated by $\mathcal{S} \cup \mathcal{I}$;

(b) the formula $\mu'((X \setminus Y) \cup Z) = \mu(X)$ yields the measure μ' on E which extends the original measure μ;

(c) $\mu'(Z) = 0$ whenever $Z \in \mathcal{I}$.

Moreover, check that if the initial measure μ is invariant (quasi-invariant) under some group G of transformations of E and the σ-ideal $\mathcal{I}$ is also invari-ant under G, then the obtained measure μ' turns out to be invariant (quasi-invariant) under G, too.

19. Let E be a ground set, μ be a nonzero σ-finite measure defined on some σ-algebra of subsets of E, and let T be any subset of E.

Demonstrate that:

(a) there exists a measure ν on E extending μ and satisfying the relation $T \in \text{dom}(\nu)$;

(b) if T is not measurable with respect to the completion of μ, then there are at least two distinct measures μ' and μ'' on E which both extend μ and satisfy the relation

$$T \in \mathrm{dom}(\mu') \cap \mathrm{dom}(\mu'').$$

To show the validity of (a), assume without loss of generality that the set $E \setminus T$ is μ-thick in E, i.e., $\mu_*(T) = 0$ (this assumption can be justified by considering a μ-measurable kernel T_0 of T and replacing T by $T \setminus T_0$). Then apply the previous exercise to the σ-ideal $\mathcal{I}$ consisting of all subsets of T.

To show the validity of (b), reduce the argument to the case

$$\mu_*(T) = 0, \quad \mu^*(T) > 0$$

and then apply (a) separately to the sets T and $T_1 \setminus T$, where T_1 stands for a μ-measurable hull of T.

20. Let E be an arbitrary uncountable set.

Prove that there exist a probability diffused measure μ on E and a countable family

$$\{X_1, \ X_2, \ \ldots, \ X_n, \ \ldots \}$$

of subsets of E such that, for every measure $\nu \in \mathcal{M}(\mu)$, at least one member of $\{X_1, X_2, ..., X_n, ...\}$ is nonmeasurable with respect to ν.

For this purpose, utilize the existence of a universal measure zero subspace of $\mathbf{R}$ of cardinality ω_1.

21. Give an example of two subsets X and Y of $\mathbf{R}$ such that:

(a) both X and Y are absolutely nonmeasurable with respect to the class $\mathcal{CBM}_0(\mathbf{R})$;

(b) the product set $X \times Y$ is not absolutely nonmeasurable with respect to the class $\mathcal{CBM}_0(\mathbf{R} \times \mathbf{R})$.

22. Sorgenfrey's topology $\mathcal{T}$ is defined as follows (see, e.g., [49]). The family of all half-open subintervals of $\mathbf{R}$ having the form

$$[a, b[\quad (a \in \mathbf{R}, b \in \mathbf{R})$$

is taken as one of the bases of $\mathcal{T}$ and the obtained space $(\mathbf{R}, \mathcal{T})$ is called the Sorgenfrey line. This space serves as a counterexample to many seemingly true statements of general and set-theoretic topology.

Supposing that the cardinal number $\mathbf{c}$ is measurable in the Ulam sense, verify that:

(a) there exists a subset of $\mathbf{R}$ absolutely nonmeasurable with respect to the class $\mathcal{CBM}_0((\mathbf{R}, \mathcal{T}))$;

(b) there exists no subset of $\mathbf{R} \times \mathbf{R}$ absolutely nonmeasurable with respect to the class $\mathcal{CBM}_0((\mathbf{R}, \mathcal{T}) \times (\mathbf{R}, \mathcal{T}))$.

23*. Demonstrate that every Luzin subset X of $\mathbf{R}$ is universal measure zero. More generally, prove that if a mapping

$$f : X \to \mathbf{R}$$

has the Baire property, then the set $f(X)$ is universal measure zero.

For this purpose, first of all reduce the argument to the case when f is a continuous mapping. Let $\{x_n : n < \omega\}$ be a countable everywhere dense subset of X and let μ be any finite diffused Borel measure on $f(X)$. Fix a real $\varepsilon > 0$ and, for each $n < \omega$, find an open neighborhood U_n of $f(x_n)$ such that

$$\mu(U_n) < \varepsilon/2^{n+1}.$$

Further, since f is continuous, for every $n < \omega$ there exists a neighborhood V_n of x_n such that $f(V_n) \subset U_n$. Therefore,

$$\mu(f(\cup\{V_n : n < \omega\})) \leq \mu(\cup\{U_n : n < \omega\}) \leq \varepsilon.$$

Since X is a Luzin set, the difference $X \setminus \cup\{V_n : n < \omega\}$ is at most countable and, consequently, $\mu(f(X)) \leq \varepsilon$. Keeping in mind the arbitrary smallness of ε, conclude that $f(X)$ is a universal measure zero subset of $\mathbf{R}$ (in particular, $f(X)$ is a totally imperfect subset of $\mathbf{R}$).

Also, check that every real-valued function on X possessing the Baire property is, in fact, a Borel function.

Let now Y be a Sierpiński set in $\mathbf{R}$ and let μ denote the measure on Y induced by the Lebesgue measure λ. Let $g : Y \to \mathbf{R}$ be an arbitrary function measurable with respect to μ.

Prove that the set $g(Y)$ does not contain any uncountable universal measure zero subspace. Infer from this fact that $g(Y)$ is totally imperfect in $\mathbf{R}$.

24. Let $(G, +)$ be a commutative group and let μ be a σ-finite G-quasi-invariant measure on G.

Show that, for any μ-measurable set X, there exists a countable family $\{g_i : i \in I\}$ of elements of G such that the set

$$X' = \cup\{g_i + X : i \in I\}$$

is μ-almost G-invariant in G, i.e.,

$$(\forall g \in G)(\mu((g + X')\triangle X') = 0).$$

For this purpose, define the required family $\{g_i : i \in I\}$ by using the method of transfinite recursion.

25*. Let $(G, +)$ be a commutative group and let X be a subset of E. Prove that these two assertions are equivalent:

(a) X is a G-absolutely negligible set;

(b) for any countable family $\{g_n \; : \; n < \omega\}$ of elements of G, there exists a countable family $\{h_m \; : \; m < \omega\}$ of elements of the same G such that

$$\cap\{h_m + (\cup\{g_n + X \; : \; n < \omega\}) \; : \; m < \omega\} = \emptyset.$$

Argue as follows. Let X satisfy (a) and suppose to the contrary that (b) is not true. Then there exists a countable family $\{g_n \; : \; n < \omega\}$ of elements of G, such that, for the set

$$X' = \cup\{g_n + X \; : \; n < \omega\}$$

and for an arbitrary countable family $\{h_m \; : \; m < \omega\}$ of elements of G, the relation

$$\cap\{h_m + X' \; : \; m < \omega\} \neq \emptyset$$

is valid. Denote $X'' = G \setminus X'$ and consider the G-invariant σ-ideal $\mathcal{I}$ of subsets of E, generated by the one-element family $\{X''\}$. Obviously, one can define a complete probability G-invariant measure μ on G such that $\mathcal{I} = I(\mu)$. In particular, the relations

$$X'' \in \mathrm{dom}(\mu), \quad X' \in \mathrm{dom}(\mu),$$

$$\mu(X'') = 0, \quad \mu(X') = 1$$

are true. According to (a), there exists a G-quasi-invariant measure ν on G extending μ and such that

$$X \in \mathrm{dom}(\nu), \quad \nu(X) = 0.$$

Consequently, one has

$$\nu(X') = \nu(\cup\{g_n + X \; : \; n < \omega\}) = 0,$$

which contradicts the relation

$$\nu(X') = \mu(X') = 1.$$

This contradiction establishes the implication (a) $\Rightarrow$ (b).

Suppose now that a set X satisfies (b). Let μ be an arbitrary σ-finite G-quasi-invariant measure on E. Denote by $\mathcal{J}$ the G-invariant σ-ideal of subsets of G, generated by the one-element family $\{X\}$. Taking (b) into account and applying Exercise 24 infer that, for each set $Y \in \mathcal{J}$, the equality $\mu_*(Y) = 0$ is valid. By virtue of Exercise 18, the measure μ can be extended to some G-quasi-invariant measure ν on G such that $\mathcal{J} \subset \mathrm{dom}(\nu)$ and $\nu(Y) = 0$ for all sets $Y \in \mathcal{J}$. In particular, $X \in \mathrm{dom}(\nu)$ and $\nu(X) = 0$. Therefore (a) is satisfied, which establishes the implication (b) $\Rightarrow$ (a).

Finally, observe that if the initial measure μ is G-invariant, then the extended measure ν is G-invariant, too.

26. Prove that there exists a translation invariant measure μ on $\mathbf{R}$ for which the following two conditions are fulfilled:

(a) μ is an extension of the Lebesgue measure λ;

(b) if X is a Sierpiński subset of $\mathbf{R}$, then $X \in \mathrm{dom}(\mu)$ and $\mu(X) = 0$.

For this purpose, keep in mind the fact that the inner Lebesgue measure of any Sierpiński set is equal to zero. Then apply to λ and to the σ-ideal generated by all Sierpiński sets the result of Exercise 18.

Remark 9. Assuming **CH**, it can be proved that there exists a countable family of Hamel bases of $\mathbf{R}$ which collectively cover the set $\mathbf{R} \setminus \{0\}$ (see [51], [243]). Also, it was shown that any Hamel basis of $\mathbf{R}$ is an $\mathbf{R}$-absolutely negligible set (see [120]). These two circumstances directly imply that if **CH** holds, then there is no nonzero σ-finite translation quasi-invariant measure ν on $\mathbf{R}$ such that all members of the above-mentioned countable family of Hamel bases are ν-measurable. In addition, it can be demonstrated that there exists a Vitali subset (a Bernstein subset) of $\mathbf{R}$ which is absolutely nonmeasurable with respect to the class of all nonzero σ-finite translation quasi-invariant measures on $\mathbf{R}$ (cf. Chapter 9).

27. Verify that the existence of an uncountable universal measure zero subspace of $\mathbf{R}$ cannot be established within **ZF** & **DC** set theory.

For this purpose, take into account Solovay's model of this theory, in which any uncountable subset of $\mathbf{R}$ contains a nonempty perfect set (see [253]).

6. Radon measures and nonmeasurable sets

Various deep interconnections between general topology and measure theory are recognized and widely known at present (see, e.g., [17], [20], [64], [76], [89], [95], [137], [265]). Connections or relationships of such a kind are very fruitful for further development of these two mathematical disciplines. Undoubtedly, the concept of quasi-compactness occupies the central place in general topology and it should be noticed that some direct analogues of this concept can be also met in contemporary measure theory. For instance, it suffices to recall the notion of compact measures first introduced by Marczewski (see [178]). This notion was motivated by concrete problems and questions of measure theory and probability theory. For example, Kolmogorov's extension theorem from the theory of stochastic processes might be indicated in this context (cf. [17], [20], [26], [199], [265]).

Here we would like to touch upon an important concept of a Radon measure on a topological space, which may be treated as a natural generalization of the classical Lebesgue measure on **R** and which found a lot of applications in diverse branches of functional analysis, convex analysis, optimization, probability theory and stochastic processes, etc. Of course, we do not intend to discuss all aspects of the theory of Radon measures. We primarily are interested in those questions of this theory which are concerned, more or less, with the existence of nonmeasurable sets (cf. the preceding three chapters of this book).

In particular, the reader will see below our special interest and inclination toward the problem of the existence of nonmeasurable sets with respect to a nonzero σ-finite diffused Radon measure given on a topological space.

For an extensive and thorough presentation of the material concerning various important properties of Radon measures, we refer the reader to [17], [20], [64], [76]. Notice, by the way, that the general theory of Radon measures was first developed by Bourbaki's school, mostly for locally compact topological spaces. Afterwards, other authors continued their research work in this direction for substantially more general classes of topological spaces.

It is reasonable to begin this chapter with a precise definition of a Radon measure on a topological space and with the naturally associated definition of Radon spaces. In the sequel, we will restrict our consideration to those measures which are either finite or σ-finite.

Let E be a Hausdorff topological space and let μ be a σ-finite measure defined on some σ-algebra of subsets of E.

We shall say that μ is a Radon measure if, for every μ-measurable set X, we have the equality

$$\mu(X) = \sup\{\mu(K) : K \in \mathrm{dom}(\mu),\ K \subset X,\ K \text{ is compact}\}.$$

In other words, μ is a Radon measure if, for every μ-measurable set X, there exists a sequence $\{K_n : n < \omega\}$ of compact subsets of X such that

$$(\forall n < \omega)(K_n \in \mathrm{dom}(\mu)),$$

$$\mu(X) = \mu(\cup\{K_n : n < \omega\}).$$

For a finite measure defined on some σ-algebra of subsets of E, the above definition can be replaced by the following one.

A finite measure ν on a Hausdorff space E is Radon if and only if, for any ν-measurable set X and for any real $\varepsilon > 0$, there exists a compact set $K \subset X$ such that $K \in \mathrm{dom}(\nu)$ and $\nu(X) < \nu(K) + \varepsilon$.

We shall say that a Hausdorff topological space E is a Radon space if every σ-finite measure defined on the Borel σ-algebra $\mathcal{B}(E)$ of E turns out to be a Radon measure.

It is easy to see that a Hausdorff space E is a Radon space if and only if every probability measure defined on the Borel σ-algebra $\mathcal{B}(E)$ turns out to be a Radon measure.

Example 1. The Lebesgue measure λ_n on the Euclidean n-dimensional space $\mathbf{R}^n$ provides a standard example of a Radon measure. More generally, let X be a Polish topological space and let μ be the completion of a σ-finite Borel measure on X. Then, according to an old theorem of Ulam, μ turns out to be a Radon measure (see Exercise 12). Therefore, any Polish topological space is a Radon space. Furthermore, let Y be a complete metric space whose topological weight is not measurable in the Ulam sense (see Appendix 1), and let ν be the completion of a σ-finite Borel measure on Y. Then, analogously to Ulam's theorem mentioned above, ν turns out to be a Radon measure. Therefore, Y is a Radon space, too.

Example 2. Let E be an analytic (i.e., Suslin) subset of a Polish space X, equipped with the induced topology. According to the classical result of Luzin, E is a Radon space (for the proof, see [17], [33], [103], [115], [199] or Appendix 5). It readily follows from this fact that the members of the σ-algebra $\mathcal{S}(X)$ generated by all analytic subsets of X are absolutely measurable sets with respect to the class of completions of all σ-finite Borel measures on X. However, the σ-algebra of all absolutely measurable sets with respect to the same class of measures may be substantially wider than $\mathcal{S}(X)$. For instance, under the Continuum Hypothesis, the σ-algebra of all absolutely measurable sets with respect to the class of completions of all σ-finite Borel measures on $\mathbf{R}$ has among its members all Luzin subsets of $\mathbf{R}$ which are universal measure zero (see Exercise 23 from Chapter 5) and, consequently, are Radon spaces. However, no Luzin set possesses the Baire property, while all members from $\mathcal{S}(\mathbf{R})$ have this property (cf. [103], [115], [152] or Appendix 5).

The following simple result shows that certain uncountable families of open sets behave nicely with respect to Radon measures.

Theorem 1. *Let E be a Hausdorff topological space and let μ be a σ-finite Radon measure on E such that $\mathcal{B}(E) \subset \mathrm{dom}(\mu)$. Suppose that a family $\{U_i : i \in I\}$ of open subsets of E is given, which is filtered (i.e., directed) with respect to the standard inclusion relation $\subset$. Then the equality*

$$\mu(\cup\{U_i : i \in I\}) = \sup\{\mu(U_i) : i \in I\}$$

holds true.

Proof. Since μ is a Radon measure and $\mathcal{B}(E) \subset \mathrm{dom}(\mu)$, we may write

$$\mu(\cup\{U_i : i \in I\}) = \sup\{\mu(K) : K \subset \cup\{U_i : i \in I\}, \ K \text{ is compact}\}.$$

Let K be any compact subset of $\cup\{U_i : i \in I\}$. Then the family $\{U_i : i \in I\}$ is an open covering of K, so there exist finitely many sets

$$U_{i_1}, \ U_{i_2}, \ \ldots, \ U_{i_m}$$

such that

$$K \subset U_{i_1} \cup U_{i_2} \cup \ldots \cup U_{i_m}.$$

Remembering that $\{U_i : i \in I\}$ is a filtered family of sets, we can find an index $i \in I$ satisfying the inclusion

$$U_{i_1} \cup U_{i_2} \cup \ldots \cup U_{i_m} \subset U_i$$

and, consequently, satisfying the relations

$$K \subset U_i, \quad \mu(K) \leq \mu(U_i).$$

This circumstance directly implies the inequality

$$\mu(\cup\{U_i : i \in I\}) \leq \sup\{\mu(U_i) : i \in I\}.$$

The opposite inequality

$$\sup\{\mu(U_i) : i \in I\} \leq \mu(\cup\{U_i : i \in I\})$$

is trivially valid and so we obtain the desired equality

$$\mu(\cup\{U_i : i \in I\}) = \sup\{\mu(U_i) : i \in I\}.$$

Theorem 1 has thus been proved.

Remark 1. In the literature, the fact expressed by Theorem 1 is sometimes called the τ-smoothness (or τ-additivity) of Radon measures (cf. [17], [76], [265]).

As usual, we denote by $\mathcal{CBM}(E)$ the class of completions of all σ-finite Borel measures on a topological space E. Let us indicate that if some subset Y of E is absolutely (universally) measurable with respect to the class $\mathcal{CBM}(E)$ and $Z \in \mathcal{B}(Y)$, then Z is also absolutely measurable with respect to $\mathcal{CBM}(E)$. This almost trivial fact will be essentially used below.

Theorem 2. *Let X be a Hausdorff topological space and let Y be a Radon subspace of X. Then Y is absolutely measurable with respect to $\mathcal{CBM}(X)$.*

Proof. Let μ' be an arbitrary nonzero measure from the class $\mathcal{CBM}(X)$. By definition, this means that there exists a nonzero σ-finite Borel measure μ on X such that μ' coincides with the completion of μ. Without loss of generality, we may assume that μ is a probability measure, i.e., $\mu(X) = 1$, and that Y is μ-thick in X, i.e., $\mu^*(Y) = 1$, where μ^* denotes the outer measure associated with μ (indeed, if $\mu^*(Y) < 1$, then we can replace X by a μ-measurable hull of Y).

Now, we define a probability Borel measure ν on Y by the following formula:

$$\nu(B \cap Y) = \mu(B) \qquad (B \in \mathcal{B}(X)).$$

The μ^*-thickness of Y implies that this definition of ν is correct (cf. Exercise 22 from Chapter 3). Further, since Y is a Radon space, the measure ν must be Radon. So there exists a σ-compact set $Z \subset Y$ such that

$$1 = \mu^*(Y) = \nu(Y) = \nu(Z).$$

The set Z is a Borel subset of the space X (being a countable union of closed subsets of X). Consequently, Z is μ-measurable and, by the definition of ν, we have

$$1 = \nu(Y) = \nu(Z) = \mu(Z) \leq \mu_*(Y) \leq 1,$$

where μ_* denotes the inner measure associated with μ. So we finally obtain the equalities

$$1 = \mu^*(Y) = \mu_*(Y),$$

which directly imply the universal measurability of Y with respect to the class $\mathcal{CBM}(X)$.

This completes the proof of Theorem 2.

Under natural additional assumptions on a topological space X, the converse statement to Theorem 2 can readily be established.

Theorem 3. *Let X be a Radon topological space and let Y be a subspace of X absolutely measurable with respect to the class $\mathcal{CBM}(X)$. Then Y is a Radon space as well.*

Proof. Let μ be an arbitrary Borel probability measure on Y. We must demonstrate that μ is a Radon measure. In order to do this, take any set $Z \in \mathcal{B}(Y)$. Since Y is absolutely measurable with respect to the class $\mathcal{CBM}(X)$, the set Z is also absolutely measurable with respect to the class $\mathcal{CBM}(X)$.

Let us introduce a Borel probability measure ν on X defined by the formula

$$\nu(A) = \mu(A \cap Y) \qquad (A \in \mathcal{B}(X)).$$

Since X is a Radon space, ν must be a Radon measure on X. Obviously, for ν we have the equalities

$$\nu^*(Y) = \nu_*(Y) = 1,$$

$$\nu^*(Z) = \nu_*(Z), \quad \mu(Z) = \nu^*(Z).$$

So we can find a set $B \in \mathcal{B}(X)$ such that

$$B \subset Z, \ \nu(B) = \nu_*(Z) = \nu^*(Z).$$

On the other hand, the definition of ν and the relation $B \subset Z \subset Y$ yield

$$\nu(B) = \mu(B) \leq \mu(Z).$$

Remembering that ν is a Radon measure, we may write $\nu(B) = \nu(P)$, where P is a σ-compact subset of X entirely contained in B (hence in Z). Clearly, P is also a σ-compact subset of Z and $\mu(P) = \nu(P)$. We finally obtain that

$$\mu(Z) = \nu^*(Z) = \nu_*(Z) = \nu(B) = \nu(P) = \mu(P) \leq \mu(Z),$$

$$\mu(Z) = \mu(P),$$

which implies that μ is a Radon probability measure. Thus Y is a Radon topological space, which finishes the proof of Theorem 3.

Natural set-theoretical operations over Radon spaces can be considered (such as topological sums, topological products, inductive and projective limits, etc.). Some of them preserve the class of Radon spaces, while others do not preserve this class (for more details, see exercises of the present chapter).

Example 3. One of the most natural topological operation is taking a continuous image of a given space. But it turns out that even in a class of quite good topological spaces a continuous image of a Radon space can be a non-Radon space. The standard example of this sort is provided by co-analytic (co-Suslin) subsets of the real line $\mathbf{R}$. Indeed, according to the result of Luzin, all analytic and co-analytic subsets of $\mathbf{R}$ are absolutely measurable with respect to the class $\mathcal{CBM}(\mathbf{R})$, so they are Radon spaces. On the other hand, under the Constructibility Axiom $\mathbf{V} = \mathbf{L}$ of Gödel (see, e.g., [10], [42], [103], [148]), there exists a set $X \subset \mathbf{R}$ having the following properties:

(*) X is a co-analytic set, i.e., $\mathbf{R} \setminus X$ is an analytic set;

(**) some continuous image of X is a Lebesgue nonmeasurable subset of $\mathbf{R}$.

This profound result of Gödel indicates that, in general, one cannot assert that a continuous image of a Radon space is also a Radon space. Under Martin's Axiom and the negation of the Continuum Hypothesis, i.e., in the theory

$$\mathbf{ZFC} \ \& \ \mathbf{MA} \ \& \ (\neg \mathbf{CH}),$$

the situation is much better, at least for continuous images of co-analytic sets (in this connection, see Exercise 25 for the present chapter).

Now, we would like to recall one important notion of measurability of those real-valued functions which are defined on a topological space (cf. [17], [20], [133], [199], [265]).

Let E be a Hausdorff topological space and let μ be a σ-finite measure on E.

We shall say that a function $f : E \to \mathbf{R}$ is measurable in the Luzin sense (with respect to μ) if there exists a sequence $\{K_n : n < \omega\}$ of compact subsets of E such that:

(1) $(\forall n < \omega)(K_n \in \text{dom}(\mu))$ and $\mu(E \setminus \cup\{K_n : n < \omega\}) = 0$;

(2) for each index $n < \omega$, the restriction $f|K_n$ is continuous.

Example 4. By virtue of Luzin's classical C-property (see, e.g., [17], [133], [197], [203]), every λ-measurable function $f : \mathbf{R} \to \mathbf{R}$ is measurable in the Luzin sense with respect to λ, where λ denotes, as usual, the standard Lebesgue measure on $\mathbf{R}$.

The next statement is a far-going generalization of the above-mentioned widely known example.

Theorem 4. *Let E be a Hausdorff topological space and let μ be a σ-finite Radon measure on E. Denote by μ' the completion of μ. Then every μ'-measurable function*

$$f : E \to \mathbf{R}$$

is measurable in the Luzin sense with respect to μ'.

Proof. Without loss of generality, we may suppose that μ is a probability measure and the set $\text{ran}(f)$ is contained in the half-open unit interval $[0, 1[$.

Actually, it suffices to show the validity of the following assertion: for any real $\varepsilon > 0$, there exists a compact set $K_\varepsilon \in \text{dom}(\mu)$ such that $\mu(K_\varepsilon) \geq 1 - \varepsilon$ and the restriction $f|K_\varepsilon$ is continuous.

The argument presented below imitates the proof of Luzin's C-property of all λ-measurable real-valued functions and, in fact, is fairly standard.

Fix a nonzero natural number n and consider the sets

$$X_{i,n} = \{x \in E : i/n \leq f(x) < (i+1)/n\} \quad (i = 0, 1, ..., n-1),$$

all of which are pairwise disjoint and μ'-measurable. These sets collectively cover the whole space E. Since μ is a Radon measure, there are compact sets

$$K_{i,n} \in \text{dom}(\mu) \quad (i = 0, 1, ..., n-1)$$

such that

$$K_{i,n} \subset X_{i,n}, \quad \mu(X_{i,n} \setminus K_{i,n}) < \varepsilon/n2^n \quad (i = 0, 1, ..., n-1).$$

For each natural index $i \in \{0, 1, ..., n-1\}$, let $f_{i,n}$ denote the constant function on $K_{i,n}$ whose range coincides with the singleton $\{i/n\}$, and let f_n be the common extension of all these $f_{i,n}$ $(i = 0, 1, ..., n-1)$. One can easily verify the validity of the following relations:

(a) the function f_n is defined on the compact set

$$K_n = \cup\{K_{i,n} : 0 \leq i < n\} \in \text{dom}(\mu)$$

and is continuous at all points of K_n;

(b) $\mu(K_n) \geq 1 - \varepsilon/2^n$;

(c) the set $K_\varepsilon = \cap\{K_n : 1 \leq n < \omega\}$ is compact, belongs to $\text{dom}(\mu)$ and $\mu(K_\varepsilon) \geq 1 - \varepsilon$;

(d) the sequence of functions

$$\{f_n|K_\varepsilon : 1 \leq n < \omega\}$$

converges uniformly to the function $f|K_\varepsilon$;

(e) $f|K_\varepsilon$ is continuous (as the limit of a uniformly convergent sequence of continuous functions).

Theorem 4 has thus been proved.

Let us introduce one notion due to Gnedenko and Kolmogorov [78], which plays a significant role in modern probability theory and the theory of stochastic processes.

Let E be a base (ground) set, $\mathcal{S}$ be a σ-algebra of subsets of E, and let μ be a probability measure whose domain coincides with $\mathcal{S}$.

The triplet $(E, \mathcal{S}, \mu)$ is usually called a probability measure space (or, briefly, a probability space).

A random variable on $(E, \mathcal{S}, \mu)$ is any real-valued function on E measurable with respect to μ.

A probability space $(E, \mathcal{S}, \mu)$ is said to be perfect if, for any random variable $f : E \to \mathbf{R}$, there exists a Borel subset T of $\mathbf{R}$ such that

$$T \subset \operatorname{ran}(f), \quad \mu(f^{-1}(T)) = 1.$$

From Theorem 4 we readily get the next important statement.

Theorem 5. *Suppose that E is a Hausdorff topological space and μ is a Radon probability measure on E. Then $(E, \operatorname{dom}(\mu), \mu)$ turns out to be a perfect probability space. Consequently, if E is a Radon topological space, then, for every Borel probability measure μ on E, the probability space $(E, \mathcal{B}(E), \mu)$ turns out to be a perfect space.*

Proof. Let μ be a Radon probability measure on a Hausdorff space E and let $f : E \to \mathbf{R}$ be a random variable on the probability measure space $(E, \operatorname{dom}(\mu), \mu)$. According to Theorem 4, this f is measurable in the Luzin sense with respect to μ, i.e., there exists a countable family $\{K_n : n < \omega\}$ of compact subsets of E such that:

(1) $(\forall n < \omega)(K_n \in \operatorname{dom}(\mu))$ and $\mu(E \setminus \cup\{K_n : n < \omega\}) = 0$;

(2) for each $n < \omega$, the restriction $f|K_n$ is continuous.

Let us denote

$$T = \cup\{f(K_n) : n < \omega\}.$$

Since all $f(K_n)$ $(n < \omega)$ are compact subsets of $\mathbf{R}$ (as continuous images of compact sets), T is a σ-compact set and hence is of type F_σ in $\mathbf{R}$. In addition, one can easily see that

$$1 \geq \mu(f^{-1}(T)) \geq \mu(\cup\{K_n : n < \omega\}) = 1,$$

so $\mu(f^{-1}(T)) = 1$, which yields the required result and finishes the proof of Theorem 5.

Theorem 6. *Let $(E, \mathcal{S}, \mu)$ be a probability space and let*

$$f : E \to \mathbf{R}$$

be a random variable on E such that:

(1) $\mu(f^{-1}(t)) = 0$ for each point $t \in \mathbf{R}$;

(2) the range of f does not contain any uncountable closed subset of $\mathbf{R}$.

Then the space $(E, \mathcal{S}, \mu)$ is not perfect.

Proof. Suppose to the contrary that $(E, \mathcal{S}, \mu)$ is a perfect probability space. Then, by virtue of the definition, there exists a Borel subset T of $\mathbf{R}$ such that

$$T \subset \operatorname{ran}(f), \quad \mu(f^{-1}(T)) = 1.$$

In view of condition (1), T cannot be countable, so is uncountable. But then, according to the well-known theorem of Alexandrov and Hausdorff (see, e.g., [10], [103], [115], [152], [191] or Exercise 5 from Chapter 20), T must contain an uncountable closed subset of $\mathbf{R}$, which contradicts condition (2).

The obtained contradiction completes the proof of Theorem 6.

Theorem 7. *Let $(E, \mathcal{S}, \mu)$ be a perfect probability space with a nonatomic measure μ. Then $\mathcal{S}$ differs from the power set $\mathcal{P}(E)$, i.e., there exists a subset of E nonmeasurable with respect to μ.*

Proof. Suppose to the contrary that $\mathcal{S} = \mathcal{P}(E)$. Then every real-valued function on E may be treated as a random variable on the probability space $(E, \mathcal{S}, \mu)$. Since our μ does not have atoms, there exists a partition $\{X_i : i \in I\}$ of E such that:

(a) $\operatorname{card}(I) \le \mathbf{c}$;

(b) $\mu(X_i) = 0$ for each index $i \in I$.

The existence of a partition $\{X_i : i \in I\}$ of E with the properties (a) and (b) can be justified by a fairly standard argument (see Exercise 9).

Let T be a totally imperfect subset of $\mathbf{R}$ with $\operatorname{card}(T) = \operatorname{card}(I)$. Obviously, this T can be realized as a certain subset of a Bernstein set in $\mathbf{R}$ (see, e.g., [33], [147], [152], [188], [190], [203] or Chapter 3 where some information about Bernstein sets is presented).

Let ϕ be a bijection acting from I onto T. We define a real-valued function f on E as follows: $f(x) = \phi(i)$ if and only if $x \in X_i$.

By virtue of our assumption, f is a random variable on the space $(E, \mathcal{S}, \mu)$ and $\mu(f^{-1}(t)) = 0$ for each point $t \in \mathbf{R}$. On the other hand, it is clear that the range of f coincides with T and does not contain any uncountable closed subset of $\mathbf{R}$. This yields a contradiction with Theorem 6 and finishes the proof.

As a corollary of Theorem 7, one can deduce the following important result (see [184]).

Theorem 8. *Let E be a Hausdorff topological space and let μ be a nonzero σ-finite diffused Radon measure on E. Then $\operatorname{dom}(\mu)$ differs from $\mathcal{P}(E)$, i.e., there exists a subset of E nonmeasurable with respect to μ.*

Proof. We may assume, without loss of generality, that μ is a Radon diffused probability measure on E. Taking into account Theorems 5, 6 and 7, it suffices to demonstrate that μ does not possess atoms, i.e., there exists no μ-measurable set X such that $\mu(X) > 0$ and, for any μ-measurable set $Y \subset X$, one has

$$\mu(Y) = 0 \quad \vee \quad \mu(X \setminus Y) = 0.$$

Suppose to the contrary that such an atom X does exist. Since μ is a Radon measure, we may assume that X is compact. Consider the family

$$\mathcal{F} = \{Y \subset X : Y \text{ is compact and } \mu(Y) = \mu(X)\}.$$

Obviously, this family is centered. Consequently, the set

$$X_0 = \cap \{Y : Y \in \mathcal{F}\}$$

is nonempty and compact as well. Moreover, using again the fact that μ is a Radon measure, it is not difficult to check that

$$\mu(X_0) = \mu(X) > 0.$$

(In this connection, see Theorem 1 and Exercise 1.) In particular, X_0 must be uncountable in view of the diffuseness of μ. Let y and z be any two distinct points from X_0. According to Urysohn's theorem (see, e.g., [49] or [152]), there exists a continuous function

$$\phi : X_0 \to [0,1]$$

such that $\phi(y) = 1$ and $\phi(z) = 0$. Let us put

$$Y = \phi^{-1}([0,1/2]), \quad Z = \phi^{-1}([1/2,1]).$$

Both Y and Z are compact subsets of X_0 and

$$X_0 = Y \cup Z, \quad y \notin Y, \quad z \notin Z.$$

Therefore, we get

$$0 < \mu(X) = \mu(X_0) \leq \mu(Y) + \mu(Z),$$

$$\mu(Y) = \mu(X) \quad \vee \quad \mu(Z) = \mu(X).$$

We may assume that $\mu(Y) = \mu(X)$. So $Y \in \mathcal{F}$ and, according to the definition of X_0, we must have $X_0 \subset Y$, which contradicts the circumstance that Y is a proper subset of X_0 (since $y \in X_0 \setminus Y$).

The obtained contradiction finishes the proof of Theorem 8.

Remark 2. For further extensions and generalizations of Theorem 8 and some related results, see [24], [31], [34], [63], [216], [274].

EXERCISES

1. Let E be a Hausdorff topological space.

Check that every nonzero σ-finite Radon measure on E is equivalent to some Radon probability measure on E. Infer from this fact that the following two assertions are equivalent:

(a) every Borel probability measure on E is Radon;
(b) E is a Radon topological space.

Also, verify that any finite Radon measure μ on E is outer regular in the sense that, for every μ-measurable set X, the equality

$$\mu(X) = \inf\{\mu(U) : U \in \text{dom}(\mu),\ X \subset U \text{ and } U \text{ is open in } E\}$$

holds true. Keeping in mind this equality and Theorem 1, infer that if μ is a finite Radon measure with $\text{dom}(\mu) = \mathcal{B}(E)$ and a family $\{K_i : i \in I\}$ of compact subsets of E is filtered by the reverse inclusion relation $\supset$, then

$$\mu(\cap\{K_i : i \in I\}) = \inf\{\mu(K_i) : i \in I\}.$$

The latter fact turns out to be useful in various applications of Radon measures (cf. the proof of Theorem 8).

2. Let E be a Hausdorff topological space, $\mathcal{S}$ be an algebra of subsets of E, and let μ be a σ-finite measure on $\mathcal{S}$ satisfying the following condition:

(i) for each set $X \in \mathcal{S}$ with $\mu(X) < +\infty$ and for any real $\varepsilon > 0$, there exists a compact set $K \subset X$ such that $K \in \mathcal{S}$ and $\mu(X \setminus K) < \varepsilon$.

Let $\sigma(\mathcal{S})$ denote the σ-algebra generated by $\mathcal{S}$ and let μ' be the measure on $\sigma(\mathcal{S})$ extending μ by Carathéodory's classical theorem (see, e.g., [17], [89], [199]).

Check that the analogous condition holds true for $\sigma(\mathcal{S})$ and μ', i.e.,

(ii) for each set $X \in \sigma(\mathcal{S})$ with $\mu(X) < +\infty$ and for any real $\varepsilon > 0$, there exists a compact set $K \subset X$ such that $K \in \sigma(\mathcal{S})$ and $\mu'(X \setminus K) < \varepsilon$.

In order to demonstrate this fact, keep in mind the circumstance that $\sigma(\mathcal{S})$ coincides with the monotone class generated by $\mathcal{S}$ (see again [17], [89], [199]).

3. Let E be a topological space all closed subsets of which are of type G_δ (or, equivalently, all open subsets of which are of type F_σ) and let μ be a σ-finite Borel measure on E.

Show that μ is inner regular with respect to the family of all closed sets in E, i.e., for any Borel set $X \subset E$, the equality

$$\mu(X) = \sup\{\mu(F) : F \subset X,\ F \text{ is closed}\}$$

holds true. Deduce from this fact that if $\mu(E) < +\infty$, then

$$\mu(X) = \inf\{\mu(G) : X \subset G,\ G \text{ is open}\}.$$

In other words, μ is outer regular with respect to the family of all open sets in E.

4*. Let E be a Hausdorff topological space, $\mathcal{S}$ be an algebra of subsets of E, and let

$$\mu : \mathcal{S} \to [0, +\infty[$$

be a finitely additive functional. Suppose that the following condition is fulfilled: for each set $X \in \mathcal{S}$ and for any real $\varepsilon > 0$, there exists a compact set $K \subset X$ such that $K \in \mathcal{S}$ and $\mu(X \setminus K) < \varepsilon$.

Demonstrate that μ is countably additive, so μ is a finite Radon measure on E with $\mathrm{dom}(\mu) = \mathcal{S}$ (A.D. Alexandrov's theorem).

Argue as follows. Let $\{X_n : n < \omega\}$ be a countable disjoint family of members of $\mathcal{S}$ such that
$$\cup\{X_n : n < \omega\} \in \mathcal{S}.$$

It suffices to establish the inequality
$$\mu(\cup\{X_n : n < \omega\}) \leq \sum\{\mu(X_n) : n < \omega\}.$$

Fix a real $\varepsilon > 0$ and take a compact set $K \subset \cup\{X_n : n < \omega\}$ satisfying the relations
$$K \in \mathcal{S},$$
$$\mu(\cup\{X_n : n < \omega\}) \leq \mu(K) + \varepsilon.$$

Further, take a countable family $\{U_n : n < \omega\}$ of open sets in E satisfying the relations
$$U_n \in \mathcal{S}, \quad X_n \subset U_n \quad \mu(U_n) \leq \mu(X_n) + \varepsilon/2^{n+1} \quad (n < \omega).$$

By virtue of the compactness of K, there exists $m < \omega$ such that
$$K \subset U_0 \cup U_1 \cup \ldots \cup U_m.$$

Deduce from the above inclusion that
$$\mu(\cup\{X_n : n < \omega\}) \leq \mu(K) + \varepsilon \leq \sum\{\mu(U_n) : n \leq m\} + \varepsilon,$$

whence it immediately follows that
$$\mu(\cup\{X_n : n < \omega\}) \leq \sum\{\mu(X_n) : n < \omega\} + 2\varepsilon,$$

which yields the required result in view of the arbitrary smallness of ε.

5. Let I be a finite set of indices and let J be a countable set of indices. Suppose that $\{\mu_i : i \in I\}$ is a family of σ-finite Radon measures and $\{\nu_j : j \in J\}$ is a family of probability Radon measures.

Show that the product measures
$$\mu = \otimes\{\mu_i : i \in I\}, \quad \nu = \otimes\{\nu_j : j \in J\}$$

are also Radon.

For this purpose, utilize the fact formulated in Exercise 2.

Remark 3. Assuming the Continuum Hypothesis, it can be proved that there exist two compact Radon spaces X and Y such that the compact product space $X \times Y$ is not a Radon space (see [267]).

6. Let I be a nonempty set of indices, $\{E_i : i \in I\}$ be a family of compact topological spaces and let, for each index $i \in I$, a Radon probability measure μ_i be given on the space E_i.

Prove that the product probability measure

$$\mu = \otimes\{\mu_i : i \in I\}$$

is also Radon.

For this purpose, utilize again the fact formulated in Exercise 2.

Remark 4. If $\{\mu_i : i \in I\}$ is an uncountable family of Radon probability measures given on Polish spaces, then one cannot assert that the product measure $\mu = \otimes\{\mu_i : i \in I\}$ is necessarily Radon (for more details, see Exercise 22 from Chapter 18).

7. Let E be a Hausdorff topological space and let μ be a σ-finite Radon measure defined on the Borel σ-algebra of E.

Show that, for an arbitrary function

$$f : E \to \mathbf{R},$$

the following three assertions are equivalent:

(a) f is measurable in the Luzin sense (with respect to μ);

(b) there exists a countable disjoint family $\{K_i : i \in I\}$ of compact subsets of E such that

$$\mu(E \setminus \cup\{K_i : i \in I\}) = 0$$

and the restriction $f|K_i$ is continuous for each index $i \in I$;

(c) for any Borel set $X \subset E$ with $\mu(X) > 0$, there exists a compact set $K \subset X$ such that $\mu(K) > 0$ and the restriction $f|K$ is continuous.

8. Let E be a base set, μ be a nonzero σ-finite measure defined on some σ-algebra of subsets of E, and let $\{X_i : i \in I\}$ be a partition of E into μ-measure zero sets.

Supposing that $\operatorname{card}(I)$ is not measurable in the Ulam sense (see Appendix 1), demonstrate that there exists a set $J \subset I$ such that the corresponding union $\cup\{X_j : j \in J\}$ is not μ-measurable.

9*. Let μ be a nonzero nonatomic σ-finite measure on a base set E.

Show that there exists a partition $\{Z_i : i \in I\}$ of E satisfying these two conditions:

(a) $\operatorname{card}(I) \leq \mathbf{c}$;

(b) each set Z_i $(i \in I)$ is of μ-measure zero.

For this purpose, first establish the following auxiliary fact: for any real $\varepsilon > 0$ and for every μ-measurable subset X of E, there exists a countable disjoint family $\{X_j : j \in J\}$ of μ-measurable subsets of X such that

$$X = \cup\{X_j : j \in J\},$$

$$\mu(X_j) < \varepsilon \quad (j \in J).$$

Then construct by recursion a sequence of countable families of pairwise disjoint sets

$$\{X_{j_1,j_2,\ldots,j_k} : j_1 \in \mathbf{N}, j_2 \in \mathbf{N}, \ldots, j_k \in \mathbf{N}\} \quad (k \in \mathbf{N}),$$

which fulfill the next three conditions:

$$X_{j_1} \supset X_{j_1,j_2} \supset \cdots \supset X_{j_1,j_2,\ldots,j_k} \supset \cdots ,$$

$$\cup\{X_{j_1,j_2\ldots,j_k} : j_1 \in \mathbf{N}, j_2 \in \mathbf{N}, \ldots, j_k \in \mathbf{N}\} = E \quad (k \in \mathbf{N}),$$

$$\mu(X_{j_1,j_2,\ldots j_k}) < 1/(k+1).$$

Finally, take the nonempty intersections of the form

$$X_{j_1} \cap X_{j_1,j_2} \cap \ldots \cap X_{j_1,j_2,\ldots,j_k} \cap \cdots$$

as members Z_i $(i \in I)$ of the required partition of E.

10. Let I be a set of indices with cardinality nonmeasurable in the Ulam sense, let $\{E_i : i \in I\}$ be a family of Radon topological spaces, and let E denote the topological sum of $\{E_i : i \in I\}$.

Show that E is also a Radon topological space.

Deduce from this fact that if X is an infinite discrete topological space and X^* is Alexandrov's one-point compactification of X, then the following two assertions are equivalent:

(a) $\mathrm{card}(X)$ is not measurable in the Ulam sense;

(b) X^* is a Radon space.

Conclude from (a) and (b) that the existence of a compact Radon space, whose cardinality is arbitrarily large and all subsets of which are Borel, is consistent with **ZFC** set theory.

11. Observe that every universal measure zero Hausdorff topological space is a Radon space.

Let I be a set of indices whose cardinality is not measurable in the Ulam sense and let $\{E_i : i \in I\}$ be a family of universal measure zero spaces.

Verify that the topological sum of $\{E_i : i \in I\}$ is also universal measure zero space.

Let $\{E_n : n \in \{1, 2, \ldots, k\}\}$ be a finite family of universal measure zero spaces.

Check that the product space

$$E = \prod\{E_n : n \in \{1, 2, ..., k\}\}$$

is also a universal measure zero space.

Finally, present an example of a countable family $\{X_n : n < \omega\}$ of universal measure zero spaces such that the product space

$$X = \prod\{X_n : n < \omega\}$$

is not universal measure zero.

12*. Give a direct proof of Ulam's theorem stating that any Polish topological space E is a Radon space.

For this purpose, consider an arbitrary Borel probability measure μ on E. For every real $\varepsilon > 0$ and for any natural number $n \geq 1$, find a finite family $\{B_i : i \in I_n\}$ of closed balls in E such that:

(a) the diameters of all balls B_i ($i \in I_n$) are strictly less than $1/n$;

(b) $\mu(\cup\{B_i : i \in I_n\}) \geq 1 - \varepsilon/2^n$.

Then define

$$K_n = \cup\{B_i : i \in I_n\} \qquad (1 \leq n < \omega),$$

$$K = \cap\{K_n : 1 \leq n < \omega\}$$

and check that K is a compact subset of E with $\mu(K) \geq 1 - \varepsilon$.

Now, if $X \in \mathcal{B}(E)$, then $\mu(X \setminus F) \leq \varepsilon$ for some set $F \subset X$ which is closed in E (see Exercise 3). Therefore,

$$\mu(X \setminus (F \cap K)) \leq \mu(X \setminus F) + \mu(X \setminus K) \leq 2\varepsilon.$$

To finish the argument, it suffices to notice that the set $F \cap K$ is compact.

13*. Let E be an arbitrary complete metric space.

Demonstrate that the following two conditions are equivalent:

(a) E is a Radon space;

(b) the topological weight of E is not measurable in the Ulam sense.

For this purpose, apply the previous exercise and the fact that under condition (b), for any σ-finite Borel measure μ on E, there exists a closed separable subset $F = F(\mu)$ of E satisfying the equality $\mu(E \setminus F) = 0$.

14*. Let E be a Hausdorff topological space such that all compact subsets of E are at most countable.

Observe that there is no nonzero σ-finite diffused Radon measure on E.

Deduce from this observation that the locally compact topological space $[0, \omega_1[$ does not admit any nonzero σ-finite diffused Radon measure.

Equip the closed interval $[0, \omega_1]$ with the order topology. In this manner, we obtain the compact space which may also be regarded as Alexandrov's one-point compactification of $[0, \omega_1[$.

For these two spaces, verify that:

(a) there exists a Borel diffused two-valued probability measure ν on $[0, \omega_1[$ (the so-called Dieudonné measure on $[0, \omega_1[$) and hence there exists a Borel diffused two-valued probability measure ν' on $[0, \omega_1]$ (the Dieudonné measure on $[0, \omega_1]$);

(b) both ν and ν' are not Radon measures, so $[0, \omega_1[$ and $[0, \omega_1]$ are not Radon spaces.

Let μ be an arbitrary Borel probability diffused measure on $[0, \omega_1]$.

Prove that $\mu(F) = 1$ for any closed uncountable set F in $[0, \omega_1]$ and, consequently, μ coincides with the above-mentioned Dieudonné measure ν'.

For this purpose, take into account the fact that ω_1 is not measurable in the Ulam sense (see Appendix 1).

Conclude that on the compact space $[0, \omega_1]$ there exists no nonatomic probability Borel measure.

15. Show that every σ-finite Radon measure μ defined on the Borel σ-algebra of a compact space E has a smallest closed support, i.e., there exists a least (with respect to the inclusion relation) closed subset $K = K(\mu)$ of E such that $\mu(E \setminus K) = 0$.

Observe that this K is unique and possesses the property that $\mu(U(x)) > 0$ for each point $x \in K$ and for any open neighborhood $U(x)$ of x.

Check that the Dieudonné measure ν' on the compact space $[0, \omega_1]$ does not possess a minimal closed support (with respect to the inclusion relation).

16*. Prove Henry's theorem stating that every σ-finite Radon measure ν on a Hausdorff topological space E can be extended to a measure ν' on E which also is Radon and contains in its domain the family of all Borel subsets of E.

Argue as follows. First of all, reduce the argument to the case where a Radon probability measure ν is given on E. Then consider the family of all finitely additive non-negative functionals μ which are defined on algebras of subsets of E, possess the Radon property (see condition (i) of Exercise 2) and extend ν. Verify that:

(a) this family has a maximal element ν' with respect to the inclusion relation;

(b) ν' is a Radon probability measure on E;

(c) all closed subsets of E belong to $\mathrm{dom}(\nu')$, so all Borel subsets of E belong to $\mathrm{dom}(\nu')$.

For checking (a), use the Kuratowski-Zorn lemma.

For establishing (b), use A.D. Alexandrov's theorem formulated in Exercise 4 of this chapter.

For establishing (c), assume that some closed set $F \subset E$ does not belong to $\mathrm{dom}(\nu')$ and suppose, without loss of generality, that $\nu'_*(E \setminus F) = 0$. Then, applying Exercise 18 from Chapter 5 to ν' and the σ-ideal generated by $\{E \setminus F\}$, extend ν' to a Radon probability measure ν'' such that $\nu''(E \setminus F) = 0$. This leads to a contradiction with the maximality of ν', so yields the required result.

17*. Let E be an arbitrary Hausdorff topological space.

Prove that the following two assertions are equivalent:

(a) there exists a nonzero σ-finite diffused Radon measure μ defined on the Borel σ-algebra of E;

(b) there exists a compact subset of E which can be continuously mapped onto the closed unit interval $[0, 1]$.

Argue as follows. Let (a) be satisfied and let K be a compact subset of E such that

$$0 < \mu(K) < +\infty.$$

According to Exercise 15, there exists a smallest closed support K_0 of the restriction of μ to K. Infer that all elements of K_0 are its condensation points and, by using ordinary recursion, construct a dyadic system of uncountable closed subsets of K_0. Then conclude that K_0 contains a compact subset which can be continuously mapped onto the Cantor space $\{0, 1\}^\omega$ and, therefore, can be continuously mapped onto $[0, 1]$. This yields (b) and the validity of the implication (a) $\Rightarrow$ (b).

To demonstrate the validity of the converse implication (b) $\Rightarrow$ (a), suppose that (b) is satisfied. Let F be a compact subset of E such that some

$$h : F \to [0, 1]$$

is a continuous surjection. Introduce the σ-algebra

$$\mathcal{S} = \{h^{-1}(B) : B \in \mathcal{B}([0, 1])\}$$

and put

$$\mu(h^{-1}(B)) = \lambda(B) \qquad (B \in \mathcal{B}([0, 1])).$$

Check that μ is a Radon nonatomic probability measure on $\mathcal{S}$. By using Exercise 16, extend μ to a Radon diffused probability measure defined on the Borel σ-algebra $\mathcal{B}(F)$. This circumstance readily yields (a) and so proves the implication (b) $\Rightarrow$ (a).

18*. Let E be a locally compact topological space and let μ be a σ-finite τ-smooth Borel measure on E (see the equality in Theorem 1).

Prove that μ is a Radon measure.

For this purpose, reduce the argument to the case where μ is a finite measure and then verify step by step that:

(a) if U is any open set in E, then there exists a σ-compact set $X \subset U$ such that $\mu(U) = \mu(X)$;

(a) if F is any closed set in E, then there exists a σ-compact set $Y \subset F$ such that $\mu(F) = \mu(Y)$;

(a) if B is any Borel set in E, then there exists a σ-compact set $Z \subset B$ such that $\mu(B) = \mu(Z)$.

19^*. Let $(S, \leq)$ be a Suslin line (see Appendix 2).

Demonstrate that any σ-finite Borel measure μ on S has a separable support. In other words, show that there exists a closed separable subset F of S such that $\mu(S \setminus F) = 0$.

For this purpose, assume without loss of generality that $\mu(\triangle) > 0$ for every non-degenerate open interval in S and then utilize the fact that the topological product $S \times S$ does not satisfy Suslin's condition, i.e., the countable chain condition (see Exercise 9 of Appendix 3).

Deduce from the stated above that any σ-finite Borel measure on S is Radon, which means that S is a Radon topological space.

Moreover, check that if $\{S_i : i \in I\}$ is a countable family of Suslin lines, then the topological product $\prod\{S_i : i \in I\}$ is also a Radon space.

20. Observe that every two-valued probability measure space is a perfect space and conclude from this fact that if $\mathrm{card}(E)$ is a two-valued measurable cardinal number, then there exists a perfect space of the form $(E, \mathcal{P}(E), \mu)$, where μ is a two-valued diffused probability measure with $\mathrm{dom}(\mu) = \mathcal{P}(E)$.

Consequently, Theorem 8 of this chapter does not admit a generalization (within **ZFC** set theory) to the class of all perfect probability spaces.

21^*. Let E be a Hausdorff topological space and let μ be a σ-finite Radon measure defined on the Borel σ-algebra of E. Suppose, in addition, that an ω_1-sequence $\{F_\xi : \xi < \omega_1\}$ of closed subsets of E is given such that

$$\mu(F_\xi) = 0 \quad (\xi < \omega_1).$$

By assuming Martin's Axiom with the negation of the Continuum Hypothesis, show that

$$\mu_*(\cup\{F_\xi : \xi < \omega_1\}) = 0.$$

For this purpose, utilize the fact that every σ-finite Radon measure defined on the Borel σ-algebra of a compact space has a support, i.e., the least (with respect to the inclusion relation) closed subset of the space, outside of which the measure is identically equal to zero (see Exercise 15). Also, keep in mind the circumstance that the above-mentioned support is compact and satisfies the Suslin condition, so a topological version of Martin's Axiom can be applied to it (see Appendix 3).

22. Let E be a complete metric space with $\mathrm{card}(E) = \mathbf{c}$ and without isolated points.

Assuming that $\mathbf{c}$ is not measurable in the Ulam sense, show that there exists a subset B of E which is absolutely nonmeasurable with respect to the class $\mathcal{CBM}_0(E)$ of completions of all nonzero σ-finite diffused Borel measures on E.

For this purpose, take as B an appropriate analogue of a Bernstein subset of $\mathbf{R}$.

23^*. Assuming that the cardinal $\mathbf{c}$ is measurable in the Ulam sense, verify that there exists no subset of the space $\mathbf{R^c}$ which is absolutely nonmeasurable with respect to the class $\mathcal{CBM}_0(\mathbf{R^c})$.

In order to obtain the required result, identify $\mathbf{c}$ with the least ordinal number α such that $\text{card}(\alpha) = \mathbf{c}$ and consider in $\mathbf{R^c}$ the family of characteristic functions

$$\mathcal{D} = \{\chi_{[0,\xi]} : \xi < \alpha\}.$$

Check that $\mathcal{D}$ is a discrete set of cardinality $\mathbf{c}$, representable as the difference of two closed subsets of $\mathbf{R^c}$, so $\mathcal{D}$ is a Borel set in $\mathbf{R^c}$. By using this $\mathcal{D}$, define a Borel probability measure μ on $\mathbf{R^c}$ whose completion μ' coincides with the power set of $\mathbf{R^c}$.

In particular, taking into account Theorem 8 of the present chapter, conclude from the above result that $\mathbf{R^c}$ is not a Radon space.

24. Let E be a Hilbert space (over $\mathbf{R}$) whose orthogonal basis is of cardinality $\mathbf{c}$, and suppose that $\mathbf{c}$ is measurable in Ulam's sense.

Demonstrate that:

(a) there exists no subset of E which is absolutely nonmeasurable with respect to the class $\mathcal{CBM}_0(E)$;

(b) there exists a set $X \subset E$ such that X is absolutely nonmeasurable with respect to the class of completions of all nonzero σ-finite diffused Radon measures on E.

25*. Assume Martin's Axiom with the negation of the Continuum Hypothesis. Let E be a Polish space and let $X \subset E$ be a continuous image of some co-analytic subset of a Polish space.

Prove that X is a Radon topological space.

For this purpose, utilize the classical result of Luzin and Sierpiński (see [10], [33], [103], [115], [152], [191] or Appendix 5), according to which X is representable as the union of an ω_1-sequence of Borel subsets of E.

26. Let E be a compact universal measure zero space and let

$$f : E \to E'$$

be a surjection such that the f-pre-images of all open sets in E' are G_δ-subsets of E.

Show that either E' is universal measure zero or E' is not a Radon space. For this purpose, utilize Henry's extension theorem (see Exercise 16).

Let X be a discrete topological space with $\operatorname{card}(X) = \omega_1$, let $x \notin X$ and let $X^* = X \cup \{x\}$ denote Alexandrov's one-point compactification of X. Consider any bijection

$$h : X^* \to [0, \omega_1]$$

such that $h(x) = \omega_1$.

Verify that the h-pre-images of all open sets in $[0, \omega_1]$ are G_δ-subsets of X^* and observe that the space X^* is universal measure zero, while the space $[0, \omega_1]$ is not Radon.

27. Let Y be a Hausdorff universal measure zero space and let Z be a Radon space.

Check that the topological product $Y \times Z$ is a Radon space.

7. Real-valued step functions with strange measurability properties

In this chapter we would like to analyze Sierpiński's example of a real-valued Lebesgue measurable function on $\mathbf{R}$ which is not bounded from above by any real-valued Borel function on $\mathbf{R}$. As Sierpiński indicates in his important and extensive article [232], the question of the existence of such a real-valued Lebesgue measurable function was raised by Luzin. In this context, several other real-valued step functions on $\mathbf{R}$ with analogous or somewhat similar properties will be discussed below.

As we have already mentioned in the Preface and in Chapter 1, the above-mentioned article [232] by Sierpiński is remarkable in various respects and may be regarded as a starting point of set-theoretic real analysis and, more generally, as a starting point of the so-called reverse mathematics. In [232], for the first time, Sierpiński specified and underlined the role of the Axiom of Choice in diverse areas of mathematics, especially, in classical point set theory, Lebesgue measure theory and mathematical analysis.

Among many concrete interesting and important mathematical results presented in [232], it was demonstrated therein (with the aid of $\mathbf{AC}$) that there exists a Lebesgue measurable real-valued function f on $\mathbf{R}$, for which there is no Borel function $\phi : \mathbf{R} \to \mathbf{R}$ satisfying the relation

$$f(x) \leq \phi(x) \qquad (x \in \mathbf{R}).$$

Here we are going to consider some natural generalizations and extensions of this result of Sierpiński. More precisely, in our further considerations we will be dealing with various measurable real-valued step functions on $\mathbf{R}$, for which certain analogues of Sierpiński's result hold true.

Let E be a base (ground) set.

Recall that a function $g : E \to \mathbf{R}$ is a step function on E if the range of g (denoted, as usual, by $\mathrm{ran}(g)$) is at most countable.

First of all, it worth noticing that Sierpiński's function $f : \mathbf{R} \to \mathbf{R}$ indicated above can be assumed to be a step function on $\mathbf{R}$. Indeed, let $\mathbf{Z}$ stand for the set of all integers and let

$$X_n = f^{-1}([n, n+1[) \qquad (n \in \mathbf{Z}).$$

The family $\{X_n : n \in \mathbf{Z}\}$ forms a countable disjoint covering of $\mathbf{R}$, all members of which are Lebesgue measurable sets. Let us introduce a new function

$$f^* : \mathbf{R} \to \mathbf{R}$$

by putting: $f^*(x) = n + 1$ if and only if $x \in X_n$.

Obviously, f^* is a Lebesgue measurable step function which is not bounded from above by any real-valued Borel function on $\mathbf{R}$, because the original function f has the same property and the inequality $f \leq f^*$ is valid.

The construction of f given in [232] starts with the existence of a Vitali type set in $\mathbf{R}$ which, as we know, is nonmeasurable in the Lebesgue sense (see, e.g., [17], [33], [77], [89], [133], [197], [203], [268] or Chapter 3; more detailed information about Vitali sets may be found in Chapter 9).

A certain generalization of Sierpiński's result can be obtained by using an approach substantially different from Sierpiński's method.

First, let us remind two dual notions from the theory of partially ordered sets (cf. [18]).

Let $(X, \preceq)$ be an arbitrary partially ordered set.

A subset Y of X is said to be coinitial in X if, for each element $x \in X$, there exists an element $y \in Y$ such that $y \preceq x$.

A subset Z of X is said to be cofinal in X if, for each element $x \in X$, there exists an element $z \in Z$ such that $x \preceq z$.

The following statement is purely set-theoretical and its nondifficult proof is within the framework of **ZFC** set theory.

Theorem 1. *Let E be an infinite set, $(F, \preceq)$ be a nonempty linearly ordered set without the least and greatest elements, and let $\mathcal{G}$ be a family of functions acting from E into F such that*

$$\mathrm{card}(\mathcal{G}) \leq \mathrm{card}(E).$$

Further, let F^* be a subset of F which is simultaneously cofinal and coinitial in F.

Then there exists a function $f : E \to F^*$ satisfying the following two conditions:

(1) there is no function $g \in \mathcal{G}$ such that

$$(\forall x \in E)(g(x) \preceq f(x));$$

(2) there is no function $h \in \mathcal{G}$ such that

$$(\forall x \in E)(f(x) \preceq h(x)).$$

In particular, f does not belong to the given family $\mathcal{G}$.

Proof. Since E is infinite, we may express E in the form

$$E = \{x_i : i \in I\} \cup \{x'_i : i \in I\},$$

where $\{x_i : i \in I\}$ and $\{x'_i : i \in I\}$ are two injective disjoint families of elements of E (consequently, by virtue of **AC**, we have $\mathrm{card}(I) = \mathrm{card}(E)$). Then we may represent $\mathcal{G}$ in the form

$$\mathcal{G} = \{g_i : i \in I\}.$$

By the way, we do not assume that the latter representation $\{g_i : i \in I\}$ is necessarily an injective family.

Now, for each index $i \in I$, define $f(x_i) \in F^*$ and $f(x'_i) \in F^*$ so that

$$f(x_i) \prec g_i(x_i) \quad g_i(x'_i) \prec f(x'_i).$$

Such a choice of the values $f(x_i)$ and $f(x'_i)$ is possible, because of the assumption that the set F^* is cofinal and coinitial in F (in this connection, we would like to recall that F and, consequently, F^* are without the least and greatest elements).

It can easily be checked that the obtained function $f : E \to F^*$ is as required, which completes the proof of Theorem 1.

Remark 1. The above argument essentially relies on the Axiom of Choice. However, an effective analogue (i.e., within **ZF** theory) of Theorem 1 can be formulated in the case where the following two conditions are fulfilled:

(a) the set F^* is countably infinite;

(b) the set E admits an effective partition $\{E', E''\}$ into two subsets of E such that

$$\mathrm{card}(E') = \mathrm{card}(E'') = \mathrm{card}(E).$$

Let λ denote the standard Lebesgue measure on the real line $\mathbf{R}$. The next statement is readily implied by Theorem 1.

Theorem 2. *There exists a function*

$$f : \mathbf{R} \to \mathbf{R}$$

satisfying the following three conditions:

(1) f is a λ-measurable step function;

(2) there is no Borel function $\phi : \mathbf{R} \to \mathbf{R}$ such that

$$(\forall x \in \mathbf{R})(f(x) \geq \phi(x));$$

(3) there is no Borel function $\psi : \mathbf{R} \to \mathbf{R}$ such that

$$(\forall x \in \mathbf{R})(f(x) \leq \psi(x)).$$

Proof. Let $\mathbf{c}$ denote the cardinality of the continuum. Take any λ-measure zero set $E \subset \mathbf{R}$ with $\operatorname{card}(E) = \mathbf{c}$. For example, the role of E can be played by the classical Cantor set $C \subset \mathbf{R}$. Obviously, C admits an effective partition into two subsets, each of which is of cardinality continuum. Let us put

$$(F, \preceq) = (\mathbf{R}, \le),$$

$$F^* = \mathbf{Z},$$

and let $\mathcal{G}$ be the family of all real-valued Borel functions on E. As is well known, $\operatorname{card}(\mathcal{G}) = \mathbf{c}$.

We may apply Theorem 1 to these E, F, F^*, and $\mathcal{G}$. So there exists a function

$$f : E \to F^*$$

satisfying conditions (1) and (2) of Theorem 1. Clearly, we can extend f to a real-valued step function on $\mathbf{R}$, e.g., by putting $f(x) = 0$ for all points $x \in \mathbf{R} \backslash E$. Preserving the same notation for the extended in such a manner function, we obtain the required Lebesgue measurable step function $f : \mathbf{R} \to \mathbf{R}$. Moreover, the definition of f directly implies that f is equal to zero for λ-almost all points of $\mathbf{R}$.

Theorem 2 has thus been proved.

Remark 2. It is useful to compare the above result with the following two widely known and basic statements of classical real analysis (see, e.g., [17], [96], [133], [197]):

(i) if a function $g : \mathbf{R} \to \mathbf{R}$ is Lebesgue measurable and bounded, then there exist two Borel bounded functions

$$\phi : \mathbf{R} \to \mathbf{R}, \quad \psi : \mathbf{R} \to \mathbf{R}$$

such that

$$\phi(x) \le g(x) \le \psi(x) \qquad (x \in \mathbf{R})$$

and $\phi(x) = \psi(x)$ for λ-almost all points $x \in \mathbf{R}$ (i.e., ϕ and ψ are λ-equivalent functions, so they both are λ-equivalent to g);

(ii) for an arbitrary λ-measurable function $h : \mathbf{R} \to \mathbf{R}$, there exists a function

$$\chi : \mathbf{R} \to \mathbf{R}$$

of Baire class 2 such that h and χ are λ-equivalent (Vitali's theorem).

Remark 3. Let $\mathcal{B}(C, \mathbf{R})$ denote the family of all Borel real-valued functions on the Cantor set C. The proof of Theorem 2 makes use of the inequality

$$\mathcal{B}(C, \mathbf{R}) \le \mathbf{c}.$$

As was demonstrated by Sierpiński many years ago, the written inequality implies the existence of a Lebesgue nonmeasurable subset of $\mathbf{R}$ (this implication holds within $\mathbf{ZF}$ & $\mathbf{DC}$ theory, where $\mathbf{DC}$ abbreviates the Axiom of Dependent Choice; see Exercise 4 from Chapter 3 of the present book). So, one may conclude that the proof of Theorem 2 given above substantially relies on an uncountable form of the Axiom of Choice. It seems that the assertion of this theorem cannot be proved without the aid of an appropriate uncountable version of $\mathbf{AC}$.

Under certain additional set-theoretic assumptions, Theorem 2 admits further extensions. To give such an extension, we need several auxiliary notions.

The important notion of a Luzin set on the real line $\mathbf{R}$ was mentioned in Exercise 15 from Chapter 4. We now recall the very similar and also important notion of a generalized Luzin set on $\mathbf{R}$.

A subset X of $\mathbf{R}$ is a generalized Luzin set if $\operatorname{card}(X) = \mathbf{c}$ and the inequality $\operatorname{card}(X \cap Y) < \mathbf{c}$ holds true for every first category set Y in $\mathbf{R}$.

It is well known that, by adding Martin's Axiom ($\mathbf{MA}$) to $\mathbf{ZFC}$ set theory, it becomes possible to construct a generalized Luzin set X on $\mathbf{R}$, and any such construction of X is based on the method of transfinite recursion. This fact is completely analogous to the existence of Luzin sets on $\mathbf{R}$ under the Continuum Hypothesis (cf. Exercise 15 from Chapter 4 and Exercise 9 from this chapter). Actually, Luzin sets and generalized Luzin sets possess many parallel properties.

Recall that a measure μ given on a σ-algebra of subsets of a ground set E is diffused (or continuous) if all singletons in E are μ-measurable and μ vanishes at all of them.

Under Martin's Axiom, generalized Luzin sets turn out to be utterly small from the point of view of topological measure theory.

More precisely, under $\mathbf{MA}$, every generalized Luzin set X has universal measure zero, i.e., if μ is any σ-finite diffused Borel measure on $\mathbf{R}$, then the equality $\mu^*(X) = 0$ holds, where μ^* denotes the outer measure on $\mathbf{R}$ associated with μ.

For the sake of brevity, in the present chapter a function $f : \mathbf{R} \to \mathbf{R}$ will be called absolutely (or universally) measurable if, for every nonzero σ-finite diffused Borel measure μ on $\mathbf{R}$, this f is measurable with respect to the completion μ' of μ.

As the reader can observe, here the absolute (universal) measurability of f is regarded as the absolute (universal) measurability of f with respect to the class $\mathcal{CBM}_0(\mathbf{R})$ of completions of all nonzero σ-finite diffused Borel measures on $\mathbf{R}$ (cf. Chapter 5).

Remark 4. There are many examples of Lebesgue measurable real-valued functions on $\mathbf{R}$ which are not absolutely measurable. On the other hand, there are absolutely measurable functions on $\mathbf{R}$ which are not Borel. In particular, the characteristic function of any analytic non-Borel subset of $\mathbf{R}$ is an absolutely measurable non-Borel function (cf. Example 2 from Chapter 6).

Theorem 3. *Assume Martin's Axiom. There exists a function*

$$f : \mathbf{R} \to \mathbf{R}$$

satisfying the following three conditions:
(1) *f is an absolutely measurable step function;*
(2) *there is no Borel function $\phi : \mathbf{R} \to \mathbf{R}$ such that*

$$(\forall x \in \mathbf{R})(f(x) \geq \phi(x));$$

(3) *there is no Borel function $\psi : \mathbf{R} \to \mathbf{R}$ such that*

$$(\forall x \in \mathbf{R})(f(x) \leq \psi(x)).$$

Proof. The argument is quite similar to the proof of Theorem 2. We take an arbitrary generalized Luzin set $X \subset \mathbf{R}$ and put $E = X$. We also put

$$(F, \preceq) = (\mathbf{R}, \leq),$$

$$F^* = \mathbf{Z},$$

$$\mathcal{G} = \mathcal{B}(X, \mathbf{R}),$$

where $\mathcal{B}(X, \mathbf{R})$ stands for the family of all Borel functions acting from X into $\mathbf{R}$. Clearly, we have

$$\operatorname{card}(\mathcal{G}) = \mathbf{c}.$$

Theorem 1 is applicable to the just introduced E, F, F^*, and $\mathcal{G}$. So there exists a function

$$f : X \to \mathbf{Z}$$

satisfying conditions (1) and (2) of Theorem 1. This f is a step function and can trivially be extended to the step function defined on the whole $\mathbf{R}$ by putting $f(x) = 0$ for all $x \in \mathbf{R} \setminus X$. It is not difficult to verify that the extended step function $f : \mathbf{R} \to \mathbf{R}$ is absolutely measurable and, consequently, is as required.

For our further purposes, we need the following auxiliary proposition from general set theory.

Lemma 1. *Let E be an infinite set and let $\{X_i : i \in I\}$ be a family of subsets of E satisfying these two conditions:*
(1) $\operatorname{card}(I) \leq \operatorname{card}(E)$;
(2) $\operatorname{card}(X_i) = \operatorname{card}(E)$ *for each index $i \in I$.*
Then there exists a disjoint family of sets $\{Y_i : i \in I\}$ such that $Y_i \subset X_i$ and $\operatorname{card}(Y_i) = \operatorname{card}(E)$ for any index $i \in I$.

Lemma 1 is due to Sierpiński (its proof may be found, e.g., in [152], [154] or [243]; see also Exercise 4 for this chapter).

We shall say that a function $h : \mathbf{R} \to \mathbf{R}$ is relatively measurable if there exists a nonzero σ-finite diffused Borel measure μ on $\mathbf{R}$ such that h is μ'-measurable, where μ' stands, as earlier, for the completion of μ.

Again, the reader can easily observe that here the relative measurability of a function h is regarded as the relative measurability of h with respect to the class $\mathcal{CBM}_0(\mathbf{R})$ of completions of all nonzero σ-finite diffused Borel measures on $\mathbf{R}$ (cf. Chapter 5).

Remark 5. A certain characterization of all relatively measurable functions was given earlier in this book. Namely, recall that a function $h : \mathbf{R} \to \mathbf{R}$ is relatively measurable if and only if there exist a Lebesgue measurable function $g : \mathbf{R} \to \mathbf{R}$ and a Borel isomorphism $\phi : \mathbf{R} \to \mathbf{R}$ such that $h = g \circ \phi$ (see Theorem 1 from Chapter 5). It readily follows from this characterization that there exist many relatively measurable functions which are nonmeasurable in the Lebesgue sense.

Remark 6. It worth noticing in connection with Remark 5 that there exists a function $h : \mathbf{R} \to \mathbf{R}$ satisfying the following three conditions:

(a) $\mathrm{ran}(h) \subset \mathbf{Q}$, where $\mathbf{Q}$ denotes the field of all rational numbers (in particular, h is a step function on $\mathbf{R}$);

(b) h is relatively measurable;

(c) h is an endomorphism of the additive group $(\mathbf{R}, +)$.

To obtain such an h, consider a nonempty perfect subset P of $\mathbf{R}$ linearly independent over the field $\mathbf{Q}$ (the existence of P is a well-known fact of classical point set theory; cf. [115], [190], [193], [268]). Let $\{e_i : i \in I\}$ stand for some Hamel basis of $\mathbf{R}$ containing P. We define $h : \mathbf{R} \to \mathbf{Q}$ as follows. Every real number x admits a unique representation in the form

$$x = q_{i_1} e_{i_1} + q_{i_2} e_{i_2} + \ldots + q_{i_n} e_{i_n},$$

where $n = n(x)$ is a natural number, $\{i_1, i_2, \ldots, i_n\}$ is a finite injective family of indices from I, and $\{q_{i_1}, q_{i_2}, \ldots, q_{i_n}\}$ is a finite family of nonzero rational numbers. We put

$$h(x) = q_{i_1} + q_{i_2} + \ldots + q_{i_n}.$$

Obviously, h is an additive function acting from $\mathbf{R}$ into $\mathbf{Q}$, so conditions (a) and (c) are valid. Further, the restriction $h|P$ is identically equal to 1. Let μ be a Borel diffused probability measure on $\mathbf{R}$ concentrated on P, i.e., $\mu(\mathbf{R} \setminus P) = 0$, and let μ' denote the completion of μ. It is clear that h turns out to be measurable with respect to μ', which implies that h is relatively measurable and thus condition (b) is satisfied, too.

Remark 7. As is well known, any endomorphism g of the additive group $(\mathbf{R}, +)$ into itself either is trivial (i.e., linear over the field $\mathbf{R}$) or is nonmeasurable in the Lebesgue sense. Furthermore, if g is nonmeasurable in the Lebesgue sense, then g is not bounded from above (from below) by any Lebesgue measurable function $\phi : \mathbf{R} \to \mathbf{R}$. In particular, a relatively measurable endomorphism h of $(\mathbf{R}, +)$ indicated in Remark 6 is not bounded from above (from below) by any Lebesgue measurable real-valued function on $\mathbf{R}$.

Lemma 2. *If g is a relatively measurable function on $\mathbf{R}$, then there exists an uncountable compact set $K \subset \mathbf{R}$ such that the restriction $g|K$ is continuous and, consequently, bounded.*

Proof. Since g is relatively measurable, there exists a measure ν which is the completion of some nonzero σ-finite diffused Borel measure on $\mathbf{R}$ and for which g turns out to be ν-measurable. According to a well-known result of topological measure theory, ν is a Radon measure (see Chapter 6). So, for the ν-measurable function g, Luzin's classical C-property is valid. This means that there exists a compact set $K \subset \mathbf{R}$ with $\nu(K) > 0$ such that the restriction $g|K$ is continuous. The diffuseness of ν and the inequality $\nu(K) > 0$ imply that K is necessarily uncountable, which finishes the proof.

Theorem 4. *There exists a function*

$$f : \mathbf{R} \to \mathbf{R}$$

satisfying the following three conditions:
(1) f is a step function;
(2) there is no relatively measurable function $\phi : \mathbf{R} \to \mathbf{R}$ such that

$$(\forall x \in \mathbf{R})(f(x) \geq \phi(x));$$

(3) there is no relatively measurable function $\psi : \mathbf{R} \to \mathbf{R}$ such that

$$(\forall x \in \mathbf{R})(f(x) \leq \psi(x)).$$

Proof. Let $\{K_i : i \in I\}$ be an injective family consisting of all uncountable compact subsets of $\mathbf{R}$. By virtue of Lemma 1, there exists a disjoint family of sets $\{Y_i : i \in I\}$ such that:
(a) $Y_i \subset K_i$ for each index $i \in I$;
(b) $\mathrm{card}(Y_i) = \mathbf{c}$ for each index $i \in I$.
For any $i \in I$, let f_i denote some surjection of Y_i onto the set $\mathbf{Z}$ of all integers. Since the domains of f_i $(i \in I)$ are pairwise disjoint, we may extend all these functions to a step function f defined on the whole real line $\mathbf{R}$. Namely, we may put $f(x) = 0$ for each point $x \in \mathbf{R} \setminus \cup \{Y_i : i \in I\}$.

We now assert that f is as required. Indeed, take any relatively measurable function

$$\phi : \mathbf{R} \to \mathbf{R}$$

and suppose for a moment that

$$(\forall x \in \mathbf{R})(f(x) \geq \phi(x)).$$

Then, according to Lemma 2, there exists an uncountable compact subset K of $\mathbf{R}$ such that the restriction $\phi|K$ is bounded from below. Consequently, the restriction $f|K$ is also bounded from below. But this circumstance contradicts the definition of f. Indeed, $K = K_i$ for some index $i \in I$ and

$$\mathbf{Z} = \mathrm{ran}(f|Y_i) \subset \mathrm{ran}(f|K_i) = \mathrm{ran}(f|K),$$

which is impossible. The obtained contradiction shows that (2) holds true for f. The analogous argument yields that (3) is valid for f, too.

Theorem 4 has thus been proved.

Remark 8. Let $\{X_i : i \in I\}$ be a family of subsets of $\mathbf{R}$ satisfying the following two conditions:

(a) $\mathrm{card}(I) \leq \mathbf{c}$;

(b) $\mathrm{card}(X_i) = \mathbf{c}$ for each index $i \in I$.

Similarly to the proof of Theorem 4, it can be demonstrated that there exists a function $f : \mathbf{R} \to \mathbf{R}$ such that $\mathrm{ran}(f|X_i) = \mathbf{R}$ for each $i \in I$.

In particular, if the family of all nonempty perfect subsets of $\mathbf{R}$ is taken as $\{X_i : i \in I\}$, then we have:

(*) for any λ-measurable set $X \subset \mathbf{R}$ with $\lambda(X) > 0$, the set $f(X)$ coincides with $\mathbf{R}$;

(**) for any second category set $Y \subset \mathbf{R}$ possessing the Baire property, the set $f(Y)$ coincides with $\mathbf{R}$.

Remark 9. It is not difficult to check that the following two assertions are equivalent within **ZF** set theory:

(a) for any function $g : \mathbf{R} \to \mathbf{R}$, there exists an uncountable set $X \subset \mathbf{R}$ such that the restriction $g|X$ is bounded;

(b) $\mathbf{R}$ cannot be represented as the union of a countable family of countable subsets of $\mathbf{R}$.

Recall that there are models of **ZF** theory in which both assertions (a) and (b) are false (see [57] or [102]). Clearly, in those models all subsets of $\mathbf{R}$ become Borel and, consequently, all functions acting from $\mathbf{R}$ into $\mathbf{R}$ become Borel functions. So it is impossible to prove within **ZF** set theory that there exists a Lebesgue measurable real-valued function on $\mathbf{R}$ which is not a Borel function.

Remark 10. Let μ be a σ-finite measure on a ground set E and let $\mathcal{M}(\mu)$ denote the class of all those measures on E which extend μ. Consider an arbitrary function $f : E \to \mathbf{R}$. It can be proved (cf. [2]) that there exist a measure $\nu \in \mathcal{M}(\mu)$ and two ν-measurable step functions

$$\phi : E \to \mathbf{R}, \quad \psi : E \to \mathbf{R}$$

satisfying the inequalities

$$\phi(x) \le f(x) \le \psi(x) \qquad (x \in E).$$

In particular, putting $E = \mathbf{R}$ and $\mu = \lambda$, we may conclude that the graph of any function $f : \mathbf{R} \to \mathbf{R}$ lies between the graphs of two real-valued step functions on $\mathbf{R}$, each of which is measurable with respect to some measure $\nu \in \mathcal{M}(\lambda)$.

EXERCISES

1. Preserve the notation of Theorem 1 and prove an effective analogue (i.e., within **ZF** set theory) of this theorem in the case where the following two conditions are simultaneously satisfied:

(a) the set F^* is countably infinite;

(b) the infinite set E admits an effective partition $\{E', E''\}$ into two subsets of E such that $\mathrm{card}(E') = \mathrm{card}(E'') = \mathrm{card}(E)$.

2. Give an effective example of a Borel isomorphism between the real line $\mathbf{R}$ and the Cantor space $C \subset \mathbf{R}$.

For this purpose, take into account the fact that C is a subset of $\mathbf{R}$ and that C contains a subset homeomorphic to the Baire canonical space $\mathbf{N}^{\mathbf{N}}$, where the set $\mathbf{N}$ of all natural numbers is assumed to be equipped with the discrete topology. Then observe that the Baire space $\mathbf{N}^{\mathbf{N}}$ is homeomorphic to the set $\mathbf{R} \setminus \mathbf{Q}$ of all irrational numbers and finally utilize an appropriate version of the Cantor–Bernstein theorem (see Appendix 1).

Let $\phi : \mathbf{R} \to C$ be a Borel isomorphism. Define a mapping

$$\Phi : \mathcal{B}(\mathbf{R}, \mathbf{R}) \to \mathcal{B}(C, \mathbf{R})$$

by putting

$$\Phi(g) = g \circ \phi^{-1} \qquad (g \in \mathcal{B}(\mathbf{R}, \mathbf{R})).$$

Check that Φ is a bijection between $\mathcal{B}(\mathbf{R}, \mathbf{R})$ and $\mathcal{B}(C, \mathbf{R})$, so one has

$$\mathrm{card}(\mathcal{B}(\mathbf{R}, \mathbf{R})) = \mathrm{card}(\mathcal{B}(C, \mathbf{R}))$$

and this equality holds within **ZF** set theory.

Remark 11. As was indicated in Exercise 4 of Chapter 3, the inequality

$$\mathrm{card}(\mathcal{B}(\mathbf{R}, \mathbf{R})) \le \mathbf{c}$$

implies, within **ZF** & **DC** theory, the existence of a Lebesgue nonmeasurable subset of $\mathbf{R}$. In view of Exercise 2, the same may be asserted on the inequality

$$\mathrm{card}(\mathcal{B}(C, \mathbf{R})) \le \mathbf{c},$$

i.e., this second inequality also implies, within **ZF** & **DC** theory, the existence of a Lebesgue nonmeasurable point set.

The latter implication can be obtained in another manner, by considering C as a compact topological group endowed with the completion of Haar probability measure on C, which is canonically isomorphic to the restriction of λ to $\mathcal{B}([0,1])$.

3. Let $\{Z_i : i \in I\}$ be a countable partition of $\mathbf{R}$ into universally measurable sets with respect to the class $\mathcal{CBM}(\mathbf{R})$ of completions of all σ-finite Borel measures on $\mathbf{R}$, and let

$$f : \mathbf{R} \to \mathbf{R}$$

be a function such that all the restrictions $f|Z_i$ $(i \in I)$ are Borel.

Verify that f is absolutely measurable.

Notice that a special case of this fact was implicitly utilized in the proof of Theorem 3.

4*. As was mentioned in this chapter, the following result is due to Sierpiński and is known as Sierpiński's lemma on disjoint subsets.

Let E be an infinite set and let $\{X_i : i \in I\}$ be a family of subsets of E satisfying these two conditions:

(a) $\mathrm{card}(I) \leq \mathrm{card}(E)$;

(b) $\mathrm{card}(X_i) = \mathrm{card}(E)$ for each index $i \in I$.

Demonstrate that there exists a disjoint family of sets $\{Y_i : i \in I\}$ such that $Y_i \subset X_i$ and $\mathrm{card}(Y_i) = \mathrm{card}(E)$ for any $i \in I$.

For this purpose, assume (without loss of generality) that $\mathrm{card}(I) = \mathrm{card}(E)$ and identify I with the half-open interval $[0, w_\alpha[$, where the cardinal number w_α is equinumerous with $\mathrm{card}(E)$. Then, by using the method of transfinite recursion, define an injective double family

$$\{x_{\xi,\eta} : \xi < w_\alpha, \ \eta < w_\alpha\}$$

of elements of E such that

$$x_{\xi,\eta} \in X_\xi \qquad (\xi < w_\alpha, \ \eta < w_\alpha).$$

Finally, put

$$Y_\xi = \{x_{\xi,\eta} : \eta < w_\alpha\} \qquad (\xi < w_\alpha)$$

and verify that the family $\{Y_\xi : \xi < w_\alpha\}$ is as required.

Remark 12. The result of Exercise 4 substantially strengthens the following well-known statement of **ZFC** set theory: for any infinite set E, the equality

$$\mathrm{card}(E \times E) = \mathrm{card}(E)$$

holds true. It worth noticing here that this statement is equivalent, within **ZF** set theory, to the Axiom of Choice (see Appendix 2).

5. Let g be an endomorphism of the additive group $(\mathbf{R}, +)$ into itself and assume that g is nonmeasurable in the Lebesgue sense.

Show that g is not bounded from above (from below) by any Lebesgue measurable function $\phi : \mathbf{R} \to \mathbf{R}$.

For this purpose, suppose otherwise, i.e., ϕ bounds from above the given g. Then $g(x) \leq t$ for some real t and for all points x from some λ-measurable set X of strictly positive λ-measure. Keeping in mind this circumstance and applying the Steinhaus property of X stating that the difference set

$$X - X = \{x - x' : x \in X, \ x' \in X\}$$

is a neighborhood of zero of $\mathbf{R}$ (see Exercise 1 from Chapter 3), infer that g must be bounded on a neighborhood of zero, so g turns out to be a trivial solution of the Cauchy functional equation, which contradicts the nonmeasurability of g.

Obtain the same result by using another method, namely, by starting with Lebesgue's classical theorem on density points of λ-measurable subsets of $\mathbf{R}$ with strictly positive λ-measure.

6. Give detailed proofs of the statement formulated in Remark 8.

7*. Let μ be a σ-finite measure on a ground set E and let $\{X_i : i \in I\}$ be a disjoint family of subsets of E.

Demonstrate that there exists a measure ν on E extending μ and such that

$$\{X_i : i \in I\} \subset \mathrm{dom}(\nu).$$

For this purpose, suppose (without loss of generality) that the given family $\{X_i : i \in I\}$ is a partition of E and consider two possible cases.

(1) The partition $\{X_i : i \in I\}$ is at most countable, i.e., we have

$$\mathrm{card}(I) \leq \mathrm{card}(\mathbf{N}) = \omega.$$

Let $\{t_i : i \in I\}$ be an injective family of real numbers and let $f : E \to \mathbf{R}$ be a step function such that $\mathrm{ran}(f|X_i) = \{t_i\}$ for any $i \in I$. Clearly, it suffices to show that there exists a measure on E extending μ for which f becomes measurable.

Since $\mathrm{card}(I) \leq \omega$, one may assume that either $I = \{1, 2, ..., m\}$ or $I = \mathbf{N} = \omega$. Under this assumption, put:

$X'_n = $ a μ-measurable hull of X_n, where $n \in I$;
$Y_n = X'_n \setminus (\cup\{X'_k : k < n\})$, where $n \in I$.

Obviously, the family of sets $\{Y_n : n \in I\}$ is a disjoint covering of E. Define a new function

$$f_0 : E \to \mathbf{R}$$

by putting: $f_0(x) = t_n$ if and only if $x \in Y_n$.

Since all sets Y_n ($n \in I$) are μ-measurable, it immediately follows from the definition of f_0 that f_0 is a μ-measurable function. Further, check the equality

$$\mu_*(\{x \in E : f(x) \neq f_0(x)\}) = 0,$$

where μ_* denotes, as usual, the inner measure associated with μ. For checking the above equality, suppose otherwise, i.e., suppose that

$$\mu_*(\{x \in E : f(x) \neq f_0(x)\}) > 0.$$

Then there exists an index $n \in I$ such that

$$\mu_*(Y_n \cap \{x \in E : f(x) \neq f_0(x)\}) > 0.$$

On the other hand, it is easy to verify the following inclusion:

$$Y_n \cap \{x \in E : f(x) \neq f_0(x)\} \subset X_n' \setminus X_n,$$

which gives a contradiction with the definition of the measurable hull X_n' of X_n. The contradiction obtained establishes the validity of the required relation

$$\mu_*(\{x \in E : f(x) \neq f_0(x)\}) = 0.$$

Now, by virtue of Marczewski's method of extending σ-finite measures (see Exercise 18 from Chapter 5), the set

$$X = \{x \in E : f(x) \neq f_0(x)\}$$

can be made measurable with respect to some measure ν on E extending μ and, moreover, ν can be chosen so that $\nu(X) = 0$. Consequently, f becomes measurable with respect to this ν.

(2) The partition $\{X_i : i \in I\}$ of E is uncountable.

Using the σ-finiteness of μ which implies the validity of the so-called countable chain condition, it is not difficult to show that there exists a set $J \subset I$ satisfying the following two conditions:

(a) $\mathrm{card}(I \setminus J) \leq \omega$;

(b) for any countable set $J_0 \subset J$, the equality

$$\mu_*(\cup\{X_j : j \in J_0\}) = 0$$

holds true.

Starting with (b) and applying again Marczewski's method, first make all the sets X_j ($j \in J$) be measurable with respect to some measure μ' on E extending μ. Notice, by the way, that $\mu'(X_j) = 0$ for all indices $j \in J$. Finally, apply the previous case to the countable disjoint family $\{X_i : i \in I \setminus J\}$ and extend μ' to a measure μ'' on E such that

$$\{X_i : i \in I \setminus J\} \subset \mathrm{dom}(\mu'').$$

Conclude from the above that the measure $\nu = \mu''$ turns out to be an extension of μ and satisfies the inclusion $\{X_i : i \in I\} \subset \text{dom}(\nu)$, which yields the required result.

8. Give a proof of the assertion formulated in Remark 10.
For this purpose, keep in mind the result of Exercise 7.

9*. Assuming Martin's Axiom (**MA**), demonstrate that there exist generalized Luzin sets on **R**.
For this purpose, identify **c** with the initial ordinal number of cardinality continuum and denote by $\{X_\xi : \xi < \mathbf{c}\}$ the family of all those first category subsets of **R** which are of type F_σ.
Applying the method of transfinite recursion, define an injective family of points

$$\{x_\xi : \xi < \mathbf{c}\} \subset \mathbf{R}$$

as follows. Suppose that, for an ordinal $\zeta < \mathbf{c}$, the partial injective family of points $\{x_\xi : \xi < \zeta\}$ has already been defined and consider the set

$$X_\zeta' = (\cup\{X_\xi : \xi < \zeta\}) \cup \{x_\xi : \xi < \zeta\}.$$

By virtue of Martin's Axiom (see Theorem 4 from Appendix 3), the set X_ζ' is of first category. Consequently, there exists a point $x \in \mathbf{R} \setminus X_\zeta'$. Put $x_\zeta = x$.
Proceeding in this manner, the family $\{x_\xi : \xi < \mathbf{c}\}$ will be constructed.
In view of this construction, the following relations are valid:
(a) $\{x_\xi : \xi < \mathbf{c}\}$ is an injective family;
(b) for any two ordinals $\xi < \mathbf{c}$ and $\zeta < \mathbf{c}$, such that $\xi < \zeta$, the point x_ζ does not belong to the set X_ξ.
Finally, define $X = \{x_\xi : \xi < \mathbf{c}\}$ and check that X is a generalized Luzin subset of **R**.

10*. Assuming again Martin's Axiom, show that:
(a) if a subset X of a generalized Luzin set has cardinality continuum, then X is a generalized Luzin set, too;
(b) no generalized Luzin set possesses the Baire property;
(c) the σ-ideal $\mathcal{I}$ generated by the family of all generalized Luzin sets in **R** is invariant under the group of transformations of **R** which preserve the σ-ideal of all first category sets in **R** (in particular, $\mathcal{I}$ is invariant under the group of all homeomorphisms of **R**);
(d) any generalized Luzin set $L \subset \mathbf{R}$ has universal measure zero;
(e) no generalized Luzin set can coincide with a Vitali subset of **R** or with a Bernstein subset of **R**.
For (d), argue as follows. First, demonstrate under **MA** that if X is a subset of **R** with $\text{card}(X) < \mathbf{c}$, then X has universal measure zero. Then take into account the fact that any σ-finite diffused Borel measure on **R** (more generally,

on a separable metric space E), is concentrated on a first category subset of $\mathbf{R}$ (of E).

11. Let $(E, \leq)$ be an infinite well-ordered set whose cardinality is regular, and let E^E denote the family of all mappings acting from E into itself.

Introduce a partial pre-ordering $\preceq$ in E^E by putting

$$f \preceq g \Leftrightarrow \operatorname{card}(\{x \in E : g(x) < f(x)\}) < \operatorname{card}(E)$$

for any two functions $f \in E^E$ and $g \in E^E$.

Let $\mathcal{F}$ be a subset of $(E^E, \preceq)$ cofinal in E^E.

Verify the validity of the inequality

$$\operatorname{card}(\mathcal{F}) > \operatorname{card}(E).$$

Consider the important particular case

$$(E, \leq) = (\mathbf{N}, \leq),$$

where $\leq$ is the canonical well-ordering of $\mathbf{N}$.

Assuming the regularity of the cardinal $\mathbf{c}$, consider also another particular case of $(E, \leq)$, where $\operatorname{card}(E) = \mathbf{c}$ and $\leq$ is some well-ordering of E.

Remark 13. In connection with the last part of Exercise 11, we would like to recall that Martin's Axiom readily implies the regularity of $\mathbf{c}$ (see Appendix 3). However, there are various models of **ZFC** set theory in which $\mathbf{c}$ is a singular cardinal, e.g., it may happen that $\mathbf{c} = \omega_{\omega_1}$ (see [103], [148]).

12*. Let $\mathbf{R}^{\mathbf{R}}$ denote the vector space (over $\mathbf{R}$) of all real-valued functions on $\mathbf{R}$.

Prove that there exists a vector subspace $\mathcal{W}$ of $\mathbf{R}^{\mathbf{R}}$ satisfying the following three conditions:

(a) $\operatorname{card}(\mathcal{W}) = 2^{\mathbf{c}}$;

(b) each element of $\mathcal{W}$ is a step function;

(c) every nonzero element of $\mathcal{W}$ is a function nonmeasurable in the Lebesgue sense.

In order to demonstrate the existence of a desired $\mathcal{W}$, start with a family $\{X_j : j \in J\}$ of subsets of $\mathbf{R}$, which is independent in the set-theoretical sense (see Appendix 1) and, moreover, has the following two properties:

(i) $\operatorname{card}(J) = 2^{\mathbf{c}}$;

(ii) for each natural number n and for every injective finite sequence

$$\{j_1, j_2, \ldots, j_n\}$$

of indices from J, any set of the form

$$X'_{j_1} \cap X'_{j_2} \cap \ldots \cap X'_{j_n},$$

where
$$X'_{j_m} = X_{j_m} \quad \vee \quad X'_{j_m} = \mathbf{R} \setminus X_{j_m} \qquad (1 \leq m \leq n),$$
is a λ-thick subset of $\mathbf{R}$.

Then consider the vector space $\mathcal{W}$ over $\mathbf{R}$, generated by the family of the characteristic functions of all members of $\{X_j : j \in J\}$, and verify that such a $\mathcal{W}$ is as required.

8. A partition of the real line into continuum many thick subsets

In this chapter some classical constructions of Lebesgue nonmeasurable sets on the real line $\mathbf{R}$ are envisaged from the point of view of the thickness of those sets. It is shown, within $\mathbf{ZF}$ & $\mathbf{DC}$ theory, that the existence of a Lebesgue nonmeasurable subset of $\mathbf{R}$ implies the existence of a partition of $\mathbf{R}$ into continuum many thick sets with respect to the Lebesgue measure.

To begin our presentation, let us recall that very soon after Lebesgue's invention (in 1902) of his measure λ on the real line $\mathbf{R}$, the three constructions of extraordinary point sets in $\mathbf{R}$ have followed. They were done, respectively, by Vitali [266], Hamel [90], and Bernstein [14].

An important by-product of each of those constructions is the statement of the existence of a Lebesgue nonmeasurable subset of $\mathbf{R}$.

In this connection, it is reasonable to stress here that those constructions differ essentially from each other. Namely, recall that:

(a) in [266] Vitali takes a selector V of the quotient set $\mathbf{R}/\mathbf{Q}$, where $\mathbf{Q}$ denotes the field of all rational numbers, and shows that V cannot be measurable with respect to any measure on $\mathbf{R}$ which extends λ and is translation invariant;

(b) in [90] Hamel considers $\mathbf{R}$ as a vector space over $\mathbf{Q}$ and establishes the existence of a basis for this space; this fact allows him to define a nontrivial endomorphism of the additive group $(\mathbf{R}, +)$ which is nonmeasurable in the Lebesgue sense;

(c) in [14] Bernstein utilizes the method of transfinite recursion and defines a subset B of $\mathbf{R}$ such that both sets B and $\mathbf{R} \setminus B$ meet every nonempty perfect set in $\mathbf{R}$; so both B and $\mathbf{R} \setminus B$ turn out to be nonmeasurable with respect to λ.

All the above-mentioned constructions are based on appropriate uncountable forms of the Axiom of Choice ($\mathbf{AC}$), which were radically rejected by Lebesgue in that time. Many years later, it was demonstrated by Solovay [253] that some uncountable version of $\mathbf{AC}$ is absolutely necessary for obtaining Lebesgue nonmeasurable point sets in $\mathbf{R}$.

Denote by $\mathbf{c}$ the cardinality of the continuum. In [168] Luzin and Sierpiński have extended Bernstein's construction for obtaining a partition of the unit interval $[0, 1]$ (or, equivalently, of $\mathbf{R}$) into continuum many Lebesgue nonmeasurable sets. Actually, they proved the following statement.

Theorem 1. *The real line* **R** *admits a partition* $\{B_i : i \in I\}$ *such that:*
(1) $\mathrm{card}(I) = \mathbf{c}$;
(2) *every set* B_i $(i \in I)$ *meets any nonempty perfect subset of* **R***.*
   In particular, all B_i $(i \in I)$ *are Bernstein subsets of* **R** *and so are nonmea
surable in the Lebesgue sense.*

Further generalization of Bernstein's construction looks as follows (see, e.g.
[128], [133], [137]).

Theorem 2. *There exists a covering* $\{B_j : j \in J\}$ *of the real line* **R** *with
its subsets, satisfying these three conditions:*
   (1) $\mathrm{card}(J) > \mathbf{c}$;
   (2) *every set* B_j $(j \in J)$ *meets each nonempty perfect set in* **R***;*
   (3) *the family* $\{B_j : j \in J\}$ *is almost disjoint, i.e.,* $\mathrm{card}(B_j \cap B_{j'}) < \mathbf{c}$ *fo
any two distinct indices* $j \in J$ *and* $j' \in J$.

The conditions (2) and (3) of Theorem 2 readily imply that every set B_j $(j \in$
J) is a Bernstein subset of **R**.
   The role of Bernstein sets in general topology, the theory of Boolean algebras
measure theory and real analysis is well known (see, for instance, [14], [76], [96]
[128], [152], [203]). In classical measure theory, the significance of these sets
is primarily caused by providing various counterexamples for seemingly valid
statements in real analysis and by constructions of measures lacking various
regularity properties (cf. [17], [89], [128], [199]).
   Let E be a ground set and let μ be a measure defined on some σ-algebra o
subsets of E.
   Recall that μ is said to be diffused (or continuous) if all singletons in E
belong to the domain of μ and μ vanishes at all of them.
   A set $Z \subset E$ is said to be μ-thick in E if the equality $\mu_*(E \setminus Z) = 0$ hold
true, where μ_* denotes the inner measure associated with μ.

Remark 1. Let $\mathcal{CBM}_0(\mathbf{R})$ denote the class of completions of all nonzero
σ-finite diffused Borel measures on **R**. It is not difficult to see that if B is any
Bernstein set in **R** and μ is any measure from the class $\mathcal{CBM}_0(\mathbf{R})$, then both E
and $\mathbf{R} \setminus B$ are μ-thick subsets of **R** and, consequently, they are nonmeasurable
with respect to μ. Actually, this property completely characterizes Bernstein
sets in **R** (see Exercise 1 from Chapter 5).

We have already mentioned three classical constructions, each of which gives
an example of a λ-nonmeasurable set in **R**. Moreover, Bernstein's construction
directly yields the partition $\{B, \mathbf{R} \setminus B\}$ of **R** into two λ-thick subsets.
   In this connection, let us demonstrate that Hamel's construction directly
leads to a partition of **R** into countably many λ-thick subsets of **R**.
   For this purpose, let us consider **R** as a vector space over the field **Q**. Let
$\{e_i : i \in I\}$ be a Hamel basis for this space containing 1, i.e., $e_{i_0} = 1$ for some

index $i_0 \in I$. Denote by V the vector space over $\mathbf{Q}$ generated by the family $\{e_i : i \in I \setminus \{i_0\}\}$, i.e., we have

$$V = \text{span}_{\mathbf{Q}}\{e_i : i \in I \setminus \{i_0\}\}.$$

It is not difficult to see that V is a special kind of a Vitali set in $\mathbf{R}$. Actually, V is a selector of $\mathbf{R}/\mathbf{Q}$ but the choice of this selector is done so carefully that V turns out to be able to carry the vector structure over $\mathbf{Q}$ induced by $\mathbf{R}$. We now assert that V is λ-thick in $\mathbf{R}$. Indeed, suppose otherwise, i.e., there exists a λ-measurable set $P \subset \mathbf{R}$ such that

$$\lambda(P) > 0, \quad P \cap V = \emptyset.$$

It is easy to see that V is everywhere dense in $\mathbf{R}$ (because any uncountable subgroup of $(\mathbf{R}, +)$ is necessarily everywhere dense in $\mathbf{R}$). Consequently, we may take a countable family $\{v_j : j \in J\} \subset V$ which is everywhere dense in $\mathbf{R}$, too. Obviously, for this family, we may write

$$V \cap (\{v_j : j \in J\} + P) = \emptyset.$$

Taking into account the metrical transitivity (ergodicity) of λ with respect to any everywhere dense subset of $\mathbf{R}$ (see Exercise 2 from Chapter 3), we get

$$\lambda(\mathbf{R} \setminus (\{v_j : j \in J\} + P)) = 0.$$

Therefore, $\lambda(V) = 0$, which is impossible in view of the translation invariance of λ and of the relations

$$\mathbf{R} = \mathbf{Q} + V = \cup\{q + V : q \in \mathbf{Q}\}, \quad \lambda(\mathbf{R}) = +\infty.$$

The obtained contradiction yields the desired result.

We thus come to the countable partition $\{q + V : q \in \mathbf{Q}\}$ of $\mathbf{R}$ into λ-thick sets. It follows from this fact that, for any natural number $n \geq 2$, there exists a partition $\{A_1, A_2, ..., A_n\}$ of $\mathbf{R}$ into λ-thick sets, and so all A_k ($1 \leq k \leq n$) are nonmeasurable with respect to λ.

Remark 2. In general, Vitali's construction does not lead to a λ-thick subset of $\mathbf{R}$. Indeed, fix a real $\varepsilon > 0$ and take an arbitrary nonempty open interval Δ in $\mathbf{R}$ with $\lambda(\Delta) < \varepsilon$. For any $x \in \mathbf{R}$, the set $x + \mathbf{Q}$ is everywhere dense in $\mathbf{R}$, so has nonempty intersection with Δ. This circumstance immediately implies that there exists a Vitali set W entirely contained in Δ and, consequently, $\lambda^*(W) < \varepsilon$, where λ^* denotes the outer measure associated with λ. We thus see that there are Vitali sets in $\mathbf{R}$ with arbitrarily small outer Lebesgue measure. Several other unexpected and extraordinary properties of Vitali sets are discussed in [139] (see also Chapter 9 of this book).

Our goal now is to obtain (within a certain weak fragment of set theory) a partition of $\mathbf{R}$ into continuum many λ-thick subsets of $\mathbf{R}$, by starting with a partition $\{A, A'\}$ of $\mathbf{R}$ consisting of some two λ-thick sets in $\mathbf{R}$.

As shown above, under an uncountable form of the Axiom of Choice, Bernstein's and Hamel's constructions give such a partition $\{A, A'\}$ of $\mathbf{R}$.

We need the following two auxiliary propositions which both belong to $\mathbf{ZF}$ & $\mathbf{DC}$ theory, where $\mathbf{DC}$ stands, as usual, for the Principle of Dependent Choices (see [93], [102], [103] or Chapter 2). This principle is stronger than the Axiom of Countable Choice ($\mathbf{CC}$) and much weaker than $\mathbf{AC}$. Moreover, according to Solovay's famous result [253], under the assumption of the existence of a strongly inaccessible cardinal number, there is a model of $\mathbf{ZF}$ & $\mathbf{DC}$, in which all subsets of $\mathbf{R}$ are measurable in the Lebesgue sense.

Lemma 1. *Let E_1 and E_2 be two Polish spaces, let μ_1 be a Borel probability diffused measure on E_1, and let μ_2 be a Borel probability diffused measure on E_2. Then there exists a Borel isomorphism*

$$\phi : E_1 \to E_2$$

which is simultaneously an isomorphism between μ_1 and μ_2, i.e., we have the equality $\mu_2(\phi(X)) = \mu_1(X)$ for every Borel subset X of E_1.

This lemma is well known (for the proof, within $\mathbf{ZF}$ & $\mathbf{DC}$ theory, see [33], [115] or Exercise 5).

Lemma 2. *Let $\{E_n : n = 1, 2, ..., k, ...\}$ be a countable family of separable metric spaces and let, for each natural number $n \geq 1$, the space E_n be equipped with a probability Borel measure μ_n. Further, let us denote:*

$$E = \prod \{E_n : n = 1, 2, ..., k, ...\},$$

$$\mu = \otimes \{\mu_n : n = 1, 2, ..., k, ...\}.$$

Suppose also that a sequence of sets $X_n \subset E_n$ $(n = 1, 2, ..., k, ...)$ is given. The following two assertions are equivalent:
(1) the product set $X = \prod \{X_n : n = 1, 2, ..., k, ...\}$ is μ-thick in E;
(2) the set X_n is μ_n-thick in E_n for each index $n \in \{1, 2, ..., k, ...\}$.

Proof. The implication (1) $\Rightarrow$ (2) is almost trivial. So we will focus our attention on the converse implication (2) $\Rightarrow$ (1). Suppose that (2) is satisfied.

Since E is a separable metric space, the probability product measure μ is defined on the Borel σ-algebra of E and, in addition to this, μ is inner regular. The latter means that, for each Borel set $Z \subset E$, the equality

$$\mu(Z) = \sup\{\mu(F) : F \subset Z, \ F \text{ is closed in } E\}$$

is valid (see Exercise 3 from Chapter 6). Therefore, it suffices to demonstrate that $X \cap P \neq \emptyset$ for any closed set $P \subset E$ with $\mu(P) > 0$.

For this purpose, we shall construct by ordinary recursion an element

$$y = (y_1, y_2, \ldots, y_n, \ldots) \in X \cap P.$$

Suppose that, for a natural number n, the finite sequence

$$(y_1, y_2, ..., y_n) \in X_1 \times X_2 \times ... \times X_n$$

has already been defined so that the inequality

$$\nu_n(P(y_1, y_2, ..., y_n)) > 0$$

holds true, where

$$\nu_n = \otimes\{\mu_m : m = n+1, n+2, ...\},$$

$$P(y_1, y_2, ..., y_n) = \{(x_{n+1}, x_{n+2}, ...) : (y_1, y_2, ..., y_n, x_{n+1}, x_{n+2}, ...) \in P\}.$$

According to classical Fubini's theorem, the set of all those elements x_{n+1} from E_{n+1} which satisfy the inequality

$$\nu_{n+1}(P(y_1, y_2, ..., y_n, x_{n+1})) > 0$$

is μ_{n+1}-measurable and has strictly positive μ_{n+1}-measure in E_{n+1}. Since the set X_{n+1} is μ_{n+1}-thick in E_{n+1}, there exists a point $y_{n+1} \in X_{n+1}$ such that

$$\nu_{n+1}(P(y_1, y_2, ..., y_n, y_{n+1})) > 0.$$

We thus see that our recursion works and, after countably many steps, yields the sequence

$$y = (y_1, y_2, ..., y_n, ...) \in X.$$

Observe now that, by virtue of the definition of y, every neighborhood of y has common elements with the set P. Since P is closed, we immediately conclude that $y \in P$, so $y \in P \cap X$.

This completes the proof of Lemma 2 (let us underline once more that the argument presented above is done within **ZF** & **DC** theory).

Remark 3. In Lemma 2, the assumption that all spaces E_n are separable and metrizable is not necessary. The conclusion of this lemma remains valid under much weaker assumptions, but the given formulation suffices for our further purposes.

Remark 4. Preserving the notation of Lemma 2, let Z be an arbitrary μ-thick set in E. Then it is easy to see that, for every natural number $n \geq 1$, the set $\text{pr}_n(Z)$ is μ_n-thick in E_n. The converse assertion is not true, in general. Indeed, simple examples show that the equalities $\text{pr}_n(Z) = E_n$ may be valid simultaneously for all natural numbers $n \geq 1$ but, at the same time, the set Z may be of μ-measure zero.

Remark 5. Let $k \geq 1$ be a natural number, $\{E_n : n = 1, 2, ..., k\}$ be a finite family of ground sets and let, for each natural number $n \in \{1, 2, ..., k\}$, the set E_n be equipped with a probability measure μ_n. Further, let us denote:

$$E = \prod \{E_n : n = 1, 2, ..., k\},$$

$$\mu = \otimes \{\mu_n : n = 1, 2, ..., k\}.$$

Suppose also that a finite sequence of sets $X_n \subset E_n$ $(n = 1, 2, ..., k)$ is given. Then the following two assertions are equivalent:

(a) the product set $X = \prod \{X_n : n = 1, 2, ..., k\}$ is μ-thick in E;

(b) the set X_n is μ_n-thick in E_n for each index $n \in \{1, 2, ..., k\}$.

We thus see that in the case of a finite sequence of probability measure spaces (or, more generally, of nonzero σ-finite measure spaces) the analogue of Lemma 2 is valid in **ZF** & **DC** theory without assuming any regularity properties of the measures.

Now, we are ready to present the main result of this chapter (in what follows we will denote by λ the restriction of the Lebesgue measure on $\mathbf{R}$ to the unit interval $[0, 1]$).

Theorem 3. *Working in* **ZF** & **DC**, *suppose that there exists a partition* $\{A, A'\}$ *of the unit interval* $[0, 1]$ *into two subsets such that*

$$\lambda^*(A) = \lambda^*(A') = 1.$$

Then there exists a partition $\{Z_i : i \in I\}$ *of the same interval, which satisfies the following two conditions:*

(1) $\mathrm{card}(I) = \mathbf{c};$

(2) $\lambda^*(Z_i) = 1$ *for each index* $i \in I$.

Proof. Let $\mathbf{N}$ denote the set of all natural numbers. Consider the Hilbert cube $E = [0, 1]^{\mathbf{N}}$ equipped with the probability product measure

$$\mu = \lambda \otimes \lambda \otimes ... \otimes \lambda \otimes ... \, .$$

Take any subset K of $\mathbf{N}$ and put: $A_{n,K} = A$ if $n \in K$, and $A_{n,K} = A'$ if $n \in \mathbf{N} \setminus K$.

Further, for the same K, introduce the corresponding product set

$$Y_K = \prod \{A_{n,K} : n \in \mathbf{N}\}.$$

Proceeding in this manner, we come to the partition $\{Y_K : K \subset \mathbf{N}\}$ of the Hilbert cube E.

By virtue of Lemma 2, all members Y_K ($K \subset \mathbf{N}$) of this partition are μ-thick in E.

Let $\phi : E \to [0,1]$ be a Borel isomorphism which simultaneously is an isomorphism of μ onto λ (the existence of ϕ follows from Lemma 1).

Obviously, $\{\phi(Y_K) : K \subset \mathbf{N}\}$ is a partition of $[0,1]$ into continuum many λ-thick subsets of $[0,1]$. So we may put

$$\{Z_i : i \in I\} = \{\phi(Y_K) : K \subset \mathbf{N}\}.$$

This finishes the proof of Theorem 3.

We immediately obtain from the above theorem that there exists a partition of the real line $\mathbf{R}$ into continuum many λ-thick sets in $\mathbf{R}$. Indeed, it suffices to consider any homeomorphism

$$h \; : \;]0,1[\; \to \mathbf{R}$$

which transforms the σ-ideal of all λ-measure zero subsets of $]0,1[$ onto the σ-ideal of all λ-measure zero subsets of $\mathbf{R}$.

As we have already mentioned at the beginning of this chapter, nontrivial endomorphisms of the additive group $(\mathbf{R},+)$ were first exhibited in [90] and all of them turned out to be nonmeasurable in the Lebesgue sense. In connection with this fact, it is worth noticing that some of such endomorphisms can be measurable with respect to certain measures belonging to the class $\mathcal{CBM}_0(\mathbf{R})$ pointed out in Remark 1. Namely, there exists a function

$$f : \mathbf{R} \to \mathbf{R}$$

satisfying the following three conditions:

(a) the range $\mathrm{ran}(f)$ of f is contained in the field $\mathbf{Q}$ (consequently, $\mathrm{ran}(f)$ is at most countable);

(b) f is measurable with respect to some measure from the class $\mathcal{CBM}_0(\mathbf{R})$;

(c) f is a nontrivial endomorphism of the additive group $(\mathbf{R},+)$.

One construction of such an f was outlined in Remark 6 of Chapter 7. Recall that the above-mentioned construction exploits the techniques of Hamel bases.

Some other applications of Hamel bases to problems and questions of real analysis may be found in [133] and [147]; see also Chapters 12, 13, and 14 of this book.

Remark 6. It can be shown that:

(a) there exists a subset of $\mathbf{R}$ which is simultaneously a Vitali set and a Bernstein set;

(b) there exists a subset of $\mathbf{R}$ which is simultaneously a Hamel basis and a Bernstein set;

(c) there exists no subset of $\mathbf{R}$ which is simultaneously a Hamel basis and a Vitali set.

Notice also that from the definitions of Bernstein sets, Sierpiński sets and Luzin sets easily follows that all of them are totally imperfect in $\mathbf{R}$. For Hamel bases and Vitali sets, we do not have an analogous result, because there is a Hamel basis H (a Vitali set V) containing a nonempty perfect subset of $\mathbf{R}$. The existence of such H and V is readily implied by one general theorem of Mycielski (see [115], [193] or [268]).

Remark 7. Let μ be the completion of a nonzero σ-finite diffused Borel measure on $\mathbf{R}$. By using Lemma 1, it is not difficult to prove within $\mathbf{ZF}$ & $\mathbf{DC}$ theory that if there exists a μ-nonmeasurable subset of $\mathbf{R}$, then there exists a partition of $\mathbf{R}$ into two μ-thick subsets (see Exercise 19). So, taking into account Lemma 1 and Theorem 3, we may conclude that the following four assertions are equivalent in $\mathbf{ZF}$ & $\mathbf{DC}$ theory:

(a) there exists a μ-nonmeasurable subset of $\mathbf{R}$;

(b) there exists a partition of $\mathbf{R}$ into two μ-thick subsets;

(c) there exists a partition of $\mathbf{R}$ into continuum many μ-thick subsets;

(d) there exists a function $g : \mathbf{R} \to \mathbf{R}$ such that $\operatorname{ran}(g|X) = \mathbf{R}$ for every μ-measurable set $X \subset \mathbf{R}$ with $\mu(X) > 0$.

In this context, the transfinite construction given in [168] becomes superfluous. At the same time, it seems that the natural analogue of Theorem 2 cannot be deduced within $\mathbf{ZF}$ & $\mathbf{DC}$ theory by assuming that there exists a λ-nonmeasurable subset of $\mathbf{R}$.

Remark 8. Consider the theory

$$\mathbf{ZF} \ \& \ \mathbf{DC} \ \& \ (\omega_1 \leq \mathbf{c}),$$

where ω_1 denotes, as usual, the least uncountable cardinal. It was proved in this theory that there exists a λ-nonmeasurable subset of $\mathbf{R}$ (see [229] and [214]). Consequently, within the same theory, there exists a partition of $\mathbf{R}$ into continuum many λ-thick subsets.

EXERCISES

1^*. Let E be an infinite ground set and let $\{X_i : i \in I\}$ be a family of subsets of E satisfying the following two conditions:

(a) $\operatorname{card}(I) \leq \operatorname{card}(E)$;

(b) $\operatorname{card}(X_i) = \operatorname{card}(E)$ for each index $i \in I$.

Demonstrate that there exists a family $\{Z_j : j \in I\}$ of subsets of E such that:

(c) $Z_j \cap Z_{j'} = \emptyset$ for any two distinct indices j and j' from I (in other words, $\{Z_j : j \in I\}$ is a disjoint family of sets);

(d) $\operatorname{card}(X_i \cap Z_j) = \operatorname{card}(E)$ for any two indices $i \in I$ and $j \in I$.

Argue as follows. According to Sierpiński's lemma on disjoint subsets (see Exercise 4 from Chapter 7), for a given family $\{X_i : i \in I\}$, there exists a disjoint family of sets $\{Y_i : i \in I\}$ such that:

(i) $Y_i \subset X_i$ for any index $i \in I$;

(ii) $\mathrm{card}(Y_i) = \mathrm{card}(X_i) = \mathrm{card}(E)$ for any index $i \in I$.

Since every set Y_i $(i \in I)$ is infinite, it can be represented in the form

$$Y_i = \cup\{U_{i,j} : j \in I\},$$

where all the sets $U_{i,j}$ $(j \in I)$ are pairwise disjoint and satisfy the equalities

$$\mathrm{card}(U_{i,j}) = \mathrm{card}(Y_i) = \mathrm{card}(E).$$

Now, define

$$Z_j = \cup\{U_{i,j} : i \in I\} \qquad (j \in I)$$

and check that the family of sets $\{Z_j : j \in I\}$ is as required.

2. Starting with the result of Exercise 1, give a proof of Theorem 1. For this purpose, put

$$E = \mathbf{R}, \quad I = \mathbf{c},$$

and take the family of all uncountable closed subsets of $\mathbf{R}$ as $\{X_i : i \in I\}$. Now, Exercise 1 guarantees the existence of a disjoint family $\{Z_j : j \in I\}$ of subsets of $\mathbf{R}$ such that

$$\mathrm{card}(X_i \cap Z_j) = \mathbf{c} \qquad (i \in I, \ j \in I).$$

Verify that this $\{Z_j : j \in I\}$ allows one to produce a partition of $\mathbf{R}$ into continuum many Bernstein subsets of $\mathbf{R}$.

3*. Let E be an infinite ground set and let $\{X_i : i \in I\}$ be a family of subsets of E satisfying the following two conditions:

(a) $\mathrm{card}(I) \leq \mathrm{card}(E)$;

(b) $\mathrm{card}(X_i) = \mathrm{card}(E)$ for each index $i \in I$.

Demonstrate that there exists a family $\{Z_j : j \in J\}$ of subsets of E such that:

(i) $\mathrm{card}(J) > \mathrm{card}(E)$;

(ii) $\mathrm{card}(Z_j \cap Z_{j'}) < \mathrm{card}(E)$ for any two distinct indices j and j' from J (in other words, $\{Z_j : j \in J\}$ is an almost disjoint family of subsets of E);

(iii) $\mathrm{card}(X_i \cap Z_j) = \mathrm{card}(E)$ for any $i \in I$ and $j \in J$.

Argue as follows. First, try to define a bijective mapping

$$\phi : E \rightarrow E \times E$$

satisfying the following relation: for each index $i \in I$, the cardinality of the set

$$\{x \in E : \mathrm{card}((\{x\} \times E) \cap \phi(X_i)) = \mathrm{card}(E)\}$$

is equal to $\mathrm{card}(E)$.

Then use the method of transfinite recursion and construct a family of functions $\{f_j : j \in J\}$ such that:

(1) $\mathrm{card}(J)$ is the least cardinal number strictly greater than $\mathrm{card}(E)$;

(2) every function f_j $(j \in J)$ acts from E into itself;

(3) $\mathrm{card}(\mathrm{Gr}(f_j) \cap \phi(X_i)) = \mathrm{card}(E)$, where $i \in I$, $j \in J$ and the symbol $\mathrm{Gr}(f_j)$ denotes, as usual, the graph of f_j;

(4) if j and j' are any two distinct indices from J, then

$$\mathrm{card}(\mathrm{Gr}(f_j) \cap \mathrm{Gr}(f_{j'})) < \mathrm{card}(E).$$

Finally, introduce the family

$$\{Z_j : j \in J\} = \{\phi^{-1}(\mathrm{Gr}(f_j)) : j \in J\}$$

and verify that this family is as required.

4. Starting with the result of Exercise 3, give a proof of Theorem 2.
For this purpose, put again

$$E = \mathbf{R}, \quad I = \mathbf{c},$$

and take the family of all uncountable closed subsets of $\mathbf{R}$ as $\{X_i : i \in I\}$. Now, there exists an almost disjoint family $\{Z_j : j \in J\}$ of subsets of $\mathbf{R}$ for which the relations (i), (ii), (iii) of Exercise 3 are fulfilled. Check that this family produces the required almost disjoint covering of $\mathbf{R}$ by Bernstein sets.

5. Give a proof of Lemma 1.
For this purpose, first show that any Borel diffused probability measure on a Polish topological space is Borel isomorphic to the restriction of the Lebesgue measure λ to $\mathcal{B}([0,1])$ (cf. Exercise 3 from Chapter 3).

6. Preserving the notation of Lemma 2, let Z be an arbitrary μ-thick set in the product space E.
Verify that, for every natural number $n \geq 1$, the set $\mathrm{pr}_n(Z)$ is μ_n-thick in the space E_n.

7. Preserving the notation of Lemma 2, give an example of a μ-measure zero set Z in E for which the equalities $\mathrm{pr}_n(Z) = E_n$ are valid simultaneously for all natural numbers $n \geq 1$.

8. Prove the statement formulated in Remark 5.

9. Let X be any Bernstein subset of $\mathbf{R}$ and let μ be a σ-finite diffused Borel measure on $\mathbf{R}$.
Show that there exist two measures ν_1 and ν_2 on $\mathbf{R}$ such that:

(a) both ν_1 and ν_2 are extensions of μ;

(b) $X \in \mathrm{dom}(\nu_1)$ and $\nu_1(X) = 0$;
(c) $X \in \mathrm{dom}(\nu_2)$ and $\nu_2(\mathbf{R} \setminus X) = 0$.

10*. Demonstrate that there exists a Bernstein set $B \subset \mathbf{R}$ which is almost translation invariant, i.e.,

$$(\forall h \in \mathbf{R})(\mathrm{card}((h + B) \triangle B) < \mathbf{c}).$$

For this purpose, construct such a B by using the method of transfinite recursion.

In addition, check that if μ is an arbitrary σ-finite translation invariant (translation quasi-invariant) measure on $\mathbf{R}$, then there exists a translation invariant (translation quasi-invariant) measure ν on $\mathbf{R}$ extending μ and satisfying the relation $B \in \mathrm{dom}(\nu)$.

11. Verify that there exists no subset of $\mathbf{R}$ which is simultaneously a Hamel basis and a Vitali set.

For this purpose, establish the following property of any Hamel basis H: the real line $\mathbf{R}$ cannot be covered by countably many translates of H.

On the other hand, keep in mind the circumstance that if V is an arbitrary Vitali set in $\mathbf{R}$, then

$$V + \mathbf{Q} = \{V + q : q \in \mathbf{Q}\} = \mathbf{R}.$$

The two above-mentioned facts indicate that V cannot be a Hamel basis for $\mathbf{R}$.

Remark 9. As was mentioned at the end of Chapter 5, any Hamel basis H in $\mathbf{R}$ is an $\mathbf{R}$-absolutely negligible set (see [120]). This fact essentially strengthens the property of H indicated in Exercise 11.

12*. Demonstrate that there exists no Borel function $f : \mathbf{R}^\omega \to \mathbf{R}$ satisfying the following conditions:
(a) $\mathrm{ran}(x) = \mathrm{ran}(y) \Rightarrow f(x) = f(y)$ for any two sequences $x \in \mathbf{R}^\omega$ and $y \in \mathbf{R}^\omega$;
(b) $f(x) \notin \mathrm{ran}(x)$ for each $x \in \mathbf{R}^\omega$.

Argue as follows. Suppose to the contrary that such a Borel function f does exist. Equip $\mathbf{R}$ with the discrete topology and denote the obtained discrete space by $\mathbf{R}_*$. Further, consider the topological product space

$$E = \mathbf{R}_* \times \mathbf{R}_* \times \dots \times \mathbf{R}_* \times \dots,$$

where the number of factors is countably infinite, and check that E is a complete metric space, hence, a Baire space as well. The same function f regarded as a mapping acting from E into $\mathbf{R}$ remains Borel and is invariant under the group Γ canonically associated with the group of all permutations of ω. To say more precisely, a transformation $g : E \to E$ belongs to Γ if and only if there exists a permutation $\phi : \omega \to \omega$ such that

$$\mathrm{pr}_n(g(x)) = x_{\phi(n)} \qquad (n < \omega, \ x \in E).$$

Verify that Γ acts topologically transitively in E (see Exercise 23 from Chapter 18). Therefore, f is constant on some co-meager set $Z \subset E$. Let $\{t\} = f(Z)$ and let

$$U = \{t\} \times \mathbf{R}_* \times \mathbf{R}_* \times ... \times \mathbf{R}_* \times ... .$$

Since U is a nonempty open set in E, the relation $U \cap Z \neq \emptyset$ must be valid. Take any element $z \in U \cap Z$ and check that $t = f(z) \in \mathrm{ran}(z)$. However, the last relation contradicts condition (b). The obtained contradiction yields the required result.

Remark 10. It is useful to compare the above exercise with Exercise 21 from Chapter 4.

13. Show that the Bernstein construction can be effectively carried out whenever the real line $\mathbf{R}$ can be well-ordered.

For this purpose, take into account the fact that there is a canonical bijection between $\mathbf{R}$ and the family of all uncountable closed subsets of $\mathbf{R}$.

14. Let $\mathcal{I}_0$ denote the σ-ideal of all Lebesgue measure zero subsets of $\mathbf{R}$ (i.e., $\mathcal{I}_0 = \mathcal{I}_\lambda$) and let $\mathcal{I}_1$ stand for the σ-ideal of all first category subsets of $\mathbf{R}$.

The quotient Boolean algebra $\mathcal{B}(\mathbf{R})/\mathcal{I}_0$ is usually called the Solovay algebra and the quotient Boolean algebra $\mathcal{B}(\mathbf{R})/\mathcal{I}_1$ is usually called the Cohen algebra.

Verify that:

(a) both $\mathcal{B}(\mathbf{R})/\mathcal{I}_0$ and $\mathcal{B}(\mathbf{R})/\mathcal{I}_1$ satisfy the countable chain condition;

(b) both $\mathcal{B}(\mathbf{R})/\mathcal{I}_0$ and $\mathcal{B}(\mathbf{R})/\mathcal{I}_1$ are complete Boolean algebras.

For checking (a), use the facts that λ is a σ-finite measure and $\mathbf{R}$ has a countable base. For checking (b), apply (a).

15*. Let the symbol $\mathcal{P}(\mathbf{R})$ denote the Boolean algebra of all subsets of $\mathbf{R}$.

Demonstrate that the following two assertions are true:

(a) the quotient Boolean algebra $\mathcal{P}(\mathbf{R})/\mathcal{I}_0$ is not complete;

(b) the quotient Boolean algebra $\mathcal{P}(\mathbf{R})/\mathcal{I}_1$ is not complete.

For establishing assertion (a), consider an arbitrary partition $\{Z_i : i \in I\}$ of $\mathbf{R}$ such that $\mathrm{card}(I) = \mathbf{c}$ and all sets Z_i $(i \in I)$ are λ-thick in $\mathbf{R}$ (see Theorem 1 of this chapter). Let $[Z_i]$ denote the element of $\mathcal{P}(\mathbf{R})/\mathcal{I}_0$ corresponding to Z_i. Show that the family $\{[Z_i] : i \in I\}$ of elements of $\mathcal{P}(\mathbf{R})/\mathcal{I}_0$ does not have supremum in $\mathcal{P}(\mathbf{R})/\mathcal{I}_0$.

For establishing assertion (b), apply a similar argument.

16*. Let a function $f : \mathbf{R} \to \mathbf{R}$ be given which is mid-point convex, i.e., the inequality

$$f(x/2 + y/2) \leq f(x)/2 + f(y)/2$$

holds true for any two points x and y from $\mathbf{R}$. Suppose, in addition, that f is measurable in the Lebesgue sense.

Prove Sierpiński's theorem stating that f is convex in the usual sense (and, in particular, f is continuous at all points of $\mathbf{R}$).

For this purpose, use the Steinhaus property (see Exercise 1 from Chapter 3) and first establish that f turns out to be bounded from above on some non-degenerate subinterval of $\mathbf{R}$ (cf. [133]; the analogous result is valid for all mid-point convex functions having the Baire property).

17*. Identify the cardinality continuum $\mathbf{c}$ with the least ordinal number α such that $card(\alpha) = \mathbf{c}$ and demonstrate that these two assertions are equivalent:

(a) $\mathbf{R}$ can be represented as the union of an increasing (by inclusion) α-sequence of Lebesgue measure zero sets;

(b) every subset of $\mathbf{R}$ can be expressed as the union of an increasing (by inclusion) α-sequence of Lebesgue measurable sets.

Argue as follows. First of all, observe that the implication (a) $\Rightarrow$ (b) is trivial. So only the converse implication (b) $\Rightarrow$ (a) needs to be established. Suppose (b) and consider any Vitali subset V of $\mathbf{R}$. Since V does not possess the Steinhaus property, the inner λ-measure of V is equal to zero. According to (b), one may write $V = \cup\{X_\xi : \xi < \alpha\}$, where $\{X_\xi : \xi < \alpha\}$ is some increasing (by inclusion) α-sequence of λ-measurable sets. Notice now that $\lambda(X_\xi) = 0$ for each ordinal $\xi < \alpha$ and put

$$Y_\xi = \cup\{X_\xi + q : q \in \mathbf{Q}\}.$$

Then $\{Y_\xi : \xi < \alpha\}$ is an increasing (by inclusion) α-sequence of λ-measure zero sets which collectively cover $\mathbf{R}$, i.e., (a) is fulfilled.

18*. Verify that:

(i) both assertions (a) and (b) of Exercise 17 are consistent with $\mathbf{ZFC}$ set theory;

(ii) none of those assertions can be proved within the same theory.

For this purpose, first suppose the Continuum Hypothesis and check that $\mathbf{R}$ is representable as the union of an increasing (by inclusion) ω_1-sequence $\{X_\xi : \xi < \omega_1\}$ of countable sets. Since the equality $\lambda(X_\xi) = 0$ trivially holds for each ordinal $\xi < \omega_1$, one may conclude the validity of (i).

On the other hand, assume that $\mathbf{c}$ coincides with the least cardinal number measurable in the Ulam sense. It is known (see, e.g., [128]) that in this case $\mathbf{c}$ is regular and there exists a subset Z of $\mathbf{R}$ satisfying the relations

$$card(Z) < \mathbf{c}, \quad Z \notin dom(\lambda).$$

Deduce from the existence of Z that $\mathbf{R}$ cannot be represented as the union of an increasing (by inclusion) $\mathbf{c}$-sequence of λ-measure zero sets.

19. Let μ be the completion of a nonzero σ-finite diffused Borel measure on **R**.

Work in **ZF** & **DC** theory and show that if there exists a μ-nonmeasurable subset of **R**, then there exists a partition of **R** into two μ-thick subsets of **R**.

For this purpose, take into account Lemma 1 of the present chapter.

Also, check the equivalence of assertions (a) and (d) in Remark 7.

20. Supposing that **c** is a regular cardinal number, deduce Theorem 1 from Theorem 2.

21. Assuming the Continuum Hypothesis, prove that:

(a) there exists a partition of **R** into continuum many Sierpiński sets, all of which are λ-thick and almost translation invariant;

(b) there exists a partition of **R** into continuum many Luzin sets, all of which are thick in the sense of category and almost translation invariant.

For proving (a) and (b), use the method of transfinite induction.

22. Under the Continuum Hypothesis, give an example of a translation invariant measure μ on **R** satisfying the following two conditions:

(a) μ is an extension of λ;

(b) there exists a Sierpiński set $S \subset \mathbf{R}$ such that $\mu(\mathbf{R} \setminus S) = 0$.

23. Let X be an arbitrary λ-thick Sierpiński subset of **R** and let λ_X denote the measure on X induced by λ (see Exercise 22 from Chapter 3).

Verify that the completion of the product measure $\lambda_X \otimes \lambda_X$ is not isomorphic to λ_X.

9. Measurability properties of Vitali sets

The main goal of this chapter is to demonstrate that some Vitali subsets of the real line $\mathbf{R}$ can be measurable with respect to certain translation quasi-invariant measures on $\mathbf{R}$ extending the standard Lebesgue measure. So, according to the general concept introduced earlier (see Chapter 5), we may say that some Vitali sets turn out to be relatively measurable with respect to the class of all those translation quasi-invariant measures on $\mathbf{R}$ which extend the Lebesgue measure. On the other hand, we will also show that there exist Vitali sets which are nonmeasurable with respect to every nonzero σ-finite translation quasi-invariant measure on $\mathbf{R}$. In other words, those Vitali sets turn out to be absolutely nonmeasurable with respect to the class of all nonzero σ-finite translation quasi-invariant measure on $\mathbf{R}$.

First, let us say a few words about some paradoxical sets and functions in mathematics. Among various mathematical objects with exotic features there are those which play a seminal role in the long process of development of mathematics. Many examples of such objects can be presented: continuous nowhere differentiable functions and their connection with Brownian motion; the Cantor set and Sierpiński's carpet (both of them are typical representatives of the so-called fractals); the Knaster–Kuratowski fan; Antoine's necklace, Alexander's horned sphere, and Milnor's spheres; the Hopf fibration; Carathéodory–Gale polytopes; etc.

Notice that the above-mentioned sets and functions were constructed effectively, i.e., without the aid of the Axiom of Choice. More delicate set-theoretic techniques based on this axiom allow one to prove the existence of other mathematical objects of a paradoxical character, for instance, a Hamel basis of the real line $\mathbf{R}$, a Bernstein subset of $\mathbf{R}$, a non-principal ultrafilter on the set $\mathbf{N}$ of all natural numbers, a Banach–Tarski decomposition of the unit ball, and so on. Undoubtedly, Vitali sets belong to this second collection. Here we would like to discuss them and their properties from the measure-theoretic point of view adopted in Chapter 5.

Dealing with classical Lebesgue measure theory or reading any textbook of this theory, we necessarily meet certain point sets of bad descriptive structure, the existence of which leads to the very important conclusion that there are Lebesgue nonmeasurable sets in $\mathbf{R}$, i.e., the standard Lebesgue measure λ on $\mathbf{R}$ is not defined for all subsets of $\mathbf{R}$ (in contrast to the outer Lebesgue measure, usually denoted by λ^*). In other words, we unavoidably have some subsets of $\mathbf{R}$ which do not belong to $\mathrm{dom}(\lambda)$.

Recall that the first ingenious example of such a subset is due to Vitali [266]. His construction is rather simple and can be explained in a few phrases. Namely, Vitali starts with the subgroup $\mathbf{Q}$ of $\mathbf{R}$, consisting of all rational numbers, and takes the corresponding quotient set $\mathbf{R}/\mathbf{Q}$ constituted of all pairwise disjoint translates of $\mathbf{Q}$. Obviously, each member of $\mathbf{R}/\mathbf{Q}$ is countably infinite and

$$\mathrm{card}(\mathbf{R}/\mathbf{Q}) = \mathrm{card}(\mathbf{R}).$$

However, it should be pointed out here that the above equality needs uncountable forms of the Axiom of Choice (see, e.g., Exercise 4 from Chapter 3). By using this axiom, Vitali fixed a selector V of $\mathbf{R}/\mathbf{Q}$ (selectors of such a kind are nowadays called Vitali sets) and then proved that V is not Lebesgue measurable and does not possess the Baire property (this topological property of subsets of $\mathbf{R}$, rather similar to the measurability property, is thoroughly discussed in [203]; see also [190] where a unified approach to these two concepts is developed). Moreover, it readily follows from Vitali's argument that V cannot belong to the domain of a translation invariant measure on $\mathbf{R}$ extending λ.

This result of Vitali turned out to be extremely fruitful for further investigations in classical measure theory and point set theory. First of all, it stimulated the appearance of much more delicate constructions of exotic sets in Euclidean spaces of higher dimension: Hausdorff's decomposition of a sphere, the Banach–Tarski paradox, von Neumann's paradox for the plane, etc. Extensive information about this topic is presented in [162], [195], [268], [272] and in many other works. Secondly, Vitali's result significantly influenced the formulation of the so-called general measure problem, in the statement of which the requirement of translation invariance of a measure is omitted. This problem directly leads to the theory of large cardinals, which is a cornerstone for the foundations of contemporary mathematics (see, e.g., [47], [110]).

The simplicity of Vitali's construction and its special place in Lebesgue measure theory inspire further study of various extraordinary properties of Vitali sets. In this context, a lot of natural questions arise and we are going to discuss some of them here. To begin, let us suppose that a nonempty finite family $\{V_i : 1 \leq i \leq m\}$ of Vitali sets is given and let us formulate the following three questions concerning this family.

Question 1. Can one assert that $\cup\{V_i : 1 \leq i \leq m\}$, like V, is nonmeasurable with respect to every translation invariant measure on $\mathbf{R}$ extending λ?

Question 2. Can $\cup\{V_i : 1 \leq i \leq m\}$ contain a λ-measurable subset of strictly positive measure?

Question 3. Can $\cup\{V_i : 1 \leq i \leq m\}$ contain a subset of the form $U \setminus K$, where U is a nonempty open set in $\mathbf{R}$ and K is a first category subset of $\mathbf{R}$?

In the material presented below, we will show that the answer to Question 1 is positive, but both Questions 2 and 3 have negative answers. We do not know a relatively simple argument which solves Question 1. The main difficulty arising here is that even two Vitali sets V_1 and V_2 can be constructed such that

$$\operatorname{card}(((V_1 \cup V_2) + q) \cap (V_1 \cup V_2)) = \operatorname{card}(\mathbf{R})$$

for each $q \in \mathbf{Q}$ (see Exercise 1 for this chapter). So our further reasoning will be based on the classical theorem of Banach [6] stating the existence of a non-negative, finitely additive, translation invariant, normalized functional defined on the family $\mathcal{BS}(\mathbf{R})$ of all bounded subsets of $\mathbf{R}$. Let us recall several notions, which we need in our further considerations.

A nonempty family $\mathcal{R}$ of subsets of $\mathbf{R}$ is a ring if the relations $X \in \mathcal{R}$ and $Y \in \mathcal{R}$ imply $X \cup Y \in \mathcal{R}$ and $X \setminus Y \in \mathcal{R}$.

A ring $\mathcal{R}$ of subsets of $\mathbf{R}$ is an algebra if $\mathbf{R} \in \mathcal{R}$.

A nonempty family $\mathcal{I}$ of subsets of $\mathbf{R}$ is an ideal if $\mathbf{R} \notin \mathcal{I}$ and the relations $X \in \mathcal{I}, Y \in \mathcal{I}, Z \subset Y$ imply $X \cup Y \in \mathcal{I}$ and $Z \in \mathcal{I}$.

A canonical example of an ideal is the above-mentioned family $\mathcal{BS}(\mathbf{R})$.

A ring (algebra, ideal) is called a σ-ring (σ-algebra, σ-ideal) if it is closed under countable unions of its members.

Canonical examples of σ-ideals are the family of all λ-measure zero sets in $\mathbf{R}$ and the family of all first category subsets of $\mathbf{R}$.

Theorem 1. *Let $\mathcal{R}$ be a translation invariant ring of subsets of $\mathbf{R}$, satisfying the relations $\mathcal{R} \subset \mathcal{BS}(\mathbf{R})$ and $[0, 1[\ \in \mathcal{R}$, and let $\nu : \mathcal{R} \to [0, +\infty[$ be a finitely additive translation invariant functional such that $\nu([0, 1[) = 1$. Then there exists a finitely additive translation invariant functional*

$$\nu' : \mathcal{BS}(\mathbf{R}) \to [0, +\infty[$$

such that ν' is an extension of ν.

The proof of this remarkable theorem of Banach can be found in many articles, textbooks, and monographs. See especially [268], where a more general result, for Euclidean space $\mathbf{R}^n$ equipped with an appropriate transformation group, is presented (cf. also Exercises 2, 3, 4 for this chapter).

Lemma 1. *Let ν be as in Theorem 1 and let a set $X \in \mathcal{BS}(\mathbf{R})$ have the following property: there exists a bounded infinite sequence $\{h_k : k \in \mathbf{N}\}$ of elements of $\mathbf{R}$ such that the family $\{X + h_k : k \in \mathbf{N}\}$ is disjoint.*
If $X \in \operatorname{dom}(\nu)$, then necessarily $\nu(X) = 0$.

Proof. Suppose to the contrary that $X \in \operatorname{dom}(\nu)$ and $\nu(X) > 0$. Consider the set $\cup\{X + h_k : k \in \mathbf{N}\}$. This set is obviously bounded, so it is contained in some interval $[a, b[$, where a and b are integers. Now, for any natural number l, we must have

$$l\nu(X) = \nu(\cup\{X + h_k : k < l\}) \leq \nu([a, b[) = b - a.$$

But this is impossible for sufficiently large natural numbers l.

The obtained contradiction finishes the proof.

Lemma 2. *Let X be a bounded subset of a Vitali set V. Then X has the property indicated in Lemma 1.*

Proof. In view of the definition of V, the family of sets $\{V + q : q \in \mathbf{Q}\}$ is disjoint (recall that V is a selector of $\mathbf{R}/\mathbf{Q}$). Consequently, the family of sets $\{X + q : q \in \mathbf{Q}\}$ is disjoint, too. So the family $\{q : q \in \mathbf{Q}, |q| \leq 1\}$ can play the role of a family $\{h_k : k \in \mathbf{N}\}$ participating in the formulation of Lemma 1.

Now, we are able to formulate and prove the following statement.

Theorem 2. *If $\{V_i : 1 \leq i \leq m\}$ is a nonempty finite family of Vitali sets, then $\cup\{V_i : 1 \leq i \leq m\}$ is absolutely nonmeasurable with respect to the class of all translation invariant measures on $\mathbf{R}$ extending λ, i.e., for every translation invariant measure μ extending λ, the set $\cup\{V_i : 1 \leq i \leq m\}$ turns out to be nonmeasurable with respect to μ.*

Proof. Suppose to the contrary that there exists a translation invariant extension μ of λ such that $\cup\{V_i : 1 \leq i \leq m\} \in \mathrm{dom}(\mu)$. Since

$$\cup\{V_i : 1 \leq i \leq m\} + \mathbf{Q} = \mathbf{R},$$

we readily infer that $\mu(\cup\{V_i : 1 \leq i \leq m\}) > 0$. Consequently, for some interval $[a, b[\subset \mathbf{R}$ whose endpoints are integers, we also have

$$\mu([a, b[\cap (\cup\{V_i : 1 \leq i \leq m\})) > 0.$$

Let now ν denote the restriction of μ to $\mathcal{BS}(\mathbf{R}) \cap \mathrm{dom}(\mu)$. For this ν, there exists a functional ν' as described in Theorem 1. Obviously, we have

$$0 < \nu([a, b[\cap (\cup\{V_i : 1 \leq i \leq m\})) = \nu'(\cup\{[a, b[\cap V_i : 1 \leq i \leq m\}).$$

Therefore, $\nu'([a, b[\cap V_i) > 0$ for at least one integer $i \in [1, m]$. But $[a, b[\cap V_i$ is a bounded subset of V_i and, by virtue of Lemma 2, it has the property described in Lemma 1. According to Lemma 1, the equality $\nu'([a, b[\cap V_i) = 0$ must be valid, so we obtain a contradiction which completes the proof.

The same method works for solving Questions 2 and 3 as well. Indeed, concerning Question 2, suppose that a finite family $\{V_i : 1 \leq i \leq m\}$ of Vitali sets is such that

$$Y \subset \cup\{V_i : 1 \leq i \leq m\}$$

for some λ-measurable set Y with $\lambda(Y) > 0$. Without loss of generality, we may assume that Y is bounded. Let ν denote the restriction of λ to the family of all bounded λ-measurable sets and let ν' be again as in Theorem 1. Obviously, we have

$$0 < \nu(Y) = \nu'(Y) = \nu'(\cup\{Y \cap V_i : 1 \leq i \leq m\}),$$

whence it follows that $\nu'(Y \cap V_i) > 0$ for some integer $i \in [1, m]$. But, as we already know, the last inequality is impossible (by the way, a similar argument shows that if μ is any translation invariant measure on $\mathbf{R}$ extending λ, then no μ-measurable set Y with $\mu(Y) > 0$ is contained in $\cup\{V_i : 1 \leq i \leq m\}$).

Concerning Question 3, suppose that a finite family $\{V_i : 1 \leq i \leq m\}$ of Vitali sets is such that

$$U \setminus Z \subset \cup\{V_i : 1 \leq i \leq m\}$$

for a nonempty open set $U \subset \mathbf{R}$ and for a first category set $Z \subset \mathbf{R}$. Without loss of generality, we may assume that $[a, b[\subset U$ where $a \in \mathbf{R}$, $b \in \mathbf{R}$ and $a < b$. Consequently, we have

$$[a, b[\setminus Z \subset \cup\{V_i : 1 \leq i \leq m\}.$$

Clearly, in this inclusion Z can be replaced by $Z \cap [a, b[$, so we may additionally suppose that Z is bounded. Now, consider the ring $\mathcal{T}$ of all finite unions of bounded half-open subintervals of $\mathbf{R}$ (of course, we mean here that these subintervals are closed on the left and open on the right). Also, consider the ideal $\mathcal{I}$ of all bounded first category subsets of $\mathbf{R}$. Let $\mathcal{R}$ denote the ring generated by $\mathcal{T} \cup \mathcal{I}$. All members of $\mathcal{R}$ are representable in the form $(A \setminus B) \cup C$, where A belongs to $\mathcal{T}$ and B and C belong to $\mathcal{I}$. We put

$$\nu((A \setminus B) \cup C) = \lambda(A).$$

A straightforward verification shows that ν is well defined on $\mathcal{R}$, because no nonempty member of $\mathcal{T}$ is of first category in $\mathbf{R}$. Indeed, if we have

$$(A \setminus B) \cup C = (A' \setminus B') \cup C',$$

then, denoting by $\triangle$ the symmetric difference of sets, we easily come to the relation

$$A \triangle A' \subset B \cup B' \cup C \cup C',$$

whence it follows that $A \triangle A' = \emptyset$ and, therefore, $A = A'$. Also, ν satisfies the assumptions of Theorem 1 and extends the restriction of λ to $\mathcal{T}$. Notice that $\nu(Z) = 0$, so

$$\nu([a, b[\setminus Z) = \nu([a, b[) = b - a > 0.$$

Let ν' be again as in Theorem 1 for the just described ν. Then we can write

$$0 < \nu([a, b[\setminus Z) = \nu'([a, b[\setminus Z) \leq \nu'(\cup\{[a, b[\cap V_i : 1 \leq i \leq m\}).$$

Consequently, $\nu'([a, b[\cap V_i) > 0$ for at least one integer $i \in [1, m]$ and once again we get a contradiction with Lemma 1.

Remark 1. Fortunately, Questions 2 and 3 can be answered in the negative without appealing to the profound Theorem 1. Sketches of the corresponding arguments are outlined in Exercises 5 and 6, respectively, and enable one to obtain slightly more general results.

Obviously, Theorem 2 established above strengthens the theorem of Vitali stating that any Vitali set V is absolutely nonmeasurable with respect to the class of all those measures on $\mathbf{R}$ which extend λ and are translation invariant. Thus, from the point of view of this class, Vitali sets and nonempty finite unions of them are extremely bad objects.

In many questions of mathematical analysis the translation invariance of a measure is not necessary and it suffices to require a much weaker property of a measure, namely, the so-called translation quasi-invariance. We would like to recall the definition of this weaker property.

Let ν be a measure defined on a translation invariant σ-algebra of subsets of $\mathbf{R}$. We say that ν is translation quasi-invariant if, for each $X \in \mathrm{dom}(\nu)$ and each $h \in \mathbf{R}$, the equalities $\nu(X) = 0$ and $\nu(h + X) = 0$ are equivalent.

Consequently, the translation quasi-invariance of a nonzero complete measure ν on $\mathbf{R}$ implies that the σ-ideal of all ν-measure zero sets is preserved under the action of any translation of $\mathbf{R}$.

The standard method to get finite translation quasi-invariant measures from a given σ-finite translation invariant measure μ on $\mathbf{R}$ is the well-known Radon-Nikodym operation. We take an arbitrary μ-integrable function

$$f : \mathbf{R} \to \ ]0, +\infty[$$

and put

$$\nu(X) = \int_X f(t) d\mu(t) \qquad (X \in \mathrm{dom}(\mu)).$$

Clearly, ν is a finite translation quasi-invariant measure whose domain coincides with the domain of μ and, in general, ν does not need to be translation invariant (notice that μ and ν are equivalent measures).

It is worth pointing out, however, that not every σ-finite quasi-invariant measure can be obtained in this simple manner (for more details, see [272]; cf. also Exercise 10).

Dealing with translation quasi-invariant extensions of λ, we detect that some Vitali sets behave better, from the measure-theoretic point of view, than other ones. To show this, let us first observe that among the selectors of $\mathbf{R}/\mathbf{Q}$ we can encounter certain subgroups of the additive group $\mathbf{R}$. Indeed, if we treat $\mathbf{R}$ as a vector space over the field $\mathbf{Q}$, then we may apply a well-known theorem from linear algebra which states that a vector subspace $\mathbf{Q}$ of $\mathbf{R}$ admits a complementary subspace in $\mathbf{R}$, i.e., we come to the representation

$$\mathbf{R} = \mathbf{Q} + H \qquad (\mathbf{Q} \cap H = \{0\}),$$

where H is some vector space over $\mathbf{Q}$. Actually, H is a hyperplane in $\mathbf{R}$ complementary to the "line" $\mathbf{Q}$ in $\mathbf{R}$. Consequently, H is also a subgroup of $\mathbf{R}$ such that no translation invariant measure on $\mathbf{R}$ extending λ can make H measurable with respect to it, because H is a particular case of a Vitali set (see Theorem 2; cf. Chapter 8).

On the other hand and somewhat surprisingly, it turns out that H becomes measurable with respect to a suitable translation quasi-invariant extension of λ. The following statement contains a stronger result which indicates that there are many possibilities for obtaining such translation quasi-invariant extensions of λ.

Theorem 3. *There exist continuum many measures μ on $\mathbf{R}$ which extend λ, are quasi-invariant under the group of all translations of $\mathbf{R}$, and satisfy the relation $H \in \mathrm{dom}(\mu)$.*

Proof. First, we would like to observe that the set H is λ-thick in $\mathbf{R}$, i.e., $\lambda_*(\mathbf{R} \setminus H) = 0$ where λ_* denotes the inner measure associated with λ (for more details, see Chapter 8). Now, we introduce the countable disjoint family of sets

$$\{H_k : k \in \mathbf{N}\} = \{q + H : q \in \mathbf{Q}\}.$$

Obviously, for any $h \in \mathbf{R}$, the family $\{h + H_k : k \in \mathbf{N}\}$ coincides with the family $\{H_{\psi(k)} : k \in \mathbf{N}\}$, where ψ is some permutation of $\mathbf{N}$. We thus derive that $\{H_k : k \in \mathbf{N}\}$ is a countable translation invariant partition of $\mathbf{R}$ into λ-thick sets. Further, consider the class of sets

$$\mathcal{S} = \{\cup\{H_k \cap X_k : k \in \mathbf{N}\} : (\forall k \in \mathbf{N})(X_k \in \mathrm{dom}(\lambda))\}.$$

It can readily be verified that $\mathcal{S}$ is a translation invariant σ-algebra of subsets of $\mathbf{R}$. Let $k \in \mathbf{N}$. Putting $X_k = \mathbf{R}$ and $X_m = \emptyset$ for each $m \in \mathbf{N} \setminus \{k\}$, we see that all H_k belong to $\mathcal{S}$ and, in particular, $H \in \mathcal{S}$.

Fix a sequence $\{a_k : k \in \mathbf{N}\}$ of strictly positive real numbers such that

$$\sum_{k \in \mathbf{N}} a_k = 1.$$

Then take an arbitrary set $\cup\{H_k \cap X_k : k \in \mathbf{N}\}$ from $\mathcal{S}$ and put

$$\mu(\cup\{H_k \cap X_k : k \in \mathbf{N}\}) = \sum_{k \in \mathbf{N}} a_k \lambda(X_k).$$

In this manner we come to a certain functional μ with $\mathrm{dom}(\mu) = \mathcal{S}$. Indeed, μ is well-defined on $\mathcal{S}$ because of the λ-thickness of all sets H_k. For the same reason, the functional μ is countably additive, so μ is a σ-finite measure on $\mathcal{S}$ (see Exercise 9). If $X \in \mathrm{dom}(\lambda)$, then

$$\mu(X) = \mu(\cup\{H_k \cap X : k \in \mathbf{N}\}) = \sum_{k \in \mathbf{N}} a_k \lambda(X) = \lambda(X),$$

which shows that μ extends λ. Further, if we have

$$\mu(\cup\{H_k \cap X_k : k \in \mathbf{N}\}) = \sum_{k \in \mathbf{N}} a_k \lambda(X_k) = 0,$$

then taking into account the inequalities $a_k > 0$ for all $k \in \mathbf{N}$, we get

$$\lambda(X_k) = 0 \quad (k \in \mathbf{N}),$$

which implies for any $h \in \mathbf{R}$ that

$$\mu(h + \cup\{H_k \cap X_k : k \in \mathbf{N}\}) = \mu(\cup\{H_{\psi(k)} \cap (h + X_k) : k \in \mathbf{N}\}) =$$

$$\sum_{k \in \mathbf{N}} a_{\psi(k)} \lambda(h + X_k) = 0,$$

where ψ is again some permutation of $\mathbf{N}$ (of course, depending on h). Therefore, the measure μ is quasi-invariant under the group of all translations of $\mathbf{R}$. Moreover, a similar argument yields that μ is also quasi-invariant under all central symmetries of $\mathbf{R}$, and hence μ is quasi-invariant under the group of all isometric transformations of $\mathbf{R}$.

Finally, since μ depends on a choice of a sequence of strictly positive real numbers $\{a_k : k \in \mathbf{N}\}$ with $\sum\{a_k : k \in \mathbf{N}\} = 1$ and there are continuum many such sequences, we conclude that there exist at least continuum many translation quasi-invariant extensions of λ for which the Vitali set H becomes measurable. Of course, it is important to emphasize here that different choices of $\{a_k : k \in \mathbf{N}\}$ produce different extensions of λ.

Theorem 3 has thus been proved.

In connection with this theorem, the following question arises.

Question 4. Does there exist a Vitali set which is absolutely nonmeasurable with respect to the class of all translation quasi-invariant measures on $\mathbf{R}$ extending λ?

In other words, we are asking whether there exists a Vitali set which is nonmeasurable with respect to every translation quasi-invariant extension of λ. The answer to this question is positive. Moreover, a much stronger result is contained in the next statement.

Theorem 4. *There exists a Vitali set V absolutely nonmeasurable with respect to the class of all nonzero σ-finite translation quasi-invariant measures on $\mathbf{R}$, i.e., for every nonzero σ-finite translation quasi-invariant measure μ on $\mathbf{R}$, we have $V \notin \mathrm{dom}(\mu)$.*

We thus see that among various Vitali sets there are those which are relatively good for translation quasi-invariant measures and there are those which are ultimately bad for the same measures. Here we only sketch the argument, leaving the details to the reader.

The crucial role is played by the following auxiliary proposition.

Lemma 3. *Let $(G, +)$ be a vector space over $\mathbf{Q}$ represented in the form*

$$G = G_0 + G_1 \quad (G_0 \cap G_1 = \{0\}),$$

where G_0 and G_1 are also vector spaces over $\mathbf{Q}$, the space G_0 is countably infinite, and $\mathrm{card}(G_1)$ is equal to the least uncountable cardinal ω_1. Then there exists a set $X \subset G$ such that:

(1) $G_0 + X = G$;

(2) $(g + X) \cap (h + X)$ is countable for any two distinct vectors $g \in G_1$ and $h \in G_1$.

For a detailed proof of Lemma 3, see Chapter 11 of [128] where a more general result is presented.

Notice that, in view of (1) of Lemma 3, we may additionally suppose that X is a selector of G/G_0. Actually, the set X is of somewhat paradoxical character, because countably many translates of X cover G and, at the same time, uncountably many translates of X form an almost disjoint family of sets.

Now, let us give the main idea of the proof of Theorem 4. As soon as Lemma 3 is established, we proceed in the following manner. We again treat $\mathbf{R}$ as a vector space over $\mathbf{Q}$ and represent it in the form of a direct sum of three vector subspaces G_0, G_1 and G_2, i.e.,

$$\mathbf{R} = G_0 + G_1 + G_2,$$

where $G_0 = \mathbf{Q}$ and G_1 are as in Lemma 3, and $(G_0 + G_1) \cap G_2 = \{0\}$. Denote $G = G_0 + G_1$ and let $X \subset G$ be a selector of G/G_0 satisfying (2) of Lemma 3. Now, it is not difficult to check that $V = X + G_2$ is a Vitali subset of $\mathbf{R}$ and this V is absolutely nonmeasurable with respect to the class of all nonzero σ-finite translation quasi-invariant measures on $\mathbf{R}$.

Remark 2. It seems somewhat surprising, but it is much easier to construct absolutely nonmeasurable sets in infinite-dimensional vector spaces over $\mathbf{R}$. For example, let E be an infinite-dimensional separable Hilbert space over $\mathbf{R}$ and let B denote its unit ball. As is well known, there exists no nonzero σ-finite translation quasi-invariant Borel measure on E (see, e.g., [250]), and this fact can be established within $\mathbf{ZF}$ & $\mathbf{DC}$ set theory. Further, it can be deduced within the same theory that B is absolutely nonmeasurable with respect to the class of all nonzero σ-finite translation quasi-invariant measures on E (actually, it suffices to check that the family of all translates of B generates the Borel σ-algebra of E). So we have a concrete subset of E which possesses very nice geometric properties but is extremely bad from the viewpoint of the above-mentioned class of measures. By using the absolute nonmeasurability of B, one can obtain, with the aid of uncountable forms of the axiom of choice, an absolutely nonmeasurable subset of $\mathbf{R}$ which essentially differs from any Vitali set (see Exercise 13).

Some additional information about Vitali sets, interesting from the measure-theoretic point of view, is presented in [32], [33], [143], [252], [272]. Notice that Vitali sets are connected with algebraic (in fact, group-theoretic) properties of λ and, first of all, with the translation invariance of λ. There are essentially different constructions of Lebesgue nonmeasurable subsets of $\mathbf{R}$ motivated by other structural properties of λ (see, e.g., [14], [24], [27], [30], [63], [71], [98], [128], [168], [184], [190], [216], [232], [235], [262]). Notice also that the class of translation invariant measures on $\mathbf{R}$ extending λ is quite large. In particular, there are even nonseparable measures belonging to this class (for more details, see [95], [107], [137], [144], [205]).

EXERCISES

1. Show that there exist two Vitali sets V_1 and V_2 on $\mathbf{R}$ such that the intersection $((V_1 \cup V_2) + q) \cap (V_1 \cup V_2)$ is of cardinality continuum for each $q \in \mathbf{Q}$.

2. Check that in order to answer Questions 1, 2 and 3, it suffices to assume that functional ν' in the formulation of Theorem 1 is invariant under all rational translations of $\mathbf{R}$ (in short, $\mathbf{Q}$-invariant).

Conclude that the union of any nonempty finite family of Vitali sets is absolutely nonmeasurable with respect to the class of all $\mathbf{Q}$-invariant measures on $\mathbf{R}$ extending λ.

Remark 3. Let $(\mathbf{Z}, +)$ denote, as usual, the group of all integers, and let $l_{\mathbf{Z}}$ be the Banach space of all real-valued bounded functions on $\mathbf{Z}$ (equipped with the standard sup-norm). Further, denote by $c_{\mathbf{Z}}$ the vector subspace of $l_{\mathbf{Z}}$ consisting of all those $\{x_n : n \in \mathbf{Z}\} \in l_{\mathbf{Z}}$ for which $\lim_{|n| \to \infty} x_n$ exists. Obviously, we have the linear functional lim on $c_{\mathbf{Z}}$ whose norm is equal to 1. The classical Hahn-Banach theorem easily implies the existence of a so-called Banach limit, i.e., the existence of a linear functional

$$\mathrm{Lim} : l_{\mathbf{Z}} \to \mathbf{R}$$

which extends lim, is shift invariant, and whose norm is also equal to 1 (see, for instance, [146], where such a Banach limit is constructed by using a nontrivial ultrafilter on $\mathbf{N}$; the case of $\mathbf{Z}$ is completely analogous, so the use of the Hahn–Banach theorem can be avoided here). The functional Lim produces a certain nonnegative, finitely additive, shift invariant functional η defined on the family of all subsets of $\mathbf{Z}$, vanishing at all singletons and such that $\eta(\mathbf{Z}) = 1$. Indeed, for any set $A \subset \mathbf{Z}$, it suffices to put $\eta(A) = \mathrm{Lim}(\chi_A)$, where χ_A denotes the characteristic function (i.e., indicator) of A. By starting with this η, one can deduce the existence of a nonnegative finitely additive $\mathbf{Q}$-invariant functional θ defined on the family of all subsets of $\mathbf{Q}$ and such that $\theta(\mathbf{Q}) = 1$. This circumstance allows to obtain a relatively simple proof of an analogue of Theorem 1 for a $\mathbf{Q}$-invariant ν'. The next two exercises outline the corresponding argument.

3. Check that the additive group $(\mathbf{Q}, +)$ can be represented as the union of an increasing (by inclusion) sequence of its subgroups, each of which is isomorphic to $(\mathbf{Z}, +)$.

Infer from this fact, with the aid of the functional Lim, that there exists a functional θ as described in Remark 3.

4*. Let $\mathcal{R}$ be a $\mathbf{Q}$-invariant ring of bounded subsets of $\mathbf{R}$ such that $[0, 1[\in \mathcal{R}$, and let $\nu : \mathcal{R} \to [0, +\infty[$ be a finitely additive $\mathbf{Q}$-invariant functional satisfying the equality $\nu([0, 1[) = 1$.

Demonstrate that there exists a finitely additive $\mathbf{Q}$-invariant functional

$$\nu' : \mathcal{BS}(\mathbf{R}) \to [0, +\infty[$$

which is an extension of ν.

Argue in the following manner. First, apply the Hahn–Banach theorem and extend ν to a nonnegative finitely additive functional defined on the family $\mathcal{BS}(\mathbf{R})$. Denote the extended functional by the symbol ν_0. Let θ be as in Exercise 3. This θ canonically produces a monotone linear functional Φ defined on the family $\mathcal{BF}(\mathbf{Q}, \mathbf{R})$ of all real-valued bounded functions on $\mathbf{Q}$ (to obtain Φ, uniformly approximate any function from $\mathcal{BF}(\mathbf{Q}, \mathbf{R})$ by appropriate step functions with finite ranges, and then consider their θ-integrals). The $\mathbf{Q}$-invariance of θ implies the analogous $\mathbf{Q}$-invariance of Φ, i.e., for every function $f \in \mathcal{BF}(\mathbf{Q}, \mathbf{R})$ and every $q \in \mathbf{Q}$, we have $\Phi(f) = \Phi(f_q)$, where f_q is defined by

$$f_q(r) = f(q + r) = f(r + q) \quad (r \in \mathbf{Q}).$$

Now, take any bounded set $X \subset \mathbf{R}$ and introduce the function $h_X \in \mathcal{BF}(\mathbf{Q}, \mathbf{R})$ by the formula

$$h_X(r) = \nu_0(X + r) \quad (r \in \mathbf{Q}).$$

Finally, by putting $\nu'(X) = \Phi(h_X)$, obtain the required $\mathbf{Q}$-invariant functional ν' extending the initial functional ν.

Remark 4. As was shown by Foreman and Wehrung, the Hahn–Banach theorem implies (within $\mathbf{ZF}$ theory) the existence of a Lebesgue nonmeasurable subset of $\mathbf{R}^3$. This result was strengthened by Pawlikowski who demonstrated that the Hahn–Banach theorem even implies the Banach–Tarski paradox in $\mathbf{R}^3$ (for details, see Fund. Math., v. 138, n. 1, 1991, pp. 13–19 and 21–22).

5*. Give a direct proof of the fact that the union of a finite family of Vitali sets has inner λ-measure zero. Moreover, try to establish the validity of the following more general assertion:

Let a set $W \subset \mathbf{R}$ be such that its intersection with every member of $\mathbf{R}/\mathbf{Q}$ is finite. Then $\lambda_*(W) = 0$.

For this purpose, utilize the strong form of the Steinhaus property of λ-measurable sets, which states that if X is an arbitrary λ-measurable set with $\lambda(X) < +\infty$, then

$$\lim_{h \to 0} \lambda((h + X) \triangle X) = 0.$$

Remark 5. In general, the hint suggested in Exercise 5 becomes useless if λ is replaced by a translation invariant measure on $\mathbf{R}$ extending λ. Indeed, it has been proved that there exists a measure μ on $\mathbf{R}$ which extends λ and is translation invariant, but does not possess the Steinhaus property. Moreover, for this μ, there are a μ-measurable set X with $\mu(X) > 0$ and a sequence $\{h_k : k \in \mathbf{N}\}$ of elements of $\mathbf{R}$ such that

$$\lim_{k \to +\infty} h_k = 0, \quad (\forall k \in \mathbf{N})((h_k + X) \cap X = \emptyset).$$

More detailed information about μ and its extraordinary properties can be found in [126] and [137] (cf. also Chapter 18).

6. Give a direct proof of the fact that no finite union of Vitali sets contains a subset of the form $U \setminus Z$, where U is a nonempty open set in $\mathbf{R}$ and Z is a first category subset of $\mathbf{R}$. Analogously to Exercise 5, try to establish the validity of the following more general assertion:

Let $W \subset \mathbf{R}$ be such that its intersection with every member of $\mathbf{R}/\mathbf{Q}$ is finite. Then W does not contain a subset of the above-mentioned form.

For this purpose, observe that every translate of $\mathbf{Q}$ has infinitely many common points with U, so there always exists a Vitali set contained in U and disjoint from W. Then take into account the fact that no Vitali set is of first category in $\mathbf{R}$.

7. By starting with the fact that there exists a bounded Vitali set (see the hint above), give an example of a countable family $\{V_i : i \in \mathbf{N}\}$ of Vitali subsets of $\mathbf{R}$ such that $\cup\{V_i : i \in \mathbf{N}\}$ contains a λ-measurable set of strictly positive measure and the complement of $\cup\{V_i : i \in \mathbf{N}\}$ has strictly positive outer λ-measure.

8. By using Theorem 1, prove that there exists a functional

$$\nu' : \mathcal{BS}(\mathbf{R}) \to [0, +\infty[$$

satisfying the following three conditions:

(a) $\nu'([0, 1[) = 1$;

(b) ν' is finitely additive and translation invariant;

(c) for some λ-measurable set Z with $\lambda(Z) > 0$, we have $\nu'(Z) = 0$.

In order to prove this fact, consider a bounded nowhere dense set $Z \in \mathrm{dom}(\lambda)$ with $\lambda(Z) > 0$ and apply to Z an argument similar to that which answers Question 3.

Remark 6. Recall that the ring $\mathcal{J}(\mathbf{R})$ of Jordan measurable subsets of $\mathbf{R}$ consists of all those bounded sets in $\mathbf{R}$ whose characteristic functions are integrable in the Riemann sense. The restriction of λ to $\mathcal{J}(\mathbf{R})$ is usually called the classical Jordan measure. The result of Exercise 8 shows, in particular, that there exists a nonnegative finitely additive translation invariant extension of the Jordan measure which is defined on the ring of all bounded λ-measurable sets, but differs from the restriction of λ to this ring.

9. Check in detail that the functional μ described in the proof of Theorem 3 is well-defined and turns out to be a measure on $\mathbf{R}$.

10. Let $h : \mathbf{R} \to \mathbf{R}$ be a mapping representable in the form

$$h(x) = ax + b \quad (x \in \mathbf{R}),$$

where a and b are fixed real numbers and a differs from $0, 1, -1$. Let G denote the group of transformations of $\mathbf{R}$ generated by h and all translations of $\mathbf{R}$.

Demonstrate that λ is quasi-invariant with respect to all transformations from G but there exists no σ-finite G-invariant measure μ on $\mathbf{R}$ such that λ can be obtained by applying to μ the Radon–Nikodym operation.

For this purpose, utilize the uniqueness property of λ stating that any σ-finite translation invariant measure defined on $\mathrm{dom}(\lambda)$ is proportional to λ (see, for instance, [118], [123]).

11. Verify that the Vitali set V described before Remark 2 is absolutely nonmeasurable with respect to the class of all nonzero σ-finite translation quasi-invariant measures on $\mathbf{R}$.

12. A function $f : \mathbf{R} \to \mathbf{R}$ is called a Vitali type function if the range of f is a Vitali set and $f(x) - x \in \mathbf{Q}$ for each $x \in \mathbf{R}$.

Show that:

(a) there exists a Vitali type function which is measurable with respect to a certain translation quasi-invariant measure on $\mathbf{R}$ extending λ;

(b) there exists a Vitali type function which is nonmeasurable with respect to every nonzero σ-finite translation quasi-invariant measure on $\mathbf{R}$.

13*. Starting with the fact that the unit ball B in a separable infinite-dimensional Hilbert space E (over $\mathbf{R}$) is absolutely nonmeasurable with respect to the class of all nonzero σ-finite translation quasi-invariant measures on E, demonstrate that there exists a set $Z \subset \mathbf{R}$ which is absolutely nonmeasurable with respect to the class of all nonzero σ-finite translation quasi-invariant measures on $\mathbf{R}$.

For this purpose, consider both E and $\mathbf{R}$ as vector spaces over $\mathbf{Q}$ and verify that they are isomorphic to each other. Let $\psi : E \to \mathbf{R}$ be an isomorphism. Put $Z = \psi(B)$ and check that Z is as required. Moreover, taking into account the fact that no disjoint family of balls can cover E, show that Z cannot be a Vitali subset of $\mathbf{R}$.

14. Denote by the symbol $[\omega]^\omega$ the family of all infinite subsets of ω and define an equivalence relation $S(X, Y)$ on this family by putting

$$S(X, Y) \Leftrightarrow \operatorname{card}(X \triangle Y) < \omega.$$

Identifying any infinite subset of ω with the corresponding point of the Cantor space $\{0, 1\}^\omega$, prove that every selector of the equivalence relation $S(X, Y)$ is nonmeasurable with respect to the completion ν' of the Haar measure ν on $\{0, 1\}^\omega$.

Taking into account the fact that ν' is isomorphic to the Lebesgue measure on $[0, 1]$, conclude that the existence of a selector of $S(X, Y)$ needs uncountable forms of the Axiom of Choice.

15*. Verify the validity of the following two assertions:
(a) there exist two Vitali sets V and W in $\mathbf{R}$ such that $V + W \neq \mathbf{R}$;
(b) there exist two Vitali sets V' and W' such that $V' + W' = \mathbf{R}$.

Argue as follows. For (a), keep in mind the fact that there is a Vitali subset V of $\mathbf{R}$ which simultaneously is a vector space over the field $\mathbf{Q}$ and $V \cap \mathbf{Q} = \{0\}$. Putting $W = V$, check that $V + W = V \neq \mathbf{R}$.

For (b), consider the same V as above and take another Vitali set U in $\mathbf{R}$ such that $\mathbf{Q} \subset U - V$. Then define $V' = U$ and $W' = V$ and check that

$$V' + W' = U + V = (U + V) + V = (U - V) + V \supset \mathbf{Q} + V = \mathbf{R},$$

which yields the required result.

10. A relationship between the measurability and continuity of real-valued functions

In the present chapter we would like to discuss deep relationships between the following two fundamental concepts of mathematical analysis: measurability and continuity. For this purpose, some nontrivial examples are given below, which underline close connections between measurable and continuous real-valued functions, and the reader will see how those connections can be described in terms of absolutely nonmeasurable functions and universal measure zero sets (cf. Chapter 5 where these notions are introduced and examined).

In particular, a variant of Luzin's C-property is formulated and proved here for a class of measures significantly wider than the class of σ-finite Borel measures or their completions. In this context, absolutely nonmeasurable functions and Sierpiński–Zygmund type functions are considered and compared to each other.

At first sight, the concept of measurability seems to be of a more general character than the concept of continuity. However, it frequently turns out that the measurability of real-valued functions is tightly linked with the continuity of their restrictions to appropriate nonsmall subsets of their domains. There are many important examples of connections of this type, for instance, Luzin's classical theorem concerning the above-mentioned C-property of all Lebesgue measurable real-valued functions on $\mathbf{R}$ (see, e.g., [17], [21], [22], [133], [197], [203]).

Such connections are primarily implied by various topological regularity properties of those measure spaces on which measurable real-valued functions are given. Recall, for example, perfect probability spaces, Radon spaces, and Prokhorov spaces. These classes of spaces are extensively investigated and discussed in the literature, e.g., in [17], [20], [64], [78], [213], [265] (see also Chapter 6).

To be more precise, the main goal of the present chapter is to express adequately nontrivial relationships between the measurability and continuity in terms of absolutely nonmeasurable functions and universal measure zero sets.

Notice that, from the measure-theoretical point of view, absolutely nonmeasurable functions are very bad and universal measure zero sets are very small. In this chapter, we also consider totally discontinuous functions (or, equivalently, functions of Sierpiński–Zygmund type). Totally discontinuous functions are extremely bad from the topological standpoint and also turn out to be closely connected with absolutely nonmeasurable real-valued functions.

To begin our presentation, let us first recall that if E is an arbitrary topological space, then the symbol $\mathcal{B}(E)$ denotes the family of all Borel subsets of E, i.e., $\mathcal{B}(E)$ is the Borel σ-algebra of E.

The topological spaces considered below are assumed to be such that all singletons in them turn out to be Borel.

If X is a subset of a topological space E, then $\mathrm{cl}(X)$ stands for the closure of X.

For a partial function f acting from a topological space E into a metric space, the symbol $\mathrm{osc}_f(x)$ denotes the oscillation of f at a point $x \in E$; in other words, the real number $\mathrm{osc}_f(x)$ is defined by the equality

$$\mathrm{osc}_f(x) = \inf\{\mathrm{diam}(f(U(x))) : U(x) \text{ is a neighborhood of } x\},$$

where $\mathrm{diam}(f(U(x)))$ stands for the diameter of $f(U(x))$.

Of course, it may happen that $\mathrm{osc}_f(x) = +\infty$ for each point $x \in E$.

If h is a function acting from a topological space E into a topological space E', then the symbol $C(h)$ denotes the set of all continuity points of h.

It is not difficult to check that if E' is a metrizable topological space, then the set

$$C(h) = \{x \in E : \mathrm{osc}_h(x) = 0\}$$

is of type G_δ in E.

Recall that a measure μ given on some σ-algebra of subsets of a base (ground) set E is diffused (or continuous) if $\mu(\{x\}) = 0$ for all $x \in E$.

All measures considered below are assumed to be diffused.

If μ is a measure, then the symbol μ_* denotes, as usual, the inner measure associated with μ and the symbol μ^* stands for the outer measure associated with μ.

A topological space X is called universal measure zero (in the classical sense) if there exists no nonzero σ-finite diffused Borel measure on X (see Chapter 5).

It is easy to see that the following two assertions are equivalent:

(i) X is universal measure zero;

(ii) for any topological space E, containing X as a subspace, and for any Borel diffused probability measure μ on E, the equality $\mu^*(X) = 0$ holds true.

Moreover, the family of all universal measure zero sets in a topological space E forms a certain σ-ideal of subsets of E. This σ-ideal is proper if E itself is not a universal measure zero space.

As we have already mentioned (see Chapter 5), there exist uncountable universal measure zero subsets of $\mathbf{R}$. In particular, this important fact of descriptive set theory was first established by Luzin and it does not need any additional set-theoretical axioms. Notice that at present there are several other constructions of uncountable universal measure zero subsets of $\mathbf{R}$, which exploit essentially different ideas and approaches (cf. [83], [208], [211], [273]).

Also, recall that a set $X \subset \mathbf{R}$ is a Luzin set if X is uncountable and meets every first category subset of $\mathbf{R}$ in (at most) countably many points.

Under the Continuum Hypothesis, there exist Luzin subsets of $\mathbf{R}$ (see, e.g., [33], [147], [152], [188], [190], [203] or Exercise 15 from Chapter 4).

Every Luzin set X in $\mathbf{R}$ has universal measure zero. Indeed, any σ-finite diffused Borel measure μ on X is concentrated on some first category subset $P = P(\mu)$ of X (see Exercise 23 from Chapter 5). But P is at most countable in view of the definition of Luzin sets, so we get the equalities

$$\mu(X) = \mu(P) = 0,$$

which directly imply that X is universal measure zero.

Let E be a base (ground) set. Throughout this chapter the symbol $\mathcal{M}(E)$ denotes the class of all nonzero σ-finite diffused measures on E (their domains may be various σ-algebras of subsets of E).

It is easy to check that the following two assertions are equivalent:

(1) E is uncountable;

(2) $\mathcal{M}(E) \neq \emptyset$.

Let E be a topological space. In our further considerations, we denote by $\mathcal{M}_1(E)$ the class of all those complete diffused probability measures μ on E which satisfy the following condition:

(*) for each set $X \in \mathrm{dom}(\mu)$, there exists a set $Y \in \mathcal{B}(E)$ such that $\mu(X \triangle Y) = 0$.

The reader can easily verify that any complete measure μ on E satisfying the condition (*) is an extension of some Borel measure on E (of course, this Borel measure depends on μ).

In the sequel, we need the following three definitions.

The first of them generalizes the classical notion of a universal measure zero subset of the real line $\mathbf{R}$ (or, equivalently, of an uncountable Polish topological space).

Definition 1. Let E be a ground set, $\mathcal{M}$ be a class of measures on E, and let X be a subset of E. We shall say that X is universal measure zero with respect to the class $\mathcal{M}$ if, for every measure $\mu \in \mathcal{M}$, the equality $\mu^*(X) = 0$ is valid.

Obviously, the standard (classical) definition of a universal measure zero set corresponds to the case where E coincides with $\mathbf{R}$ (or with some uncountable Polish topological space) and $\mathcal{M}$ coincides with the class of all nonzero σ-finite diffused Borel measures on E.

The next definition generalizes the notion of Sierpiński–Zygmund functions.

Definition 2. Let E be a set, $\mathcal{T}$ be a class of topologies on E, and let f be a real-valued function defined on E. We shall say that f is totally discontinuous with respect to $\mathcal{T}$ (or f is a Sierpiński–Zygmund type function with respect to $\mathcal{T}$) if, for any topology $T \in \mathcal{T}$ and for any set $X \subset E$ with $\mathrm{card}(X) = \mathrm{card}(E)$ the restriction $f|X$ is not continuous with respect to the induced topology

$$T|X = \{U \cap X : U \in T\}.$$

Also, it is reasonable to recall here one more definition which was introduced in Chapter 5 of this book.

Definition 3. Let E be a set, $\mathcal{M}$ be a class of measures on E, and let f be a real-valued function defined on E. We say that f is relatively measurable with respect to $\mathcal{M}$ if there exists at least one measure $\mu \in \mathcal{M}$ such that f turns out to be μ-measurable.

Naturally, in accordance with Definition 3, we say that a function $g : E \to \mathbf{R}$ is absolutely nonmeasurable with respect to $\mathcal{M}$ if, for every measure $\mu \in \mathcal{M}$ this g turns out to be nonmeasurable with respect to μ.

Let us give several examples illustrating the above-mentioned concepts.

Example 1. Let E be a set, μ be a σ-finite measure on E, and let $\mathcal{M}(\mu)$ denote the class of all those measures on E which extend μ. It is easy to see that, for a set $X \subset E$, the following two assertions are equivalent:
(a) $\mu^*(X) = 0$;
(b) X is universal measure zero with respect to $\mathcal{M}(\mu)$.

Example 2. Let E be a topological space and let Y be a subset of E. Taking into account the circumstance that every measure from the class $\mathcal{M}_1(E)$ is an extension of some Borel diffused probability measure on E, one can readily check that these two assertions are equivalent:
(a) Y is universal measure zero (in the classical sense);
(b) Y is universal measure zero with respect to $\mathcal{M}_1(E)$.

Example 3. Let E be a separable metric space of cardinality $\mathbf{c}$, let T denote the topology of E induced by its metric, and let $\mathcal{T} = \{T\}$. In this case, as actually was demonstrated by Sierpiński and Zygmund [246], there exists a totally discontinuous function with respect to $\mathcal{T}$. For the proof, see [246] or [152]. In this context, we would like to notice that a much stronger result is valid. Namely, it can be established that if E is a ground set with $\mathrm{card}(E) = \mathbf{c}$ and $\mathcal{T}$ is a family of topologies on E such that $\mathrm{card}(\mathcal{T}) \leq \mathbf{c}$ and each topology from $\mathcal{T}$ has a countable base, then there exists a totally discontinuous function with respect to $\mathcal{T}$.

Example 4. Assuming the Continuum Hypothesis (or the more general Martin's Axiom), it can be proved that if E is an uncountable Polish space, then any Sierpiński–Zygmund function on E is absolutely nonmeasurable with respect to the class $\mathcal{M}_1(E)$ (see Exercise 4 from Chapter 5 and Remark 4 below).

Example 5. Let E be an uncountable Polish space and let B be a Bernstein subset of E (see, e.g., [96], [147], [152], [188], [190], [203] for the definition and basic properties of Bernstein sets; cf. also Chapters 3 and 5). It meets no difficulty to show that the characteristic function

$$\chi_B : E \to \{0,1\}$$

is absolutely nonmeasurable with respect to the class $\mathcal{CBM}_0(E)$ of the completions of all nonzero σ-finite diffused Borel measures on E. On the other hand, the same χ_B turns out to be measurable with respect to a certain measure from the class $\mathcal{M}_1(E)$ (see Exercise 3 from Chapter 5).

Comparing Example 4 with Example 5, we may conclude that from the measure-theoretical point of view the characteristic functions of Bernstein sets are much better than Sierpiński–Zygmund type functions.

Let E be an uncountable set and let $f : E \to \mathbf{R}$ be a function.

The natural question arises: what conditions are necessary and sufficient for treating f as a random variable on E?

More precisely, we are interested whether there exists a probability space $(E, \mathrm{dom}(\mu), \mu)$ such that f becomes a random variable on this space (recall that μ is assumed to be a diffused probability measure). The worst situation is when such a space $(E, \mathrm{dom}(\mu), \mu)$ (or, equivalently, when a diffused probability measure μ on E) does not exist at all.

In this case, we may say that f is an absolutely nonmeasurable function on E and, actually, this means that f is absolutely nonmeasurable with respect to the class $\mathcal{M}(E)$, because any nonzero σ-finite measure on E is equivalent to an appropriate probability measure on the same E.

It is natural to try to find some suitable characterization of absolutely nonmeasurable functions with respect to the class $\mathcal{M}(E)$ of all nonzero σ-finite diffused measures on E. We already know one of such characterizations and we would like to recall it here.

Theorem 1. *Let $f : E \to \mathbf{R}$ be a function. Then f is absolutely nonmeasurable with respect to the class $\mathcal{M}(E)$ if and only if the following two conditions are satisfied:*

(a) $\mathrm{card}(f^{-1}(t)) \leq \omega$ *for every point $t \in \mathbf{R}$;*
(b) *the set $\mathrm{ran}(f)$ is universal measure zero.*

For the proof of Theorem 1, see Chapter 5.

It immediately follows from this theorem that if $\mathrm{card}(E) > \mathbf{c}$, then every real-valued function on E can be regarded as a random variable on $(E, \mathcal{S}, \mu)$ for some σ-algebra $\mathcal{S}$ of subsets of E and for some probability diffused complete measure μ on $\mathcal{S}$. So, speaking of absolutely nonmeasurable real-valued functions, we may restrict our considerations to the case where $\operatorname{card}(E) \leq \mathbf{c}$ and, in fact, we may put $E = \mathbf{R}$ without essential loss of generality.

It is not difficult to prove the next statement.

Theorem 2. *Let E be an uncountable Polish space and let $f : E \to \mathbf{R}$ be a function absolutely nonmeasurable with respect to the class of completions of all Borel diffused probability measures on E. Then the following two assertions are valid:*

(1) *$C(f)$ is at most countable;*

(2) *if E contains no isolated points, then $C(f)$ is nowhere dense.*

Proof. Let us show that assertion (1) holds true. Suppose to the contrary that $C(f)$ is uncountable. According to a well-known fact from general topology (see, for instance, [49] or [152]), the set $C(f)$ is of type G_δ and, in particular, is Borel in E. Consequently, there exists a Borel diffused probability measure μ on E such that

$$\mu(C(f)) = 1.$$

Let μ' denote the completion of μ. It can readily be verified that f turns out to be measurable with respect to μ'. But this circumstance contradicts the absolute nonmeasurability of f with respect to the class of completions of all Borel diffused probability measures on E. The obtained contradiction establishes the validity of assertion (1).

Now, let us show that assertion (2) holds true. Suppose to the contrary that there exists a nonempty open set $U \subset E$ such that the set

$$P = C(f) \cap U$$

is everywhere dense in U. By virtue of (1), P is at most countable. At the same time, taking into account the descriptive structure of $C(f)$, we infer that P must be of type G_δ in the Polish topological space U. Since U does not contain isolated points, we immediately come to a contradiction with the Baire theorem on category, applied to U. The obtained contradiction establishes the validity of assertion (2) and thus finishes the proof of Theorem 2.

It is useful to compare Theorem 2 with the next statement.

Theorem 3. *Assuming the Continuum Hypothesis, there exists a function $f : \mathbf{R} \to \mathbf{R}$ such that:*

(1) *$\operatorname{card}(C(f)) = \omega$;*

(2) *f is absolutely nonmeasurable with respect to the class $\mathcal{M}(\mathbf{R})$.*

Proof. Denote, as usual, by $\mathbf{N}$ the set of all natural numbers and by $\mathbf{Z}$ the set of all integers. For any $n \in \mathbf{Z}$ and for any $k \in \mathbf{N}$, a Luzin set $L(n, k) \subset \mathbf{R}$

can be constructed such that $n \in L(n,k)$ and the diameter of $L(n,k)$ is strictly less than $1/(k+2)$. Consider the family of half-open intervals

$$\{[n - 1/2, n + 1/2[\ : \ n \in \mathbf{Z}\}.$$

Obviously, this family is a disjoint covering of the real line $\mathbf{R}$. For every integer n, denote by $\{U(n,k) : k \in \mathbf{N}\}$ a family of subintervals of $[n - 1/2, n + 1/2[$ satisfying the following relations:

(a) $U(n,0) = [n - 1/2, n + 1/2[$;
(b) n is an interior point of all intervals $U(n,k)$, where $k \in \mathbf{N}$;
(c) for each $k \in \mathbf{N}$, we have the inclusion

$$U(n, k + 1) \subset U(n, k)$$

and the length of $U(n, k+1)$ is strictly less than the length of $U(n,k)$;
(d) the length of $U(n,k)$ tends to zero as k tends to infinity.

Now, we introduce the required function f as follows. First of all, we put $f(n) = n$ for all $n \in \mathbf{Z}$. Then we fix $n \in \mathbf{Z}$ and define the restriction of f to the interval $[n - 1/2, n + 1/2[$. For this purpose, it suffices to determine the restriction of f to any set of the form $U(n,k) \setminus U(n, k + 1)$. Since we have the equality

$$\mathrm{card}(U(n,k) \setminus U(n, k + 1)) = \mathbf{c},$$

there exists a bijection

$$f_{n,k} : U(n,k) \setminus U(n, k + 1) \to L(n,k).$$

The common extension of all partial functions

$$f_{n,k} \quad (n \in \mathbf{Z}, \ k \in \mathbf{N})$$

yields the desired f. Indeed, keeping in mind Theorem 1, it is not difficult to verify that f is continuous at all points of $\mathbf{Z}$ and, simultaneously, is absolutely nonmeasurable with respect to the class $\mathcal{M}(\mathbf{R})$.

This completes the proof of Theorem 3.

Remark 1. The direct analogue of Theorem 3 can be proved under Martin's Axiom instead of the Continuum Hypothesis. The argument remains almost the same and we only must replace Luzin sets by so-called generalized Luzin sets which also are universal measure zero (assuming $\mathbf{MA}$).

Remark 2. An additional set-theoretical assumption in the formulation of Theorem 3 is necessary, because it is impossible to establish within $\mathbf{ZFC}$ set theory the existence of functions acting from $\mathbf{R}$ into $\mathbf{R}$ and absolutely nonmeasurable with respect to the class $\mathcal{M}(\mathbf{R})$ (see Remark 4 of Chapter 5).

For an arbitrary topological space E, let us indicate two simple and readily verified properties of the class $\mathcal{M}_1(E)$.

Lemma 1. *Let E be a topological space and let $\mu \in \mathcal{M}_1(E)$. The following two assertions are valid:*

(1) if B is a Borel subset of E such that $\mu(B) > 0$, then the measure ν on E defined by

$$\nu(X) = \mu(X \cap B)/\mu(B) \qquad (X \in dom(\mu))$$

also belongs to the class $\mathcal{M}_1(E)$;

(2) if S is a σ-algebra of subsets of E containing $\mathcal{B}(E)$ and contained in $dom(\mu)$, then the completion of the restriction of μ to S also belongs to $\mathcal{M}_1(E)$.

The proof of this proposition is quite easy, so we omit it here and leave it to the reader.

Lemma 2. *Let E be a topological space and let*

$$f : E \to \mathbf{R}$$

be a partial continuous function. Then f can be extended to a partial continuous function

$$f^* : E \to \mathbf{R}$$

such that $dom(f^*)$ is a subset of E representable in the form $A \cap B$, where A is a closed subset of E and B is of type G_δ in E (consequently, $dom(f^*)$ is a Borel set in E).

Proof. The argument is fairly standard and exploits the local compactness (or completeness) of $\mathbf{R}$. We denote

$$A = \mathrm{cl}(dom(f)), \quad B = \{x \in E : osc_f(x) = 0\}.$$

Clearly, A is closed in E and B is of type G_δ. For each point $x \in A \cap B$, consider the family $\{U_\xi(x) : \xi \in \Xi\}$ of all neighborhoods of x. The associated family of closed sets $\{\mathrm{cl}(f(U_\xi(x))) : \xi \in \Xi\}$ is centered in $\mathbf{R}$ and has members with arbitrarily small diameters. This circumstance implies that

$$\cap\{\mathrm{cl}(f(U_\xi(x))) : \xi \in \Xi\} = \{y\}$$

for some uniquely determined point $y \in \mathbf{R}$. Define $f^*(x) = y$. In this manner, we get the function

$$f^* : A \cap B \to \mathbf{R}.$$

From the definition of f^*, one can readily deduce the inclusion

$$f^*(U) \subset \mathrm{cl}(f(U))$$

for any open set $U \subset E$. Therefore, f^* extends f and is continuous as well.
This completes the proof of Lemma 2.

In fact, Lemma 2 is a certain version of Lavrentiev's classical theorem on extensions of partial continuous functions (cf. [49], [152] and Exercise 12 for this chapter). Notice also that if every closed subset of E is of type G_δ, then $\mathrm{dom}(f^*)$ in Lemma 2 can be chosen to be of type G_δ. Obviously, Lemma 2 remains valid for any complete metric space F instead of $\mathbf{R}$.

Lemma 3. *Let E be a ground set, μ be a σ-finite measure on E, and let $\mathcal{I}$ be a σ-ideal of subsets of E such that*

$$(\forall X \in \mathcal{I})(\mu_*(X) = 0).$$

Then μ can be extended to a measure μ' on E such that $\mathrm{dom}(\mu')$ coincides with the σ-algebra generated by $\mathrm{dom}(\mu) \cup \mathcal{I}$, and $\mu'(X) = 0$ for every set $X \in \mathcal{I}$.

Lemma 3 is well known and goes back to Marczewski's method of extending σ-finite measures by using certain σ-ideals of sets (cf. [174], [176] and Exercise 18 from Chapter 5).

Lemma 4. *Let E be a topological space, μ be a measure from the class $\mathcal{M}_1(E)$ and let $\mathcal{I}$ be a σ-ideal of subsets of E such that*

$$(\forall X \in \mathcal{I})(\mu_*(X) = 0).$$

Let μ' denote the extension of μ obtained with the aid of this σ-ideal as described in Lemma 3. Then μ' also belongs to the class $\mathcal{M}_1(E)$.

Proof. Since μ is complete, μ' is complete, too. Let X be an arbitrary μ'-measurable set. According to the definition of μ', there exists a μ-measurable set Y such that

$$\mu'(X \triangle Y) = 0.$$

Further, since $\mu \in \mathcal{M}_1(E)$, we can find a Borel set $Z \subset E$ such that

$$\mu(Y \triangle Z) = 0.$$

Consequently, we have

$$0 \le \mu'(X \triangle Z) \le \mu'(X \triangle Y) + \mu'(Y \triangle Z) = 0,$$

which implies that $\mu' \in \mathcal{M}_1(E)$ and finishes the proof of the lemma.

Remark 3. The previous auxiliary proposition directly shows that the class $\mathcal{M}_1(E)$ is closed under the operation of extending measures described in Lemma 3. In general, the class $\mathcal{M}_1(E)$ is sufficiently wide (assuming, of course, that E is not a universal measure zero space). For example, if $\mathrm{card}(E)$ is not measurable

in the Ulam sense, then any measure from $\mathcal{M}_1(E)$ admits a proper extensio●
belonging to the same class.

Lemma 5. *Let E be a topological space in which every closed set is of typ●*
G_δ (or, equivalently, every open set is of type F_σ). Then any Borel probabilit●
measure μ on E is regular, i.e., for each set $X \in \mathcal{B}(E)$, we have the equalities

$$\mu(X) = \sup\{\mu(F) : F \subset X, \ F \text{ is closed}\} = \inf\{\mu(G) : X \subset G, \ G \text{ is open}\}.$$

This auxiliary proposition is well known (see, e.g., [17], [89], [199], [203] an●
Exercise 3 from Chapter 6).

Now, we are able to formulate and prove the main statement of this chapter●

Theorem 4. *Let E be a topological space such that every closed subset o●*
E is of type G_δ. Let $f : E \to \mathbf{R}$ be a function. The following two relations ar●
equivalent:

(1) f is absolutely nonmeasurable with respect to the class $\mathcal{M}_1(E)$;
(2) for each set $X \subset E$, the continuity of the restriction $f|X$ implies that X●
is universal measure zero with respect to the class $\mathcal{M}_1(E)$.

Proof. (1) $\Rightarrow$ (2). Let relation (1) be satisfied. Suppose to the contrar●
that (2) is not valid. Then there exists a set $Y \subset E$ such that:

(a) the restriction $f|Y$ is continuous;

(b) Y is not universal measure zero with respect to $\mathcal{M}_1(E)$.

As we already know (see Example 1), relation (b) implies that Y is no●
universal measure zero in the classical sense. Consequently, there exists at leas●
one Borel diffused probability measure μ on Y. From the existence of μ it easil●
follows that there exists a Borel diffused probability measure ν on the whol●
space E, for which we have $\nu^*(Y) = 1$. According to (a) and Lemma 2, the●
function $f|Y$ can be extended to a continuous function

$$f^* : Y^* \to \mathbf{R},$$

where Y^* is a Borel subset of E. We may also consider f^* as a Borel functio●
on the whole E (by putting $f^*(x) = 0$ for all points $x \in E \setminus Y^*$). In this way
we get a ν-measurable real-valued function f^* on E. Furthermore, in view o●
the obvious inclusions

$$Y \subset Y^* \subset E,$$

we infer that

$$\nu(Y^*) = \nu^*(Y) = 1$$

and, consequently, $\nu_*(Y^* \setminus Y) = 0$. It can readily be verified that

$$\{x \in E : f^*(x) \neq f(x)\} \subset (Y^* \setminus Y) \cup (E \setminus Y^*).$$

Therefore, we have

$$\nu_*(\{x \in E : f^*(x) \neq f(x)\}) = 0.$$

Applying Lemmas 3 and 4 to the σ-ideal $\mathcal{I}$ generated by the set

$$\{x \in E : f^*(x) \neq f(x)\},$$

we derive that there exists a measure $\nu' \in \mathcal{M}_1(E)$ extending ν and such that

$$\nu'(\{x \in E : f^*(x) \neq f(x)\}) = 0.$$

Finally, taking in view the circumstance that the function f^* is ν-measurable (hence, automatically, ν'-measurable), we conclude that f also turns out to be ν'-measurable. But this conclusion contradicts our assumption that f is absolutely nonmeasurable with respect to the class $\mathcal{M}_1(E)$.

$(2) \Rightarrow (1)$. Let relation (2) be satisfied. We must demonstrate that f is absolutely nonmeasurable with respect to $\mathcal{M}_1(E)$. To show this fact, we use a certain modification of the fairly standard argument (cf. the proof of Theorem 4 from Chapter 6).

First of all, we may assume without loss of generality that the function f is non-negative and bounded, e.g.,

$$0 \leq f(x) < 1 \quad (x \in E).$$

Suppose to the contrary that there exists a measure $\mu \in \mathcal{M}_1(E)$ such that f is μ-measurable. Then, for any natural number $n > 0$, the sets

$$X_k = \{x \in E : k/n \leq f(x) < (k+1)/n\} \quad (k \in \{0, 1, ..., n-1\})$$

turn out to be μ-measurable and collectively cover E. Let us associate to each set X_k a Borel set $Y_k \subset E$ such that $\mu(X_k \triangle Y_k) = 0$. Since all sets X_k are pairwise disjoint, we get

$$\mu(Y_k \cap Y_r) = 0 \quad (\{k, r\} \subset \{0, 1, ..., n-1\}, \ k \neq r).$$

By virtue of Lemma 5, these equalities imply that, for any real $\varepsilon \in \ ]0, 1[$, there are pairwise disjoint closed sets $F_k \subset E$ satisfying the relations

$$F_k \subset Y_k \quad (k \in \{0, 1, ..., n-1\}), \quad \mu(E \setminus (F_0 \cup F_1 \cup \cdots \cup F_{n-1})) < \varepsilon/2^n.$$

Denoting

$$Z_n = (F_0 \cap X_0) \cup (F_1 \cap X_1) \cup \cdots \cup (F_{n-1} \cap X_{n-1}),$$

we come to the inequality

$$\mu(E \setminus Z_n) < \varepsilon/2^n.$$

Further, we can readily define a continuous step function

$$f_n : Z_n \to [0,1]$$

such that

$$(\forall x \in Z_n)(|f(x) - f_n(x)| \leq 1/n).$$

Therefore, the sequence of continuous step functions $\{f_n : n \in \mathbf{N} \setminus \{0\}\}$ uniformly converges to f on the set

$$Z = \cap\{Z_n : n \in \mathbf{N} \setminus \{0\}\},$$

which shows that the restriction $f|Z$ is continuous, too.

In addition to the stated above, the inequality

$$\mu(Z) \geq 1 - \varepsilon$$

holds true, according to which Z is not universal measure zero (because ε was taken strictly less than 1). So we get a contradiction with our assumption that relation (2) is valid.

The contradiction obtained gives the required result and finishes the proof.

Obviously, Theorem 4 admits the following equivalent formulation:

Theorem 4'. *Let E be a topological space in which every closed subset is of type G_δ and let $f : E \to \mathbf{R}$ be a function. This f is relatively measurable with respect to the class $\mathcal{M}_1(E)$ if and only if there exists a set $X \subset E$ which is not universal measure zero with respect to $\mathcal{M}_1(E)$ and for which the restriction $f|X$ is continuous.*

Remark 4. The theorem just presented characterizes the functions absolutely nonmeasurable with respect to the class $\mathcal{M}_1(E)$ in terms of their continuous restrictions. As has already been mentioned, if E is an uncountable Polish space, then there exist Sierpiński–Zygmund type functions f acting from E into $\mathbf{R}$. Theorem 4 implies that if the Continuum Hypothesis (or Martin's Axiom) holds, then any such f is absolutely nonmeasurable with respect to the class $\mathcal{M}_1(E)$.

Thus, under certain set-theoretical assumptions, every Sierpiński–Zygmund type function on $\mathbf{R}$ turns out to be absolutely nonmeasurable with respect to the class $\mathcal{M}_1(\mathbf{R})$. In this context, it is reasonable to give here a simple example which shows that, in general, the family of all absolutely nonmeasurable functions with respect to $\mathcal{M}_1(\mathbf{R})$ is essentially wider than the family of all Sierpiński–Zygmund functions on $\mathbf{R}$.

Example 6. Assuming the Continuum Hypothesis, a function $f : \mathbf{R} \to \mathbf{R}$ can easily be constructed in such a manner that the following two relations would be satisfied:

(a) f is injective;

(b) ran(f) is a Luzin subset of $\mathbf{R}$.

Let $X \subset \mathbf{R}$ be an arbitrary uncountable universal measure zero set and let

$$g : \mathbf{R} \to \mathbf{R}$$

denote the function extending $f|(\mathbf{R} \setminus X)$ and coinciding with the identity mapping on X. By virtue of Theorem 1, this g is absolutely nonmeasurable with respect to the class $\mathcal{M}(\mathbf{R})$ (hence with respect to the class $\mathcal{M}_1(\mathbf{R})$ as well) but, at the same time, g is not a Sierpiński–Zygmund function.

Remark 5. It follows from Example 6 that (under **CH**) the family of all Sierpiński–Zygmund functions acting from $\mathbf{R}$ into $\mathbf{R}$ and the family of all functions absolutely nonmeasurable with respect to $\mathcal{M}_1(\mathbf{R})$ differ from each other. More precisely, the first family is properly contained in the second one. On the other hand, let Z be a Sierpiński subset of $\mathbf{R}$ equipped with the induced topology. Then no uncountable subset of Z is universal measure zero, so the family of all Sierpiński–Zygmund type functions acting from Z into $\mathbf{R}$ coincides with the family of all those functions acting from Z into $\mathbf{R}$ which are absolutely nonmeasurable with respect to the class $\mathcal{M}_1(Z)$.

Remark 6. Dealing with some classes of probability diffused measures on $\mathbf{R}$, which are essentially bigger than the class $\mathcal{M}_1(\mathbf{R})$, one may detect that certain Sierpiński–Zygmund functions can be measurable with respect to concrete measures from those classes. Indeed, if a Sierpiński–Zygmund function $f : \mathbf{R} \to \mathbf{R}$ is such that

$$\text{card}(f^{-1}(t_0)) > \omega$$

for a point $t_0 \in \mathbf{R}$, then, according to Theorem 1, f is not absolutely nonmeasurable with respect to the class $\mathcal{M}(\mathbf{R})$. Moreover, it can be proved that there exists a Sierpiński–Zygmund function $h : \mathbf{R} \to \mathbf{R}$ which is measurable with respect to some translation-invariant measure on $\mathbf{R}$ extending the classical Lebesgue measure λ (for a detailed explanation, see [132] or [137]).

In this context, it should also be mentioned that, by virtue of one deep result of Roslanowski and Shelah [220], there is a model of set theory in which every function $\phi : \mathbf{R} \to \mathbf{R}$ admits a continuous restriction to a subset of $\mathbf{R}$ having strictly positive outer Lebesgue measure. In that model all functions acting from $\mathbf{R}$ into $\mathbf{R}$ (including Sierpiński–Zygmund ones) turn out to be relatively measurable with respect to the class $\mathcal{M}_1(\mathbf{R})$.

EXERCISES

1. Give an example of a topological space E in which some singletons are not Borel subsets of E.

2. Check that, for an arbitrary ground set E, the following two assertions are equivalent:

(a) $\operatorname{card}(E) \geq \omega_1$;

(b) the class $\mathcal{M}(E)$ of all nonzero σ-finite diffused (i.e., continuous) measures on E is nonempty.

3. Let E be a ground set and let μ be a measure on E. Denote by $\mathcal{M}(\mu)$ the class of all those measures on E which extend μ. Let X be a subset of E.

Verify that the following two assertions are equivalent:

(a) $\mu^*(X) = 0$;

(b) X is universal measure zero with respect to $\mathcal{M}(\mu)$.

4. Let E be a topological space and let Y be a subset of E.

Check that these two assertions are equivalent:

(a) Y is universal measure zero (in the classical sense);

(b) Y is universal measure zero with respect to the class $\mathcal{M}_1(E)$.

5*. Let E be a ground set with $\operatorname{card}(E) = \mathbf{c}$ and let $\mathcal{T}$ be a family of topologies on E such that:

(a) $\operatorname{card}(\mathcal{T}) \leq \mathbf{c}$;

(b) each topology from $\mathcal{T}$ has a countable base.

Demonstrate that there exists a function $f : E \to \mathbf{R}$ totally discontinuous with respect to the family $\mathcal{T}$. This means the following:

for any set $X \subset E$ with $\operatorname{card}(X) = \operatorname{card}(E)$ and for any topology $T \in \mathcal{T}$, the restriction $f|X$ is discontinuous with respect to the induced topology $T|X$.

For this purpose, apply an argument similar to that of Sierpiński and Zygmund (see [152], [246]) and construct the required function f by using the method of transfinite recursion.

6. Prove the direct analogue of Theorem 3 assuming Martin's Axiom (instead of the Continuum Hypothesis).

For this purpose, in the corresponding argument replace Luzin sets by generalized Luzin sets which also have universal measure zero (under **MA**).

7. Complete some details in the proof of Lemma 2, namely, in the final part of the argument where it is claimed that the function f^* is continuous and extends f.

8. Give a proof of Lemma 3.

9. Let E be a topological space such that $\operatorname{card}(E)$ is not measurable in the Ulam sense.

Demonstrate that any measure from the class $\mathcal{M}_1(E)$ admits a proper extension belonging to the same class.

10. Let h be a function acting from a topological space E into a metric space E' and let the symbol $C(h)$ denote the set of all continuity points of h.

Verify that the set $C(h)$ is of type G_δ in E.

11. As was pointed out in this chapter, there is a model of **ZFC** set theory, in which every function $\phi : \mathbf{R} \to \mathbf{R}$ admits a continuous restriction to some subset of $\mathbf{R}$ having strictly positive outer Lebesgue measure.

Check that, in the above-mentioned model, all functions acting from $\mathbf{R}$ into $\mathbf{R}$ (including Sierpiński–Zygmund ones) turn out to be relatively measurable with respect to the class $\mathcal{M}_1(\mathbf{R})$.

12. Let E be a topological space satisfying the first countability axiom and let E' be a complete metric space. Suppose that a partial function

$$f : E \to E'$$

is given which is continuous on its domain.

Prove Lavrentiev's classical theorem stating that there exists a partial function

$$f^* : E \to E'$$

for which the following three relations are valid:
(a) f^* is an extension of f;
(b) the domain of f^* is of type G_δ in E;
(c) f^* is continuous on $\mathrm{dom}(f^*)$.

In order to establish the existence of f^* satisfying (a), (b) and (c), argue as in the proof of Lemma 2.

13*. Let E and E' be any two complete metric spaces and let

$$h : E \to E'$$

be a partial homeomorphism, i.e., h establishes a homeomorphism between the subspace $\mathrm{dom}(h)$ of E and the subspace $h(E)$ of E'.

Prove Lavrentiev's other classical theorem (on extensions of homeomorphisms), stating that there exists a partial function

$$h^* : E \to E'$$

for which the following three relations are valid:
(a) h^* is an extension of h;
(b) h^* is a partial homeomorphism;
(c) the set $\mathrm{dom}(h^*)$ is of type G_δ in E and the set $\mathrm{ran}(h^*)$ is of type G_δ in E'.

For this purpose, apply the result of Exercise 12 to both partial continuous functions $h : E \to E'$ and $h^{-1} : E' \to E$.

14. Let E be a ground space and let $\mathcal{M}_1$ and $\mathcal{M}_2$ be two classes of σ-finite measures on E. Denote by $\mathcal{U}_1$ (by $\mathcal{U}_2$) the class of all those subsets of E which are universal measure zero with respect to $\mathcal{M}_1$ (with respect to $\mathcal{M}_2$).

Check the validity of the implication

$$\mathcal{M}_1 \subset \mathcal{M}_2 \Rightarrow \mathcal{U}_2 \subset \mathcal{U}_1.$$

15*. Assuming the Continuum Hypothesis, prove that:

(a) the σ-ideal generated by all Luzin subsets of $\mathbf{R}$ does not possess a base of cardinality not exceeding $\mathbf{c}$;

(b) the σ-ideal generated by all Sierpiński subsets of $\mathbf{R}$ does not possess a base of cardinality not exceeding $\mathbf{c}$;

(c) the σ-ideal of all universal measure zero subsets of $\mathbf{R}$ does not possess a base of cardinality not exceeding $\mathbf{c}$.

Also, assume Martin's Axiom and establish (c) with two direct analogues of (a) and (b), respectively, for the σ-ideal generated by all generalized Luzin sets and for the σ-ideal generated by all generalized Sierpiński sets (which are defined similarly to generalized Luzin sets).

For this purpose, in each of the above cases (a), (b) and (c) suppose to the contrary that there exists a base of a σ-ideal, having cardinality not exceeding $\mathbf{c}$. Then, by using the method of transfinite recursion, construct a set belonging to the σ-ideal and not contained in any member of this base.

16. Show that, for every function $f : \mathbf{R} \to \mathbf{R}$, the following two conditions are equivalent:

(a) the graph $\mathrm{Gr}(f)$ of f is totally imperfect in the plane $\mathbf{R}^2$ (i.e., $\mathrm{Gr}(f)$ does not contain a nonempty perfect subset of $\mathbf{R}^2$);

(b) f is an absolutely nonmeasurable function with respect to the class of completions of all nonzero σ-finite diffused Borel measures on $\mathbf{R}$.

Apply the above-mentioned equivalence of (a) and (b) to Sierpiński–Zygmund functions and to the characteristic functions of Bernstein subsets of $\mathbf{R}$.

11. A relationship between absolutely nonmeasurable functions and Sierpiński–Zygmund type functions

Here we continue our study of properties of absolutely nonmeasurable real-valued functions (with respect to various classes of measures) and we are going to compare them with properties of Sierpiński–Zygmund type real-valued functions.

Let $\mathbf{R}$ denote the real line, E be a base (ground) set, and let $\mathcal{M}$ be a class of σ-finite measures on E.

Recall (see Chapter 5) that a function $f : E \to \mathbf{R}$ is absolutely (or universally) measurable with respect to the class $\mathcal{M}$ if f is measurable with respect to each measure from $\mathcal{M}$.

Accordingly, we say that a set $X \subset E$ is absolutely (or universally) measurable with respect to the class $\mathcal{M}$ if the characteristic function of X is absolutely (or universally) measurable with respect to $\mathcal{M}$.

It is not difficult to see that the following two assertions are equivalent:

(a) a function $f : E \to \mathbf{R}$ is universally measurable with respect to $\mathcal{M}$;

(b) for any point $t \in \mathbf{R}$, the set $f^{-1}(]-\infty, t])$ is universally measurable with respect to $\mathcal{M}$.

Also, a straightforward verification shows that the family of all those sets in E which are universally measurable with respect to the class $\mathcal{M}$ is a certain σ-algebra of subsets of E (depending on $\mathcal{M}$).

We also recall that f is relatively measurable with respect to the class $\mathcal{M}$ if there exists at least one measure $\mu \in \mathcal{M}$ such that f turns out to be μ-measurable.

Accordingly, we say that a set $X \subset E$ is relatively measurable with respect to $\mathcal{M}$ if the characteristic function of X is relatively measurable with respect to $\mathcal{M}$.

Finally, recall that f is absolutely nonmeasurable with respect to the class $\mathcal{M}$ if f is nonmeasurable with respect to every measure $\mu \in \mathcal{M}$.

Accordingly, we say that a set $X \subset E$ is absolutely nonmeasurable with respect to $\mathcal{M}$ if the characteristic function of X is absolutely nonmeasurable with respect to $\mathcal{M}$.

These notions were discussed in Chapter 5 and were specified for some concrete classes $\mathcal{M}$ of measures on E (see, e.g., Examples 1–5 from Chapter 5).

Also, one important particular case was considered in Chapter 10, when a ground set E is a topological space and $\mathcal{M}$ is a certain class of extensions of Borel probability measures on E. More precisely, let E be a topological space in which all singletons are Borel subsets of E. As in Chapter 10, we denote by $\mathcal{M}_1(E)$ the class of all complete diffused probability measures μ on E satisfying the following natural condition:

(*) for any set $X \in \mathrm{dom}(\mu)$, there exists a set $Y \in \mathcal{B}(E)$ such that the equality $\mu(X \triangle Y) = 0$ holds true.

Notice that condition (*) is realizable in many situations. In particular, such a situation occurs when one extends a given Borel diffused probability measure μ_0 on E by using a σ-ideal of subsets of E, all members of which have inner measure zero with respect to μ_0. The extension μ obtained in this manner trivially satisfies condition (*).

We recall that a set $Z \subset E$ is universal measure zero if, for every Borel diffused probability measure μ on E, the equality $\mu^*(Z) = 0$ is fulfilled, where μ^* denotes the outer measure associated with μ.

Example 1. Let E be a topological space and let $f : E \to \mathbf{R}$ be a function. Suppose that there exist a universal measure zero set $Z \subset E$ and a Borel function $g : E \to \mathbf{R}$ such that

$$f|(E \setminus Z) = g|(E \setminus Z).$$

One can check that, in this case, f is universally measurable with respect to the class $\mathcal{M}_1(E)$.

However, one cannot assert that, for any real-valued function $f : E \to \mathbf{R}$ universally measurable with respect to $\mathcal{M}_1(E)$, there exist a universal measure zero set $Z \subset E$ and a Borel real-valued function $h : E \to \mathbf{R}$ such that

$$f|(E \setminus Z) = h|(E \setminus Z).$$

The next example serves to confirm this fact.

Example 2. Consider the case where E is an uncountable Polish topological space and assume that all uncountable co-analytic sets in E possess the perfect subset property. It is well known that this assumption is consistent with the standard axioms of **ZFC** set theory (see, for instance, [103], [115] or Appendix 5). Let A be a non-Borel analytic subset of E and let

$$f = \chi_A : E \to \mathbf{R}$$

denote the characteristic function of A. Then, according to a well-known result of classical descriptive set theory, f is absolutely (universally) measurable with respect to the class of completions of all σ-finite Borel measures on E, so f is also universally measurable with respect to the class $\mathcal{M}_1(E)$. At the same time, it is not difficult to verify that there exist no universal measure zero set $Z \subset E$ and a Borel function $h : E \to \mathbf{R}$ for which

$$f|(E \setminus Z) = h|(E \setminus Z).$$

For a certain class of topological spaces E, a characterization of those real-valued functions on E which are absolutely nonmeasurable with respect to $\mathcal{M}_1(E)$ can be given in terms of universal measure zero subsets of E.

More precisely, we have the following statement.

Theorem 1. *Let E be a topological space in which every closed set is a G_δ-subset of E (equivalently, every open set is an F_σ-subset of E) and let $f : E \to \mathbf{R}$ be a function. These two assertions are equivalent:*

(1) f is absolutely nonmeasurable with respect to the class $\mathcal{M}_1(E)$;

(2) for any set $X \subset E$, the continuity of the restriction $f|X$ implies that X is universal measure zero.

For the proof of this theorem, see Chapter 10. Obviously, the equivalence of (1) and (2) can be formulated in another form:

A function $f : E \to \mathbf{R}$ is relatively measurable with respect to the class $\mathcal{M}_1(E)$ if and only if there exists a set $Y \subset E$ which is not universal measure zero and for which the restriction $f|Y$ is continuous.

This statement may be regarded as a weak analogue of Luzin's basic theorem concerning the C-property of all real-valued Lebesgue measurable functions.

Let E be a topological space in which all singletons are Borel subsets of E.

In our further considerations, we shall denote by the symbol $\mathcal{AN}(E)$ the family of all those real-valued functions on E which are absolutely nonmeasurable with respect to the class $\mathcal{M}_1(E)$.

We shall say that $g : E \to \mathbf{R}$ is a Sierpiński–Zygmund type function if there exists no uncountable subset Z of E such that $g|Z$ is continuous.

Below, the family of all Sierpiński–Zygmund type functions on E will be denoted by the symbol $\mathcal{SZ}(E)$.

The next statement may be regarded as a consequence of Theorem 1.

Theorem 2. *Let E be a topological space in which every closed set is a G_δ-subset of E. The following two assertions are valid:*

(1) any Sierpiński–Zygmund type function $g : E \to \mathbf{R}$ is absolutely nonmeasurable with respect to the class $\mathcal{M}_1(E)$;

(2) if there exists no uncountable universal measure zero subset of E, then any real-valued function on E absolutely nonmeasurable with respect to the class $\mathcal{M}_1(E)$ is a Sierpiński–Zygmund type function.

Proof. Let $g : E \to \mathbf{R}$ be a Sierpiński–Zygmund type function. Take any set $X \subset E$ for which the restriction $g|X$ is continuous. By virtue of the definition of

Sierpiński–Zygmund type functions, X must be countable and hence universal measure zero. We thus see that g automatically satisfies (2) of Theorem 1, which yields the absolute nonmeasurability of g with respect to the class $\mathcal{M}_1(E)$.

Assume now that there exists no uncountable universal measure zero subset of E and take an arbitrary function $f : E \to \mathbf{R}$ absolutely nonmeasurable with respect to $\mathcal{M}_1(E)$. We must show that f is a Sierpiński–Zygmund type function. Indeed, supposing otherwise, we come to an uncountable set $X \subset E$ for which the restriction $f|X$ is continuous. Since this X is not universal measure zero, we infer that (2) of Theorem 1 does not hold. Therefore, f cannot be absolutely nonmeasurable with respect to the class $\mathcal{M}_1(E)$, which contradicts the definition of f.

Theorem 2 has thus been proved.

Remark 1. Assuming the Continuum Hypothesis, the class of Sierpiński–Zygmund functions acting from $\mathbf{R}$ into itself coincides with the class $\mathcal{SZ}(\mathbf{R})$. Theorem 1 directly implies that, under **CH**, every Sierpiński–Zygmund type function on $\mathbf{R}$ is absolutely nonmeasurable with respect to the class $\mathcal{M}_1(\mathbf{R})$. On the other hand, let us mention once more that in [220] a model of set theory is presented in which every function $f : \mathbf{R} \to \mathbf{R}$ possesses a continuous restriction to some subset of $\mathbf{R}$ of strictly positive outer Lebesgue measure. Therefore, in that model, there are no absolutely nonmeasurable functions with respect to the class $\mathcal{M}_1(\mathbf{R})$.

Below, we will use the standard notions of Sierpiński sets and Luzin sets (see, e.g., [33], [147], [152], [188], [190], [203], and Chapter 4).

Recall that every Luzin set in $\mathbf{R}$ is universal measure zero. Recall also that if X is a Sierpiński set in $\mathbf{R}$, then any uncountable subset of X has strictly positive outer Lebesgue measure. Consequently, X does not contain uncountable universal measure zero subsets. Now, it follows from Theorem 2 that the family of all real-valued functions on X which are absolutely nonmeasurable with respect to the class $\mathcal{M}_1(X)$ coincides with the family of all Sierpiński–Zygmund type functions on X, i.e., the equality

$$\mathcal{AN}(X) = \mathcal{SZ}(X)$$

holds true. Taking into account these remarks, it would be interesting to find a general topological characterization of all those spaces E for which

$$\mathcal{AN}(E) = \mathcal{SZ}(E).$$

The following two examples seem to be relevant in connection with the notions introduced above.

Example 3. Let $\{X_i : i \in I\}$ be a family of topological spaces such that $\operatorname{card}(I) \leq \omega$ and let
$$\mathcal{AN}(X_i) = \mathcal{SZ}(X_i) \qquad (i \in I).$$

Assume, in addition, that every space X_i $(i \in I)$ possesses the property that any open set in X_i is an F_σ-subset of X_i. Denote by X the topological sum of the family $\{X_i : i \in I\}$. Then the equality

$$\mathcal{AN}(X) = \mathcal{SZ}(X)$$

holds, too. Indeed, it suffices to show the inclusion

$$\mathcal{AN}(X) \subset \mathcal{SZ}(X).$$

Let $f \in \mathcal{AN}(X)$ and suppose to the contrary that $f \notin \mathcal{SZ}(X)$. Then there exists an uncountable set $Y \subset X$ such that the restriction $f|Y$ is continuous. Clearly, for some index $j \in I$, the set $Y \cap X_j$ is also uncountable. Since $\mathcal{AN}(X_j) = \mathcal{SZ}(X_j)$ and $f|(Y \cap X_j)$ is continuous, we infer that $f|X_j \notin \mathcal{AN}(X_j)$. Consequently, there exists a measure $\mu \in \mathcal{M}_1(X_j)$ for which $f|X_j$ turns out to be μ-measurable. But this fact readily implies that there exists a measure $\nu \in \mathcal{M}_1(X)$ such that f is ν-measurable. So we obtain a contradiction which yields the required result.

The above example shows that the class of those spaces E which satisfy $\mathcal{AN}(E) = \mathcal{SZ}(E)$ and in which all open sets are of type F_σ is closed under taking countable topological sums. But the same class is not closed under finite topological products.

Example 4. Assume the Continuum Hypothesis. Let X be a Sierpiński subset of $\mathbf{R}$. We consider X as a subspace of $\mathbf{R}$ endowed with the Sorgenfrey topology (see [49] or Exercise 22 from Chapter 5). Since the Sorgenfrey topology is hereditarily Lindelöf, the Borel σ-algebra of X is the same as in the usual Euclidean topology. This also implies that X does not contain uncountable universal measure zero subsets. Moreover, every open set in X is of type F_σ. Therefore, according to Theorems 1 and 2, we have

$$\mathcal{AN}(X) = \mathcal{SZ}(X).$$

Let us consider the set

$$Y = 1 - X = \{1 - x : x \in X\}$$

which is also a Sierpiński subset of $\mathbf{R}$. So we may write

$$\mathcal{AN}(Y) = \mathcal{SZ}(Y).$$

Now, take the topological product $Z = X \times Y$. In this product space the set

$$D = \{(x, 1 - x) : x \in X\}$$

is uncountable, closed and discrete. By Ulam's classical theorem stating the non-real-valued-measurability of the least uncountable cardinal ω_1 (see, e.g.,

[10], [103], [203], [263] or Appendix 1), D is universal measure zero. Let L be a Luzin subset of $\mathbf{R}$ and let

$$g : Z \setminus D \to L$$

be a bijection. We define a function

$$f : Z \to \mathbf{R}$$

as follows: if $z \in D$, then $z = (x, 1 - x)$ and we put $f(z) = x$; if $z \in Z \setminus D$, then we put $f(z) = g(z)$.

It is not difficult to check that f is absolutely nonmeasurable with respect to the class $\mathcal{M}_1(Z)$, but the same f is not a Sierpiński–Zygmund type function on Z.

Remark 2. The function f of Example 4 is absolutely nonmeasurable with respect to a much wider class of measures on Z, namely, with respect to the class $\mathcal{M}(Z)$ of all nonzero σ-finite diffused measures on Z (by virtue of Theorem 2 from Chapter 5). We thus see that Example 4 also presents a somewhat stronger result.

Theorem 2 of this chapter shows that, in order to distinguish the family of all absolutely nonmeasurable functions on a space E and the family of all Sierpiński–Zygmund type functions on E, it is necessary to have an uncountable universal measure zero subset of E. Actually, this condition turns out to be sufficient as well.

Lemma 1. *Let E be a topological space, U be an uncountable universal measure zero subset of E, and let the relation $\mathcal{AN}(E) \neq \emptyset$ be satisfied. Then we have*

$$\mathcal{AN}(E) \neq \mathcal{SZ}(E).$$

Proof. Indeed, since $\mathcal{AN}(E) \neq \emptyset$, we may pick a function $h \in \mathcal{AN}(E)$. Now, define $h^* : E \to \mathbf{R}$ as follows: $h^*(x) = 0$ for all $x \in U$ and $h^*(x) = h(x)$ for all $x \in E \setminus U$.

Then it is easy to see that h^* also belongs to the class $\mathcal{AN}(E)$, but h^* is not a Sierpiński–Zygmund function.

As was pointed out in preceding sections of this book, there are several constructions (within **ZFC** set theory) of uncountable universal measure zero subsets of the real line (clearly, those constructions are applicable to any uncountable Polish space). It should be noticed, however, that the existence of a universal measure zero subset of $\mathbf{R}$ of cardinality continuum ($= \mathbf{c}$) cannot be established within **ZFC** theory. In this connection, see [188] and the references therein.

The question of the existence of an uncountable universal measure zero subspace is also interesting for an arbitrary topological space E in which all singletons are Borel subsets of E.

Let us consider the situation where a given topological space E is nonseparable but has the property that any Borel probability measure μ on E possesses a separable support, i.e., there exists a closed separable subset $F = F(\mu)$ of E such that $\mu(F) = 1$.

Lemma 2. *Assume the Continuum Hypothesis. Let E be a nonseparable topological space of cardinality $\mathbf{c}$ such that every Borel probability measure on E possesses a separable support. Then there exists a universal measure zero set $Z \subset E$ with $\mathrm{card}(Z) = \mathbf{c}$.*

The proof of this lemma can be carried out by using the standard transfinite construction of Luzin (or Sierpiński) type, applied to the σ-ideal generated by the family of all separable subsets of E (cf. Chapter 4).

Remark 3. Lemma 2 is of interest only for those topological spaces E which are not metrizable. Indeed, if E is metrizable and nonseparable, then it contains a closed discrete subset Y of cardinality ω_1. Obviously, this Y has universal measure zero (because of Ulam's theorem mentioned earlier).

The next auxiliary statement is a version of the classical Sierpiński–Zygmund theorem (cf. [152], [246]).

Lemma 3. *Let E be a topological space such that*

$$\mathrm{card}(E) = \mathrm{card}(\mathcal{B}(E)) = \mathbf{c}.$$

Then there exists a function $f : E \to \mathbf{R}$ having the following property: for every set $X \subset E$ with $\mathrm{card}(X) = \mathbf{c}$, the restriction $f|X$ is not continuous.

Proof. Denote by $\mathcal{G}$ the family of all real-valued functions g such that g is defined on a Borel subset of E, is continuous and

$$\mathrm{card}(\mathrm{dom}(g)) = \mathbf{c}.$$

In view of the assumption of the lemma, we have the equality

$$\mathrm{card}(\mathcal{G}) = \mathbf{c}.$$

So, applying the standard diagonal argument, we are able to construct a function $f : E \to \mathbf{R}$ satisfying the following condition: for any function $g \in \mathcal{G}$, the cardinality of the set

$$\{x \in \mathrm{dom}(g) : g(x) = f(x)\}$$

is strictly less than $\mathbf{c}$.

Let us check that f has the property indicated in the formulation of the lemma. For this purpose, take an arbitrary set $X \subset E$ with $\mathrm{card}(X) = \mathbf{c}$ and suppose to the contrary that $h = f|X$ is continuous. Let $\mathrm{cl}(X)$ denote the closure of X and let

$$A = \{x \in E : \mathrm{osc}_h(x) = 0\},$$

where the symbol $\mathrm{osc}_h(x)$ stands, as usual, for the oscillation of h at x. The set A is of type G_δ, so the set

$$B = \mathrm{cl}(X) \cap A$$

is Borel in E and includes X (because h is continuous on X).

Take any point $x \in B$ and consider the family of closed sets

$$\{\mathrm{cl}(h(U)) : U \text{ is a neighborhood of } x\}.$$

This family is centered and contains members with arbitrarily small diameters. Consequently, the intersection of this family is a singleton, say $\{y\}$. We define $g(x) = y$. Now, it is not difficult to check that the obtained function

$$g : B \to \mathbf{R}$$

is a continuous extension of $h = f|X$. In particular, we have

$$\mathrm{card}(\{x \in \mathrm{dom}(g) : g(x) = f(x)\}) \geq \mathrm{card}(X) = \mathbf{c},$$

which yields a contradiction and finishes the proof.

Remark 4. It directly follows from Lemma 3 that if a topological space E satisfies the condition

$$\mathrm{card}(E) = \mathrm{card}(\mathcal{B}(E)) = \mathbf{c},$$

then there are many Sierpiński–Zygmund functions on E; more precisely, the family of all such functions has cardinality $2^{\mathbf{c}}$.

Example 5. Assume the Continuum Hypothesis. Let S be a Suslin line, i.e., S is a Dedekind complete dense linearly ordered set without endpoints, satisfying the Suslin condition and not containing a countable everywhere dense subset. Then there exists a function

$$f : S \to \mathbf{R}$$

which is absolutely nonmeasurable with respect to the class $\mathcal{M}_1(S)$, but is not a Sierpiński–Zygmund type function on S. The existence of such a f is based on the following facts:

(a) the cardinality of S is equal to $\mathbf{c}$;

(b) every Borel probability measure on S possesses a separable support.

Assertion (a) is well known (see Exercise 5 from Appendix 2). For the proof of assertion (b), see Exercise 19 from Chapter 6. Further, let us check that $\mathcal{AN}(S) \neq \emptyset$. For this purpose, take a Luzin set $L \subset \mathbf{R}$ and consider any bijection

$$h : S \to L.$$

This h is absolutely nonmeasurable with respect to the class $\mathcal{M}(S)$ of all nonzero σ-finite diffused measures on S (see Theorem 2 from Chapter 5) and, consequently, h is absolutely nonmeasurable with respect to the class $\mathcal{M}_1(S)$. Now, according to Lemmas 1 and 2, there exists a function

$$f : S \to \mathbf{R}$$

which is absolutely nonmeasurable with respect to $\mathcal{M}_1(S)$ and, at the same time, is not a Sierpiński–Zygmund type function on S. Notice by the way that

$$\mathrm{card}(\mathcal{B}(S)) = \mathbf{c},$$

so there are many Sierpiński–Zygmund type functions on S (by virtue of Lemma 3). Notice also that S does not contain uncountable discrete subsets.

In Theorem 1, the implication $(1) \Rightarrow (2)$ does not need the assumption that every open set in a topological space E is an F_σ-subset of E. But the proof of the reverse implication $(2) \Rightarrow (1)$, given in Chapter 10, essentially exploits this assumption.

Below we present two examples which show that in certain situations the same equivalence $(1) \Leftrightarrow (2)$ can be true without the above-mentioned assumption.

Example 6. On the real line $\mathbf{R}$ consider the family of sets

$$\mathcal{T} = \{U \setminus D \ : \ U \text{ is open in } \mathbf{R} \text{ and } \mathrm{card}(D) \leq \omega\}.$$

As is well known, $\mathcal{T}$ is a topology on $\mathbf{R}$ much stronger than the standard Euclidean topology of $\mathbf{R}$ (see Exercise 27 from Chapter 1). The topological space $\mathbf{R}^* = (\mathbf{R}, \mathcal{T})$ has the property that

$$\mathcal{B}(\mathbf{R}^*) = \mathcal{B}(\mathbf{R}),$$

i.e., the Borel sets in $\mathbf{R}^*$ coincide with the Borel sets in $\mathbf{R}$. This circumstance implies that:

(i) the Borel real-valued functions on $\mathbf{R}^*$ are identical with the Borel real-valued functions on $\mathbf{R}$;

(ii) the universal measure zero sets in $\mathbf{R}^*$ are identical with the universal measure zero sets in $\mathbf{R}$;

(iii) $\mathcal{M}_1(\mathbf{R}^*) = \mathcal{M}_1(\mathbf{R})$.

It follows from Theorem 1 and the above-mentioned properties (i)–(iii) that, for any function $f : \mathbf{R}^* \to \mathbf{R}$, the equivalence (1) $\Leftrightarrow$ (2) of Theorem 1 holds true. On the other hand, a simple argument leads to the conclusion that not every closed subset of $\mathbf{R}^*$ is of type G_δ. For instance, the set $\mathbf{Q}$ of all rational numbers is closed in $\mathbf{R}^*$ and is not of type G_δ (this fact readily follows from the Baire category theorem applied to $\mathbf{R}$).

To present another example, consider the closed interval of ordinal numbers

$$E' = [0, \omega_1],$$

equipped with its order topology. This E' is a compact space and carries the so-called Dieudonné probability measure λ which is defined on the Borel σ-algebra of E', is diffused, two-valued, and takes values 1 on all closed uncountable (equivalently, closed unbounded) subsets of $E = [0, \omega_1[$, where E is endowed with the induced topology.

One can describe the family of all universal measure zero subsets of E. It turns out that they precisely are the so-called nonstationary sets in E (see Appendix 1).

Lemma 4. *The class of all universal measure zero subsets of the topological space $E = [0, \omega_1[$ coincides with the class of all nonstationary subsets of E.*

Proof. Let X be an arbitrary subset of E. If X is stationary, i.e., has nonempty intersection with every closed unbounded subset of E, then the outer Dieudonné measure of X is equal to 1. Therefore, by using Marczewski's standard method of extending σ-finite measures (see Exercise 18 from Chapter 5), one can extend the Dieudonné measure λ to a certain measure $\mu \in \mathcal{M}_1(E)$ such that $X \in \text{dom}(\mu)$ and $\mu(X) = 1$. So, in this case, X cannot be universal measure zero.

It remains to show that any nonstationary subset Y of E is universal measure zero. We may assume, without loss of generality, that Y is an open set in E and is representable in the form

$$Y = \cup\{Y_\xi : \xi < \omega_1\},$$

where all Y_ξ ($\xi < \omega_1$) are open, bounded, and pairwise disjoint subintervals of E. In particular, $\text{card}(Y_\xi) \leq \omega$ for each ordinal $\xi < \omega_1$.

Now, suppose to the contrary that there exists a Borel probability diffused measure μ on E such that $\mu(Y) > 0$. Then, for every set $\Xi \subset [0, \omega_1[$, the partial union $\cup\{Y_\xi : \xi \in \Xi\}$ is μ-measurable, because it is open in E. So we may define

$$\nu(\Xi) = \mu(\cup\{Y_\xi : \xi \in \Xi\}) \quad (\Xi \subset [0, \omega_1[).$$

The obtained functional ν is a nonzero finite diffused measure whose domain coincides with the family of all subsets of $[0, \omega_1[$. But this contradicts Ulam's theorem stating that ω_1 is not a real-valued measurable cardinal (see Appendix 1).

The contradiction obtained finishes the proof.

Theorem 3. *In the locally compact space $E = [0, \omega_1[$ the closed set of all limit ordinals is not of type G_δ, but the equivalence (1) $\Leftrightarrow$ (2) of Theorem 1 holds true for E.*

Proof. The first assertion of Theorem 3 is easy and probably well known. So we leave its checking to the reader.

To verify the validity of the second assertion, take any function

$$f : E \to \mathbf{R}.$$

Actually, we only have to show the validity of the implication (2) $\Rightarrow$ (1) for this f. Suppose that (2) is satisfied and suppose to the contrary that (1) is not valid. Then there exists a measure $\mu \in \mathcal{M}_1(E)$ such that f turns out to be μ-measurable. By virtue of Lemma 4, all nonstationary subsets of E are universal measure zero with respect to $\mathcal{M}_1(E)$. This implies that the restriction of μ to the Borel σ-algebra of E is identical with the Dieudonné measure λ. Since λ is two-valued and $\mu \in \mathcal{M}_1(E)$, we readily infer that μ is also two-valued. Further, a fairly standard argument shows that f is constant on some μ-measurable set Z with $\mu(Z) = 1$. Consequently, the restriction $f|Z$ turns out to be continuous, which contradicts (2).

The contradiction obtained completes the proof.

From Theorem 3 we readily get the following example.

Example 7. For the compact topological space $E' = [0, \omega_1]$, the equivalence (1) $\Leftrightarrow$ (2) of Theorem 1 holds true. The space E' may be treated as Alexandrov's one-point compactification of $E = [0, \omega_1[$. Obviously, in this E' the closed singleton $\{\omega_1\}$ is not of type G_δ.

In connection with the presented results, it would be interesting to give a full characterization of all those topological spaces E for which the equivalence (1) $\Leftrightarrow$ (2) of Theorem 1 is fulfilled.

EXERCISES

1. Let E be a base set and let $\mathcal{M}$ be a class of measures on E.

Check that, for a function $f : E \to \mathbf{R}$, the following two conditions are equivalent:

(a) $f : E \to \mathbf{R}$ is absolutely measurable with respect to $\mathcal{M}$;

(b) for any $t \in \mathbf{R}$, the set $f^{-1}(]-\infty, t])$ is absolutely measurable with respect to $\mathcal{M}$.

2. Verify the validity of the assertion formulated in Example 1.

3. Verify the validity of the assertion formulated in Example 2.

4. Let $\{X_i : i \in I\}$ be a family of topological spaces such that in every X_i ($i \in I$) any open set is representable as the union of countably many closed sets, and let X denote the topological sum of $\{X_i : i \in I\}$.

Show that in the space X any open set is also representable as the union of countably many closed sets.

5. Prove the assertion formulated in Remark 2.

6. Give a detailed proof of Lemma 2.

7. Show that if a metric space E is nonseparable, then E contains an uncountable closed discrete subset.

Also, give an example of a nonseparable Hausdorff topological space which does not contain any uncountable discrete subspace.

8. Let $E = [0, \omega_1[$ be equipped with its standard order topology and let $f : E \to \mathbf{R}$ be a continuous function.

Show that f is eventually constant, i.e., there exists an ordinal $\xi < \omega_1$ such that the restriction $f|[\xi, \omega_1[$ is constant.

9. Let E be as in the previous exercise and let a function $g : E \to \mathbf{R}$ be measurable with respect to the Dieudonné measure λ.

Verify that there exist a set $X \subset E$ such that $\lambda(X) = 1$ and the restriction $g|X$ is constant.

10. Prove the assertion formulated in Remark 4.

11*. Assume the Continuum Hypothesis and let S be as in Example 5 of this chapter.

Demonstrate that the space S does not contain any uncountable discrete subspace.

For this purpose, keep in mind the fact that the cardinality of the Borel σ-algebra $\mathcal{B}(S)$ is equal to $\mathbf{c}$.

12. Check that in the space $\mathbf{R}^*$ of Example 6 the closed set $\mathbf{Q}$ is not of type G_δ.

13*. Below, the symbol $(\mathbf{T}, +)$ stands for the one-dimensional unit torus or equivalently, the circle group, i.e.,

$$(\mathbf{T}, +) = (\mathbf{S}_1, \cdot) \subset (\mathbf{C}, \cdot),$$

where $(\mathbf{C}, \cdot)$ is the set of all complex numbers endowed with the standard multiplication operation. Both $(\mathbf{T}, +)$ and $(\mathbf{R}, +)$ are considered as commutative groups with respect to their standard addition operations.

A function $g : \mathbf{R} \to \mathbf{T}$ is called a Sierpiński–Zygmund function if the restriction of g to any subset of $\mathbf{R}$ of cardinality continuum is discontinuous.

A set $X \subset \mathbf{T}$ ($X \subset \mathbf{R}$) is a generalized Sierpiński subset of $\mathbf{T}$ (of $\mathbf{R}$) if $\operatorname{card}(X) = \mathbf{c}$ and $\operatorname{card}(X \cap Y) < \mathbf{c}$ for each first category set $Y \subset \mathbf{T}$ ($Y \subset \mathbf{R}$).

Prove that, under Martin's Axiom, there exists a homomorphism

$$g : (\mathbf{R}, +) \to (\mathbf{T}, +)$$

satisfying the following three conditions:

(a) for every set $Z \subset \mathbf{R}$ with $\operatorname{card}(Z) = \mathbf{c}$, the restriction $g|Z$ is not a Borel mapping; in particular, g is a Sierpiński–Zygmund function;

(b) the graph of g is thick in the product space $\mathbf{R} \times \mathbf{T}$;

(c) the range of g is a generalized Sierpiński subset of $\mathbf{T}$.

In the process of constructing the required g, utilize the method of transfinite recursion (cf. [137]).

Taking into account the existence of g, prove that, under Martin's Axiom, there exists a Sierpiński–Zygmund function $f : \mathbf{R} \to \mathbf{R}$ such that:

(d) f is relatively measurable with respect to the class of all translation-invariant measures on $\mathbf{R}$ extending the Lebesgue measure λ;

(e) for every set $X \subset \mathbf{R}$ of cardinality $\mathbf{c}$, the restriction of f to X is a function relatively measurable with respect to the class $\mathcal{M}(X)$.

Argue as follows. First, it is easy to construct a Borel isomorphism

$$h : \mathbf{T} \to \mathbf{R}$$

such that both h and h^{-1} preserve all Lebesgue measure zero sets, i.e., for any Lebesgue measure zero subset A of $\mathbf{T}$, the set $h(A)$ is of Lebesgue measure zero in $\mathbf{R}$ and, conversely, for any Lebesgue measure zero subset B of $\mathbf{R}$, the set $h^{-1}(B)$ is of Lebesgue measure zero in $\mathbf{T}$.

Then define $f = h \circ g$, where g is as described above.

Check that the following relations are fulfilled:

(i) for every set $Y \subset \mathbf{R}$ with $\operatorname{card}(Y) = \mathbf{c}$, the restriction $f|Y$ is not a Borel mapping;

(ii) the range of f is a generalized Sierpiński subset of $\mathbf{R}$;

(iii) f is measurable with respect to some translation-invariant measure on $\mathbf{R}$ extending the Lebesgue measure λ.

In order to show the validity of (i), suppose to the contrary that $f|Y$ is Borel measurable for some $Y \subset \mathbf{R}$ with $\operatorname{card}(Y) = \mathbf{c}$. Then, taking into account that h is a Borel isomorphism, obtain that

$$g|Y = (h^{-1} \circ f)|Y$$

is a Borel mapping, which contradicts the definition of g. Therefore, relation (i) is valid for f. In particular, f is a Sierpiński–Zygmund function.

Further, for checking that $\operatorname{ran}(g)$ is a generalized Sierpiński subset of $\mathbf{T}$, keep in mind that the Borel isomorphism h transforms all Lebesgue measure zero subsets of $\mathbf{T}$ onto all Lebesgue measure zero subsets of $\mathbf{R}$, whence it follows that the set

$$\mathrm{ran}(f) \quad (= \mathrm{ran}(h \circ g))$$

is a generalized Sierpiński subset of $\mathbf{R}$. Consequently, relation (ii) holds for f.

Since the graph of g is thick in the product space $\mathbf{R} \times \mathbf{T}$, there exists a translation-invariant extension μ of the Lebesgue measure λ such that g turns out to be measurable with respect to μ (cf. [137]). Therefore, the composition

$$h \circ g = f$$

is also measurable with respect to μ, i.e. relation (iii) is satisfied.

It remains to verify the validity of assertion (e). For this purpose, take any set $X \subset \mathbf{R}$ with $\mathrm{card}(X) = \mathbf{c}$. Only two cases are possible.

(1) $\mathrm{card}(\mathrm{ran}(f|X)) < \mathbf{c}$. In this case, there exists a point $t \in \mathrm{ran}(f|X)$ whose pre-image with respect to $f|X$ is uncountable. So $f|X$ cannot be absolutely nonmeasurable with respect to the class $\mathcal{M}(X)$.

(2) $\mathrm{card}(\mathrm{ran}(f|X)) = \mathbf{c}$. In this case, the set $\mathrm{ran}(f|X)$ is a generalized Sierpiński subset of $\mathbf{R}$. Since the outer Lebesgue measure of any generalized Sierpiński set is always strictly positive, $\mathrm{ran}(f|X)$ is not universal measure zero. Consequently, $f|X$ cannot be absolutely nonmeasurable with respect to the class $M(X)$.

Remark 5. The result established in Exercise 13 shows that, under Martin's Axiom, there exist Sierpiński–Zygmund functions on $\mathbf{R}$ all restrictions of which to the subsets of $\mathbf{R}$ of cardinality continuum are not too bad from the measure-theoretical viewpoint.

12. Sums of absolutely nonmeasurable injective functions

As was pointed out in Remark 4 of Chapter 5, the existence of absolutely nonmeasurable real-valued functions with respect to the class of all nonzero σ-finite continuous (i.e., diffused) measures on the real line $\mathbf{R}$ is not provable within **ZFC** set theory, so the help of additional set-theoretical assumptions becomes indispensable. Therefore, speaking of such functions, we should enrich **ZFC** theory by other axioms.

In the present chapter, we will exploit Martin's Axiom (see Appendix 3). Recall that, for the sake of brevity, Martin's axiom is usually denoted by **MA**.

Actually, in what follows we do not need the full power of this axiom. For our further purposes, it suffices to use some consequences of **MA**. Among those consequences, there are important statements concerning the structure of the two classical σ-ideals of subsets of $\mathbf{R}$: the σ-ideal of all Lebesgue measure zero subsets of $\mathbf{R}$ and the σ-ideal of all first category subsets of $\mathbf{R}$.

More precisely, as was shown by Martin and Solovay [181], the following two assertions are valid in **ZFC** & **MA** theory.

(M) Let μ be the completion of a σ-finite diffused Borel measure on $\mathbf{R}$ and let $\{X_i : i \in I\}$ be a family of μ-measure zero subsets of $\mathbf{R}$ such that $\mathrm{card}(I) < \mathbf{c}$; then the set $\cup\{X_i : i \in I\}$ is also of μ-measure zero (in particular, if X is a subset of $\mathbf{R}$ with $\mathrm{card}(X) < \mathbf{c}$, then $\mu(X) = 0$).

(C) Let $\{Y_i : i \in I\}$ be a family of first category subsets of $\mathbf{R}$ such that $\mathrm{card}(I) < \mathbf{c}$; then the set $\cup\{Y_i : i \in I\}$ is also of first category in $\mathbf{R}$ (in particular, if Y is a subset of $\mathbf{R}$ with $\mathrm{card}(Y) < \mathbf{c}$, then Y is of first category in $\mathbf{R}$).

In our considerations below the usage of statements (M) and (C) is completely sufficient for obtaining the main result of this chapter. Notice that in Appendix 3 it is demonstrated how both statements (M) and (C) can be deduced from Martin's Axiom.

Here we continue our discussion of various types of absolutely nonmeasurable functions. By assuming Martin's Axiom (or only the two consequences of **MA** indicated above), we are going to prove in the present chapter that every function acting from the real line $\mathbf{R}$ into itself is representable as the sum of two absolutely nonmeasurable injective functions. A similar result will be obtained for any endomorphism of the additive group $(\mathbf{R}, +)$.

It is a well-known fact that any function acting from $\mathbf{R}$ into itself is representable as a sum of two injective functions (see [165] or Sierpiński's remarkable monograph [243]). The main goal of this chapter is to show that, under Martin's Axiom, every function acting from $\mathbf{R}$ into itself can be expressed as a sum of two very bad (from the measure-theoretical viewpoint) injective functions. Fortunately, a certain algebraic version of this result is valid, too. Namely, assuming Martin's Axiom, it is proved below that every endomorphism of the additive group $\mathbf{R}$ can be written as a sum of two very bad (again, from the measure-theoretical viewpoint) injective endomorphisms of $\mathbf{R}$.

In our further consideration, we need some auxiliary notions and statements.

As usual, the symbol $\mathbf{Q}$ denotes the field of all rational numbers. This classical field is a necessary object here, because throughout the present chapter we are going to treat the real line $\mathbf{R}$ as a vector space over $\mathbf{Q}$.

If X and Y are any two subsets of $\mathbf{R}$, then $X + Y$ denotes the vector sum (or Minkowski's sum) of these subsets, i.e.,

$$X + Y = \{x + y : x \in X, \ y \in Y\}.$$

Analogously, if $X \subset \mathbf{R}$ and $T \subset \mathbf{R}$, then

$$TX = \{tx : t \in T, \ x \in X\}.$$

$\mathrm{span}_{\mathbf{Q}}(X) = $ the vector space over $\mathbf{Q}$ generated by a set $X \subset \mathbf{R}$.

Recall that a set $X \subset \mathbf{R}$ is a Luzin subset of $\mathbf{R}$ if X is uncountable and $\mathrm{card}(X \cap Y) \leq \omega$ for every first category set $Y \subset \mathbf{R}$.

Various properties of Luzin sets are discussed in [33], [147], [152], [188], [190], [203]; see also Exercises 15, 16 from Chapter 4 and Exercise 23 from Chapter 5.

Analogously, a set $X \subset \mathbf{R}$ is said to be a generalized Luzin subset of $\mathbf{R}$ if $\mathrm{card}(X) = \mathbf{c}$ and $\mathrm{card}(X \cap Y) < \mathbf{c}$ for every first category set $Y \subset \mathbf{R}$.

Let μ be a nonzero σ-finite measure defined on a σ-algebra of subsets of a given nonempty set E. As usual, we denote by $\mathrm{dom}(\mu)$ the σ-algebra of all μ-measurable sets and by $\mathcal{I}(\mu)$ the σ-ideal generated by the family of all μ-measure zero sets. The symbol μ^* stands for the outer measure associated with μ.

For $E \neq \emptyset$, we denote by $\mathcal{M}(E)$ the class of all nonzero σ-finite continuous (diffused) measures on E. Let us emphasize once more that the domains of measures from $\mathcal{M}(E)$ are, in general, various σ-algebras of subsets of E.

Any function $f : E \to \mathbf{R}$ which is absolutely nonmeasurable with respect to the class $\mathcal{M}(E)$ can be regarded as an extremely nonmeasurable real-valued function on E. In order to characterize such functions, the classical notion of a universal measure zero subset of $\mathbf{R}$ is needed.

Let $Z \subset \mathbf{R}$. We recall that Z is a universal measure zero set if, for any σ-finite continuous Borel measure μ on $\mathbf{R}$, we have $\mu^*(Z) = 0$. Equivalently, we may say that $Z \subset \mathbf{R}$ is a universal measure zero set if there exists no nonzero σ-finite continuous Borel measure on Z (where Z is assumed to be endowed with the induced topology).

Some properties of universal measure zero sets have already been discussed in preceding sections of this book (see, e.g., Exercise 11 from Chapter 6 or Lemma 4 from Chapter 11).

The following statement yields a characterization of absolutely nonmeasurable functions with respect to the class $\mathcal{M}(E)$.

Theorem 1. *For an arbitrary function*

$$f : E \to \mathbf{R},$$

these two assertions are equivalent:

(1) f is absolutely nonmeasurable with respect to $\mathcal{M}(E)$;

(2) the range of f is a universal measure zero subset of $\mathbf{R}$ and, for each point $t \in \mathbf{R}$, the set $f^{-1}(t)$ is at most countable.

The proof of the equivalence of (1) and (2) meets no difficulties and may be found in Chapter 5 of this book.

It directly follows from Theorem 1 that if a function $f : E \to \mathbf{R}$ is injective and the range of f is a universal measure zero set, then f is absolutely nonmeasurable with respect to $\mathcal{M}(E)$.

Remark 1. It is easy to see that if $\mathrm{card}(E) > \omega$, then there exist no subsets of E which are absolutely nonmeasurable with respect to the class $\mathcal{M}(E)$. Theorem 1 also implies that if $\mathrm{card}(E) > \mathbf{c}$, then there exist no functions on E which are absolutely nonmeasurable with respect to $\mathcal{M}(E)$. More precisely, the existence of an absolutely nonmeasurable function with respect to $\mathcal{M}(E)$ is equivalent to the existence of a universal measure zero set $Z \subset \mathbf{R}$ with $\mathrm{card}(Z) = \mathrm{card}(E)$. Consequently, the following two assertions are equivalent:

(a) there exists a function $f : \mathbf{R} \to \mathbf{R}$ absolutely nonmeasurable with respect to the class $\mathcal{M}(\mathbf{R})$;

(b) there exists a universal measure zero set $Z \subset \mathbf{R}$ with $\mathrm{card}(Z) = \mathbf{c}$.

Remark 2. Several classical constructions (within **ZFC** theory) of uncountable universal measure zero subsets of $\mathbf{R}$ are known. We have already mentioned in Chapter 5 that those constructions belong to Hausdorff, Luzin, Sierpiński, Marczewski, and other authors. According to them, every nonempty perfect set $P \subset \mathbf{R}$ contains an uncountable universal measure zero subset. It was also shown that there exists a model of **ZFC** theory in which the Continuum Hypothesis fails to be true and every universal measure zero subset of $\mathbf{R}$ has cardinality less than or equal to ω_1, where ω_1 stands for the least uncountable cardinal number (for more details, see [188] and references therein).

In this context, it should be recalled once more that every Luzin subset of $\mathbf{R}$ and, under Martin's Axiom, every generalized Luzin subset of $\mathbf{R}$ are universal measure zero sets in $\mathbf{R}$ (cf. Exercise 23 from Chapter 5).

Actually, the reader is already familiar with the above material, because he or she has some information about universal measure zero sets and absolutely nonmeasurable functions from preceding sections of the book. From now on, we will be dealing with those uncountable universal measure zero sets which are endowed with an additional algebraic structure.

It turns out that uncountable universal measure zero subsets of $\mathbf{R}$ can carry the structure of a vector space over the field $\mathbf{Q}$ of all rational numbers. For instance, by assuming the Continuum Hypothesis, it was shown that there exists a Luzin set which is a vector space over $\mathbf{Q}$. Analogously, under Martin's Axiom, there exists a generalized Luzin set which also is a vector space over $\mathbf{Q}$.

A more general result was obtained in [52] by Erdös, Kunen, and Mauldin. It looks as follows.

Theorem 2. *Assuming Martin's Axiom, there exist two generalized Luzin sets L_1 and L_2 in $\mathbf{R}$ such that:*

(1) both L_1 and L_2 are vector spaces over $\mathbf{Q}$;

(2) the additive group $(\mathbf{R}, +)$ is a direct sum of L_1 and L_2, i.e., the relations

$$\mathbf{R} = L_1 + L_2, \quad L_1 \cap L_2 = \{0\}$$

hold true.

For our further purposes, we need some substantially stronger version of Theorem 2. Namely, we are going to prove the following statement.

Theorem 3. *Under Martin's Axiom, there exist two generalized Luzin sets L_1 and L_2 in $\mathbf{R}$ such that:*

(1) both L_1 and L_2 are vector spaces over $\mathbf{Q}$;

(2) $\mathbf{R} = L_1 + L_2$;

(3) $\operatorname{card}(L_1 \cap L_2) = \mathbf{c}$.

Proof. As usual, we may identify $\mathbf{c}$ with the least ordinal number having the same cardinality (cf. Appendix 1 where it is indicated that in contemporary set theory such an identification is implied by von Neumann's definition of infinite cardinal numbers).

Let $\{t_\xi : \xi < \mathbf{c}\}$ be a transfinite sequence consisting of all points of $\mathbf{R}$ and let $\{F_\xi : \xi < \mathbf{c}\}$ be a transfinite sequence consisting of all nowhere dense closed subsets of $\mathbf{R}$. We are going to construct three $\mathbf{c}$-sequences

$$\{X_\xi : \xi < \mathbf{c}\}, \quad \{Y_\xi : \xi < \mathbf{c}\}, \quad \{Z_\xi : \xi < \mathbf{c}\}$$

satisfying the following conditions:

(a) if $\xi < \zeta < \mathbf{c}$, then $X_\xi \subset X_\zeta$, $Y_\xi \subset Y_\zeta$, and $Z_\xi \subset Z_\zeta$;

(b) for every ordinal $\xi < \mathbf{c}$, the sets X_ξ, Y_ξ, and Z_ξ are vector subspaces of $\mathbf{R}$ over $\mathbf{Q}$;

(c) for every ordinal $\xi < \mathbf{c}$, we have the inequalities

$$\operatorname{card}(X_\xi) \le \operatorname{card}(\xi) + w, \quad \operatorname{card}(Y_\xi) \le \operatorname{card}(\xi) + w, \quad \operatorname{card}(Z_\xi) \le \operatorname{card}(\xi) + w;$$

(d) if $\xi < \zeta < \mathbf{c}$, then

$$(X_\xi + Y_\xi) \cap F_\xi = (X_\zeta + Y_\zeta) \cap F_\xi,$$

$$(Y_\xi + Z_\xi) \cap F_\xi = (Y_\zeta + Z_\zeta) \cap F_\xi;$$

(e) $t_\xi \in X_\xi + Y_\xi + Z_\xi$.

To obtain the desired three $\mathbf{c}$-sequences, we use the method of transfinite recursion. First of all, we put

$$X_0 = Y_0 = Z_0 = \{0\}.$$

Suppose that, for an ordinal $\xi < \mathbf{c}$, the partial ξ-sequences

$$\{X_\eta : \eta < \xi\}, \quad \{Y_\eta : \eta < \xi\}, \quad \{Z_\eta : \eta < \xi\}$$

of the required vector subspaces of $\mathbf{R}$ (over $\mathbf{Q}$) have already been constructed, and put

$$X'_\xi = \cup\{X_\eta : \eta < \xi\},$$
$$Y'_\xi = \cup\{Y_\eta : \eta < \xi\},$$
$$Z'_\xi = \cup\{Z_\eta : \eta < \xi\}.$$

Obviously, X'_ξ, Y'_ξ, and Z'_ξ are vector spaces over $\mathbf{Q}$ and

$$\operatorname{card}(X'_\xi + Y'_\xi + Z'_\xi) < \operatorname{card}(\xi) + w < \mathbf{c}.$$

By virtue of Martin's Axiom (or of its consequence (C) mentioned at the beginning of this chapter), the set

$$X'_\xi + Y'_\xi + Z'_\xi + \mathbf{Q}(\cup\{F_\eta : \eta \le \xi\})$$

is of first category in $\mathbf{R}$, so there exists an element $y_\xi \in \mathbf{R}$ such that

$$y_\xi \notin X'_\xi + Y'_\xi + Z'_\xi + \mathbf{Q}(\cup\{F_\eta : \eta \le \xi\}).$$

Consequently, we have $y_\xi \notin Y'_\xi$ and

$$((\mathbf{Q} \setminus \{0\})y_\xi + Y'_\xi) \cap (\mathbf{Q}(\cup\{F_\eta : \eta \le \xi\})) = \emptyset.$$

Now, let us denote $u_\xi = t_\xi - y_\xi$ and observe that both sets

$$X'_\xi + Y'_\xi + \mathbf{Q}y_\xi + \mathbf{Q}(\cup\{F_\eta : \eta \leq \xi\}),$$

$$Y'_\xi + \mathbf{Q}y_\xi + Z'_\xi + \mathbf{Q}(\cup\{F_\eta : \eta \leq \xi\})$$

are of first category in $\mathbf{R}$. Therefore, there exist two elements $x_\xi \in \mathbf{R}$ and $z_\xi \in \mathbf{R}$ satisfying the relations

$$x_\xi + z_\xi = u_\xi,$$

$$x_\xi \notin X'_\xi + Y'_\xi + \mathbf{Q}y_\xi + \mathbf{Q}(\cup\{F_\eta : \eta \leq \xi\}),$$

$$z_\xi \notin Y'_\xi + \mathbf{Q}y_\xi + Z'_\xi + \mathbf{Q}(\cup\{F_\eta : \eta \leq \xi\}).$$

Taking this circumstance into account, let us define the following three vector spaces over $\mathbf{Q}$:

$$X_\xi = X'_\xi + \mathbf{Q}x_\xi,$$

$$Y_\xi = Y'_\xi + \mathbf{Q}y_\xi,$$

$$Z_\xi = Z'_\xi + \mathbf{Q}z_\xi.$$

Proceeding in this manner, we obtain the desired $\mathbf{c}$-sequences

$$\{X_\xi : \xi < \mathbf{c}\}, \quad \{Y_\xi : \xi < \mathbf{c}\}, \quad \{Z_\xi : \xi < \mathbf{c}\}.$$

Finally, we define

$$L_1 = \cup\{X_\xi + Y_\xi : \xi < \mathbf{c}\},$$

$$L_2 = \cup\{Y_\xi + Z_\xi : \xi < \mathbf{c}\}.$$

Now, it is not difficult to check that L_1 and L_2 are as required.

Indeed, conditions (a), (b), (c) are fulfilled by virtue of our construction, and condition (d) can be verified by transfinite induction. Actually, it suffices to show the validity of the equalities

$$(X_\eta + Y_\eta) \cap F_\xi = (X_{\eta+1} + Y_{\eta+1}) \cap F_\xi,$$

$$(Y_\eta + Z_\eta) \cap F_\xi = (Y_{\eta+1} + Z_{\eta+1}) \cap F_\xi$$

for any ordinal $\eta \in [\xi, \mathbf{c}[$. These equalities easily follow from the construction of X_ξ, Y_ξ, and Z_ξ. So both L_1 and L_2 turn out to be generalized Luzin sets in $\mathbf{R}$.

Further, relation (1) trivially holds and relation (2) is implied by the fact that

$$t_\xi = x_\xi + y_\xi + z_\xi = (x_\xi + y_\xi/2) + (y_\xi/2 + z_\xi),$$

$$x_\xi + y_\xi/2 \in L_1, \quad y_\xi/2 + z_\xi \in L_2$$

for every ordinal $\xi < \mathbf{c}$.

Relation (3) is a straightforward consequence of the injectivity of $\{y_\xi : \xi < \mathbf{c}\}$ and of the inclusion

$$\{y_\xi : \xi < \mathbf{c}\} \subset L_1 \cap L_2.$$

Theorem 3 has thus been proved.

Remark 3. Theorem 2 readily follows from Theorem 3. Indeed, suppose that there are two generalized Luzin sets $L_1 \subset \mathbf{R}$ and $L_2 \subset \mathbf{R}$ such as in Theorem 3. Denote

$$L = L_1 \cap L_2$$

and observe that L is a generalized Luzin set, too. In addition, since L is a vector subspace of L_1, there exists a complementary vector space L' for L, i.e., we have

$$L + L' = L_1, \quad L \cap L' = \{0\}.$$

Now, it is clear that $\mathbf{R}$ is a direct sum of L' and L_2. Since L_2 is a generalized Luzin set and

$$L' + L_2 = \mathbf{R},$$

we may assert (under Martin's Axiom or under its consequence (M)) that L' is necessarily of cardinality continuum and, being a subset of L_1, this L' turns out to be a generalized Luzin set as well.

Theorem 4. *Assuming Martin's Axiom, for any function*

$$f : \mathbf{R} \to \mathbf{R},$$

there exist two injective functions $f_1 : \mathbf{R} \to \mathbf{R}$ and $f_2 : \mathbf{R} \to \mathbf{R}$ which are absolutely nonmeasurable with respect to the class $\mathcal{M}(\mathbf{R})$ and for which the equality $f = f_1 + f_2$ holds true.

Proof. Let $f : \mathbf{R} \to \mathbf{R}$ be a function. We start with two generalized Luzin sets L_1 and L_2 satisfying Theorem 3. Identify again $\mathbf{c}$ with the initial ordinal number of cardinality continuum, and let $\{t_\xi : \xi < \mathbf{c}\}$ be a bijective enumeration of all points of $\mathbf{R}$. We are going to construct the required functions

$$f_1 : \mathbf{R} \to \mathbf{R}, \quad f_2 : \mathbf{R} \to \mathbf{R}$$

by means of transfinite recursion. Suppose that, for an ordinal $\xi < \mathbf{c}$, the injective partial functions

$$f_1 : \{t_\eta : \eta < \xi\} \to \mathbf{R}, \quad f_2 : \{t_\eta : \eta < \xi\} \to \mathbf{R}$$

have already been defined so that

$$\mathrm{ran}(f_1) \subset L_1, \quad \mathrm{ran}(f_2) \subset L_2,$$

$$f_1(t_\eta) + f_2(t_\eta) = f(t_\eta) \quad (\eta < \xi).$$

Let us denote
$$v = f(t_\xi), \quad U = L_1 \cap L_2.$$
In view of the equality
$$\mathbf{R} = L_1 + L_2,$$
there are two elements $l_1 \in L_1$ and $l_2 \in L_2$ such that $v = l_1 + l_2$. Evidently, we may write
$$v = (l_1 - u) + (l_2 + u) \quad (u \in U).$$
Since $\operatorname{card}(U) = \mathbf{c}$, it follows from the above relation that there exists an element $u \in U$ such that
$$l_1 - u \notin \{f_1(t_\eta) : \eta < \xi\}, \quad l_2 + u \notin \{f_2(t_\eta) : \eta < \xi\}.$$
Let us put
$$f_1(t_\xi) = l_1 - u, \quad f_2(t_\xi) = l_2 + u.$$
Then we get
$$f_1(t_\xi) \in L_1, \quad f_2(t_\xi) \in L_2, \quad f(t_\xi) = v = f_1(t_\xi) + f_2(t_\xi).$$

Proceeding in this manner, we come to the injective functions f_1 and f_2 which are defined on the whole $\mathbf{R}$, the range of f_1 is contained in L_1, the range of f_2 is contained in L_2, and $f = f_1 + f_2$. By virtue of Theorem 1, f_1 and f_2 are absolutely nonmeasurable functions with respect to the class $\mathcal{M}(\mathbf{R})$.

This completes the proof of Theorem 4.

Theorem 5. *Assume Martin's Axiom and let $n \geq 2$ be a natural number. Then, for every function*
$$f : \mathbf{R} \to \mathbf{R},$$
there exist functions
$$f_1 : \mathbf{R} \to \mathbf{R}, \quad f_2 : \mathbf{R} \to \mathbf{R}, \quad \dots, \quad f_n : \mathbf{R} \to \mathbf{R}$$
such that:

(1) each function f_i ($i \in \{1, 2, ..., n\}$) is injective and absolutely nonmeasurable with respect to the class $\mathcal{M}(\mathbf{R})$;

(2) $f = f_1 + f_2 + ... + f_n$.

Keeping in mind Theorem 4, the proof of Theorem 5 directly follows by induction on n.

By using an argument similar to the proof of Theorem 4, the next statement can be established.

Theorem 6. *Assume Martin's Axiom. Let $f : \mathbf{R} \to \mathbf{R}$ be an endomorphism of the additive group $(\mathbf{R}, +)$. Then there exist two monomorphisms $f_1 : \mathbf{R} \to \mathbf{R}$*

and $f_2 : \mathbf{R} \to \mathbf{R}$ which are absolutely nonmeasurable with respect to $\mathcal{M}(\mathbf{R})$ and satisfy the equality $f = f_1 + f_2$.

Proof. We argue analogously to the proof of Theorem 4. Let $f : \mathbf{R} \to \mathbf{R}$ be an arbitrary endomorphism of the additive group $\mathbf{R}$. Identify again $\mathbf{c}$ with the initial ordinal number of cardinality continuum, and let $\{e_\xi : \xi < \mathbf{c}\}$ denote some Hamel basis of $\mathbf{R}$.

Further, let L_1 and L_2 be again two generalized Luzin sets such as indicated in Theorem 3. We are going to define two monomorphisms

$$f_1 : \mathbf{R} \to \mathbf{R}, \quad f_2 : \mathbf{R} \to \mathbf{R}$$

by means of transfinite recursion. Suppose that, for an ordinal $\xi < \mathbf{c}$, the injective partial homomorphisms

$$f_1 : \mathrm{span}_{\mathbf{Q}}(\{e_\eta : \eta < \xi\}) \to \mathbf{R}, \quad f_2 : \mathrm{span}_{\mathbf{Q}}(\{e_\eta : \eta < \xi\}) \to \mathbf{R}$$

have already been defined so that

$$\mathrm{ran}(f_1) \subset L_1, \quad \mathrm{ran}(f_2) \subset L_2,$$

$$f_1(e_\eta) + f_2(e_\eta) = f(e_\eta) \quad (\eta < \xi).$$

Let us denote

$$v = f(e_\xi), \quad U = L_1 \cap L_2,$$

$$P_1 = f_1(\mathrm{span}_{\mathbf{Q}}(\{e_\eta : \eta < \xi\})),$$

$$P_2 = f_2(\mathrm{span}_{\mathbf{Q}}(\{e_\eta : \eta < \xi\})).$$

Clearly, we have the inequalities

$$\mathrm{card}(P_1) \le \mathrm{card}(\xi) + \omega, \quad \mathrm{card}(P_2) \le \mathrm{card}(\xi) + \omega.$$

In view of the equality

$$\mathbf{R} = L_1 + L_2,$$

there exist two elements $l_1 \in L_1$ and $l_2 \in L_2$ such that $v = l_1 + l_2$. Again, we may write

$$v = (l_1 - u) + (l_2 + u) \quad (u \in U).$$

Since $\mathrm{card}(U) = \mathbf{c}$, it follows from the above relation that there exists an element $u \in U$ such that

$$l_1 - u \notin P_1, \quad l_2 + u \notin P_2.$$

Let us put

$$f_1(e_\xi) = l_1 - u, \quad f_2(e_\xi) = l_2 + u.$$

Then we get

$$f_1(e_\xi) \in L_1, \quad f_2(e_\xi) \in L_2, \quad f(e_\xi) = v = f_1(e_\xi) + f_2(e_\xi).$$

Proceeding in this manner, we come to the injective homomorphisms f_1 and f which are defined on the whole $\mathbf{R}$, the range of f_1 is contained in L_1, the range of f_2 is contained in L_2, and $f = f_1 + f_2$. By virtue of Theorem 1, both f_1 and f_2 are absolutely nonmeasurable endomorphisms of $(\mathbf{R}, +)$.

This finishes the proof of Theorem 6.

Theorem 7. *Assume Martin's Axiom and let $n \geq 2$ be a natural number. Then, for every additive function*

$$f : \mathbf{R} \to \mathbf{R},$$

there exist functions

$$f_1 : \mathbf{R} \to \mathbf{R}, \quad f_2 : \mathbf{R} \to \mathbf{R}, \quad \ldots, \quad f_n : \mathbf{R} \to \mathbf{R}$$

such that:

(1) each function f_i ($i \in \{1, 2, ..., n\}$) is injective, additive, and absolutely nonmeasurable with respect to the class $\mathcal{M}(\mathbf{R})$;

(2) $f = f_1 + f_2 + ... + f_n$.

Taking into account Theorem 6, the proof of Theorem 7 readily follows by induction on n.

Remark 4. Let E be a set, $\mathcal{M}$ be a class of σ-finite measures on E, and let $f : E \to \mathbf{R}$ be a function absolutely measurable with respect to $\mathcal{M}$. It is easy to see that f cannot be represented as the sum of two functions, one of which is relatively measurable with respect to $\mathcal{M}$, and the other is absolutely nonmeasurable with respect to $\mathcal{M}$. In particular, no constant real-valued function on E is expressible as a sum of two functions, one of which is relatively measurable with respect to the class $\mathcal{M}(E)$, and the other is absolutely nonmeasurable with respect to $\mathcal{M}(E)$.

Remark 5. Consider any function $g : \mathbf{R} \to \mathbf{R}$ whose graph is thick in the plane $\mathbf{R}^2$, with respect to the standard two-dimensional Lebesgue measure λ_2 on $\mathbf{R}^2$. This means that the graph of g meets every λ_2-measurable subset of $\mathbf{R}^2$ with strictly positive measure. It is not difficult to show that g is relatively measurable with respect to the class $\mathcal{M}(\lambda)$ of all those measures on $\mathbf{R}$ which extend the standard Lebesgue measure λ on $\mathbf{R}$ (see, e.g., Exercise 14 from Chapter 5). Also, every function

$$f : \mathbf{R} \to \mathbf{R}$$

can be expressed as a sum

$$f = g_1 + g_2,$$

where both g_1 and g_2 have thick graphs in $\mathbf{R}^2$ with respect to λ_2 (for more details, see Exercise 5 of this chapter). In particular, one may assert that if f is absolutely nonmeasurable with respect to the class $\mathcal{M}(\mathbf{R})$ (and hence with respect to the class $\mathcal{M}(\lambda)$), then f can be expressed as a sum of two relatively measurable functions with respect to $\mathcal{M}(\lambda)$. Consequently, the sum of an absolutely nonmeasurable function with respect to $\mathcal{M}(\lambda)$ and relatively measurable function with respect to $\mathcal{M}(\lambda)$ can be a relatively measurable function with respect to $\mathcal{M}(\lambda)$.

Remark 6. The family $\mathbf{R}^{\mathbf{R}}$ of all functions acting from $\mathbf{R}$ into $\mathbf{R}$ carries the commutative group structure with respect to the standard addition operation:

$$(f + g)(x) = f(x) + g(x) \quad (x \in \mathbf{R}),$$

where $f \in \mathbf{R}^{\mathbf{R}}$ and $g \in \mathbf{R}^{\mathbf{R}}$. It is natural to ask whether there is a large subgroup of $\mathbf{R}^{\mathbf{R}}$, all elements of which (excluding the function identically equal to zero) are absolutely nonmeasurable with respect to $\mathcal{M}(\mathbf{R})$. In this direction, it will be proved in the next Chapter 13 that, under the Continuum Hypothesis, there exists a group $\mathcal{F} \subset \mathbf{R}^{\mathbf{R}}$ satisfying the relations:

(1) $\operatorname{card}(\mathcal{F}) = \mathbf{c}^+$, where $\mathbf{c}^+$ stands for the least cardinal number strictly greater than $\mathbf{c}$ (in other words, $\mathbf{c}^+$ is the successor of $\mathbf{c}$);

(2) every $f \in \mathcal{F}$ is an endomorphism of the additive group $(\mathbf{R}, +)$;

(3) every $f \in \mathcal{F} \setminus \{0\}$ is a function absolutely nonmeasurable with respect to the class $\mathcal{M}(\mathbf{R})$.

In particular, assuming $2^{\mathbf{c}} = \mathbf{c}^+$, we obtain that

$$\operatorname{card}(\mathcal{F}) = \operatorname{card}(\mathbf{R}^{\mathbf{R}}).$$

A more detailed explanation will be given in Chapter 13.

EXERCISES

1. Deduce from Theorem 2 (or from Theorem 3) that the topological product of two Luzin subsets of $\mathbf{R}$ can be a non-Luzin subset of $\mathbf{R}^2$.

To demonstrate this fact, utilize Exercise 26 of Chapter 4 and Exercise 23 of Chapter 5.

2*. Show that there exist two Bernstein sets B_1 and B_2 in $\mathbf{R}$ satisfying the following relations:

(a) B_1 is a vector space over $\mathbf{Q}$;

(b) B_2 is a vector space over $\mathbf{Q}$;

(c) $\mathbf{R}$ is a direct sum of B_1 and B_2, i.e.,

$$B_1 + B_2 = \mathbf{R}, \quad B_1 \cap B_2 = \{0\}.$$

For this purpose, try to construct the required B_1 and B_2 by using the method of transfinite recursion (cf. the proof of Theorem 3).

3. Assume Martin's Axiom and demonstrate that, for an arbitrary function

$$f : \mathbf{R} \to \;]0, +\infty[,$$

there exist two functions

$$f_1 : \mathbf{R} \to \mathbf{R}, \quad f_2 : \mathbf{R} \to \mathbf{R}$$

satisfying the following conditions:
   (a) f_1 is absolutely nonmeasurable with respect to the class $\mathcal{M}(\mathbf{R})$;
   (b) f_2 is absolutely nonmeasurable with respect to the class $\mathcal{M}(\mathbf{R})$;
   (c) $f = f_1 \cdot f_2$.

4. Let $f : \mathbf{R} \to \mathbf{R}$ be a function absolutely nonmeasurable with respect to the class $\mathcal{M}(\mathbf{R})$.
Is it true that the function $f^2 = f \cdot f$ is also absolutely nonmeasurable with respect to $\mathcal{M}(\mathbf{R})$?

5*. Prove that every function $f : \mathbf{R} \to \mathbf{R}$ can be written as $f = g_1 + g_2$, where both g_1 and g_2 are injective and have thick graphs in $\mathbf{R}^2$ with respect to the standard two-dimensional Lebesgue measure λ_2.
For this purpose, use the method of transfinite recursion and argue similarly to the proof of Theorem 3.
Conclude from the above result that if f is absolutely nonmeasurable with respect to the class $\mathcal{M}(\mathbf{R})$ (and hence with respect to the class $\mathcal{M}(\lambda)$), then f can be expressed as a sum of two relatively measurable functions with respect to $\mathcal{M}(\lambda)$.

6. Under Martin's Axiom, construct two functions

$$f : \mathbf{R} \to \mathbf{R}, \quad g : \mathbf{R} \to \mathbf{R}$$

satisfying the following conditions:
   (a) f is absolutely nonmeasurable with respect to the class $\mathcal{M}(\mathbf{R})$;
   (b) g is relatively measurable with respect to the class $\mathcal{M}(\lambda)$;
   (c) the sum $f + g$ is a relatively measurable function with respect to the class $\mathcal{M}(\lambda)$.
For this purpose, utilize the result of the previous exercise.

7. Under Martin's Axiom, construct (by transfinite recursion) two Sierpiński subsets X and Y of $\mathbf{R}$ satisfying the following relations:
   (a) both X and Y are vector spaces over the field $\mathbf{Q}$;
   (b) $X + Y = \mathbf{R}$;
   (c) $\mathrm{card}(X \cap Y) = \mathbf{c}$.

Deduce from this fact that, under the same assumption, there exist two Sierpiński subsets X' and Y' of $\mathbf{R}$ such that both of them are vector spaces over $\mathbf{Q}$, and $\mathbf{R}$ is a direct sum of X' and Y', i.e., the equalities

$$\mathbf{R} = X' + Y', \quad X' \cap Y' = \{0\}$$

hold true.

To obtain the required result, argue analogously to the proof of Theorem 3.

Conclude that the topological product of two Sierpiński subsets of $\mathbf{R}$ can be a non-Sierpiński subset of $\mathbf{R}^2$.

8. Let X be a subset of the Euclidean space $\mathbf{R}^n$, where $n \geq 1$. This X is called a strong measure zero set if, for any sequence $\{\varepsilon_k : k < \omega\}$ of strictly positive real numbers, there exists a sequence $\{\Delta_k : k < \omega\}$ of n-dimensional cubes in $\mathbf{R}^n$ which collectively cover X and $\lambda_n(\Delta_k) < \varepsilon_k$ for each $k < \omega$ (see, e.g., [133], [188]).

Prove that:

(a) every strong measure zero set is universal measure zero;

(b) every Luzin set in $\mathbf{R}^n$ has strong measure zero;

(c) the family of all strong measure zero subsets of $\mathbf{R}^n$ forms a proper σ-ideal of sets in $\mathbf{R}^n$;

(d) if $f : \mathbf{R}^n \to \mathbf{R}^m$ is a continuous mapping and X is a strong measure zero set in $\mathbf{R}^n$, then the set $f(X)$ is a strong measure zero set in $\mathbf{R}^m$.

In order to establish (a), first prove the following auxiliary proposition.

Let μ be a finite diffused Borel measure on a compact cube $[a, b]^n \subset \mathbf{R}^n$. Then, for every $\varepsilon > 0$, there exists $\delta > 0$ such that $\mu(\Delta) < \varepsilon$ whenever Δ is any n-dimensional cube in $[a, b]^n$ with diameter strictly less than δ.

9*. Assuming the Continuum Hypothesis, show that the product of two strong measure zero subsets of $\mathbf{R}$ can be a non-strong measure zero set in the Euclidean plane $\mathbf{R}^2$.

For this purpose, utilize Theorem 2 and the result of Exercise 8.

10*. Demonstrate that, under Martin's Axiom, there exists a Hamel basis in $\mathbf{R}$ which simultaneously is a generalized Luzin set (consequently, is a universal measure zero set).

For this purpose, use Theorem 2 of the present chapter and an argument similar to that given in Exercise 17 of Chapter 3.

Likewise, supposing that the Continuum Hypothesis holds, demonstrate that there exists a Hamel basis in $\mathbf{R}$ which is a Luzin subset of $\mathbf{R}$.

11*. Assuming Martin's Axiom, show that:

(a) there exists a universal measure zero set $X \subset \mathbf{R}$ and an injective continuous mapping

$$f : X \to \mathbf{R}$$

such that no subset of $f(X)$ having cardinality $\mathbf{c}$ is universal measure zero;

(b) the above-mentioned set X contains a subset Y such that the set $f(Y)$ i absolutely nonmeasurable with respect to the class of completions of all nonzero σ-finite diffused Borel measures on $f(X)$.

In order to establish the validity of (a) and (b), keep in mind the result o Exercise 10 with the fact that the topological product of any finite family o universal measure zero spaces is again universal measure zero.

On the other hand, suppose that X and Y are two topological spaces, Y i universal measure zero and $f : X \to Y$ is an injective Borel mapping.

Check that X is also universal measure zero.

12. Demonstrate that if Martin's Axiom holds, then there exists a Hame basis in $\mathbf{R}$ which simultaneously is a generalized Sierpiński set.

For this purpose, use Exercise 7 of the present chapter and an argument analogous to that given in Exercise 17 of Chapter 3.

13. A large group of absolutely nonmeasurable additive functions

The set $\mathbf{R}^{\mathbf{R}}$ of all real-valued functions defined on $\mathbf{R}$ carries several canonical mathematical structures. One of them is a natural commutative group structure on $\mathbf{R}^{\mathbf{R}}$. Namely, for any two functions $f \in \mathbf{R}^{\mathbf{R}}$ and $g \in \mathbf{R}^{\mathbf{R}}$, we have by definition

$$(f + g)(x) = f(x) + g(x) \qquad (x \in \mathbf{R}).$$

In this chapter it is proved, by assuming the Continuum Hypothesis (**CH**), that there exists a subgroup of $\mathbf{R}^{\mathbf{R}}$ whose cardinality is strictly greater than **c** and all nonzero members of which are additive functions absolutely nonmeasurable with respect to the class $\mathcal{M}(\mathbf{R})$ of all nonzero σ-finite diffused (i.e., continuous) measures on $\mathbf{R}$.

This result seems to be of interest in light of the theorem presented in Chapter 12 and stating that any function from $\mathbf{R}^{\mathbf{R}}$ is expressible as a sum of two absolutely nonmeasurable injective functions.

The existence of Lebesgue nonmeasurable sets in $\mathbf{R}$ and, accordingly, the existence of Lebesgue nonmeasurable real-valued functions on $\mathbf{R}$ are very important facts of mathematical analysis and play a seminal role for the foundations of contemporary mathematics. Moreover, as has already been mentioned in preceding sections of the book, these facts are closely connected with uncountable forms of the Axiom of Choice. In various topics and questions of real analysis, nonmeasurable sets or nonmeasurable functions turn out to be endowed with additional algebraic structure. For instance, one may be required to have a Lebesgue nonmeasurable subgroup of the additive group $(\mathbf{R}, +)$ or a Lebesgue nonmeasurable homomorphism of $(\mathbf{R}, +)$ into itself. As a rule, one needs more delicate constructions for proving the existence of such Lebesgue nonmeasurable algebraic objects. The following example illustrates this circumstance.

Example 1. As usual, denote by λ the standard Lebesgue measure on the real line $\mathbf{R}$. As is well known (see, e.g., [17], [33], [77], [190], [203]), $\mathbf{R}$ admits a partition $\{A, B\}$ such that the set A is of λ-measure zero and the set B is of first category in $\mathbf{R}$. This circumstance easily implies that there exists a subset B' of B which is not λ-measurable. Consequently, B' is of first category, but is not Lebesgue measurable. The analogous question can be posed for subgroups of the additive group $(\mathbf{R}, +)$. Namely, one may ask whether there exists a subgroup of $\mathbf{R}$ which is of first category, but is not λ-measurable. The answer is positive under the Continuum Hypothesis (or under Martin's Axiom which is much weaker than $\mathbf{CH}$), but the corresponding technique is more complicated and needs the method of transfinite induction. Moreover, assuming Martin's Axiom, there exists a subset X of $\mathbf{R}$ that is a generalized Sierpiński set and, simultaneously, a vector space over the field $\mathbf{Q}$ of all rational numbers. Obviously, this X is of first category in $\mathbf{R}$, but every subset of X of cardinality continuum is not λ-measurable. Extensive information about Sierpiński sets and generalized Sierpiński sets may be found in [33], [147], [152], [188], [190], [203] (see also Chapter 4 of the present book).

Example 2. The dual question naturally arises for subgroups of $(\mathbf{R}, +)$, namely, whether there exists a subgroup of $\mathbf{R}$ which is of λ-measure zero, but does not possess the Baire property. Again, the answer is positive under the Continuum Hypothesis (or under Martin's Axiom). Indeed, as we already know, assuming Martin's Axiom, there exists a subset Y of $\mathbf{R}$ that is a generalized Luzin set and, simultaneously, a vector space over the field $\mathbf{Q}$ of all rational numbers (see, e.g., Chapter 12). Clearly, this Y is universal measure zero (hence is of λ-measure zero) and every subset of Y having cardinality continuum does not possess the Baire property.

A number of works are devoted to algebraic properties of various families of measurable and nonmeasurable functions (see, e.g., [13], [37], [73], [74], [75], [84], [85], [129], [131], [140], [147], [198], [218]).

As has already been announced, in this chapter we are going to construct, with the aid of the Continuum Hypothesis, a family $\mathcal{F} \subset \mathbf{R}^{\mathbf{R}}$ of functions such that $\mathrm{card}(\mathcal{F})$ is strictly greater than the cardinality of the continuum, $\mathcal{F}$ itself is a group with respect to the standard addition operation, and each nonzero member from $\mathcal{F}$ is an absolutely nonmeasurable additive function.

For our further purposes, we need some auxiliary notions and statements.

Let μ be a nonzero measure defined on a σ-algebra of subsets of a nonempty set E. As usual, we denote by the symbol $\mathrm{dom}(\mu)$ the domain of μ (i.e., the σ-algebra of all μ-measurable sets) and by the symbol $\mathcal{I}(\mu)$ the σ-ideal generated by the family of all μ-measure zero sets.

Let $\mathcal{M}$ be some class of measures on E (in general, their domains are diverse σ-algebras of subsets of E) and let $f : E \to \mathbf{R}$ be a function.

Recall that f is absolutely nonmeasurable with respect to the class $\mathcal{M}$ if there exists no measure from $\mathcal{M}$ for which f turns out to be measurable.

Accordingly, we say that a set $X \subset E$ is absolutely nonmeasurable with respect to the class $\mathcal{M}$ if the characteristic function (i.e., indicator) of X is absolutely nonmeasurable with respect to $\mathcal{M}$.

As in preceding sections of this book, for a nonempty ground (base) set E, we denote by $\mathcal{M}(E)$ the class of all nonzero σ-finite continuous measures on E.

(let us underline once more that their domains may be various σ-algebras of subsets of E).

Of course, any function $f : E \to \mathbf{R}$ that is absolutely nonmeasurable with respect to the class $\mathcal{M}(E)$ may be regarded as an utterly nonmeasurable real-valued function on E. A useful description of such functions can be given with the aid of universal measure zero subsets of $\mathbf{R}$.

Let $Z \subset \mathbf{R}$. We recall that Z has universal measure zero if for any σ-finite continuous Borel measure μ on $\mathbf{R}$, the equality $\mu^*(Z) = 0$ holds, where μ^* denotes the outer measure associated with μ. Equivalently, we may say that $Z \subset \mathbf{R}$ has universal measure zero if there exists no nonzero σ-finite continuous Borel measure on Z (where Z is assumed to be endowed with the induced topology).

Some important properties of universal measure zero sets were discussed in preceding sections of this book (see, e.g., Chapter 5).

Here we are going to exploit once more the following auxiliary proposition which yields a characterization of absolutely nonmeasurable functions with respect to the class $\mathcal{M}(E)$.

Lemma 1. *For any function $f : E \to \mathbf{R}$, these two assertions are equivalent:*
(1) f is absolutely nonmeasurable with respect to $\mathcal{M}(E)$;
(2) the range of f is a universal measure zero subset of $\mathbf{R}$ and, for each point $t \in \mathbf{R}$, the set $f^{-1}(t)$ is at most countable.

The proof of this lemma is not difficult and was presented in Chapter 5 (see Theorem 2 of that chapter).

It is worth noticing that some uncountable universal measure zero subsets of $\mathbf{R}$ can carry a certain algebraic structure. The following auxiliary proposition will be useful for our purposes.

Lemma 2. *There exists (within $\mathbf{ZFC}$ theory) an uncountable universal measure zero set $Z \subset \mathbf{R}$ which simultaneously is a vector space over the field $\mathbf{Q}$ of all rational numbers.*

This lemma is well known (see, e.g., [208] where a much deeper result is presented).

Remark 1. Another way to obtain Lemma 2 is to deduce it from one general statement of metamathematical character. We would like to formulate this statement as a metatheorem.

Let $S(X)$ be a property of a subset X of a Polish space. Suppose that the following conditions are satisfied:
(a) if $S(X)$ and $Y \subset X$, then $S(Y)$;
(b) if $\{X_i : i \in I\}$ is a countable family of subsets of a Polish space and $S(X_i)$ for all $i \in I$, then $S(\cup\{X_i : i \in I\})$;
(c) if $S(X)$ and $S(Y)$, then $S(X \times Y)$;

(d) if B_1 and B_2 are two Borel subsets of Polish spaces, $h : B_1 \rightarrow B_2$ is an injective Borel mapping and X is a subset of B_1 with $S(X)$, then $S(h(X))$;

(e) there exists at least one Polish space containing an uncountable set X such that $S(X)$.

Then there exists an uncountable vector space $Z \subset \mathbf{R}$ (over the field $\mathbf{Q}$) such that $S(Z)$.

Let us sketch the proof of this metatheorem. As is well known, there exists a nonempty perfect set $P \subset \mathbf{R}$ that is linearly independent over $\mathbf{Q}$ (see, e.g., [115], [190], [193], [268]). The conditions (d) and (e) imply that there exists an uncountable set $X \subset P$ satisfying $S(X)$. Clearly, X is also linearly independent over $\mathbf{Q}$. The conditions (b), (c) and (d) imply that, for the set

$$Y = \{(q_1 x_1, ..., q_n x_n) : 0 < n < \omega, (q_1, ..., q_n) \in (\mathbf{Q} \setminus \{0\})^n, (x_1, ..., x_n) \in X^n\},$$

the property $S(Y)$ holds true. Consider an arbitrary element

$$(q_1 x_1, ..., q_n x_n) \in Y,$$

where $n > 0$ is a natural number, $(q_1, ..., q_n) \in (\mathbf{Q} \setminus \{0\})^n$ and $(x_1, ..., x_n) \in X^n$.

We shall say that this element is admissible if $x_i < x_j$ for any two natural indices $i \in [1, n]$ and $j \in [1, n]$ such that $i < j$.

Let Y' denote the set of all admissible elements of Y. In view of condition (a), we have $S(Y')$. Now, for every natural number $n > 0$, define the set T_n by the equality

$$T_n = \{(x_1, ..., x_n) \in P^n : (\forall i \in [1, n])(\forall j \in [1, n])(i < j \Rightarrow x_i < x_j)\}$$

and, in addition to this, define the set

$$T_n' = \{(q_1 x_1, ..., q_n x_n) : (x_1, ..., x_n) \in T_n, (q_1, ..., q_n) \in (\mathbf{Q} \setminus \{0\})^n\}.$$

Obviously, both T_n and T_n' are Borel subsets of the Euclidean space $\mathbf{R}^n$. Since all spaces $\mathbf{R}^n$ ($n < \omega$) are canonically embedded in the space $\mathbf{R}^\omega$, the set

$$T' = \cup\{T_n' : 0 < n < \omega\}$$

is a Borel subset of $\mathbf{R}^\omega$, and

$$Y' \subset \cup\{T_n' : 0 < n < \omega\} = T'.$$

Further, consider the Borel mapping

$$g : T' \rightarrow \mathbf{R}$$

given by the formula

$$g(y_1, ..., y_n) = y_1 + ... + y_n, \qquad (0 < n < \omega, (y_1, ..., y_n) \in T_n').$$

By virtue of the linear independence of P over $\mathbf{Q}$, this mapping g has the property that, for any point $t \in \mathbf{R}$, the set $g^{-1}(t)$ is either empty or one-element. In other words, g is an injective Borel mapping and hence is a Borel isomorphism between the sets $\mathrm{dom}(g)$ and $\mathrm{ran}(g)$ (see Appendix 5). Now, it can easily be checked that the set

$$Z = g(Y') \cup \{0\}$$

is an uncountable vector space over $\mathbf{Q}$ and the relation $S(Z)$ holds true.

The next auxiliary proposition belongs to infinite combinatorics and states the existence of a quite large almost disjoint family of subsets of a given infinite set. This proposition is crucial for obtaining the main result of the chapter.

Lemma 3. *Let Ξ be an infinite set. There exists a family $\{\Xi_j : j \in J\}$ of subsets of Ξ such that:*
(1) $\mathrm{card}(J) > \mathrm{card}(\Xi)$;
(2) $\mathrm{card}(\Xi_j) = \mathrm{card}(\Xi)$ *for each index $j \in J$;*
(3) $\mathrm{card}(\Xi_j \cap \Xi_{j'}) < \mathrm{card}(\Xi)$ *for any two distinct indices $j \in J$ and $j' \in J$.*

We omit the proof of this lemma. It is based on a fairly standard argument by transfinite recursion, and we refer the reader to the classical monograph [243] by Sierpiński, where Lemma 3 is proved in detail (see also Exercise 3 from Chapter 8 where a much stronger result is presented).

Let $(V, +)$ be a vector space (over some field of scalars) and let $\{V_j : j \in J\}$ be a family of vector subspaces of V.

We shall say that this family is admissible if, for any finite sequence

$$(j_0, j_1, j_2, \ldots, j_k) \qquad (0 \leq k < \omega)$$

of distinct indices from J, the relation

$$\mathrm{card}(V_{j_0} \cap (V_{j_1} + V_{j_2} + \ldots + V_{j_k})) < \mathrm{card}(V)$$

holds true.

The next proposition guarantees the existence of a large admissible family of vector subspaces of an uncountable vector space (over $\mathbf{Q}$).

Lemma 4. *Let V be an uncountable vector space over $\mathbf{Q}$. There exists an admissible family $\{V_j : j \in J\}$ of vector subspaces of V such that:*
(1) $\mathrm{card}(J) > \mathrm{card}(V)$;
(2) $\mathrm{card}(V_j) = \mathrm{card}(V)$ *for all $j \in J$.*

Proof. Let Ξ be a basis of V. Clearly, we have the equality

$$\mathrm{card}(\Xi) = \mathrm{card}(V).$$

Let $\{\Xi_j : j \in J\}$ be a family of subsets of Ξ satisfying the relations (1)–(3) of Lemma 3. For each index $j \in J$, let us put

$$V_j = \mathrm{span}_{\mathbf{Q}}(\Xi_j).$$

In other words, V_j denotes the vector space over $\mathbf{Q}$ generated by Ξ_j.

It is not difficult to check that the obtained family $\{V_j : j \in J\}$ of vector subspaces of V is as required.

Now, we are ready to establish the following statement.

Theorem 1. *Under the Continuum Hypothesis, there exists a family $\mathcal{F}$ of functions from $\mathbf{R}^{\mathbf{R}}$ satisfying the following relations:*

(1) $\mathrm{card}(\mathcal{F}) > \mathbf{c}$;

(2) $\mathcal{F}$ *is a vector space over $\mathbf{Q}$;*

(3) all functions from $\mathcal{F}$ are homomorphisms of the additive group $(\mathbf{R}, +)$ into itself;

(4) all nonzero functions from $\mathcal{F}$ are absolutely nonmeasurable with respect to the class $\mathcal{M}(\mathbf{R})$.

Proof. Let V be an uncountable universal measure zero set in $\mathbf{R}$ which simultaneously is a vector space over $\mathbf{Q}$. As was already mentioned, such a V does exist (see Lemma 2).

Let $\{V_j : j \in J\}$ be an admissible family of vector subspaces of V satisfying relations (1) and (2) of Lemma 4.

Consider $\mathbf{R}$ as a vector space over $\mathbf{Q}$. In view of the supposed equality $\mathbf{c} = \omega_1$, the vector space $\mathbf{R}$ is isomorphic to each vector space from the family $\{V_j : j \in J\}$.

For any index $j \in J$, denote by $f_j : \mathbf{R} \to V_j$ some isomorphism between the two vector spaces $\mathbf{R}$ and V_j.

Notice that the family of functions $\{f_j : j \in J\}$ is linearly independent over $\mathbf{Q}$. Indeed, consider any linear (over $\mathbf{Q}$) combination

$$q_0 f_{j_0} + q_1 f_{j_1} + \ldots + q_k f_{j_k},$$

where $k \geq 1$, all coefficients $q_0, q_1, \ldots, q_k$ are nonzero, and $j_0, j_1, \ldots, j_k$ are distinct indices from J. Since we have

$$\mathrm{card}(V_{j_0}) = \mathbf{c} = \omega_1, \quad \mathrm{card}(V_{j_0} \cap (V_{j_1} + \ldots + V_{j_k})) \leq \omega,$$

there exists an element

$$y \in V_{j_0} \setminus (V_{j_1} + \ldots + V_{j_k}).$$

Since $f_{j_0}(\mathbf{R}) = V_{j_0}$, we can find an element $x \in \mathbf{R}$ such that $y = f_{j_0}(x)$. Now, it readily follows that

$$(q_0 f_{j_0} + q_1 f_{j_1} + \ldots + q_k f_{j_k})(x) \neq 0,$$

so $q_0 f_{j_0} + q_1 f_{j_1} + \ldots + q_k f_{j_k}$ is not identically equal to zero.

Further, we put

$$\mathcal{F} = \operatorname{span}_\mathbf{Q}\{f_j : j \in J\}$$

and we claim that $\mathcal{F}$ is the required family of functions.

Notice that the relations (1), (2), and (3) of the theorem trivially hold by virtue of the definition of $\mathcal{F}$. So it remains to verify the validity of relation (4).

In view of Lemma 1, we must check that if f is an arbitrary function from $\mathcal{F} \setminus \{0\}$, then the set $\operatorname{ran}(f)$ is universal measure zero and the set $f^{-1}(t)$ is at most countable for every point $t \in \mathbf{R}$.

First, observe that if $f \in \mathcal{F}$, then $\operatorname{ran}(f) \subset V$, hence $\operatorname{ran}(f)$ is indeed universal measure zero.

Further, if $f \in \mathcal{F} \setminus \{0\}$, then f admits a unique representation in the form

$$f = q_0 f_{j_0} + q_1 f_{j_1} + \ldots + q_k f_{j_k},$$

where k is a natural number, $j_0, j_1, j_2, \ldots, j_k$ are distinct indices from J, and $q_0, q_1, \ldots, q_k$ are nonzero rational numbers.

It suffices to demonstrate that, for any point $t \in V$, the set $f^{-1}(t)$ is at most countable. We will show this fact by induction on k.

If $k = 0$, then $f = q_0 f_{j_0}$, where $q_0 \neq 0$. In this case f is an isomorphism between $\mathbf{R}$ and V_{j_0}, and for each $t \in V$ the set $f^{-1}(t)$ either is empty or is a singleton.

Assume that our assertion has already been proved for natural numbers strictly smaller than k and suppose to the contrary that, for a function f represented in the above-mentioned form, there exists $\tau \in V$ such that

$$\operatorname{card}(f^{-1}(\tau)) \geq \omega_1.$$

It is not difficult to see that in this case the set

$$\{x \in \mathbf{R} : q_0 f_{j_0}(x) = (-q_1 f_{j_1} - q_2 f_{j_2} - \ldots - q_k f_{j_k})(x)\}$$

is uncountable. Since the Continuum Hypothesis is assumed and the family of vector spaces $\{V_j : j \in J\}$ is admissible, the vector space

$$V_{j_0} \cap (V_{j_1} + V_{j_2} + \ldots + V_{j_k})$$

must be at most countable. But we obviously have

$$\operatorname{ran}(q_0 f_{j_0}) \subset V_{j_0},$$

$$\operatorname{ran}(-q_1 f_{j_1} - q_2 f_{j_2} - \ldots - q_k f_{j_k}) \subset V_{j_1} + V_{j_2} + \ldots + V_{j_k}.$$

These two inclusions readily imply that there exists a point

$$t' \in V_{j_0} \cap (V_{j_1} + V_{j_2} + \ldots + V_{j_k})$$

such that
$$\mathrm{card}((-q_1 f_{j_1} - q_2 f_{j_2} - \ldots - q_k f_{j_k})^{-1}(t')) \geq \omega_1,$$
which contradicts the inductive assumption on $q_1 f_{j_1} + q_2 f_{j_2} + \ldots + q_k f_{j_k}$.

The obtained contradiction finishes the proof of Theorem 1.

Remark 2. Denote by $\mathbf{c}^+$ the least cardinal number strictly greater than $\mathbf{c}$. As a straightforward consequence of Theorem 1 we obtain that if $2^{\mathbf{c}} = \mathbf{c}^+$ then $\mathrm{card}(\mathcal{F}) = \mathrm{card}(\mathbf{R}^{\mathbf{R}})$.

Remark 3. Let K be an uncountable subfield of $\mathbf{R}$ and consider $\mathbf{R}$ as a vector space over K. Then there does not exist a vector space $\mathcal{G} \subset \mathbf{R}^{\mathbf{R}}$ over K such that:

(1) $\mathrm{card}(\mathcal{G}) > \mathbf{c}$;
(2) all $g \in \mathcal{G}$ are K-linear homomorphisms of $\mathbf{R}$ into itself;
(3) all $g \in \mathcal{G} \setminus \{0\}$ are absolutely nonmeasurable functions with respect to the class $\mathcal{M}(\mathbf{R})$.

Indeed, suppose to the contrary that such a $\mathcal{G}$ does exist and take any point $x \in \mathbf{R} \setminus \{0\}$. Obviously, there are two distinct elements $g \in \mathcal{G}$ and $h \in \mathcal{G}$, for which we have $g(x) = h(x)$ and, consequently, $(g-h)(x) = 0$. This fact directly implies that
$$(g-h)(yx) = y((g-h)(x)) = 0$$
for each element $y \in K$, i.e. the set $(g-h)^{-1}(0)$ is uncountable, which contradicts the absolute nonmeasurability of $g-h$ (see Lemma 1).

As was mentioned earlier, the existence of at least one function from $\mathbf{R}^{\mathbf{R}}$ that is absolutely nonmeasurable with respect to the class $\mathcal{M}(\mathbf{R})$ necessarily needs additional set-theoretical hypotheses (see Remark 4 from Chapter 5). For the class $\mathcal{MQI}(\mathbf{R})$ of all nonzero σ-finite translation quasi-invariant measures on $\mathbf{R}$, a certain analogue of Theorem 1 can be established within **ZFC** theory. Let us formulate and prove this analogue. The reader will see that its proof is rather similar to the proof of Theorem 1.

For this purpose, we need one more auxiliary proposition.

Lemma 5. *Let $f : \mathbf{R} \to \mathbf{R}$ be a function satisfying the following two conditions:*

(1) f is additive;
(2) the range of f is an uncountable universal measure zero set.

Then f turns out to be absolutely nonmeasurable with respect to the class $\mathcal{MQI}(\mathbf{R})$.

The proof of Lemma 5 is not difficult, so we omit it here and leave to the reader (see Exercise 7 for this chapter).

Theorem 2. *There exists a family $\mathcal{F} \subset \mathbf{R}^{\mathbf{R}}$ satisfying the following relations:*

(1) $\text{card}(\mathcal{F}) > \omega_1$;

(2) $\mathcal{F}$ is a vector space over $\mathbf{Q}$;

(3) all functions from $\mathcal{F}$ are homomorphisms of the additive group $(\mathbf{R}, +)$ into itself;

(4) all nonzero functions from $\mathcal{F}$ are absolutely nonmeasurable with respect to the class $\mathcal{MQI}(\mathbf{R})$.

In particular, if $2^{\mathbf{c}} = \omega_2$, then $\text{card}(\mathcal{F}) = \text{card}(\mathbf{R}^{\mathbf{R}})$.

Proof. As in the proof of Theorem 1, we again start with an uncountable universal measure zero set V in $\mathbf{R}$ which simultaneously is a vector space over the field $\mathbf{Q}$ of all rational numbers. As was already mentioned, such a V does exist within **ZFC** set theory (see Lemma 2 or Remark 1) and we may assume, without loss of generality, that $\text{card}(V) = \omega_1$.

Treating the real line $\mathbf{R}$ also as a vector space over $\mathbf{Q}$, we can represent $\mathbf{R}$ in the form of a direct sum

$$\mathbf{R} = V + V' \qquad (V \cap V' = \{0\}),$$

where V' is some vector space over the same field $\mathbf{Q}$.

Let $\{V_j : j \in J\}$ be an admissible family of vector subspaces of V satisfying relations (1) and (2) of Lemma 4. Since $\text{card}(V) = \omega_1$, we have the inequality $\text{card}(J) > \omega_1$.

In our situation we cannot assert that the vector space $\mathbf{R}$ (over $\mathbf{Q}$) is isomorphic to each vector space from the family $\{V_j : j \in J\}$, because, in general, the cardinality continuum $\mathbf{c}$ may be strictly greater than ω_1. However, for any index $j \in J$, we are able to introduce some isomorphism

$$g_j : V \to V_j$$

between V and V_j, because of the equalities

$$\text{card}(V) = \text{card}(V_j) \qquad (j \in J).$$

Similarly to the previous case, the family of additive functions $\{g_j : j \in J\}$ is linearly independent over $\mathbf{Q}$. Indeed, consider any linear (over $\mathbf{Q}$) combination

$$q_0 g_{j_0} + q_1 g_{j_1} + \ldots + q_k g_{j_k},$$

where $k \geq 1$, all coefficients $q_0, q_1, \ldots, q_k$ are nonzero, and $j_0, j_1, \ldots, j_k$ are distinct indices from J. Since we have

$$\text{card}(V_{j_0}) = \omega_1, \quad \text{card}(V_{j_0} \cap (V_{j_1} + \ldots + V_{j_k})) \leq \omega,$$

there exists an element

$$y \in V_{j_0} \setminus (V_{j_1} + \ldots + V_{j_k}).$$

In view of the equality $g_{j_0}(V) = V_{j_0}$, we can find an element $x \in V$ such that $y = g_{j_0}(x)$. Now, it readily follows that

$$(q_0 g_{j_0} + q_1 g_{j_1} + \ldots + q_k g_{j_k})(x) \neq 0,$$

so $q_0 g_{j_0} + q_1 g_{j_1} + \ldots + q_k g_{j_k}$ is not identically equal to zero.

Further, we put

$$\mathcal{G} = \text{span}_\mathbf{Q}\{g_j : j \in J\}.$$

Actually, the argument presented in the proof of Theorem 1 shows that, for each element $v \in V$ and for any function $g \in \mathcal{G} \setminus \{0\}$, the inequality

$$\text{card}(g^{-1}(v)) \leq \omega$$

holds true. It immediately follows from this inequality that

$$\text{card}(\text{ran}(g)) = \omega_1 > \omega$$

for every $g \in \mathcal{G} \setminus \{0\}$.

Taking into account the representation of $\mathbf{R}$ in the form of a direct sum

$$\mathbf{R} = V + V' \quad (V \cap V' = \{0\}),$$

we may consider the canonical projection

$$\text{pr}_V : (\mathbf{R}, +) \to (V, +)$$

of $(\mathbf{R}, +)$ onto $(V, +)$. In other words, since each element $x \in \mathbf{R}$ admits a unique representation in the form

$$x = v + v' \quad (v \in V, \ v' \in V'),$$

we may put, by definition, $\text{pr}_V(x) = v$.

Finally, we introduce the family

$$\mathcal{F} = \text{span}_\mathbf{Q}\{g_j \circ \text{pr}_V : j \in J\}$$

of additive functions acting from $\mathbf{R}$ into V, and we claim that $\mathcal{F}$ is the required family of functions, i.e., $\mathcal{F}$ satisfies all relations (1)–(4) of Theorem 2.

Notice that the relations (1), (2) and (3) trivially hold by virtue of the definition of $\mathcal{F}$. So it remains to verify the validity of relation (4).

In view of Lemma 5, we only must check that if f is an arbitrary function from $\mathcal{F} \setminus \{0\}$, then the set $\text{ran}(f)$ is uncountable and has universal measure zero.

Firstly, observe that if $f \in \mathcal{F}$, then $\text{ran}(f) \subset V$, hence $\text{ran}(f)$ is indeed universal measure zero.

Secondly, if $f \in \mathcal{F} \setminus \{0\}$, then f admits a representation in the form

$$f = g \circ \text{pr}_V,$$

where g is some nonzero function from the family $\mathcal{G}$. Since pr_Y is a surjection and, as indicated above, the set $\mathrm{ran}(g)$ is uncountable, we infer that the set $\mathrm{ran}(f)$ is uncountable, too.

Theorem 2 has thus been proved.

Remark 4. Observe that if $2^{\mathbf{c}} = \omega_2$, then $\mathbf{c} = \omega_1$. The converse implication is not true, in general. Indeed, there are models of **ZFC** theory in which we have $\mathbf{c} = \omega_1$ but $2^{\mathbf{c}} = 2^{\omega_1} > \omega_2$ (see, for instance, [103], [148]).

Remark 5. Almost disjoint families of subsets of an infinite set play a remarkable role in many questions of set theory, infinite combinatorics, model theory, etc. For example, the existence of large almost disjoint families of subsets of a given infinite base set Ξ (cf. Lemma 3) implies that if M is an infinite model of a first-order mathematical theory, then there exists a model M' of the same theory such that $\mathrm{card}(M') > \mathrm{card}(M)$. In other words, if a first-order theory admits at least one infinite model, then it admits another model of strictly greater cardinality. Therefore such a theory cannot be categorical (for further development of this topic culminated by Morley's famous theorem, see, e.g., [29]).

EXERCISES

1. Let G be a subgroup of $(\mathbf{R}, +)$ which is of strictly positive outer λ-measure.

Demonstrate, by using Martin's Axiom, that G contains a subgroup G' which is a generalized Sierpiński set.

2. Let H be a subgroup of $(\mathbf{R}, +)$ which is of second category in $\mathbf{R}$.

Demonstrate, by using Martin's Axiom, that H contains a subgroup H' which is a generalized Luzin set.

3*. Let E be an infinite set with $\mathrm{card}(E) = \alpha$ and let β and γ be two cardinal numbers satisfying the conditions

$$1 < \beta < \gamma, \quad (\forall \delta < \gamma)(\beta^{\delta} < \alpha), \quad \sum \{\beta^{\delta} : \delta < \gamma\} = \alpha.$$

Prove that there exists a family $\mathcal{F}$ of subsets of E such that:
(i) $\mathcal{F}$ is a covering of E and $\mathrm{card}(\mathcal{F}) = \beta^{\gamma}$;
(ii) the cardinality of each member of $\mathcal{F}$ is equal to γ;
(iii) if X and Y are any two distinct members of $\mathcal{F}$, then $\mathrm{card}(X \cap Y) < \gamma$.

Argue as follows. Assume, without loss of generality, that β and γ are well-ordered by their canonical orderings (so β and γ are regarded as initial ordinal numbers) and consider the family $\mathcal{K}$ of all mappings acting from proper initial subintervals of γ into β. Check that

$$\mathrm{card}(\mathcal{K}) = \alpha.$$

Further, for any mapping h acting from γ into β, denote by $\mathcal{K}_h$ the family of all restrictions of h to proper initial subintervals of γ. Then take some bijection

$$\phi : \mathcal{K} \to E$$

and define

$$\mathcal{F} = \{\phi(\mathcal{K}_h) : h \in \beta^\gamma\}.$$

Verify that the family $\mathcal{F}$ of subsets of E satisfies the conditions (i)–(iii), so is as required.

Finally, consider the particular case

$$\alpha = \omega, \quad \beta = 2, \quad \gamma = \omega,$$

and conclude (within **ZF** set theory) that there exists a family $\mathcal{S}$ of infinite subsets of ω such that $\operatorname{card}(\mathcal{S}) = 2^\omega$ and the intersection of any two distinct members from $\mathcal{S}$ is finite.

Remark 6. Actually, the second part of Exercise 3 is concerned with the partially ordered set $(P, \preceq)$, where P is the family of all mappings acting from proper initial intervals of ω into $\{0, 1\}$, and $f \preceq g$ means that f is a restriction of g. This partially ordered set is usually called the complete binary ω-tree.

4. Suppose that all those subsets of $\mathbf{R}$ whose cardinalities are strictly less than $\mathbf{c}$ have λ-measure zero. Consider the quotient Boolean algebra $\mathcal{P}(\mathbf{R})/\mathcal{I}(\lambda)$, where $\mathcal{P}(\mathbf{R})$ denotes, as usual, the power set of $\mathbf{R}$ and $\mathcal{I}(\lambda)$ stands for the σ-ideal of all λ-measure zero sets.

Show that the Suslin number of $\mathcal{P}(\mathbf{R})/\mathcal{I}(\lambda)$ is strictly greater than $\mathbf{c}$, i.e., there are $\mathbf{c}^+$ many pairwise disjoint elements of $\mathcal{P}(\mathbf{R})/\mathcal{I}(\lambda)$.

5. Suppose that all those subsets of $\mathbf{R}$ whose cardinalities are strictly less than $\mathbf{c}$ are of first category in $\mathbf{R}$. Consider the quotient Boolean algebra $\mathcal{P}(\mathbf{R})/\mathcal{K}(\mathbf{R})$, where $\mathcal{K}(\mathbf{R})$ stands for the σ-ideal of all first category subsets of $\mathbf{R}$.

Show that the Suslin number of $\mathcal{P}(\mathbf{R})/\mathcal{K}(\mathbf{R})$ is strictly greater than $\mathbf{c}$.

6. Let E be an infinite set and let $\mathcal{B} = \mathcal{P}(E)$ denote the Boolean algebra of all subsets of E. Consider the family of sets

$$\mathcal{B}_0 = \{X \subset E : \operatorname{card}(X) < \operatorname{card}(E) \ \vee \ \operatorname{card}(E \setminus X) < \operatorname{card}(E)\}.$$

Obviously, $\mathcal{B}_0$ is a Boolean subalgebra of $\mathcal{B}$.

Demonstrate that $\mathcal{B}_0$ is not a direct summand in $\mathcal{B}$; in other words, there exists no Boolean subalgebra $\mathcal{B}_1$ of $\mathcal{B}$ such that

$$\mathcal{B} = \mathcal{B}_0 + \mathcal{B}_1 \quad (\mathcal{B}_0 \cap \mathcal{B}_1 = \{\emptyset\}).$$

For this purpose, use Lemma 3 of the present chapter.

7. Give a detailed proof of Lemma 5.

For this purpose, take into account the fact that if Γ is a subgroup of $(\mathbf{R}, +)$ satisfying the relation $\mathrm{card}(\mathbf{R}/\Gamma) > \omega$ and μ-measurable with respect to some σ-finite translation quasi-invariant measure μ on $\mathbf{R}$, then $\mu(\Gamma) = 0$.

8*. In this exercise a special construction of an Aronszajn tree is outlined by using almost disjoint subsets of ω (see Remark 9 in Appendix 1 for the definition of Aronszajn trees).

The desired Aronszajn tree $(T, \leq)$ has to be constructed by the method of transfinite recursion up to ω_1.

Namely, the role of elements of T are played by certain infinite subsets of ω. $X \leq Y$ means that $Y \subset X$. Transfinite recursion must be carried out so that the following two conditions would be satisfied:

(a) for every ordinal $\alpha < \omega_1$, the respective level T_α of T consists of countably many infinite and pairwise almost disjoint subsets of ω;

(b) if $\alpha < \beta < \omega_1$, a set X belongs to T_α, and D is a finite subset of X, then there exists a set $Y \in T_\beta$ such that $D \subset Y \subset X$.

In the process of the construction of T_α ($\alpha < \omega_1$), consider separately two cases:

(i) α is a successor ordinal, i.e., $\alpha = \gamma + 1$ for some $\gamma < \omega_1$;

(ii) α is a limit ordinal, i.e., $\alpha = \sup\{\gamma : \gamma < \alpha\}$.

Remark 7. There are several other well-known transfinite constructions of an ω_1-Aronszajn tree (see, for instance, [103], [148]). The method described above and based on almost disjoint infinite subsets of ω was suggested by Shelah.

9*. Let Z be a subset of the Euclidean space $\mathbf{R}^m$ ($m \geq 2$) such that any three distinct points of Z form either an acute-angled triangle or a right-angled triangle.

Prove that $\mathrm{card}(Z) \leq 2^m$, so Z is necessarily finite.

Let H denote an infinite-dimensional separable Hilbert space (over the field $\mathbf{R}$).

Show that there exists a set $X \subset H$ possessing the following two properties:

(a) the cardinality of X is equal to $\mathbf{c}$;

(b) any three distinct points of X form an acute-angled triangle.

Argue as follows. First of all, without loss of generality, identify H with the standard Hilbert space

$$l_2 = \{t \in \mathbf{R}^{\mathbf{N}} : \sum\{(t(n))^2 : n \in \mathbf{N}\} < +\infty\}.$$

Let $\{N_j : j \in J\}$ be a family of infinite subsets of $\mathbf{N}$ such that:

(i) $\mathrm{card}(J) = \mathbf{c}$;

(ii) $\mathrm{card}(N_j \cap N_{j'})$ is finite for any two distinct indices $j \in J$ and $j' \in J$.

Now, for each $j \in J$, define the element $x_j \in l_2$ by the formula

$$x_j(n) = (1/2^n)\chi_j(n) \qquad (n \in \mathbf{N}),$$

where χ_j denotes the characteristic function of the set $N_j \subset \mathbf{N}$.

Putting $X = \{x_j : j \in J\}$, verify that any three distinct points of X form an acute-angled triangle.

10. Let X be an uncountable subset of the Euclidean space $\mathbf{R}^m$.

Demonstrate that there exists an uncountable set $Y \subset X$ such that all distances between different points of Y are distinct.

For this purpose, use induction on m.

11*. Let H be an infinite-dimensional separable Hilbert space over $\mathbf{R}$.

Give an example of a set $X \subset H$ whose cardinality is equal to $\mathbf{c}$ and all distances between the points of X belong to some fixed countable subset of $\mathbf{R}$.

For establishing this fact, argue similarly to the hint of Exercise 9, i.e. identify H with l_2 and utilize again the existence of an almost disjoint family $\mathcal{N}$ of infinite subsets of $\mathbf{N}$ such that $\text{card}(\mathcal{N}) = \mathbf{c}$.

12. Let E be an infinite-dimensional topological vector space, $\{e_\xi : \xi \in \Xi\}$ be some basis for the vector structure of E, and suppose that there exists a local base $\{U_i : i \in I\}$ of open neighborhoods of $0 \in E$ satisfying the inequality

$$\text{card}(I) \leq \text{card}(\Xi).$$

Show that there is a vector space $\mathcal{F}$ of linear functionals on E such that:

(a) $\text{card}(\mathcal{F}) > \text{card}(\Xi)$;

(b) every nonzero functional $f \in \mathcal{F}$ is discontinuous at all points of E.

For this purpose, apply Lemma 3 to the infinite set Ξ.

Finally, consider $\mathbf{R}$ as a topological vector space over the field $\mathbf{Q}$ of all rational numbers and obtain from the above result the existence of nontrivial solutions of Cauchy's functional equation (cf. Chapter 3).

14. Additive properties of certain classes of pathological functions

The material of this chapter is concerned with some additive properties of the following three typical families of pathological functions: continuous nowhere differentiable functions, Sierpiński–Zygmund functions, and absolutely nonmeasurable functions.

Notice that many works were devoted to additive properties of various families of real-valued functions on $\mathbf{R}$ which are not necessarily bad or pathological. Among those works, we would like to point out [13], [73], [74], [75], [85], [129], [131], [140], [147], [198], [218]. Here our main goal is to show direct analogues linking the above-mentioned properties. Of course, our presentation is far from being complete. We only wish to indicate some vivid parallels and interrelations between additive properties of certain well-known classes of real-valued functions.

To be more concrete, let us consider the following three classes of functions acting from the real line $\mathbf{R}$ into itself.

(1) Continuous nowhere differentiable functions;

(2) Sierpiński–Zygmund functions, i.e., those functions whose restrictions to all subsets of $\mathbf{R}$ of cardinality continuum are discontinuous;

(3) Absolutely nonmeasurable functions, i.e., those functions which are nonmeasurable with respect to all nonzero σ-finite diffused measures on $\mathbf{R}$.

Definitely, it can be said that the functions belonging to the first class are very bad from the differential point of view, the functions belonging to the second class are very bad from the topological point of view, and the functions belonging to the third class can be regarded as very bad from the measure-theoretical point of view.

As is widely known, there are nontrivial individual examples of continuous nowhere differentiable functions. Recall that the first examples of this kind are due to Bolzano and Weierstrass (they were discovered in the second half of the 19th century).

The next important step was made by Banach [7] and Mazurkiewicz [186] in 1931. They demonstrated (independently) that, in the space $\mathcal{C}[0,1]$ of all continuous real-valued functions defined on the unit segment $[0,1]$, the family of nowhere differentiable functions is co-meager, i.e., is the complement of a first category subset of $\mathcal{C}[0,1]$ (for more details, see Exercise 5 of this chapter).

A nontrivial consequence of their result is the following fact:

Any continuous real-valued function on $\mathbf{R}$ can be represented as a sum (difference) of two continuous nowhere differentiable functions.

This fact does not follow from concrete individual constructions of a continuous nowhere differentiable function on $\mathbf{R}$, so needs an argument based on the above-mentioned result of Banach and Mazurkiewicz.

Indeed, for any real numbers a and b such that $a < b$, consider the Banach space $\mathcal{C}[a,b]$ of all continuous real-valued functions on $[a,b]$ and denote by $\mathcal{P}$ the subset of $\mathcal{C}[a,b]$ consisting of all nowhere differentiable functions on $[a,b]$. Take any function g from $\mathcal{C}[a,b]$ and observe that both sets $\mathcal{P}$ and $g + \mathcal{P}$ are co-meager in $\mathcal{C}[a,b]$. Consequently,

$$\mathcal{P} \cap (g + \mathcal{P}) \neq \emptyset$$

which directly implies the existence of two functions g_1 and g_2 from $\mathcal{P}$ such that

$$g = g_1 - g_2.$$

Now, let $f : \mathbf{R} \to \mathbf{R}$ be an arbitrary continuous function and let $\mathbf{Z}$ denote, as usual, the set of all integers. For every integer n, let f_n be the restriction of f to the segment $[n, n+1]$. According to the above, we can write

$$f_n = \phi_n - \psi_n \quad (n \in \mathbf{Z}),$$

where ϕ_n and ψ_n are some continuous and nowhere differentiable functions on $[n, n+1]$. Without loss of generality, we may assume that

$$\phi_n(n+1) = \phi_{n+1}(n+1) \quad (n \in \mathbf{Z}).$$

Keeping in mind these contact conditions for the family of continuous functions $\{\phi_n : n \in \mathbf{Z}\}$ and taking into account the trivial equalities

$$f_n(n+1) = f_{n+1}(n+1) \quad (n \in \mathbf{Z}),$$

we get the analogous contact conditions for the family of continuous functions $\{\psi_n : n \in \mathbf{Z}\}$, i.e.,

$$\psi_n(n+1) = \psi_{n+1}(n+1) \quad (n \in \mathbf{Z}).$$

Therefore, denoting by ϕ (respectively, by ψ) the common extension of all functions ϕ_n ($n \in \mathbf{Z}$) (respectively, of all functions ψ_n ($n \in \mathbf{Z}$)), we obtain that both ϕ and ψ are continuous nowhere differentiable functions on $\mathbf{R}$ and

$$f = \phi - \psi.$$

On the other hand, it was shown that there are sufficiently large vector subspaces $\mathcal{U}$ of $\mathcal{C}[0, 1]$ such that all members of $\mathcal{U} \setminus \{0\}$ are nowhere differentiable functions (see, for instance, [85]). Moreover, it was proved in the article [218] that, for every separable Banach space $\mathcal{W}$, there exists a closed vector subspace of $\mathcal{C}[0, 1]$ which is isometric to $\mathcal{W}$ and all nonzero members of which are nowhere differentiable functions.

The second type of pathological functions are Sierpiński–Zygmund functions first introduced in [246]. They are usually constructed by exploiting an appropriate well-ordering of $\mathbf{R}$ and it is clear that their existence needs uncountable forms of the Axiom of Choice, because any such function turns out to be nonmeasurable with respect to the standard Lebesgue measure λ on $\mathbf{R}$).

Furthermore, every Sierpiński–Zygmund function turns out to be nonmeasurable with respect to the completion of any nonzero σ-finite diffused Borel measure on $\mathbf{R}$. However, it was demonstrated in [132] that there exists a translation invariant measure μ on $\mathbf{R}$ extending the Lebesgue measure λ and such that some Sierpiński–Zygmund functions become measurable with respect to μ.

For various interesting properties of Sierpiński–Zygmund functions, see, e.g., [5], [37], [73], [131], [198], [212], and references given therein. See also Chapters 3 and 11 of this book.

It is not difficult to prove the following two statements (cf. [212], Proposition 1, or Chapter 12 of the present book).

Theorem 1. *Any function from $\mathbf{R}^{\mathbf{R}}$ can be represented as a sum (difference) of two Sierpiński–Zygmund injective functions.*

Theorem 2. *Any additive function from $\mathbf{R}^{\mathbf{R}}$ can be represented as a sum (difference) of two Sierpiński–Zygmund injective additive functions.*

For the sake of completeness, we give below the proof of Theorem 2. The proof of Theorem 1 can be done analogously and, in fact, is much easier.

Let $\mathbf{Q}$ denote the field of all rational numbers, ω denote the least infinite cardinal number, and let $\mathbf{c}$ denote the cardinality of the continuum. As usual, we identify $\mathbf{c}$ with the initial ordinal number equinumerous with $\mathbf{c}$.

Fix a Hamel basis $\{e_\xi : \xi < \mathbf{c}\}$ of $\mathbf{R}$.

Let $\{h_\xi : \xi < \mathbf{c}\}$ be the family of all real-valued Borel functions whose domains are the uncountable Borel subsets of $\mathbf{R}$.

Let $f : \mathbf{R} \to \mathbf{R}$ be an arbitrary additive function. We are going to construct by transfinite recursion two injective additive functions

$$f_1 : \mathbf{R} \to \mathbf{R}, \quad f_2 : \mathbf{R} \to \mathbf{R}$$

such that $f = f_1 + f_2$. For this purpose, it suffices to define recursively the values

$$f_1(e_\xi), \quad f_2(e_\xi) \quad (\xi < \mathbf{c}).$$

Suppose that, for an ordinal number $\xi < \mathbf{c}$, the two ξ-sequences of real numbers

$$\{f_1(e_\zeta) : \zeta < \xi\}, \quad \{f_2(e_\zeta) : \zeta < \xi\}$$

have already been defined so that

$$f(e_\zeta) = f_1(e_\zeta) + f_2(e_\zeta) \quad (\zeta < \xi)$$

and the corresponding partial additive functions

$$f_1 : E_\xi \to \mathbf{R}, \quad f_2 : E_\xi \to \mathbf{R}$$

are injective, where E_ξ denotes the vector space over $\mathbf{Q}$ generated by the family $\{e_\zeta : \zeta < \xi\}$, i.e.,

$$E_\xi = \mathrm{span}_{\mathbf{Q}}\{e_\zeta : \zeta < \xi\}.$$

We may assert that there exist two real numbers y_1 and y_2 satisfying the relations:

$$f(e_\xi) = y_1 + y_2, \quad y_1 \notin f_1(E_\xi), \quad y_2 \notin f_2(E_\xi),$$
$$(f_1(E_\xi) + \mathbf{Q}y_1) \cap (\cup\{h_\zeta(E_\xi + \mathbf{Q}e_\xi) : \zeta < \xi\}) = \emptyset,$$
$$(f_2(E_\xi) + \mathbf{Q}y_2) \cap (\cup\{h_\zeta(E_\xi + \mathbf{Q}e_\xi) : \zeta < \xi\}) = \emptyset.$$

Indeed, the existence of y_1 and y_2 easily follows from the inequalities

$$\mathrm{card}(E_\xi) \le \mathrm{card}(\xi) + \omega < \mathbf{c},$$

$$\mathrm{card}(\cup\{h_\zeta(E_\xi + \mathbf{Q}e_\xi) : \zeta < \xi\}) \le \mathrm{card}(\xi) + \omega < \mathbf{c}.$$

Now, we put

$$f_1(e_\xi) = y_1, \quad f_2(e_\xi) = y_2.$$

Proceeding in this manner, we obtain all the values

$$f_1(e_\xi), \quad f_2(e_\xi) \quad (\xi < \mathbf{c})$$

and, consequently, the corresponding two injective additive functions

$$f_1 : \mathbf{R} \to \mathbf{R}, \quad f_2 : \mathbf{R} \to \mathbf{R}.$$

It is not difficult to verify that both f_1 and f_2 are Sierpiński–Zygmund functions (the corresponding details are left to the reader).

Remark 1. Let $f : \mathbf{R} \to \mathbf{R}$ be a function. We shall say that f is a Sierpiński–Zygmund function in the strong sense if, for every set $X \subset \mathbf{R}$ with $\mathrm{card}(X) = \mathbf{c}$, the restriction $f|X$ is not a Borel function. Both functions f_1 and f_2 constructed above are Sierpiński–Zygmund functions in the strong sense. Notice that, under Martin's Axiom, there exist Sierpiński–Zygmund functions which are not Sierpiński–Zygmund functions in the strong sense. For more details, see Exercises 10 and 11 of this chapter.

The family $\mathbf{R}^{\mathbf{R}}$ of all functions acting from $\mathbf{R}$ into itself carries the canonical structure of a vector space over the field $\mathbf{R}$. It was demonstrated in [73] that there exists a vector subspace $\mathcal{V}$ of $\mathbf{R}^{\mathbf{R}}$ such that all members of $\mathcal{V} \setminus \{0\}$ are Sierpiński–Zygmund functions and the cardinality of $\mathcal{V}$ is strictly greater than the cardinality of the continuum.

The following statement essentially strengthens the above-mentioned result of [73].

Theorem 3. *There exists a vector subspace $\mathcal{V}$ of $\mathbf{R}^{\mathbf{R}}$ such that:*
(1) all members of $\mathcal{V}$ are additive functions;
(2) all members of $\mathcal{V} \setminus \{0\}$ are Sierpiński–Zygmund functions;
(3) the cardinality of $\mathcal{V}$ is strictly greater than the cardinality of the continuum.

Proof. As before, in the argument presented below we identify $\mathbf{c}$ with the initial ordinal number of cardinality continuum.

Let $\{e_\xi : \xi < \mathbf{c}\}$ be again a Hamel basis of $\mathbf{R}$.

Let $\{K_\xi : \xi < \mathbf{c}\}$ denote an increasing (by inclusion) transfinite sequence of subfields of $\mathbf{R}$ such that

$$K_0 = \mathbf{Q},$$

$$\cup\{K_\xi : \xi < \mathbf{c}\} = \mathbf{R},$$

$$\mathrm{card}(K_\xi) \leq \mathrm{card}(\xi) + \omega \quad (\xi < \mathbf{c}).$$

Let $\{h_\xi : \xi < \mathbf{c}\}$ be again the family of all real-valued Borel functions whose domains are the uncountable Borel subsets of $\mathbf{R}$.

Let $\mathbf{c}^+$ denote the successor of $\mathbf{c}$, i.e., the least cardinal number strictly greater than $\mathbf{c}$. We may assume that $\mathbf{c}^+$ coincides with the initial ordinal number of the same cardinality, i.e., for any ordinal number $\alpha < \mathbf{c}^+$, we have the inequality $\mathrm{card}(\alpha) < \mathbf{c}^+$.

We are going to construct by transfinite recursion a certain family

$$\{f_\alpha : \alpha < \mathbf{c}^+\}$$

of additive functions acting from $\mathbf{R}$ into $\mathbf{R}$.

Suppose that, for an ordinal $\beta < \mathbf{c}^+$, the partial family $\{f_\alpha : \alpha < \beta\}$ has already been constructed.

In order to define the additive function

$$f_\beta : \mathbf{R} \to \mathbf{R},$$

it completely suffices to determine all values

$$f_\beta(e_\xi) \quad (\xi < \mathbf{c}),$$

because $\{e_\xi : \xi < \mathbf{c}\}$ is a Hamel basis of $\mathbf{R}$.

Since $\operatorname{card}(\beta) \leq \mathbf{c}$, we may represent $\{f_\alpha : \alpha < \beta\}$ in the form of a $\mathbf{c}$-sequence $\{g_\xi : \xi < \mathbf{c}\}$. Here we do not assume that the family $\{g_\xi : \xi < \mathbf{c}\}$ is necessarily injective (moreover, if $\operatorname{card}(\beta) < \mathbf{c}$, then it is clear that the corresponding family $\{g_\xi : \xi < \mathbf{c}\}$ cannot be injective).

For every ordinal $\xi < \mathbf{c}$, let $\mathcal{V}_\xi$ denote the vector space over K_ξ generated by the family of functions $\{g_\zeta : \zeta < \xi\}$.

Also, for every ordinal $\xi < \mathbf{c}$, introduce the notation:

E_ξ = the vector space over $\mathbf{Q}$ generated by the family $\{e_\zeta : \zeta < \xi\}$;

E'_ξ = the vector space over $\mathbf{Q}$ generated by the family $\{e_\zeta : \zeta \leq \xi\}$.

Now, we define the values $f_\beta(e_\xi)$ $(\xi < \mathbf{c})$ by transfinite recursion over ξ.

Suppose that, for $\xi < \mathbf{c}$, the partial family $\{f_\beta(e_\zeta) : \zeta < \xi\}$ has already been constructed. Then we may consider the corresponding additive functional f_β on the vector space E_ξ. As usual, denote

$$\mathcal{V}_\xi(E'_\xi) = \{g(x) : g \in \mathcal{V}_\xi, \ x \in E'_\xi\}$$

and choose the value $f_\beta(e_\xi)$ so that the relation

$$(\mathcal{V}_\xi(E'_\xi) + K_\xi f_\beta(E_\xi) + (K_\xi \setminus \{0\})f_\beta(e_\xi)) \cap (\cup\{h_\zeta(E'_\xi) : \zeta < \xi\}) = \emptyset$$

would be satisfied. It is not difficult to check that such a choice of $f_\beta(e_\xi)$ is always possible, because of the inequalities

$$\operatorname{card}(\mathcal{V}_\xi) \leq \operatorname{card}(\xi) + \omega < \mathbf{c},$$

$$\operatorname{card}(E_\xi) \leq \operatorname{card}(E'_\xi) \leq \operatorname{card}(\xi) + \omega < \mathbf{c}.$$

Proceeding in this manner, we obtain the family of real numbers

$$f_\beta(e_\xi) \quad (\xi < \mathbf{c})$$

and, consequently, the associated additive function $f_\beta : \mathbf{R} \to \mathbf{R}$.

So, by virtue of the just described construction, we come to the transfinite sequence of additive functions $\{f_\alpha : \alpha < \mathbf{c}^+\}$, each of which acts from $\mathbf{R}$ into itself.

Let $\mathcal{V}$ denote the vector space over $\mathbf{R}$ generated by $\{f_\alpha : \alpha < \mathbf{c}^+\}$. We claim that $\mathcal{V}$ is as required, i.e., $\mathcal{V}$ satisfies conditions (1)–(3) of the theorem.

Indeed, condition (1) is trivially valid.

Let us show that condition (2) holds true, too. For this purpose, take any nonzero function f from $\mathcal{V}$ and any function h from the family $\{h_\xi : \xi < \mathbf{c}\}$.

The function f can be written as

$$f = t_1 f_{\alpha_1} + t_2 f_{\alpha_2} + \ldots + t_n f_{\alpha_n} + t f_\beta,$$

where n is a natural number, $t_1, t_2, \ldots, t_n, t$ are some nonzero real numbers, and $\alpha_1, \alpha_2, \ldots, \alpha_n, \beta$ are ordinals such that

$$\alpha_1 < \alpha_2 < \ldots < \alpha_n < \beta < \mathbf{c}^+.$$

Obviously, we can find an ordinal number $\xi_0 < \mathbf{c}$ such that

$$\{t_1, t_2, \ldots, t_n, t\} \subset K_\xi,$$

$$\{f_{\alpha_1}, f_{\alpha_2}, \ldots, f_{\alpha_n}\} \subset \{g_\zeta : \zeta < \xi\},$$

$$h \in \{h_\zeta : \zeta < \xi\}$$

for every ordinal ξ satisfying the inequalities $\xi_0 < \xi < \mathbf{c}$.

Now, let us consider an element $z \in \mathbf{R}$ whose representation via our Hamel basis $\{e_\xi : \xi < \mathbf{c}\}$ looks as follows:

$$z = q_1 e_{\zeta_1} + q_2 e_{\zeta_2} + \ldots + q_m e_{\zeta_m} + q e_\xi,$$

where m is a natural number, $q_1, q_2, \ldots, q_m, q$ are some nonzero rational numbers, and $\zeta_1, \zeta_2, \ldots, \zeta_m, \xi$ are some ordinal numbers such that

$$\zeta_1 < \zeta_2 < \ldots < \zeta_m < \xi, \quad \xi_0 < \xi.$$

It is not difficult to see from the definition of $f_\beta(e_\xi)$ that $f(z) \neq h(z)$. Consequently, we have

$$\mathrm{card}(\{x \in \mathbf{R} : f(x) = h(x)\}) \leq \mathrm{card}(\xi_0) + \omega < \mathbf{c},$$

which yields that f is a Sierpiński–Zygmund function. Moreover, the same f has the following much stronger property: for every set $X \subset \mathbf{R}$ with $\mathrm{card}(X) = \mathbf{c}$, the restriction of f to X is not a Borel function, i.e., f is a Sierpiński–Zygmund function in the strong sense.

It remains to check the validity of condition (3). The preceding argument shows, in particular, that if ordinals α and β satisfy the inequalities

$$\alpha < \beta < \mathbf{c}^+,$$

then the difference $f_\alpha - f_\beta$ is a Sierpiński–Zygmund function and, consequently, $f_\alpha \neq f_\beta$. It immediately follows from this observation that

$$\mathrm{card}(\mathcal{V}) = \mathbf{c}^+ > \mathbf{c},$$

i.e., condition (3) is fulfilled. Notice also that if f and f^* are any two distinct functions from $\mathcal{V}$, then the difference $f - f^*$ also belongs to $\mathcal{V}$ and is a nonzero function. According to the said above, $f - f^*$ is a Sierpiński–Zygmund function, so

$$\mathrm{card}(\{x \in \mathbf{R} : (f - f^*)(x) = 0\}) < \mathbf{c}$$

or, equivalently,
$$\text{card}(\{x \in \mathbf{R} : f(x) = f^*(x)\}) < \mathbf{c}.$$

In other words, the graphs of f and f^* are almost disjoint subsets (in fact, almost disjoint subgroups) of the Euclidean plane $\mathbf{R}^2 = \mathbf{R} \times \mathbf{R}$.

Theorem 3 has thus been proved.

Remark 2. Let $\mathcal{F}$ be a family of functions acting from $\mathbf{R}$ into $\mathbf{R}$. The cardinal number $\text{A}(\mathcal{F})$ is usually defined as the smallest cardinality of a family $\mathcal{H} \subset \mathbf{R}^{\mathbf{R}}$ for which there exists no $h \in \mathbf{R}^{\mathbf{R}}$ such that $h + \mathcal{H} \subset \mathcal{F}$. This cardinal number was investigated for concrete classes of real-valued functions on $\mathbf{R}$ (see, e.g., [212] and references therein). A more general concept was also introduced in [212]. Namely, let $\mathcal{F}_1$ and $\mathcal{F}_2$ be two subfamilies of $\mathbf{R}^{\mathbf{R}}$. Define the cardinal number $\text{Add}(\mathcal{F}_1, \mathcal{F}_2)$ as the smallest cardinality of a family $\mathcal{H} \subset \mathbf{R}^{\mathbf{R}}$ for which there exists no $h \in \mathcal{F}_1$ such that $h + \mathcal{H} \subset \mathcal{F}_2$. By using an argument somewhat similar to the proof of Theorem 3, it can be demonstrated that if $\mathcal{F}_1$ is the family of all additive real-valued functions on $\mathbf{R}$ and $\mathcal{F}_2$ is the family of all Sierpiński–Zygmund functions on $\mathbf{R}$, then

$$\text{Add}(\mathcal{F}_1, \mathcal{F}_2) > \mathbf{c}.$$

For details, see Theorem 10 (iv) from [212] and its proof. In this context, it is natural to consider the cardinal number $\text{A}(\mathcal{G})$, where $\mathcal{G}$ denotes the family of all those real-valued functions on $\mathbf{R}$ which simultaneously are additive and Sierpiński–Zygmund functions. It is not difficult to check that

$$\text{A}(\mathcal{G}) = \text{Add}(\mathcal{G}, \mathcal{G}) + 1 = 2.$$

Notice also that the equality

$$\text{A}(\mathcal{F}) = \text{Add}(\mathcal{F}, \mathcal{F}) + 1$$

holds true for any family $\mathcal{F} \subset \mathbf{R}^{\mathbf{R}}$ (see again [212], Proposition 1).

Certain analogues of Theorems 1, 2, and 3 are valid for absolutely non-measurable real-valued functions on $\mathbf{R}$. To formulate them, let us recall some notions which were introduced and examined in preceding sections of this book.

For a given nonempty set E, we denote, as usual, by $\mathcal{M}(E)$ the class of all nonzero σ-finite continuous measures on E (their domains are, in general, various σ-algebras of subsets of E).

A function $f : E \to \mathbf{R}$ which is nonmeasurable with respect to each measure from the class $\mathcal{M}(E)$ can be regarded as an extremely nonmeasurable real-valued function on E. As before, we call such an f an absolutely nonmeasurable function on E.

We already know that the notion of a universal measure zero subset of $\mathbf{R}$ is closely connected with absolutely nonmeasurable functions and turns out to be a necessary tool for their characterization.

Let $Z \subset \mathbf{R}$. We recall that Z is a universal measure zero set if, for any σ-finite continuous Borel measure μ on $\mathbf{R}$, we have $\mu^*(Z) = 0$ where μ^* denotes the outer measure associated with μ.

Equivalently, one may say that $Z \subset \mathbf{R}$ is a universal measure zero set if there exists no nonzero σ-finite continuous Borel measure on Z (where Z is assumed to be endowed with the induced topology).

As we already know, the following statement yields a useful characterization of absolutely nonmeasurable functions with respect to the class $\mathcal{M}(E)$.

Theorem 4. *For an arbitrary function*

$$f : E \to \mathbf{R},$$

these two assertions are equivalent:
(1) f is absolutely nonmeasurable with respect to $\mathcal{M}(E)$;
(2) the range of f is a universal measure zero subset of $\mathbf{R}$ and, for each point $t \in \mathbf{R}$, the set $f^{-1}(t)$ is at most countable.

The nondifficult proof of the above theorem was given in Chapter 5 of this book.

We would like to remind readers that the following three facts are straightforward consequences of Theorem 4:

(i) if a function $f : E \to \mathbf{R}$ is injective and the range of f is a universal measure zero set, then f is absolutely nonmeasurable with respect to $\mathcal{M}(E)$;

(ii) the composition of any two functions which are absolutely nonmeasurable with respect to the class $\mathcal{M}(\mathbf{R})$ is absolutely nonmeasurable with respect to the same class;

(iii) if $\mathrm{card}(E) > \mathbf{c}$, then there exist no functions on E which are absolutely nonmeasurable with respect to the class $\mathcal{M}(E)$; more precisely, the existence of an absolutely nonmeasurable function with respect to $\mathcal{M}(E)$ is equivalent to the existence of a universal measure zero set $Z \subset \mathbf{R}$ with $\mathrm{card}(Z) = \mathrm{card}(E)$.

Very important representatives of the family of all universal measure zero subsets of $\mathbf{R}$ are Luzin sets which have already been discussed in preceding chapters of this book.

Recall that $L \subset \mathbf{R}$ is a Luzin set if L is uncountable and the intersection of L with any first category subset of $\mathbf{R}$ is at most countable.

Various properties of Luzin sets are presented in the widely known textbook by Oxtoby [203] (see also [33], [133], [147], [152], [188], [190]).

A set $L' \subset \mathbf{R}$ is a generalized Luzin set if $\mathrm{card}(L') = \mathbf{c}$ and the intersection of L with any first category subset of $\mathbf{R}$ has cardinality strictly less than $\mathbf{c}$.

Repeat once more that every Luzin subset of $\mathbf{R}$ and, under Martin's Axiom, every generalized Luzin subset of $\mathbf{R}$ are universal measure zero sets in $\mathbf{R}$. These two facts are easy to prove (see Exercise 23 from Chapter 5).

The following statement was established in Chapter 12. It shows that, under **MA**, the family of all absolutely nonmeasurable functions with respect to the class $\mathcal{M}(\mathbf{R})$ is sufficiently rich in the additive group $(\mathbf{R}^{\mathbf{R}}, +)$.

Theorem 5. *Assuming Martin's Axiom, for any function*

$$f : \mathbf{R} \to \mathbf{R},$$

there exist two injective functions $f_1 : \mathbf{R} \to \mathbf{R}$ and $f_2 : \mathbf{R} \to \mathbf{R}$ which are absolutely nonmeasurable with respect to the class $\mathcal{M}(\mathbf{R})$ and for which the equality $f = f_1 + f_2$ holds true.

The next statement was also proved in Chapter 12 and shows that, under **MA**, the family of all additive absolutely nonmeasurable functions with respect to the same class $\mathcal{M}(\mathbf{R})$ is sufficiently rich in the vector space of all additive functions acting from $\mathbf{R}$ into itself.

Theorem 6. *Assume Martin's Axiom. Let $f : \mathbf{R} \to \mathbf{R}$ be any additive function. Then there exist two injective additive functions*

$$f_1 : \mathbf{R} \to \mathbf{R}, \quad f_2 : \mathbf{R} \to \mathbf{R}$$

which are absolutely nonmeasurable with respect to $\mathcal{M}(\mathbf{R})$ and satisfy the equality $f = f_1 + f_2$.

Further, assuming **CH**, it was demonstrated in Chapter 13 that there is a large subgroup of $(\mathbf{R}^{\mathbf{R}}, +)$, all nonzero members of which are absolutely nonmeasurable functions with respect to the class $\mathcal{M}(\mathbf{R})$. More precisely, we have the following statement.

Theorem 7. *Under the Continuum Hypothesis, there exists a subgroup $\mathcal{F}$ of $(\mathbf{R}^{\mathbf{R}}, +)$ satisfying the relations:*
(1) $\mathrm{card}(\mathcal{F}) = \mathbf{c}^+$;
(2) every $f \in \mathcal{F}$ is an additive function;
(3) every $f \in \mathcal{F} \setminus \{0\}$ is a function absolutely nonmeasurable with respect to the class $\mathcal{M}(\mathbf{R})$.
In particular, if $2^{\mathbf{c}} = \mathbf{c}^+$, then the equality

$$\mathrm{card}(\mathcal{F}) = \mathrm{card}(\mathbf{R}^{\mathbf{R}})$$

holds true.

A certain analogue of Theorem 7 was also obtained within **ZFC** set theory for the class of all nonzero σ-finite translation quasi-invariant measures on $\mathbf{R}$ (see Chapter 13).

Remark 3. The usage of an additional set-theoretical assumption in the formulation of Theorem 7 is necessary. Indeed, suppose that Martin's Axiom and the negation of the Continuum Hypothesis hold. Then, as is well known, we have the equalities

$$2^\omega = 2^{\omega_1} = \mathbf{c},$$

where ω_1 stands, as usual, for the least uncountable ordinal number. Let $\mathcal{G}$ be a subgroup of $(\mathbf{R}^\mathbf{R}, +)$ satisfying the analogue of condition (3) of Theorem 7, i.e., every $g \in \mathcal{G} \setminus \{0\}$ is a function absolutely nonmeasurable with respect to $\mathcal{M}(\mathbf{R})$. Fix a subset X of $\mathbf{R}$ with $\mathrm{card}(X) = \omega_1$. For any function $g \in \mathcal{G}$, consider the restriction of g to X and let $\mathrm{Gr}(g|X)$ denote the graph of this restriction. Thus, we come to the mapping

$$F : \mathcal{G} \to \mathcal{P}(\mathbf{R} \times \mathbf{R})$$

defined by the formula

$$F(g) = \mathrm{Gr}(g|X) \quad (g \in \mathcal{G}).$$

In fact, this F acts from $\mathcal{G}$ into the family of all those subsets of $\mathbf{R} \times \mathbf{R}$ whose cardinalities are equal to ω_1. Let $g \in \mathcal{G}$, $h \in \mathcal{G}$ and $F(g) = F(h)$. Then

$$\mathrm{Gr}(g|X) = \mathrm{Gr}(h|X), \quad g|X = h|X, \quad (g-h)|X = 0,$$

whence it follows, by virtue of Theorem 4, that

$$g - h = 0, \quad g = h.$$

In other words, the introduced mapping F is injective, which immediately yields

$$\mathrm{card}(\mathcal{G}) \leq \mathrm{card}((\mathbf{R} \times \mathbf{R})^{\omega_1}) = 2^{\omega_1} = \mathbf{c}.$$

So we conclude that, under Martin's Axiom and the negation of the Continuum Hypothesis, there is no large subgroup of $(\mathbf{R}^\mathbf{R}, +)$, all nonzero members of which are absolutely nonmeasurable with respect to the class $\mathcal{M}(\mathbf{R})$.

Remark 4. As has been already mentioned, in some models of set theory there are no absolutely nonmeasurable functions with respect to $\mathcal{M}(\mathbf{R})$. On the other hand, if we assume Martin's Axiom, then the class of Sierpiński–Zygmund functions on $\mathbf{R}$ and the class of absolutely nonmeasurable functions with respect to $\mathcal{M}(\mathbf{R})$ are in general position, i.e., these two classes of functions have nonempty intersection and none of them contains another one.

Finishing this chapter, let us introduce one measure-theoretical version of Sierpiński–Zygmund functions which is closely connected with the notion of absolute nonmeasurability.

Let E be a non-universal measure zero topological space, all singletons in which are Borel subsets of E.

We shall say that a function $f : E \to \mathbf{R}$ is a Sierpiński–Zygmund type function (in the measure-theoretical sense) if, for any non-universal measure zero set $X \subset E$, the restriction $f|X$ is not a Borel function.

Theorem 8. *Let E be a non-universal measure zero topological space (a singletons in which are Borel) and let*

$$f : E \to \mathbf{R}$$

be an absolutely nonmeasurable function with respect to the class $\mathcal{M}(E)$. Then there exists a non-universal measure zero subset Y of E such that the restriction $f|Y$ is an injective Sierpiński–Zygmund type function on Y.

Proof. According to Theorem 4, the set $\mathrm{ran}(f)$ is universal measure zero and the set $f^{-1}(x)$ is at most countable for every point $x \in \mathbf{R}$. Consequently there exists a countable disjoint family $\{Y_i : i \in I\}$ such that

$$\cup\{Y_i : i \in I\} = E$$

and the restriction $f|Y_i$ is injective for any index $i \in I$. Since the original space E is not universal measure zero, at least one set Y_{i_0} is not universal measure zero, either. Let $Y = Y_{i_0}$ denote one of such sets. We claim that the restriction $f|Y$ is an injective Sierpiński–Zygmund type function on Y. Indeed, suppose for a moment otherwise, i.e., suppose that there exists a non-universal measure zero set $Z \subset Y$ for which the corresponding restriction $f|Z$ is Borel. Let μ be some probability continuous Borel measure on Z. For every Borel subset B of $\mathrm{ran}(f)$, define

$$\nu(B) = \mu((f|Z)^{-1}(B)).$$

In this way, we come to the probability continuous Borel measure ν on $\mathrm{ran}(f)$ which contradicts the fact that $\mathrm{ran}(f)$ is a universal measure zero set.

The obtained contradiction finishes the proof.

EXERCISES

1. Let E and F be any two topological spaces and let $g : E \to F$ be a mapping. According to the well-known definition from general topology (see [19], [49], [152]), this g is called a closed mapping if, for each closed subset A of E, the set $g(A)$ is closed in F.

Let X be a topological space, Y be a quasi-compact topological space, and let

$$\mathrm{pr}_1 : X \times Y \to X$$

denote the first canonical projection of the product space $X \times Y$ to X, i.e., we have

$$\mathrm{pr}_1((x, y)) = x \qquad ((x, y) \in X \times Y).$$

Work in **ZF** set theory and verify that pr_1 is a closed mapping (this statement is due to Kuratowski and is known as Kuratowski's lemma on closed projection).

2. Let E and F be two topological spaces and let $g : E \to F$ be a mapping. This g is called perfect if it is continuous, closed and, for all points $y \in F$, the pre-images $g^{-1}(y)$ are quasi-compact subspaces of E.

Check the validity of the following assertions:

(a) if $f : X \to Y$ is perfect, then for every quasi-compact set $K \subset Y$, the pre-image $f^{-1}(K)$ is quasi-compact, too;

(b) the composition of any two perfect mappings is also a perfect mapping;

(c) if a mapping $f : X \to Y$ is perfect and a set $A \subset X$ is closed, then the restriction $f|A$ is a perfect mapping;

(d) if a mapping $f : X \to Y$ is perfect and B is any subset of Y, then the mapping

$$f|f^{-1}(B) : f^{-1}(B) \to B$$

is perfect, too.

By using the above-mentioned facts and Kuratowski's lemma on closed projection (see Exercise 1), demonstrate within **ZF** set theory that if X and Y are two quasi-compact spaces, then their product $X \times Y$ is also quasi-compact.

Remark 5. According to the above result, we have in **ZF** theory the theorem stating that the product of finitely many quasi-compact topological spaces is quasi-compact, too. On the other hand, it cannot be proved within the same theory that the product of countably many quasi-compact topological spaces is also quasi-compact (for more details, see [93]).

3. Let $P(z)$ and $Q(z)$ be two polynomials over the field $\mathbf{C}$ of all complex numbers, let the degree of $P(z)$ be strictly greater than the degree of $Q(z)$ and let D denote the set of all roots of $Q(z)$. Define a mapping

$$F : \mathbf{C} \setminus D \to \mathbf{C}$$

by the formula

$$F(z) = P(z)/Q(z) \qquad (z \in \mathbf{C} \setminus D).$$

Demonstrate that F is a perfect mapping.

4*. Let X be a Hausdorff space and let Y be a Hausdorff image of X under some perfect mapping.

Prove that if Y admits at least one nonzero σ-finite diffused Radon measure, then X admits a probability diffused Radon measure defined on the Borel σ-algebra of X.

For this purpose, use Exercise 17 from Chapter 6.

Conclude that if X is a compact space, Y is a Hausdorff continuous image of X and there exists a nonzero σ-finite diffused Radon measure on Y, then there exists a probability diffused Radon measure defined on the Borel σ-algebra of X.

5*. Let $C[0,1]$ denote, as usual, the Banach space of all continuous real-valued functions which are defined on the unit segment $[0,1]$, and let $\mathcal{D}$ stand for the set of all those functions $f \in C[0,1]$ which are differentiable at some point of $[0,1]$ (of course, this point depends on f).

Prove that the set $\mathcal{D}$ is of first category in $C[0,1]$ (this classical result is due to Banach and Mazurkiewicz; see their remarkable works [7] and [186]).

Argue as follows. Fix a natural number $n > 0$ and consider the set

$$\mathcal{D}(n) = \{f \in C[0,1] \; : \; \text{there exists a point } x \in [0,1] \text{ such that}$$

the absolute values of all derived numbers of f at x do not exceed $n\}$.

By using Kuratowski's lemma on closed projection, verify that the set $\mathcal{D}(n)$ is closed and nowhere dense in $C[0,1]$. Then keep in mind the inclusion

$$\mathcal{D} \subset \cup\{\mathcal{D}(n) : 0 < n < \omega\}$$

and obtain the required result.

6*. Recall that a function $g : \mathbf{R} \to \mathbf{R}$ is said to be approximately differentiable at a point $x \in \mathbf{R}$ if there exists a Lebesgue measurable set $X \subset \mathbf{R}$ (depending on x) such that x is a density point of X and there exists a (path) derivative of $g|X$ at x.

Prove that, for every continuous function $f : \mathbf{R} \to \mathbf{R}$, there exist two continuous nowhere approximately differentiable functions

$$f_1 : \mathbf{R} \to \mathbf{R}, \quad f_2 : \mathbf{R} \to \mathbf{R}$$

such that $f = f_1 + f_2$.

For this purpose, utilize the fact analogous to the result of Banach and Mazurkiewicz, which states that in the space $C[0,1]$ the family of all nowhere approximately differentiable functions is co-meager (this fact was first established by Jarnik [100]).

7. Give a detailed proof of Theorem 1, i.e., demonstrate that every function

$$f : \mathbf{R} \to \mathbf{R}$$

can be represented in the form

$$f = f_1 + f_2,$$

where both $f_1 : \mathbf{R} \to \mathbf{R}$ and $f_2 : \mathbf{R} \to \mathbf{R}$ are injective Sierpiński–Zygmund functions.

8. Let $\mathcal{F}_1$ and $\mathcal{F}_2$ be two subfamilies of $\mathbf{R}^{\mathbf{R}}$. Show that the following two conditions are equivalent:

(a) $\mathrm{Add}(\mathcal{F}_1, \mathcal{F}_2) \geq 2$;
(b) $\mathbf{R}^{\mathbf{R}} = \mathcal{F}_1 - \mathcal{F}_2$.

9. Check the equality $A(\mathcal{G}) = 2$, where $\mathcal{G}$ denotes the family of all real-valued additive functions on $\mathbf{R}$ which are simultaneously Sierpiński–Zygmund functions.

10. Under Martin's Axiom, prove that there exists a semi-continuous function $f : \mathbf{R} \to \mathbf{R}$ which is not $\mathbf{c}$-continuous, i.e., f is semi-continuous but there does not exist a family $\{f_i : i \in I\}$ of partial functions acting from $\mathbf{R}$ into $\mathbf{R}$ such that:
(a) $\mathrm{card}(I) < \mathbf{c}$;
(b) each f_i $(i \in I)$ is continuous on $\mathrm{dom}(f_i)$;
(c) the graph of f is a union of the graphs of f_i $(i \in I)$.

11*. Assuming Martin's Axiom, construct a Sierpiński–Zygmund function on $\mathbf{R}$ which is not a Sierpiński–Zygmund function in the strong sense.
For this purpose, start with the fact formulated in Exercise 10 and use the method of transfinite recursion.

12. Assume Martin's Axiom and consider the space $\mathcal{C}[0, 1]$ of all real-valued continuous functions defined on the closed unit interval $[0, 1]$. Let $\mathcal{F}$ be a co-meager subset of $\mathcal{C}[0, 1]$.
Demonstrate that there exists a family $\mathcal{G} \subset \mathcal{F}$ satisfying the following three conditions:
(a) $\mathrm{card}(\mathcal{G}) = \mathbf{c}$;
(b) $\mathcal{G}$ is a vector space over the field $\mathbf{Q}$;
(c) $\mathcal{G}$ is thick in the sense of category, i.e., $\mathcal{G}$ meets every non-meager subset of $\mathcal{C}[0, 1]$ having the Baire property.
For this purpose, construct the required $\mathcal{G}$ by the method of transfinite recursion.

Remark 6. The previous exercise shows, in particular, that if $\mathcal{F}$ consists of pathological functions of a certain type, then there exists a thick (in the sense of category) group $\mathcal{G} \subset \mathcal{F}$, all nonzero members of which are pathological functions of the same type.

13*. Recall that a topological space E is arcwise connected if, for any two points $x \in E$ and $y \in E$, there exists a continuous mapping

$$\phi : [0, 1] \to E$$

such that $\phi(0) = x$ and $\phi(1) = y$. The set $\phi([0, 1])$ is called a curve connecting the two points x and y.
Give an example of a compact subset K of $\mathbf{R}^2$ which is arcwise connected, but no nontrivial curve (i.e., no curve distinct from a singleton) in K has finite length.

For this purpose, check that the role of such a K can be played by the graph of a continuous nowhere differentiable function acting from $[0, 1]$ into $\mathbf{R}$.

Starting with the above-mentioned result, try to construct a surface S in $\mathbf{R}^3$ having the following two properties:

(a) S is homeomorphic to the unit square $[0, 1]^2$;

(b) the length of any nontrivial curve contained in S is equal to infinity.

14*. Denote by E the subset of $\mathcal{C}[0, 1]$ consisting of all convex functions on $[0, 1]$. Observe that E is closed in $\mathcal{C}[0, 1]$, so E may be regarded as a Polish topological space.

Demonstrate that the set of all everywhere differentiable functions from E is co-meager in E.

For this purpose, use an argument somewhat similar to that sketched in Exercise 5.

Remark 7. It can be proved that the set of all everywhere twice differentiable functions from E is of first category in E.

15. Absolutely nonmeasurable homomorphisms of commutative groups

In Chapter 3 we were briefly concerned with Cauchy's functional equation

$$f(x+y) = f(x) + f(y) \qquad (x \in \mathbf{R}, \ y \in \mathbf{R}).$$

Recall that in this equation f denotes an unknown function acting from $\mathbf{R}$ into itself (in fact, f is an endomorphism of the additive group $(\mathbf{R}, +)$). To resolve this equation means to find all those functions $f : \mathbf{R} \to \mathbf{R}$ which satisfy the equation or, in other words, to find all possible homomorphisms of the additive group $(\mathbf{R}, +)$ into itself. As was mentioned in Chapter 3, besides the trivial continuous solutions of Cauchy's equation, which are of the form

$$f(x) = ax \qquad (x \in \mathbf{R}),$$

where a parameter a belongs to $\mathbf{R}$, there exist many nontrivial solutions and all of the latter ones have very bad descriptive properties, i.e., all of them are nonmeasurable with respect to the Lebesgue measure $\lambda = \lambda_1$ on $\mathbf{R}$ and, in addition, do not possess the Baire property with respect to the ordinary Euclidean topology of $\mathbf{R}$ (see, e.g., [133], [147] or Exercise 10 from Chapter 3).

The above circumstance may be regarded as a starting point for studying a more general situation when we have an uncountable commutative group $(G, +)$ instead of $(\mathbf{R}, +)$ and are dealing with various homomorphisms of G into other commutative groups.

For instance, it is well known that, for any commutative locally compact topological group $(G, +)$, there are many continuous homomorphisms of $(G, +)$ into the one-dimensional unit torus (or the circle group)

$$(\mathbf{T}, +) = (\mathbf{S}_1, \cdot) \subset (\mathbf{C}, +, \cdot),$$

where $\mathbf{C}$ stands, as usual, for the field of all complex numbers and

$$\mathbf{S}_1 = \{z \in \mathbf{C} : |z| = 1\}.$$

Such homomorphisms are called characters and have one important property, namely, they separate points in G, i.e., for any two distinct elements $x \in G$ and $y \in G$, there exists a character

$$\phi : G \to \mathbf{T}$$

such that $\phi(x) \neq \phi(y)$.

A similar result holds true for so-called real characters, namely, if $(G, +)$ belongs to a certain, sufficiently wide, class of commutative locally compact groups then the family of all continuous homomorphisms from $(G, +)$ into $(\mathbf{R}, +)$ separates points of G (see, e.g., [41], [95]).

At the same time, there are several constructions of everywhere discontinuous homomorphisms acting from commutative locally compact groups into $(\mathbf{T}, +)$ (or into $(\mathbf{R}, +)$). Those homomorphisms can be constructed by the method of transfinite recursion for concrete groups $(G, +)$. For instance, it is well known that there are homomorphisms

$$\phi : (\mathbf{R}, +) \to (\mathbf{R}, +),$$

$$\psi : (\mathbf{R}, +) \to (\mathbf{T}, +)$$

such that the graph of ϕ is everywhere dense in $\mathbf{R} \times \mathbf{R}$ and the graph of ψ is everywhere dense in $\mathbf{R} \times \mathbf{T}$. It is not difficult to see that ϕ and ψ are discontinuous at all points of $\mathbf{R}$.

If a σ-compact locally compact commutative group $(G, +)$ is assumed to be endowed with the standard Haar measure ν, then everywhere discontinuous homomorphisms from $(G, +)$ into $(\mathbf{T}, +)$ turn out to be nonmeasurable with respect to the completion of ν. However, these bad homomorphisms are sometimes useful for constructing nonseparable translation invariant extensions of the Haar measure and become measurable with respect to those extensions (cf [144]). So, according to our terminology, the above-mentioned homomorphisms turn out to be relatively measurable with respect to the class of all those translation invariant measures on $(G, +)$ which extend ν.

Treating $(G, +)$ only as an abstract group, we cannot assert that a homomorphism acting from $(G, +)$ into $(\mathbf{T}, +)$ (or into $(\mathbf{R}, +)$) which has bad descriptive properties with respect to one group topology on G is also bad with respect to another group topology on G. The following simple example illustrates this interesting fact.

Example 1. Consider the real line $(\mathbf{R}, +)$ and the Euclidean plane $(\mathbf{R}^2, +)$ as abstract commutative groups. As is well known, these two groups are isomorphic to each other. Let

$$\phi : (\mathbf{R}^2, +) \to (\mathbf{R}, +)$$

be an isomorphism. Notice, by the way, that the existence of ϕ needs uncountable forms of the Axiom of Choice, because ϕ is a nonmeasurable function with respect to the ordinary two-dimensional Lebesgue measure λ_2 on $\mathbf{R}^2$ and, simultaneously, ϕ does not possess the Baire property (cf. Exercise 6 for this chapter). Briefly speaking, ϕ is bad from the point of view of the standard

Lebesgue measure λ_2 on $\mathbf{R}^2$ and, simultaneously, from the point of view of ordinary Euclidean topology on $\mathbf{R}^2$.

On the other hand, consider the bijection ϕ^{-1} and equip $(\mathbf{R}^2, +)$ with the topology $\phi^{-1}(\mathcal{T})$, where $\mathcal{T}$ is the standard Euclidean topology on $\mathbf{R}$. We thus obtain the locally compact topological group $(\mathbf{R}^2, \phi^{-1}(\mathcal{T}))$ such that ϕ turns out to be an isomorphism between $(\mathbf{R}^2, \phi^{-1}(\mathcal{T}))$ and the topological group $\mathbf{R}$, so ϕ has very good descriptive properties and, in particular, it is measurable with respect to the completion of the Haar measure on $(\mathbf{R}^2, \phi^{-1}(\mathcal{T}))$, which is the ϕ^{-1}-image of the one-dimensional Lebesgue measure λ_1 on $\mathbf{R}$. Now, denote by

$$\gamma : (\mathbf{R}, +) \to (\mathbf{T}, +)$$

the canonical epimorphism given by

$$\gamma(t) = (\cos(t), \sin(t)) \quad (t \in \mathbf{R}),$$

and take the composition

$$\chi = \gamma \circ \phi.$$

Then it is easy to see that the argument used above is applicable to the homomorphism χ, too.

Example 1 inspires the question of whether there are ultimately bad homomorphisms acting from an uncountable commutative group $(G, +)$ into $(\mathbf{R}, +)$ (or into $(\mathbf{T}, +)$). Naturally, here the ultimate badness of homomorphisms

$$\phi : (G, +) \to (\mathbf{R}, +),$$

$$\psi : (G, +) \to (\mathbf{T}, +)$$

means that ϕ and ψ should be nonmeasurable with respect to every nonzero σ-finite G-quasi-invariant measure on G.

In the present chapter we discuss this question and describe all those commutative groups $(G, +)$ for which such homomorphisms ϕ and ψ do exist.

First, let us recall several notions from measure theory.

Let E be a nonempty set and let μ be a nonzero σ-finite measure defined on a σ-algebra of subsets of E.

As usual, we denote by the symbol $\text{dom}(\mu)$ the domain of μ (i.e., the σ-algebra of all μ-measurable sets) and by the symbol $\mathcal{I}(\mu)$ the σ-ideal generated by the family of all μ-measure zero sets.

A subset X of E is thick (or massive) with respect to μ if the equality $\mu_*(E \setminus X) = 0$ holds, where μ_* denotes the inner measure associated with μ.

Let $\mathcal{M}$ be some class of measures on E (in general, their domains are various σ-algebras of subsets of E) and let ϕ be a function on E all values of which are in $\mathbf{R}$ (or in $\mathbf{T}$).

Recall (see Chapter 3) that ϕ is absolutely nonmeasurable with respect to $\mathcal{M}$ if there exists no measure from $\mathcal{M}$ for which ϕ turns out to be measurable.

Accordingly, we say that a set $X \subset E$ is absolutely nonmeasurable with respect to $\mathcal{M}$ if the characteristic function (i.e., indicator) of X is absolutely nonmeasurable with respect to $\mathcal{M}$.

Also, by the standard definition, a measure μ given on a commutative group $(G, +)$ is translation quasi-invariant if both $\mathrm{dom}(\mu)$ and $\mathcal{I}(\mu)$ are translation invariant classes of subsets of G.

If, in addition, we have

$$\mu(g + X) = \mu(X) \quad (X \in \mathrm{dom}(\mu), \ g \in G),$$

then μ is called a translation invariant measure on $(G, +)$.

In general, the class of all σ-finite translation quasi-invariant measures on $(G, +)$ is much wider than the class of all σ-finite translation invariant measures on $(G, +)$.

Notice also that if $(G, \cdot)$ is a σ-compact locally compact group (not necessarily commutative), then the left (right) Haar measure ν on $(G, \cdot)$ is left (right) translation invariant, and any σ-finite measure on $(G, \cdot)$ equivalent to ν is left (right) translation quasi-invariant.

For a given commutative group $(G, +)$, we shall denote by $\mathcal{MQI}(G)$ the class of all nonzero σ-finite translation quasi-invariant measures on G.

Example 2. Let $\mathbf{Q}$ be the field of all rational numbers. As we have already noticed several times, every Vitali subset of $\mathbf{R}$ (i.e., every selector of $\mathbf{R}/\mathbf{Q}$) is absolutely nonmeasurable with respect to all those translation invariant measures on $\mathbf{R}$ which extend the Lebesgue measure $\lambda = \lambda_1$. On the other hand, there exist Vitali sets which are measurable with respect to certain translation quasi-invariant extensions of λ (in this connection, see Chapter 9). The above-mentioned facts vividly show that the concept of the absolute nonmeasurability of $\mathbf{R}$-valued ($\mathbf{T}$-valued) functions substantially depends on a choice of an initial class $\mathcal{M}$ of measures.

For the sake of brevity, throughout this chapter we shall say that a function acting from a commutative group $(G, +)$ into $\mathbf{R}$ (into $\mathbf{T}$) is absolutely nonmeasurable if, for every measure $\mu \in \mathcal{MQI}(G)$, this function is not measurable with respect to μ.

First, we would like to give a condition for a group homomorphism ϕ acting from $(G, +)$ into $\mathbf{R}$ (or into $\mathbf{T}$), which guarantees the absolute nonmeasurability of ϕ.

For this purpose, we need again the classical notion of a universal measure zero subset of $\mathbf{R}$ (of $\mathbf{T}$).

Let $Z \subset \mathbf{R}$ (respectively, $Z \subset \mathbf{T}$). We recall that Z is universal measure zero if, for any σ-finite continuous Borel measure μ on $\mathbf{R}$ (respectively, on $\mathbf{T}$), we have $\mu^*(Z) = 0$, where μ^* denotes, as usual, the outer measure associated with μ.

Equivalently, we may say that $Z \subset \mathbf{R}$ (respectively, $Z \subset \mathbf{T}$) is universal measure zero if there exists no nonzero σ-finite continuous Borel measure on Z. Of course, here Z is assumed to be endowed with the induced topology.

Some properties of universal measure zero sets are discussed in Chapter 5 of this book.

Several classical constructions (within **ZFC** set theory) of uncountable universal measure zero subsets of $\mathbf{R}$ have already been mentioned in preceding sections of the book. Recall once more that every Luzin subset of $\mathbf{R}$ and (under Martin's Axiom) every generalized Luzin subset of $\mathbf{R}$ are universal measure zero (in this connection, see Exercise 23 from Chapter 5).

We also know that certain uncountable universal measure zero subsets of $\mathbf{R}$ can carry the commutative group structure. The following auxiliary proposition is helpful for our further purposes.

Lemma 1. *There exists, within **ZFC** theory, an uncountable universal measure zero set $Z \subset \mathbf{R}$ (respectively, $Z \subset \mathbf{T}$) which simultaneously is a vector space over the field $\mathbf{Q}$ of all rational numbers.*

There are several different proofs of Lemma 1. As was pointed out in Chapter 13, the assertion of this lemma directly follows from one general statement of metamathematical character, which is formulated as follows.

Let $S(X)$ be a property of a subset X of a Polish topological space. Suppose that the following conditions are satisfied:

(a) if $S(X)$ and $Y \subset X$, then $S(Y)$;

(b) if $\{X_i : i \in I\}$ is a countable family of subsets of a Polish space and $S(X_i)$ for all $i \in I$, then $S(\cup\{X_i : i \in I\})$;

(c) if $S(X)$ and $S(Y)$, then $S(X \times Y)$;

(d) if E and E' are Polish spaces, $\psi : E \to E'$ is an injective Borel mapping and X is a subset of E with $S(X)$, then $S(\psi(X))$;

(e) there is a Polish space which contains an uncountable set X such that $S(X)$.

Then there exists an uncountable vector space $Z \subset \mathbf{R}$ (over the field $\mathbf{Q}$) such that $S(Z)$.

Notice, in this context, that if there exists an uncountable universal measure zero set $U \subset \mathbf{R}$ which is a vector space over $\mathbf{Q}$, then there exists an uncountable universal measure zero set $U' \subset \mathbf{T}$ which is also a vector space over $\mathbf{Q}$ (for more details, see Exercise 13).

The next auxiliary proposition yields a sufficient condition for the absolute nonmeasurability of homomorphisms acting from a commutative group $(G, +)$ into $\mathbf{R}$ (or into $\mathbf{T}$).

Lemma 2. *Let ϕ be a homomorphism acting from a commutative group $(G, +)$ into $\mathbf{R}$ (into $\mathbf{T}$) such that the range of ϕ is an uncountable universal measure zero subset of $\mathbf{R}$ (of $\mathbf{T}$). Then ϕ is absolutely nonmeasurable.*

Proof. Let ϕ satisfy the assumptions of Lemma 2. We have to show that ϕ is absolutely nonmeasurable. Suppose otherwise. Then there exists a measure $\mu \in \mathcal{MQI}(G)$ such that ϕ is μ-measurable. We may assume, without loss of generality, that μ is a probability measure, i.e., $\mu(G) = 1$.

Further, denote by $\mathrm{ran}(\phi)$ the range of ϕ and observe that, for any point $t \in \mathrm{ran}(\phi)$, the set $\phi^{-1}(t)$ is a translate of the set $\phi^{-1}(0)$. It is clear that in G there are uncountably many pairwise disjoint translates of $\phi^{-1}(0)$. In view of the translation quasi-invariance of μ and the supposed μ-measurability of ϕ, all the sets

$$\phi^{-1}(t) \quad (t \in \mathrm{ran}(\phi))$$

must be of μ-measure zero. Now, for every Borel subset B of $\mathrm{ran}(\phi)$, define

$$\mu'(B) = \mu(\phi^{-1}(B)).$$

We thus come to the Borel diffused probability measure μ' on $\mathrm{ran}(\phi)$ contradicting the fact that $\mathrm{ran}(\phi)$ is universal measure zero. The obtained contradiction shows that ϕ is absolutely nonmeasurable, which completes the proof.

Let $(G, +)$ be an arbitrary commutative group. As is widely known, the family of all elements of G of finite order constitutes a subgroup G_0 of G which is usually called the torsion subgroup of G (or the periodic part of G).

If $G_0 = \{0\}$, then G is called a torsion free group.

The next lemma is of purely group-theoretical character and probably is known (at least, for algebraists). However, for the sake of completeness, we give its proof here.

Lemma 3. *Let $(G, +)$ be a commutative group, G_0 be its torsion subgroup, and suppose that the quotient group G/G_0 is uncountable. Then there exists an uncountable subgroup H of G such that the relation*

$$G_0 \cap H = \{0\}$$

holds true.

Proof. It directly follows from the Kuratowski–Zorn lemma that there exists a maximal (with respect to the inclusion relation) group $H \subset G$ satisfying the equality $G_0 \cap H = \{0\}$. It suffices to show that H is uncountable. Suppose otherwise, i.e.,

$$\mathrm{card}(H) \leq \omega$$

where ω denotes, as usual, the first infinite cardinal. Since

$$\mathrm{card}(G/G_0) > \omega,$$

we may choose an uncountable family $\{f_i : i \in I\}$ of elements from G which are pairwise incomparable modulo G_0, i.e.,

$$f_i - f_j \notin G_0 \quad (i \in I, \ j \in I, \ i \neq j).$$

Moreover, since H is at most countable, we may assume without loss of generality that $f_i \notin H$ for all indices $i \in I$.

By virtue of the maximality of H, for each index $i \in I$, there exist a natural number $m_i \neq 0$, an element $h_i \in H$ and an element $g_i \in G_0$ such that

$$g_i = m_i f_i + h_i.$$

Further, since m_i and h_i range over countable sets and f_i ranges over an uncountable set, there exists a natural number $m \neq 0$ and two distinct indices $i \in I$ and $j \in I$ for which the equalities

$$g_i = m f_i + h_i,$$

$$g_j = m f_j + h_j,$$

$$h_i = h_j$$

are fulfilled. From these equalities we immediately infer

$$g_i - g_j = m(f_i - f_j) + (h_i - h_j) = m(f_i - f_j),$$

whence it follows that $f_i - f_j$ belongs to G_0, which contradicts the definition of the family $\{f_i : i \in I\}$.

The obtained contradiction finishes the proof of Lemma 3.

Lemma 4. *Let $(G, +)$ be a commutative group and let H be any subgroup of G. Then H is not absolutely nonmeasurable.*

The proof of Lemma 4 is not difficult and may be found in Chapter 13 of [128] (see also Exercises 7 and 8 for the present chapter).

Remark 1. In connection with the previous lemma, we would like to notice that:

(i) if $\mathrm{card}(G/H) \leq \omega$, then there exists a σ-finite translation invariant measure μ on G for which we have $H \in \mathrm{dom}(\mu)$ and $\mu(H) > 0$;

(ii) if $\mathrm{card}(G/H) > \omega$, then H turns out to be a G-absolutely negligible set in G (for the definition of G-absolutely negligible sets, see Chapter 5 of this book), so any σ-finite translation invariant (translation quasi-invariant) measure ν on G admits a translation invariant (translation quasi-invariant) extension ν' on G for which we have $H \in \mathrm{dom}(\nu')$ and $\nu'(H) = 0$.

A more detailed explanation of (ii) is given in Exercise 8 for the present chapter.

Lemma 5. *Any commutative group* $(H, +)$ *can be represented in the form*

$$H = \cup \{H_n : n < \omega\},$$

where the family $\{H_n : n < \omega\}$ *is increasing by inclusion and all* H_n *are direc sums of cyclic groups.*

Lemma 5 is due to Kulikov and is known as Kulikov's theorem on the alge braic structure of commutative groups. Its proof can be found in [72] or [159].

We now are ready to formulate and prove the main statement of this chapter

Theorem 1. *Let* $(G, +)$ *be a commutative group and let* G_0 *denote the torsion subgroup of* G. *The following two assertions are equivalent:*

(1) *the quotient group* G/G_0 *is uncountable;*

(2) *there exists an absolutely nonmeasurable homomorphism acting from* $(G, +)$ *into* $(\mathbf{R}, +)$ *(or into* $(\mathbf{T}, +)$*).*

Proof. Suppose that (1) is satisfied. According to Lemma 3, there exists an uncountable subgroup H of G such that

$$G_0 \cap H = \{0\}.$$

By virtue of Lemma 5, this H can be represented in the form

$$H = \cup \{H_n : n < \omega\},$$

where the family of groups $\{H_n : n < \omega\}$ is increasing by inclusion and all H_n are direct sums of cyclic groups. Since all elements from H are of infinite order every H_n is a direct sum of infinite cyclic groups (each of them is isomorphic to the additive group $\mathbf{Z}$ of integers). Consequently, every H_n turns out to be a free commutative group. Since H is uncountable, at least one of the groups H_n is uncountable, too. We may assume, without loss of generality, that H_0 is such a group.

Further, consider an uncountable universal measure zero subset U of $\mathbf{R}$ (of $\mathbf{T}$) which is a vector space over $\mathbf{Q}$ (the existence of U is guaranteed by Lemma 1). Obviously, there exists a homomorphism ϕ from the free commutative group H_0 into U such that the range of ϕ is uncountable. Since U is a divisible commutative group, this ϕ can be extended to a homomorphism acting from the whole $(G, +)$ into U (see again [72], [159] or Exercise 1 of this chapter). For the sake of brevity, we preserve the same notation ϕ for the extended in this manner homomorphism. Now, Lemma 2 says that ϕ is absolutely nonmeasurable, so (2) is fulfilled.

Conversely, assume that (2) is satisfied. We have to show that (1) holds true, too. Suppose otherwise, i.e.,

$$\text{card}(G/G_0) \leq \omega.$$

Let ϕ be any homomorphism acting from $(G, +)$ into $(\mathbf{R}, +)$ (or into $(\mathbf{T}, +)$). It can easily be seen that the range of ϕ is at most countable. Denote by F the kernel of ϕ. Since F is a subgroup of $(G, +)$, it is not absolutely nonmeasurable in view of Lemma 4. Consequently, there exists a nonzero σ-finite translation invariant measure μ on G such that $F \in \mathrm{dom}(\mu)$. Since the range of ϕ is countable, we deduce that there are countably many pairwise disjoint translates of F which collectively cover the whole group G. Therefore, ϕ turns out to be measurable with respect to μ. In other words, we have shown that there are no absolutely nonmeasurable homomorphisms acting from $(G, +)$ into $(\mathbf{R}, +)$ (into $(\mathbf{T}, +)$). But this fact contradicts (2).

Theorem 1 has thus been proved.

As a straightforward corollary of Theorem 1, we get the next statement.

Theorem 2. *Let $(G, +)$ be a commutative group. The following two assertions are equivalent:*

(1) there exists an absolutely nonmeasurable homomorphism from $(G, +)$ into $(\mathbf{R}, +)$;

(2) there exists an absolutely nonmeasurable homomorphism from $(G, +)$ into $(\mathbf{T}, +)$.

Remark 2. The implication $(1) \Rightarrow (2)$ can be proved directly, because of the existence of a canonical surjective continuous homomorphism γ acting from $\mathbf{R}$ onto $\mathbf{T}$ and defined by

$$\gamma(t) = (\cos(t), \sin(t)) \quad (t \in \mathbf{R}).$$

Here it suffices to apply Lemma 2 and the simple fact that there is a covering of $\mathbf{R}$ by countably many compacts such that all restrictions of γ to those compacts are injective functions. However, the converse implication $(2) \Rightarrow (1)$ makes essential use of Theorem 1.

Another consequence of Theorem 1 can be formulated as follows.

Theorem 3. *Let $(G, +)$ be a commutative group, G_0 be its torsion subgroup, and suppose that $\mathrm{card}(G/G_0) > \omega$. Then there exists a homomorphism ϕ from G into $\mathbf{R}$ (into $\mathbf{T}$) having the following property: if G is regarded as a thick subgroup of a σ-compact locally compact group G' and μ is the measure induced on G by the Haar measure μ' on G', then ϕ turns out to be absolutely nonmeasurable with respect to the class of all translation quasi-invariant extensions of μ.*

We leave to the reader the proof of Theorem 3.

Remark 3. As far as we know, analogous questions for non-commutative groups were not considered in the literature. Consequently, it would be interesting to envisage absolutely nonmeasurable homomorphisms of uncountable non-commutative groups.

EXERCISES

1*. Recall that a commutative group $(H, +)$ is divisible if, for any natural number $m \neq 0$ and for any element $h \in H$, the equation $mx = h$ has a solution in H.

For example, the classical additive groups $(\mathbf{R}, +)$ and $(\mathbf{T}, +)$ are divisible. Recall also that a partial mapping

$$\phi : (G, +) \to (H, +)$$

is a partial homomorphism of a commutative group $(G, +)$ into a commutative group $(H, +)$ if there exists a subgroup G_0 of G such that $\mathrm{dom}(\phi) = G_0$ and ϕ is a homomorphism acting from $(G_0, +)$ into $(H, +)$.

Demonstrate that if $(H, +)$ is divisible and $\phi : (G, +) \to (H, +)$ is a partial homomorphism, then there exists a group homomorphism

$$\phi' : (G, +) \to (H, +)$$

extending ϕ.

For this purpose, consider the family of all those partial group homomorphisms of G to H which extend ϕ. By virtue of the Kuratowski-Zorn lemma, there exists a maximal (with respect to the inclusion relation) element ϕ' of the above-mentioned family. In order to show that

$$\mathrm{dom}(\phi') = G,$$

suppose to the contrary that

$$G \setminus \mathrm{dom}(\phi') \neq \emptyset$$

and pick an element $g \in G \setminus \mathrm{dom}(\phi')$.

Then denote by F the subgroup of G generated by $\{g\} \cup \mathrm{dom}(\phi')$.

Only two cases are possible.

(a) For every natural number $m > 0$, the element mg does not belong to $\mathrm{dom}(\phi')$.

In this case, put $\phi''(g) = 0$ and check that ϕ' can be extended to a homomorphism ϕ'' acting from F into H.

(b) There exists a natural number $m > 0$ such that $mg \in \mathrm{dom}(\phi')$.

In this case, one may assume without loss of generality that m is a least nonzero natural number satisfying $mg \in \mathrm{dom}(\phi')$. Denote

$$h' = \phi'(mg).$$

Since H is divisible, there exists an element $h \in H$ such that

$$h' = mh.$$

Then define
$$\phi''(g) = h$$
and check that ϕ' can be extended to a homomorphism ϕ'' acting from F into H.

Thus, in both cases a contradiction is obtained with the maximality of ϕ', which yields the required result.

2. Applying the result of Exercise 1, prove that every commutative group $(G, +)$ can be isomorphically embedded in the torus $(\mathbf{T}^\alpha, +)$ of topological weight α, where α is an appropriate cardinal number (depending on an initial group G).

Conclude that, any commutative group $(G, +)$ may be treated as an everywhere dense subgroup of some compact topological group.

3*. Let α be a nonzero cardinal number less than or equal to the cardinality continuum **c**.

By using the method of transfinite recursion, construct a homomorphism
$$\psi : (\mathbf{R}, +) \to (\mathbf{T}^\alpha, +)$$
such that the graph of ψ is $(\lambda \otimes \nu)$-thick in $\mathbf{R} \times \mathbf{T}^\alpha$, where λ denotes the standard Lebesgue measure on $\mathbf{R}$ and ν stands for the completion of the Haar probability measure on $\mathbf{T}^\alpha$.

Check that:
(a) ψ is discontinuous at all points of $\mathbf{R}$;
(b) ψ is nonmeasurable with respect to λ.

Now, for every $(\lambda \otimes \nu)$-measurable subset B of the product group $\mathbf{R} \times \mathbf{T}^\alpha$, define
$$B' = \{x \in \mathbf{R} : (x, \psi(x)) \in B\}$$
and introduce the family of sets
$$\mathcal{S} = \{B' : B \in \mathrm{dom}(\lambda \otimes \nu)\}.$$

Further, put
$$\mu(B') = (\lambda \otimes \nu)(B) \qquad (B' \in \mathcal{S})$$
and verify the validity of the following assertions:

(c) $\mathcal{S}$ contains $\mathrm{dom}(\lambda)$ and is a translation invariant σ-algebra of subsets of the real line $\mathbf{R}$;
(d) μ is a translation invariant measure on $\mathbf{R}$ strictly extending λ;
(e) ψ is measurable with respect to μ;
(f) if $\alpha > \omega$, then μ is a nonseparable measure.

4*. Demonstrate that all discontinuous characters on a σ-compact locally compact topological group $(G, \cdot)$ are necessarily nonmeasurable with respect to the completion of the left (right) Haar measure on G.

For this purpose, utilize the Steinhaus property of Haar measurable sets in G (cf. Exercise 1 from Chapter 3).

Remark 4. Several different versions of the classical result presented in the previous exercise will be discussed in Chapter 19.

5. Show that all discontinuous characters on a σ-compact locally compact topological group $(G, \cdot)$ do not possess the Baire property.

For this purpose, utilize a certain topological analogue of the Steinhaus property, namely, the Banach–Kuratowski–Pettis property (see [147], [152], [202]).

6. Consider $(\mathbf{R}, +)$ and $(\mathbf{R}^2, +)$ as abstract commutative groups. Let

$$\phi : (\mathbf{R}, +) \to (\mathbf{R}^2, +)$$

be a group homomorphism such that the range of ϕ contains some three non-collinear points of the plane $\mathbf{R}^2$.

Verify that at least one of the induced endomorphisms

$$\mathrm{pr}_1 \circ \phi : \mathbf{R} \to \mathbf{R},$$

$$\mathrm{pr}_2 \circ \phi : \mathbf{R} \to \mathbf{R}$$

is not measurable in the Lebesgue sense and does not possess the Baire property.

In particular, if $\mathrm{ran}(\phi)$ is not of λ_2-measure zero (or is not of first category in $\mathbf{R}^2$), then at least one of the above-mentioned induced endomorphisms of $\mathbf{R}$ is nonmeasurable in the Lebesgue sense and does not possess the Baire property.

7. Let $(G, +)$ be a commutative group and let H be a subgroup of G. Show that if

$$\mathrm{card}(G/H) \leq \omega,$$

then there exists a σ-finite translation invariant measure μ on G for which one has

$$H \in \mathrm{dom}(\mu), \quad \mu(H) > 0.$$

In particular, H is not an absolutely nonmeasurable subset of G.

Moreover, if $\mathrm{card}(G/H) < \omega$, then μ can be assumed to be a probability translation invariant measure on G.

8. Let $(G, +)$ be a commutative group and let H be a subgroup of G. Show that if

$$\mathrm{card}(G/H) > \omega,$$

then H turns out to be a G-absolutely negligible set in G.

For this purpose, use a characterization of absolutely negligible sets given in Exercise 25 of Chapter 5.

Conclude that H is not an absolutely nonmeasurable subset of G.

9. Give a detailed proof of Theorem 3 of this chapter.

10*. Let $C = \{0,1\}^\omega$ be the Cantor space regarded as a commutative compact metrizable group with respect to the standard product topology and group operation modulo 2.

By using the Continuum Hypothesis (Martin's axiom), demonstrate that C contains a Luzin subset (a generalized Luzin subset) which simultaneously is a subgroup of C.

Conclude that, under these additional axioms, there exist universal measure zero subgroups of C which are equinumerous with C.

11. Let $(G,+)$ be an arbitrary 2-divisible commutative group, i.e., all the equations

$$x + x = g \quad (g \in G)$$

are solvable in G (for example, one may take $G = \mathbf{R}$ or $G = \mathbf{T}$).

Verify that any homomorphism

$$\phi : G \to C$$

is trivial, where C is again the Cantor space treated as a compact metrizable commutative group (see the previous exercise).

Conclude from the stated above that there exist no absolutely nonmeasurable homomorphisms acting from G into C (although condition (1) of Theorem 1 may be satisfied for G).

Remark 5. Taking into account the result of Exercise 11, it would be interesting to describe all those Polish commutative groups $(H,+)$ for which a certain analogue of Theorem 1 holds true. More precisely, in that analogue the classical groups $\mathbf{R}$ and $\mathbf{T}$ should be replaced by a group $(H,+)$.

12. Let L be a generalized Luzin subset of the Cantor space $C = \{0,1\}^\omega$ and suppose, in addition, that L is simultaneously a subgroup of C.

Assuming Martin's Axiom, check that the identical embedding of L into C is a group monomorphism absolutely nonmeasurable with respect to the class of all nonzero σ-finite continuous measures on L.

Remark 6. In connection with the previous exercise and the generalized Luzin set L described therein, it should be noticed that the class $\mathcal{MQI}(L)$ of all nonzero σ-finite translation quasi-invariant measures on L is ample in the sense that, for every measure $\mu \in \mathcal{MQI}(L)$, there exists a measure $\mu' \in \mathcal{MQI}(L)$ which strictly extends μ (in this connection, see [128] or Chapter 5 of this book).

13*. Demonstrate that if there exists an uncountable universal measure zero vector space $U \subset \mathbf{R}$ over $\mathbf{Q}$, then there exists an uncountable universal measure zero vector space $U' \subset \mathbf{T}$ over $\mathbf{Q}$.

For this purpose, represent $\mathbf{T}$ in the form of a direct sum of its subgroups

$$\mathbf{T} = T_0 + W \qquad (\operatorname{card}(T_0 \cap W) = 1),$$

where T_0 denotes the torsion subgroup of $\mathbf{T}$, and consider the set $\gamma(U)$, where γ stands for the canonical epimorphism of $\mathbf{R}$ onto $\mathbf{T}$ defined by

$$\gamma(t) = (\cos(t), \sin(t)) \qquad (t \in \mathbf{R}).$$

Then, keeping in mind the equality $\operatorname{card}(T_0) = \omega$, check that the set

$$U' = W \cap \gamma(U)$$

is as required.

16. Measurable and nonmeasurable sets with homogeneous sections

In this chapter we will be dealing with those sets in the Euclidean plane $\mathbf{R}^2$, which have homogeneous (from the measure-theoretical viewpoint) horizontal and vertical linear sections. Also, we will touch upon those sets in the Euclidean space $\mathbf{R}^3$, which have homogeneous (from the same viewpoint) sections produced by planes parallel to one of the three coordinate planes of $\mathbf{R}^3$.

In particular, we will see that some of such sets can be measurable in the sense of the standard two-dimensional Lebesgue measure λ_2 on $\mathbf{R}^2$ (respectively, in the sense of the standard three-dimensional Lebesgue measure λ_3 on $\mathbf{R}^3$) and some of them can be nonmeasurable in the sense of λ_2 (respectively, in the sense of λ_3).

Needless to say, the topic we touch upon here is mainly motivated by the classical theorem of Fubini concerning double and iterated integrals of Lebesgue measurable functions of two variables. Also, it is well known that Fubini's theorem plays an utterly important role in many topics of mathematical analysis and probability theory (see, for instance, [17], [89], [96], [197], [199], [203]). This theorem has a very long history. Its prototype is Cavalieri's principle which had been applied repeatedly (mostly, at intuitive level) for calculating the areas and volumes of various geometric figures. In this context, it should be recalled that much earlier than Cavalieri, the great ancient Greek scientist Archimedes successfully utilized the above-mentioned principle in practice and, as a result, was able to obtain the beautiful formula of the volume of a three-dimensional Euclidean ball.

Of course, after introducing the rigorous concepts of general measure theory, the spectrum of applications of Cavalieri's principle became substantially wider and the principle itself was fully replaced by Fubini's theorem. Recall that, according to this theorem, if a given function $f : [0,1]^2 \to \mathbf{R}$ of two real variables is bounded and Lebesgue measurable, then there exist the corresponding iterated integrals of f and the equality

$$\int_0^1 (\int_0^1 f(x,y)dx)dy = \int_0^1 (\int_0^1 f(x,y)dy)dx$$

holds true. In fact, both sides of this equality are identical with the double Lebesgue integral $\int_0^1 \int_0^1 f(x,y)dxdy$.

On the other hand, it is also known that, for the existence of these iterated integrals and for their coincidence, the measurability of f with respect to λ_2 is not necessary. Indeed, Sierpiński [235] was able to construct, by using the method of transfinite recursion, an injective function

$$\phi : [0, 1] \to [0, 1]$$

whose graph $\mathrm{Gr}(\phi)$ is thick in $[0, 1]^2$ with respect to λ_2. This means that $\mathrm{Gr}(\phi)$ meets every Borel subset of $[0, 1]^2$ with strictly positive λ_2-measure (notice that similar constructions are presented, e.g., in [10], [77], [203]; cf. also Exercise 14 from Chapter 5). Now, denoting by g the characteristic function of $\mathrm{Gr}(\phi)$, we infer that g is not λ_2-measurable, but its iterated integrals do exist and both of them are equal to zero (hence they are equal to each other).

At the same time, by assuming the Continuum Hypothesis (**CH**):

$$\mathbf{c} = 2^\omega = \omega_1,$$

where ω_1 stands, as usual, for the least uncountable cardinal (ordinal) number, Sierpiński constructed a subset S of $[0, 1]^2$ satisfying the following relations:

(i) for every $x \in [0, 1]$, the set $S \cap (\{x\} \times [0, 1])$ is at most countable;

(ii) for every $y \in [0, 1]$, the set $([0, 1]^2 \setminus S) \cap ([0, 1] \times \{y\})$ is at most countable.

Though this construction is not difficult, it is quite ingenious. Let us briefly recall it (cf. [233] or the proof of Theorem 4 from Chapter 4).

Indeed, supposing that **CH** holds true, one may equip the unit interval $[0, 1]$ with a well-ordering $\preceq$ which is isomorphic to the ordinal number ω_1. Now, putting

$$S = \{(x, y) \in [0, 1]^2 : y \preceq x\},$$

it can readily be checked that both relations (i) and (ii) are satisfied for S.

Moreover, Sierpiński also established that the existence of a set S with the properties (i) and (ii) is equivalent to the Continuum Hypothesis (for more details, see his widely known book [243] or Exercise 11 of Chapter 4).

Thus, the Continuum Hypothesis is equivalent to the possibility of decomposing the unit square $[0, 1]^2$ into two sets

$$S_1 = S, \quad S_2 = [0, 1]^2 \setminus S$$

such that S_1 meets each vertical line of the plane at finitely or countably many points and S_2 meets every horizontal line of the plane at finitely or countably many points.

This remarkable decomposition of the square implies numerous nontrivial consequences (see, e.g., the extensive survey [247]). For example, an analogous decomposition of the product set $\omega_1 \times \omega_1$ leads to the classical theorem of Ulam [263] stating the non-real-valued measurability of ω_1. Recall that Ulam's theorem does not need additional set-theoretical axioms, so it is a result of **ZFC** set theory (cf. Exercise 2 for the present chapter). This result strengthens the theorem of Banach and Kuratowski [8] proved by them, with the aid of **CH**, one year earlier than Ulam's theorem was established.

Remark 1. In a certain sense, we may say that both sets S_1 and S_2 of Sierpiński's decomposition are absolutely nonmeasurable with respect to the class of completions of products of the form $\mu \otimes \nu$, where μ and ν are any nonzero σ-finite continuous measures on $[0, 1]$. In this context, let us especially underline that there are many nonproduct measures on $\mathbf{R}^2$ strictly extending λ_2, for which the assertion of Fubini's theorem remains true. One of such measures can be obtained if we expand the σ-ideal $\mathcal{I}(\lambda_2)$ of all λ_2-measure zero sets by adding to it all those subsets of $\mathbf{R}^2$ whose vertical and horizontal sections are at most countable (see Exercise 3).

Now, denoting by $h = \chi_S$ the characteristic function of $S \subset [0, 1]^2$, we can readily verify that h admits both iterated integrals, but one of them is equal to 0 and the other is equal to 1. This fact shows that, for a nonnegative bounded real-valued function of two variables, the existence of its iterated integrals does not always imply their coincidence. In view of these two important circumstances, it makes sense to consider the class of all those functions

$$f : [0, 1]^2 \to [0, +\infty[,$$

for which the corresponding iterated integrals exist. As was already mentioned, these iterated integrals sometimes may coincide, and sometimes not. Since any such f can be approximated by nonnegative linear combinations of character-istic functions of subsets of $[0, 1]^2$, it seems to be natural first to consider the characteristic functions of plane point sets (cf. [209]).

To continue our presentation, let us give the following simple example con-cerning unbounded subsets of the Euclidean plane $\mathbf{R}^2$.

Example 1. Let a and b be any two strictly positive real numbers. It is easy to indicate a subset Z of the plane $\mathbf{R}^2$, such that all horizontal sections of Z are line segments of length a and all vertical sections of Z are line segments of length b. In fact, Z can be taken as a strip in $\mathbf{R}^2$, the two boundary lines of which are analytically expressible in the form:

$$y = (b/a)x + c_1, \quad y = (b/a)x + c_2,$$

where c_1 and c_2 are real numbers satisfying the equality $|c_1 - c_2| = b$.

This example is absolutely elementary and geometrically visual. The natural question arises whether it is possible to construct a bounded set with somewhat analogous properties of its linear (i.e., horizontal and vertical) sections.

To be more precise, let us denote by $\lambda = \lambda_1$ the standard one-dimensional Lebesgue measure on $\mathbf{R} = \mathbf{R}^1$ and take any two real numbers a and b such that

$$0 \le a \le 1, \quad 0 \le b \le 1.$$

Then one may ask whether there exists a set $W \subset [0,1]^2$ satisfying the relations:

(1) all horizontal sections $([0,1] \times \{y\}) \cap W$, where y ranges over $[0,1]$, are of λ-measure a;

(2) all vertical sections $(\{x\} \times [0,1]) \cap W$, where x ranges over $[0,1]$, are of λ-measure b.

Clearly, the above-mentioned set $S \subset [0,1]^2$ of Sierpiński corresponds to the case $a = 1$ and $b = 0$.

For our further purposes, it will be convenient to introduce the following definition.

We shall say that $W \subset [0,1]^2$ is an (a,b)-homogeneous set in the unit square $[0,1]^2$ if both relations (1) and (2) are fulfilled for W.

It turns out that if $a = b$, then an (a,b)-homogeneous set W can be constructed within the framework of $\mathbf{ZF}$ & $\mathbf{DC}$ theory. The corresponding construction is presented below. It is based on some auxiliary notions and propositions.

We shall say that a subset Z of the square $[0,1]^2$ is saturated (in $[0,1]^2$) if both sets $[0,1] \setminus \mathrm{pr}_1(Z)$ and $[0,1] \setminus \mathrm{pr}_2(Z)$ are at most countable.

We shall say that $W \subset [0,1]^2$ is an almost (a,b)-homogeneous set in the square $[0,1]^2$ if:

(i') λ-almost all horizontal sections $([0,1] \times \{y\}) \cap W$, where $y \in [0,1]$, are of λ-measure a;

(ii') λ-almost all vertical sections $(\{x\} \times [0,1]) \cap W$, where $x \in [0,1]$, are of λ-measure b.

Lemma 1. *Let W be an almost (a,b)-homogeneous set in $[0,1]^2$. Then there exists an (a,b)-homogeneous set W^* in $[0,1]^2$ such that $\lambda_2(W^* \triangle W) = 0$.*

Consequently, W^* is λ_2-measurable if and only if W is λ_2-measurable.

This lemma is easy, so we omit its proof here and leave it to the reader (see Exercise 4).

Lemma 2. *Let a real number $a \in [0,1]$ be of the form $a = 1/2^n$, where $n > 0$ is a natural number. Then there are two families*

$$\{Q_i : 1 \le i \le 2^n\}, \quad \{Q'_j : 1 \le j \le 2^n\}$$

of open squares in $[0,1]^2$ such that:

(1) the sides of all squares Q_i $(1 \le i \le 2^n)$ and Q'_j $(1 \le j \le 2^n)$ are respectively parallel to the coordinate axes $\mathbf{R} \times \{0\}$ and $\{0\} \times \mathbf{R}$;

(2) all squares Q_i $(1 \le i \le 2^n)$ are pairwise disjoint;

(3) all squares Q'_j $(1 \le j \le 2^n)$ are pairwise disjoint;

(4) no two squares Q_i and Q'_j, where $1 \leq i \leq 2^n$ and $1 \leq j \leq 2^n$, have common points;

(5) both sets $\cup\{Q_i : 1 \leq i \leq 2^n\}$ and $\cup\{Q'_j : 1 \leq j \leq 2^n\}$ are saturated;

(6) the side lengths of all squares Q_i $(1 \leq i \leq 2^n)$ and Q'_j $(1 \leq j \leq 2^n)$ are equal to $1/2^n$.

Proof. Decompose each side of $[0,1]^2$ into 2^n many pairwise congruent segments and consider the straight lines which pass through the endpoints of these segments and are parallel to the coordinate axes. In this manner, we come to the decomposition of $[0,1]^2$ into 2^{2n} smaller pairwise congruent open squares. Let $\{Q_i : 1 \leq i \leq 2^n\}$ be the family of all those smaller squares which are arranged along one diagonal of $[0,1]^2$, and let $\{Q'_j : 1 \leq j \leq 2^n\}$ be the family of all those smaller squares which are arranged along the other diagonal of $[0,1]^2$. It is easy to check that the above-mentioned two families are as required.

Lemma 2 has thus been proved.

It directly follows from Lemma 2 that, for any $a = 1/2^n$, there exists an almost (a, a)-homogeneous subset $Z_a = \cup\{Q_i : 1 \leq i \leq 2^n\}$ of the square $[0,1]^2$ and, moreover, there is a family $\{Q'_j : 1 \leq j \leq 2^n\}$ of open squares which do not have common points with Z_a and all side lengths of which are equal to $1/2^n$. Notice also that the set $\cup\{Q'_j : 1 \leq j \leq 2^n\}$ is saturated (this fact will be exploited below).

Lemma 3. Let a real number $a \in [0,1]$ admit a finite dyadic expansion

$$a = 1/2^{n_1} + 1/2^{n_2} + ... + 1/2^{n_k},$$

where $k \geq 1$ and $(n_1, n_2, ..., n_k)$ is a strictly increasing sequence of positive natural numbers. Then there exists an almost (a, a)-homogeneous subset Z_a of the square $[0,1]^2$.

Proof. We argue by induction on k. Suppose that, for some $k \geq 1$, the existence of Z_a has already been established in such a manner that there is a family $\{Q'_j : 1 \leq j \leq 2^{n_k}\}$ of open squares which do not have common points with Z_a, all side lengths of which are equal to $1/2^{n_k}$, and the set

$$\cup\{Q'_j : 1 \leq j \leq 2^{n_k}\}$$

is saturated. Now, take any real number a' of the form

$$a' = 1/2^{n_1} + 1/2^{n_2} + ... + 1/2^{n_k} + 1/2^{n_{k+1}} = a + 1/2^{n_{k+1}},$$

where $n_{k+1} > n_k$. Every square Q'_j can be dissected into pairwise congruent smaller open squares whose side lengths are equal to $1/2^{n_{k+1}}$. We denote by D_j the union of those squares from this dissection which are arranged along one diagonal of Q'_j, and denote by D'_j the union of those squares from this dissection which are arranged along the other diagonal of Q'_j (cf. the proof of Lemma 2). Then we put

$$Z_{a'} = Z_a \cup (\cup\{D_j : 1 \le j \le 2^{n_k}\}).$$

It can readily be verified that $Z_{a'}$ is as required. At the same time, the set

$$\cup\{D'_j : 1 \le j \le 2^{n_k}\}$$

is saturated, has no common points with $Z_{a'}$, and is representable as the union of 2^{n_k+1} open squares with side lengths equal to $1/2^{n_k+1}$.

This completes the proof of Lemma 3.

Theorem 1. *For every real number $a \in [0,1]$, there exists a λ_2-measurable (a,a)-homogeneous subset Z_a^* of $[0,1]^2$.*

Proof. Take any real number $a \in [0,1]$. If $a = 0$, then we define $Z_a^* = \emptyset$. Suppose now that $0 < a \le 1$ and consider the dyadic expansion of a:

$$a = 1/2^{n_1} + 1/2^{n_2} + ... + 1/2^{n_k} + ...,$$

where $n_1 < n_2 < ... < n_k < ...$. Further, denote

$$a_k = 1/2^{n_1} + 1/2^{n_2} + ... + 1/2^{n_k} \quad (k \ge 1).$$

According to Lemma 3, for any a_k, there exists an almost (a_k, a_k)-homogeneous set $Z_{a_k} \subset [0,1]^2$. In addition, the proof of Lemma 3 shows that we may suppose the validity of the inclusions

$$Z_{a_1} \subset Z_{a_2} \subset ... \subset Z_{a_k} \subset ... \ .$$

Let us put

$$Z_a = Z_{a_1} \cup Z_{a_2} \cup ... \cup Z_{a_k} \cup ... \ .$$

Then it becomes clear that the set Z_a defined in this manner is almost (a,a)-homogeneous. Evidently, we may apply Lemma 1 to Z_a. Utilizing this lemma, we obtain an (a,a)-homogeneous set Z_a^* such that $\lambda_2(Z_a^* \triangle Z_a) = 0$. Finally, since the set Z_a is λ_2-measurable, Z_a^* is λ_2-measurable, too.

Remark 2. It is not difficult to check that the (a,a)-homogeneous set Z_a^* of Theorem 1 can be chosen to be Borel measurable. In fact, the construction of the set Z_a^* presented above is done within **ZF & DC** theory.

On the other hand, if should be underlined that if $a \ne b$, then no construction of an (a,b)-homogeneous set $Z_{a,b} \subset [0,1]^2$ can be realized in **ZF & DC** theory because according to Fubini's theorem, such a set must be nonmeasurable with respect to λ_2. Moreover, as follows from one result of Friedman [70], a set $Z_{a,b}$ with the desired properties cannot be constructed even within **ZFC** set theory.

However, by starting with Sierpiński's decomposition of the unit square $[0,1]^2$, it becomes possible to establish the following statement which relies on one consequence of Martin's Axiom (see the beginning of Chapter 12 and Appendix 3).

Theorem 2. *Suppose that all subsets of* $\mathbf{R}$ *with cardinalities strictly less than* $\mathbf{c}$ *are of λ-measure zero. Then, for any two real numbers* $a \in [0,1]$ *and* $b \in [0,1]$*, there exists an (a,b)-homogeneous subset W of the square* $[0,1]^2$.

Proof. The argument is very similar to that of Sierpiński's construction. Indeed, using a well-ordering of $[0,1]$ isomorphic to the least ordinal number equinumerous with $[0,1]$, we see that there exists a partition $\{A, B\}$ of the square $[0,1]^2$ such that:

(1) for every $x \in [0,1]$, the cardinality of $A \cap (\{x\} \times [0,1])$ is strictly less than $\mathbf{c}$;

(2) for every $y \in [0,1]$, the cardinality of $B \cap ([0,1] \times \{y\})$ is strictly less than $\mathbf{c}$.

Further, for any $y \in [0,1]$, we have the relation

$$\lambda(A \cap ([0,1] \times \{y\})) = 1,$$

so there is a λ-measurable set $F_y \subset A \cap ([0,1] \times \{y\})$ whose λ-measure is equal to a.

Analogously, for any $x \in [0,1]$, we have the relation

$$\lambda(B \cap (\{x\} \times [0,1])) = 1,$$

so there is a λ-measurable set $G_x \subset B \cap (\{x\} \times [0,1])$ whose λ-measure is equal to b.

Now, we define the set W as follows:

$$W = (\cup\{G_x : x \in [0,1]\}) \cup (\cup\{F_y : y \in [0,1]\}).$$

It is not difficult to verify that W is an (a,b)-homogeneous subset of the unit square $[0,1]^2$.

Remark 3. For an arbitrary (a,b)-homogeneous set $W \subset [0,1]^2$, denote by

$$\chi_W : [0,1]^2 \to \{0,1\}$$

the characteristic function of W. It is easy to see that there exist iterated integrals

$$\int_0^1 (\int_0^1 \chi_W(x,y)dx)dy = a, \qquad \int_0^1 (\int_0^1 \chi_W(x,y)dy)dx = b.$$

Clearly, these iterated integrals are equal to each other if and only if $a = b$. Some situations with the equality of the iterated integrals for characteristic functions

χ_Z, where $Z \subset [0,1]^2$ is not *a priori* assumed to be λ_2-measurable, are discussed in [209] (cf. also [135]).

It is natural to extend the above considerations to the case of the three-dimensional Euclidean space $\mathbf{R}^3$. However, here the following two possibilities should be taken into account.

Firstly, we may consider the sections of a given set $W \subset \mathbf{R}^3$ by straight lines parallel to one of the coordinate axes Ox, Oy and Oz.

Secondly, we may also consider the sections of $W \subset \mathbf{R}^3$ by those planes which are parallel to one of the three coordinate planes xOy, yOz, and zOx.

Let us give a simple example concerning unbounded subsets of $\mathbf{R}^3$ and analogous to Example 1 which was concerned with unbounded subsets of $\mathbf{R}^2$.

Example 2. Suppose that a, b and c are any three strictly positive real numbers. There exists a set $P \subset \mathbf{R}^3$ such that:

(1) all sections of P by the planes parallel to xOy are triangles of area a;

(2) all sections of P by the planes parallel to yOz are triangles of area b;

(3) all sections of P by the planes parallel to zOx are triangles of area c.

Moreover, an elementary argument shows that, similarly to the case of $\mathbf{R}^2$, some unbounded triangular prism can be taken as P (see Exercise 5).

If we want to obtain an analogous result for bounded sets in $\mathbf{R}^3$, then we again need to use some delicate set-theoretical techniques inspired by the Sierpiński decomposition of $[0,1]^2$. Of course, in the case of $\mathbf{R}^3$, the unit square $[0,1]^2$ should be replaced by the unit cube $[0,1]^3$.

Suppose that

$$0 \leq a \leq 1, \quad 0 \leq b \leq 1, \quad 0 \leq c \leq 1.$$

We shall say that a set $W \subset [0,1]^3$ is (a,b,c)-homogeneous with respect to its two-dimensional sections if the following conditions are satisfied:

(i) all sections of W by the planes $\{x\} \times \mathbf{R} \times \mathbf{R}$, where x ranges over $[0,1]$, have λ_2-measure a;

(ii) all sections of W by the planes $\mathbf{R} \times \{y\} \times \mathbf{R}$, where y ranges over $[0,1]$, have λ_2-measure b;

(iii) all sections of W by the planes $\mathbf{R} \times \mathbf{R} \times \{z\}$, where z ranges over $[0,1]$, have λ_2-measure c.

In terms of this definition, we can formulate and prove the statement analogous to Theorem 2.

Theorem 3. *Suppose that all subsets of $\mathbf{R}$ with cardinalities strictly less than $\mathbf{c}$ are of λ-measure zero. Let a, b and c be three real numbers such that*

$$0 \leq a \leq 1, \quad 0 \leq b \leq 1, \quad 0 \leq c \leq 1.$$

Then there exists a set $W \subset [0,1]^3$ which is (a,b,c)-homogeneous with respect to its two-dimensional sections.

Proof. The argument is very similar to the proof of Theorem 2. Let $\preceq$ be a well-ordering of $[0, 1]$ isomorphic to the least ordinal number of cardinality $\mathbf{c}$. We introduce the following three sets:

$$A = \{(x, y, z) \in [0, 1]^3 : x \preceq \min(y, z)\},$$

$$B = \{(x, y, z) \in [0, 1]^3 : y \preceq \min(x, z)\},$$

$$C = \{(x, y, z) \in [0, 1]^3 : z \preceq \min(x, y)\}.$$

It is clear that

$$A \cup B \cup C = [0, 1]^3.$$

In addition to this, the next three relations are valid:

(1) if Γ is a plane in $\mathbf{R}^3$ of the form $x = t$, where $t \in [0, 1]$, then $\lambda_2(\Gamma \cap A) = 1$, and if Γ is a plane in $\mathbf{R}^3$ of the form $y = t$ or $z = t$, where $t \in [0, 1]$, then $\lambda_2(\Gamma \cap A) = 0$;

(2) if Γ is a plane in $\mathbf{R}^3$ of the form $y = t$, where $t \in [0, 1]$, then $\lambda_2(\Gamma \cap B) = 1$, and if Γ is a plane in $\mathbf{R}^3$ of the form $x = t$ or $z = t$, where $t \in [0, 1]$, then $\lambda_2(\Gamma \cap B) = 0$;

(3) if Γ is a plane in $\mathbf{R}^3$ of the form $z = t$, where $t \in [0, 1]$, then $\lambda_2(\Gamma \cap C) = 1$, and if Γ is a plane in $\mathbf{R}^3$ of the form $x = t$ or $y = t$, where $t \in [0, 1]$, then $\lambda_2(\Gamma \cap C) = 0$.

Taking in view these relations, we may apply to the sets A, B and C the construction analogous to that of the proof of Theorem 2 and, as a result, we come to the desired $(a, b, c,)$-homogeneous set $W \subset [0, 1]^3$.

Theorem 3 has thus been proved.

Obviously, if the disjunction

$$a \neq b \quad \vee \quad b \neq c \quad \vee \quad c \neq a$$

holds, then the set W of Theorem 3 is not measurable with respect to λ_3 (because Fubini's classical theorem fails to be true for this W).

Example 3. Let a, b, and c be any three strictly positive real numbers. It is easy to see that there exists a subset Q of $\mathbf{R}^3$ such that all linear sections of Q by the lines parallel to the axis Oz are segments of length a, all linear sections of Q by the lines parallel to the axis Ox are segments of length b, and all linear sections of Q by the lines parallel to the axis Oy are segments of length c.

Actually, the role of Q can be played by the set of all points lying between certain two parallel planes in $\mathbf{R}^3$.

Keeping in mind the fact indicated in Example 3, we may introduce another notion of (a, b, c)-homogeneity of subsets of $[0, 1]^3$ for $\{a, b, c\} \subset [0, 1]$.

Namely, we shall say that a set $W \subset [0, 1]^3$ is (a, b, c)-homogeneous with respect to its linear sections if the following three conditions are satisfied:

(i′) all sections of W by the lines $\{x\} \times \{y\} \times \mathbf{R}$, where $x \in [0,1]$ and $y \in [0,1]$ have λ-measure a;

(ii′) all sections of W by the lines $\mathbf{R} \times \{y\} \times \{z\}$, where $y \in [0,1]$ and $z \in [0,1]$, have λ-measure b;

(iii′) all sections of W by the lines $\{x\} \times \mathbf{R} \times \{z\}$, where $x \in [0,1]$ and $z \in [0,1]$, have λ-measure c.

By applying the method similar to Sierpiński's construction for $[0,1]^2$, it becomes possible to show the validity of the next statement.

Theorem 4. *Let $\{a,b,c\} \subset [0,1]$. Under the assumption that every subset of $\mathbf{R}$ with cardinality strictly less than $\mathbf{c}$ is of λ-measure zero, there exists a set $W \subset [0,1]^3$ which is (a,b,c)-homogeneous with respect to its linear sections.*

Consequently, all sections of W by the planes parallel to one of the coordinate planes are, respectively, (a,b)-homogeneous, (b,c)-homogeneous and (c,a)-homogeneous.

Proof. We only sketch the argument which is completely analogous to that of the proof of Theorem 3. Again, endow the interval $[0,1]$ with a well-ordering $\preceq$ isomorphic to the least ordinal number of cardinality $\mathbf{c}$. Then define the following three sets:

$$A = \{(x,y,z) \in [0,1]^3 : \max(y,z) \preceq x\},$$

$$B = \{(x,y,z) \in [0,1]^3 : \max(x,z) \preceq y\},$$

$$C = \{(x,y,z) \in [0,1]^3 : \max(x,y) \preceq z\}.$$

It directly follows from the definition of A, B and C that

$$A \cup B \cup C = [0,1]^3.$$

In addition to this equality, we have the following relations:

(1) if l is any straight line in $\mathbf{R}^3$ parallel to one of the Oy and Oz axes, then $\lambda(A \cap l) = 0$, and if l is any straight line in $\mathbf{R}^3$ parallel to Ox axis and intersecting the cube $[0,1]^3$, then $\lambda(A \cap l) = 1$;

(2) if l is any straight line in $\mathbf{R}^3$ parallel to one of the Ox and Oz axes, then $\lambda(B \cap l) = 0$, and if l is any straight line in $\mathbf{R}^3$ parallel to Oy axis and intersecting the cube $[0,1]^3$, then $\lambda(B \cap l) = 1$;

(3) if l is any straight line in $\mathbf{R}^3$ parallel to one of the Ox and Oy axes, then $\lambda(C \cap l) = 0$, and if l is any straight line in $\mathbf{R}^3$ parallel to Oz axis and intersecting the cube $[0,1]^3$, then $\lambda(C \cap l) = 1$.

It is now clear that, applying to A, B and C the argument of the proof of Theorem 2, we come to a set $W \subset [0,1]^3$ which is (a,b,c)-homogeneous with respect to its linear sections.

This completes the proof of Theorem 4.

As has already been mentioned, if a set $\{a, b, c\} \subset [0, 1]$ contains at least two distinct elements, then the corresponding (a, b, c)-homogeneous set W is necessarily nonmeasurable with respect to λ_3 (by virtue of Fubini's theorem).

If $0 \leq a = b = c \leq 1$, then one may pose the question whether there exists a (more or less) simple construction of a λ_3-measurable set $W \subset [0, 1]^3$ which is (a, a, a)-homogeneous with respect to its linear sections. It turns out that the answer to this question is positive and the construction of such a set W is similar to the recursive construction described earlier for the two-dimensional case (see Lemmas 1, 2, 3 and Theorem 1).

A key role in the construction of such $W \subset [0, 1]^3$ is played by the following auxiliary geometric proposition.

Lemma 4. *There exist two families*

$$(K_1, K_2, ..., K_9), \quad (T_1, T_2, ..., T_9)$$

of cubes in $[0, 1]^3$ satisfying these four conditions:

(a) all cubes K_i $(i = 1, 2, ..., 9)$ and T_j $(j = 1, 2, ..., 9)$ have edges of length $1/3$, which are parallel to the corresponding edges of $[0, 1]^3$;

(b) the orthogonal projection of the set $\cup\{K_i : 1 \leq i \leq 9\}$ to any facet of $[0, 1]^3$ coincides with that facet;

(c) the orthogonal projection of the set $\cup\{T_j : 1 \leq j \leq 9\}$ to any facet of $[0, 1]^3$ coincides with that facet;

(d) the sets $K_i \cap T_j$ have no common interior points for all indices $i = 1, 2, ..., 9$ and $j = 1, 2, ..., 9$.

Moreover, the above-mentioned two families of cubes can be chosen to be symmetric to each other with respect to the center of $[0, 1]^3$.

The proof of this lemma is purely geometric and is not difficult. So we leave it to the reader.

Theorem 5. *For any number $a \in [0, 1]$, there exists (within **ZF** & **DC** theory) a λ_3-measurable set $W \subset [0, 1]^3$ which is (a, a, a)-homogeneous with respect to its linear sections.*

Proof. Starting with Lemma 4, we may carry out the construction of the required W similarly to the two-dimensional case. Namely, first we take an arbitrary number a from $[0, 1/2]$. If $a = 0$, then we define $W = \emptyset$. Suppose now that $0 < a \leq 1/2$ and represent this a in the form

$$a = 1/3^{n_1} + 1/3^{n_2} + ... + 1/3^{n_k} + ... ,$$

where $n_1 < n_2 < ... < n_k < ...$. Then we can recursively construct an increasing (by the inclusion relation) sequence $(W_1, W_2, ..., W_k, ...)$ of λ_3-measurable subsets of $[0, 1]^3$ such that all sections of W_k by the segments of the form

$$\{x\} \times \{y\} \times [0, 1], \quad [0, 1] \times \{y\} \times \{z\}, \quad \{x\} \times [0, 1] \times \{z\} \quad (x, y, z \in [0, 1])$$

are of λ-measure $1/3^{n_1} + 1/3^{n_2} + ... + 1/3^{n_k}$. Further, we put

$$W = \cup\{W_k : 0 < k < \omega\}$$

and so obtain the λ_3-measurable set W which is (a, a, a)-homogeneous.

If a satisfies the inequalities $1/2 < a \le 1$, then we take the number $a' = 1 - a$ and observe that

$$0 \le a' < 1/2.$$

According to the stated above, there exists a λ_3-measurable set $W' \subset [0, 1]^3$ corresponding to this number a'. Putting $W = [0, 1]^3 \setminus W'$, we come to the (a, a, a)-homogeneous λ_3-measurable set W corresponding to a.

This finishes the proof of Theorem 5.

Remark 4. The results presented in this chapter for the unit square $[0, 1]^2$ and for the unit cube $[0, 1]^3$ admit a natural extension to the case of the unit hypercube $[0, 1]^m \subset \mathbf{R}^m$, where $m \ge 4$ (see Exercise 8).

EXERCISES

1. Let S be a subset of $[0, 1]^2$ such that:

(i) for every $x \in [0, 1]$, the set $S \cap (\{x\} \times [0, 1])$ has cardinality strictly less than $\mathbf{c}$;

(ii) for every $y \in [0, 1]$, the set $([0, 1]^2 \setminus S) \cap ([0, 1] \times \{y\})$ has cardinality strictly less than $\mathbf{c}$.

Assuming Martin's Axiom, show that:

(a) S does not possess the Baire property in $[0, 1]^2$;

(b) S is nonmeasurable with respect to the completion of the product measure $\mu \otimes \nu$, where μ and ν are any two nonzero σ-finite diffused Borel measures on $[0, 1]$.

2*. Work in **ZF** set theory and define in the product set $\omega_1 \times \omega_1$ some analogue U of Sierpiński's set S, i.e., U must be such that all vertical sections of U are at most countable and all horizontal sections of U are co-countable.

Further, work in **ZFC** set theory and deduce from the existence of U the classical result of Ulam stating that there is no nontrivial σ-finite diffused measure whose domain coincides with the power set $\mathcal{P}(\omega_1)$ of ω_1.

Argue as follows. Let $h : \mathbf{R} \to \mathbf{R}$ be a partial function.

First, check that there exists a countably generated σ-algebra $\mathcal{F}_1$ of subsets of $\mathrm{dom}(h)$ and a countably generated σ-algebra $\mathcal{F}_2$ of subsets of $\mathrm{ran}(h)$ such that the graph $\mathrm{Gr}(h)$ of h belongs to the product σ-algebra $\mathcal{F}_1 \otimes \mathcal{F}_2$. Moreover it can be assumed that $\mathcal{F}_1$ contains all singletons of $\mathrm{dom}(h)$ and $\mathcal{F}_2$ contains all singletons of $\mathrm{ran}(h)$.

Then, keeping in mind the fact that there exists a subset of $\mathbf{R}$ with cardinality ω_1, infer from the above that, for any partial function $g : \omega_1 \to \omega_1$, there exists a countably generated σ-algebra $\mathcal{S}_1$ of subsets of $\mathrm{dom}(g)$ and a countably generated σ-algebra $\mathcal{S}_2$ of subsets of $\mathrm{ran}(g)$ such that the graph $\mathrm{Gr}(g)$ of g belongs to the product σ-algebra $\mathcal{S}_1 \otimes \mathcal{S}_2$. As above, it can be assumed that $\mathcal{S}_1$ contains all singletons of $\mathrm{dom}(g)$ and $\mathcal{S}_2$ contains all singletons of $\mathrm{ran}(g)$.

Now, by using the set U, represent the product set $\omega_1 \times \omega_1$ in the form

$$\omega_1 \times \omega_1 = (\cup\{Z_i : i \in I\}) \cup (\cup\{Z'_j : j \in J\}),$$

where $\{Z_i : i \in I\}$ is a countable family of the graphs of partial functions acting from ω_1 into itself, and $\{(Z'_j)^{-1} : j \in J\}$ is also a countable family of partial functions acting from ω_1 into itself.

Deduce from the above-mentioned representation that if Z is an arbitrary subset of $\omega_1 \times \omega_1$, then there exists a countably generated σ-algebra $\mathcal{S}$ of subsets of ω_1 such that Z belongs to the product σ-algebra $\mathcal{S} \otimes \mathcal{S}$.

Conclude, in particular, that the equality

$$\mathcal{P}(\omega_1 \times \omega_1) = \mathcal{P}(\omega_1) \otimes \mathcal{P}(\omega_1)$$

holds true, where $\mathcal{P}(\omega_1 \times \omega_1)$ (respectively, $\mathcal{P}(\omega_1)$) denotes, as usual, the power set of $\omega_1 \times \omega_1$ (respectively, the power set of ω_1).

Finally, assuming the existence of a nonzero σ-finite diffused measure μ with $\mathrm{dom}(\mu) = \mathcal{P}(\omega_1)$, apply to the product measure $\mu \otimes \mu$ and to the set U Fubini's theorem and obtain a contradiction, which shows that such a μ cannot exist.

Remark 5. Actually, Exercise 2 yields a proof of the non-real-valued-measurability of ω_1 without using Ulam's transfinite matrix. Another way to obtain the same result, again without the aid of Ulam's matrix, is to prove the existence of an uncountable universal measure zero subspace X of $\mathbf{R}$. Notice that the classical construction of X given by Luzin exploits some delicate properties of analytic (co-analytic) subsets of $\mathbf{R}$ (see, e.g., [152], [188]). Other interesting constructions of X are presented in [208], [211], [273].

3^*. Show that there exists a translation invariant measure on $\mathbf{R}^2$ which strictly extends the Lebesgue measure λ_2 and for which the assertion of Fubini's theorem remains true.

For this purpose, consider the σ-ideal $\mathcal{I}$ consisting of those subsets of $\mathbf{R}^2$, all horizontal and vertical sections of which are at most countable. Verify that:

(a) among the members of $\mathcal{I}$ there are some sets nonmeasurable with respect to λ_2;

(b) $\mathcal{I}$ is a translation invariant σ-ideal;

(c) for each set $X \in \mathcal{I}$, the inner λ_2-measure of X is equal to zero.

Taking into account relations (a)–(b) and applying Marczewski's method (see Exercise 18 from Chapter 5), conclude that there exists a translation invariant measure μ on $\mathbf{R}^2$ strictly extending λ_2 and such that the σ-algebra $\mathrm{dom}(\mu)$ is generated by $\mathrm{dom}(\lambda_2) \cup \mathcal{I}$. Check that μ is as required.

4. Give a proof of Lemma 1 within **ZF** & **DC** set theory.

5. Construct an unbounded triangular prism in $\mathbf{R}^3$ which satisfies relations (1)–(3) of Example 2.

6. Show the validity of the assertion formulated in Example 3.

7. Construct (within **ZF** set theory) two families of cubes described in Lemma 4.

8*. Try to obtain the statement analogous to Theorem 5 for the case of the m-dimensional Euclidean space $\mathbf{R}^m$, where $m \geq 4$.

For this purpose, first formulate and prove an appropriate m-dimensional analogue of Lemma 4.

9. Let $(E, \mathcal{S}, \mu)$ be a nonatomic probability measure space such that each subset X of E with $\mathrm{card}(X) < \mathrm{card}(E)$ belongs to $\mathcal{S}$ and has μ-measure zero.

Demonstrate that, for any two real numbers $a \in [0, 1]$ and $b \in [0, 1]$, there exists an (a, b)-homogeneous subset of $E \times E$.

For this purpose, argue similarly to the proof of Theorem 2.

Apply the above result to a probability measure λ' on $[0, 1]$ which satisfies the following two conditions:

(a) λ' extends the restriction of the Lebesgue measure λ to $[0, 1]$;

(b) any set $X \subset [0, 1]$ with $\mathrm{card}(X) < \mathbf{c}$ belongs to $\mathrm{dom}(\lambda')$ and $\lambda'(X) = 0$.

Notice that the existence of such a λ' does not need any additional set-theoretical hypotheses.

17. A combinatorial problem on translation invariant extensions of the Lebesgue measure

In this chapter we are going to prove that, for every natural number $k \geq 2$, there exist k many subsets of the real line $\mathbf{R}$ such that any $k - 1$ of them can be simultaneously made measurable with respect to a certain translation invariant extension of the Lebesgue measure (in general, depending on a choice of these $k-1$ subsets), but there is no nonzero σ-finite translation quasi-invariant measure on $\mathbf{R}$ for which all of these k subsets become measurable.

In connection with this result, a related open problem of combinatorial nature is posed in the end of the chapter.

Let E be a base (ground) set and let μ be a nonzero σ-finite measure defined on some σ-algebra of subsets of E.

Recall that the general measure extension problem is to extend the given μ to a maximally wide class of subsets of E. If μ is continuous (i.e., diffused), then this problem turns out to be closely connected with the theory of large cardinals and additional set-theoretical axioms (cf. [47], [103], [110], [154]).

For example, suppose that E coincides with the real line $\mathbf{R}$ and μ coincides with the standard one-dimensional Lebesgue measure λ on this line. Then, as is well known (see, e.g., [103]), the following two assertions are equivalent:

(1) there exists a measure on $\mathbf{R}$ which extends λ and whose domain coincides with the family of all subsets of $\mathbf{R}$;

(2) there exists a nonzero σ-finite continuous measure defined on the family of all subsets of $\mathbf{R}$.

On the other hand, according to the classical result of Ulam [263], if the cardinality of the continuum $\mathbf{c} = \mathrm{card}(\mathbf{R})$ is strictly less than the first uncountable weakly inaccessible cardinal, then there exists no nonzero σ-finite continuous measure defined on the family of all subsets of $\mathbf{R}$ (see Exercise 2 for the present chapter). Moreover, under Martin's Axiom, there are countably many sets

$$Z_1 \subset \mathbf{R}, \ Z_2 \subset \mathbf{R}, \ \ldots, \ Z_k \subset \mathbf{R}, \ \ldots$$

such that no nonzero σ-finite continuous measure ν on $\mathbf{R}$ can make all these Z_k $(1 \leq k < \omega)$ to be measurable with respect to ν (cf. Exercise 3 for this chapter).

Let E be again a base set, μ be a σ-finite measure defined on some σ-algebra of subsets of E, and let $\{X_1, X_2, ..., X_k\}$ be a finite family of subsets of E. It is well known that there always exists a measure μ' on E extending μ and such that all sets $X_1, X_2, ..., X_k$ are μ'-measurable (see Remark 1 from Chapter 5).

In contrast to this situation, if the original measure μ is invariant under a group G of transformations of E, then we cannot assert, in general, that there exists an extension μ' of μ which also is invariant under G and for which all given sets $X_1, X_2, ..., X_k$ are μ'-measurable. Even for $k = 1$, it may happen that a single set X_1 turns out to be nonmeasurable with respect to every G-invariant measure on E extending μ. This circumstance has already been mentioned several times in preceding sections of the book. However, we would like to recall once more the following classical example.

Example 1. If E coincides with the real line $\mathbf{R}$ and μ coincides with the standard Lebesgue measure λ on $\mathbf{R}$, then the construction of Vitali [266] yields a set $V \subset \mathbf{R}$ which is nonmeasurable with respect to every translation invariant measure on $\mathbf{R}$ extending λ; or, in other words, V turns out to be absolutely nonmeasurable with respect to the class of all translation invariant measures on $\mathbf{R}$ extending λ (cf. Chapter 9 where a much stronger result is presented).

At the same time, it was established by various authors that there exist many subsets of $\mathbf{R}$ measurable with respect to certain measures on $\mathbf{R}$ which are translation invariant and strictly extend λ (see, e.g., [37], [99], [123], [174], [176], [210], [272]). Moreover, it was proved that there exists even a nonseparable translation invariant extension ν of λ (see [69], [95], [107], [144]). Clearly, the domain of such a ν contains in itself a large class of subsets of $\mathbf{R}$ which are not measurable with respect to λ.

Many delicate problems and questions of the theory of translation invariant extensions of λ were discussed in the literature (see, for instance, [123], [137], [210], [272]). One of the problems of this type will be considered below. It has a certain combinatorial flavor (cf. also an open problem formulated before exercises of this chapter).

To begin our presentation, let us first recall the following fact.

Example 2. In [118] some sets $A_1 \subset \mathbf{R}$ and $A_2 \subset \mathbf{R}$ were constructed, which satisfy these three conditions:

(1) there exists a translation invariant measure μ_1 on $\mathbf{R}$ extending λ and such that $\mu_1(A_1) = 0$;

(2) there exists a translation invariant measure μ_2 on $\mathbf{R}$ extending λ and such that $\mu_2(A_2) = 0$;

(3) there exists no nonzero σ-finite translation invariant measure ν on $\mathbf{R}$ such that both sets A_1 and A_2 are ν-measurable.

Actually, it was demonstrated in [118] that A_1 and A_2 possess a property much stronger than condition (3), namely, for any nonzero σ-finite translation invariant measure μ on $\mathbf{R}$, the set $A_1 \cup A_2$ is nonmeasurable with respect to μ. In other words, according to our terminology, the set $A_1 \cup A_2$ is absolutely nonmeasurable with respect to the class of all nonzero σ-finite translation invariant measures on $\mathbf{R}$.

The natural question arises whether it is possible to generalize the above-mentioned Example 2 to the case of several subsets of the real line. As has already been announced, the main goal of this chapter is to establish an analogous result for a certain finite family of subsets

$$A_1, \; A_2, \; \ldots, \; A_k$$

of $\mathbf{R}$, where k is an arbitrary natural number greater than 2. In fact, it will be shown later that by combining techniques of Hamel bases with the argument used in the proof of an old theorem of Sierpiński [243] concerning a certain logical equivalent of the Continuum Hypothesis, one can get a positive answer to this question. Briefly speaking, the usage of Hamel bases enables us to avoid the Continuum Hypothesis and to obtain the desired result within **ZFC** set theory (for more details, see below).

For our further considerations, it is reasonable to recall some notions from the general theory of invariant and quasi-invariant measures.

Let E be a base (ground) set, G be a group of transformations of E, and let μ be a σ-finite measure defined on some G-invariant σ-algebra of subsets of E. This μ is said to be G-quasi-invariant if, for any transformation $g \in G$ and for any set $X \in \operatorname{dom}(\mu)$, the relation

$$\mu(X) = 0 \Leftrightarrow \mu(g(X)) = 0$$

holds true.

Obviously, the quasi-invariance of measures is a much weaker property than the ordinary invariance of measures.

In the sequel, we need one specific notion from the theory of quasi-invariant measures (cf. Exercise 16 from Chapter 5).

Let E be again a base set and let G be a group of transformations of E.

We say that a set $X \subset E$ is G-negligible in E if the following two conditions are satisfied:

(a) there exists a nonzero σ-finite G-quasi-invariant measure μ_0 on E such that $X \in \operatorname{dom}(\mu_0)$;

(b) for any σ-finite G-quasi-invariant measure μ on E, we have the implication

$$X \in \operatorname{dom}(\mu) \Rightarrow \mu(X) = 0.$$

Some properties of G-negligible sets are discussed in [118], [126], and [128]. In particular, the following auxiliary proposition is formulated therein.

Lemma 1. *Let* $(\Gamma_1, +)$ *and* $(\Gamma_2, +)$ *be two commutative groups and suppos* *that* $\phi : \Gamma_1 \to \Gamma_2$ *is a surjective homomorphism. If* Y *is a* Γ_2*-negligible subse of* Γ_2, *then* $X = \phi^{-1}(Y)$ *is a* Γ_1*-negligible subset of* Γ_1.

The proof of Lemma 1 follows directly from the definition of negligible sets so is omitted here (see Exercise 4 for this chapter).

Notice by the way that the commutativity of the groups Γ_1 and Γ_2 is no essential in the formulation of this lemma. However, below we will be dealing only with commutative groups (even with vector spaces over $\mathbf{Q}$), so the presented formulation is sufficient for our further purposes.

We also need some other auxiliary statements.

Lemma 2. *Let* $(G, +)$ *be a commutative group and let a nonempty se* $X \subset G$ *be such that* $D + X \neq G$ *for every countable set* $D \subset G$. *Denote by* $\mathcal{S}$ *the* G*-invariant* σ*-algebra of subsets of* G, *generated by* X *and the family of al countable subsets of* G. *Then there exists a continuous* G*-invariant probabilit measure* μ *on* $\mathcal{S}$ *satisfying the equality* $\mu(X) = 0$.

Further, let $\mathcal{T}$ *be a* σ*-algebra of subsets of* G *and let* ν *be a* σ*-finite measur on* $\mathcal{T}$ *such that* $\nu_*(D + X) = 0$ *for any countable set* $D \subset G$. *Let* $\mathcal{R}$ *denot the* σ*-algebra of subsets of* G, *generated by* $\mathcal{S} \cup \mathcal{T}$. *Then there exists a uniqu* σ*-finite measure* θ *on* $\mathcal{R}$ *satisfying the following four relations:*

(a) $\theta(Y \cap Z) = \mu(Y)\nu(Z)$ *for all* $Y \in \mathcal{S}$ *and* $Z \in \mathcal{T}$;
(b) *if* $\mathcal{T}$ *is* G*-invariant, then* $\mathcal{R}$ *is also* G*-invariant;*
(c) *if* ν *is* G*-invariant, then* θ *is also* G*-invariant;*
(d) *if* ν *is* G*-quasi-invariant, then* θ *is also* G*-quasi-invariant.*

Proof. The first part of this lemma is almost trivial. Indeed, it immediately follows from the assumption on X that the family of all those sets in G, which can be covered by countably many translates of X, forms a G-invariant σ-idea $\mathcal{J}$. Each member of the σ-algebra $\mathcal{S}$ either belongs to $\mathcal{J}$ or is the complement of a set belonging to $\mathcal{J}$. Now, for any set $Y \in \mathcal{S}$, put:

$\mu(Y) = 0$ if $Y \in \mathcal{J}$;
$\mu(Y) = 1$ if $G \setminus Y \in \mathcal{J}$.

It can easily be seen that the probability measure μ defined in this manne is continuous and G-invariant.

To establish the second part of Lemma 2, take the product measure $\nu \otimes \mu$ and consider the diagonal

$$\Delta = \{(g, g) : g \in G\} \subset G \times G$$

with the canonical bijection $\phi : G \to \Delta$ given by

$$\phi(g) = (g, g) \quad (g \in G).$$

Obviously, we have

$$\mathcal{R} = \phi^{-1}(\mathcal{T} \otimes \mathcal{S}).$$

Since μ is a two-valued probability measure and $\nu_*(Z) = 0$ for each $Z \in \mathcal{S}$ with $\mu(Z) = 0$, we easily infer that the diagonal Δ is $(\nu \otimes \mu)$-thick in $G \times G$. As is well known, in this case there exists a unique σ-finite measure θ on $\mathcal{R}$ satisfying the equality

$$(\nu \otimes \mu)(P) = \theta(\phi^{-1}(P)) \quad (P \in \mathcal{T} \otimes \mathcal{S}).$$

This equality directly implies relation (a). The relations (b), (c), and (d) are also readily verified.

Lemma 2 has thus been proved (cf. Exercise 18 from Chapter 5).

Lemma 3. *Let* $(H, +)$ *be a commutative group equipped with a σ-finite H-quasi-invariant measure μ and let X be a μ-measurable subset of H. Then there exists a countable subgroup H' of H such that the set*

$$X' = \cup \{h' + X : h' \in H'\}$$

is μ-almost H-invariant, i.e., for any $h \in H$, we have the equality

$$\mu((h + X')\triangle X') = 0.$$

Proof. If $\mu(X) = 0$, then there is nothing to prove. So let us consider the case when $\mu(X) > 0$ (only this case is of interest to us in the sequel). Suppose to the contrary that there exists no countable subgroup H' of H with the required property, and define by transfinite recursion an increasing ω_1-sequence of μ-measurable subsets of H. Namely, put $X_0 = X$. Assume that, for an ordinal $\xi < \omega_1$, the partial ξ-sequence $\{X_\zeta : \zeta < \xi\}$ has already been defined in such a way that every set X_ζ is of the form

$$X_\zeta = \cup \{h_{k,\zeta} + X : k < \omega\}$$

for some countable family $\{h_{k,\zeta} : k < \omega\} \subset H$. Introduce the set

$$Y_\xi = \cup \{X_\zeta : \zeta < \xi\}.$$

Clearly, Y_ξ can be represented in a similar form (because ξ is a countable ordinal). According to our assumption, there exists an element $h \in H$ such that

$$\mu((h + Y_\xi)\triangle Y_\xi) > 0,$$

which implies the disjunction

$$\mu((h + Y_\xi) \setminus Y_\xi) > 0 \quad \vee \quad \mu((-h + Y_\xi) \setminus Y_\xi) > 0.$$

Now, let us put

$$X_\xi = Y_\xi \cup (h + Y_\xi) \cup (-h + Y_\xi).$$

Proceeding in this manner, we get the ω_1-sequence $\{X_\xi : \xi < \omega_1\}$ of μ-measurable sets, which increases by inclusion and the inequality

$$\mu(X_{\xi+1} \setminus X_\xi) > 0$$

is fulfilled for every ordinal $\xi < \omega_1$. But the latter fact contradicts the σ-finiteness of μ. The obtained contradiction finishes the proof.

Remark 1. Obviously, the assertion of Lemma 3 remains true in a more general situation where a σ-finite measure μ is given on a base set E and this μ is quasi-invariant under some group G of transformations of E (in short, G-quasi-invariant). Moreover, the argument presented above shows that the assertion of the lemma holds true for any G-quasi-invariant measure μ on E which satisfies the countable chain condition and, in general, such a measure does not need to be σ-finite (see Exercise 5).

Below, having two commutative groups $(G, +)$ and $(H, +)$, we will consider their direct sum $G + H$ which, in fact, may be identified with the product group $G \times H$. Naturally, under such an identification $(G, +)$ may be regarded as the subgroup $G \times \{0\}$ of $G \times H$ and $(H, +)$ may be regarded as the subgroup $\{0\} \times H$ of $G \times H$. Analogously, having finitely many commutative groups

$$(G_1, +), \ (G_2, +), \ \ldots, \ (G_k, +),$$

we can identify their direct sum $G_1 + G_2 + \ldots + G_k$ with the product group $G_1 \times G_2 \times \ldots \times G_k$.

Lemma 4. *Let $(G, +)$ and $(H, +)$ be two commutative groups and let $\mathrm{card}(H) > \omega$. Consider the direct sum $G + H$. Let X be a subset of $G + H$ having the property that*

$$\mathrm{card}((g + H) \cap X) < \omega$$

for each element $g \in G$. Then X is a $(G + H)$-negligible subset of $G + H$.

Proof. It readily follows from the described property of X that, for every countable family

$$\{g_i + h_i : i < \omega\} \subset G + H,$$

the inequality

$$\mathrm{card}((G + H) \setminus \cup \{g_i + h_i + X : i < \omega\}) > \omega$$

is valid. In view of Lemma 2, this implies that there exists a probability continuous $(G + H)$-invariant measure μ_0 on $G + H$ such that $X \in \mathrm{dom}(\mu_0)$ and $\mu_0(X) = 0$.

Now, let μ be any σ-finite $(G + H)$-quasi-invariant measure on $G + H$ such that $X \in \mathrm{dom}(\mu)$. We have to show that $\mu(X) = 0$. Suppose to the contrary that $\mu(X) > 0$. Since our μ is H-quasi-invariant, we can find (by virtue of Lemma 3) a countable subgroup H' of H for which the set

$$X' = H' + X = \cup\{h' + X : h' \in H'\}$$

turns out to be almost H-invariant with respect to μ, i.e., the equality

$$\mu(X'\triangle(h + X')) = 0$$

holds true for each $h \in H$. Taking into account this equality, we readily infer that, for any countable family $\{h_j : j < \omega\} \subset H$, the relation

$$\mu(\cap\{h_j + X' : j < \omega\}) = \mu(X') > 0$$

is valid and, consequently,

$$\cap\{h_j + X' : j < \omega\} \neq \emptyset.$$

But, if a countable family $\{h_j : j < \omega\} \subset H$ is chosen satisfying the condition

$$h_j - h_r \notin H' \quad (j \in J,\ r \in J,\ j \neq r),$$

then, keeping in mind the definition of X, it is not difficult to verify that

$$\cap\{h_j + X' : j < \omega\} = \emptyset,$$

which yields a contradiction with the stated above.

The obtained contradiction finishes the proof.

Lemma 5. *Let $(G, \|\cdot\|)$ be a normed vector space over $\mathbf{Q}$ with $\mathrm{card}(G) > 1$ and let $\{B_m : m < \omega\}$ be a countable family of balls in G. Then there exists a disjoint countable family $\{P_j : j \in J\}$ of subsets of G satisfying the following relations:*

(1) each set P_j is a translate of some ball B_m;

(2) for any ball B_m $(m < \omega)$, there are infinitely many indices $j \in J$ such that the set P_j is a translate of B_m.

We omit a simple proof of the above assertion, because the required disjoint family $\{P_j : j \in J\}$ can easily be constructed by ordinary recursion.

Lemma 6. *Let $G \neq \{0\}$ and H be two vector spaces over $\mathbf{Q}$ and let $G + H$ be their direct sum. Suppose that a set $X \subset G + H$ is given such that the inequality*

$$\mathrm{card}(X \cap (g + H)) \leq \omega$$

holds for each element $g \in G$. Then there exists a set $Y \subset G + H$ satisfying the following two conditions:

(a) $\mathrm{card}(Y \cap (g + H)) \leq 1$ for every $g \in G$;

(b) $X \subset \cup\{g_i + Y : i \in I\}$ for some countable family $\{g_i : i \in I\} \subset G$.

Proof. We may treat G as a normed vector space over $\mathbf{Q}$. Indeed, denote by $\{z_\xi : \xi \in \Xi\}$ any Hamel basis of G. Each element $g \in G$ admits a unique representation in the form

$$g = \sum\{q_\xi z_\xi : \xi \in \Xi\},$$

where all coefficients q_ξ belong to $\mathbf{Q}$ and only finitely many of them differ from zero. Putting

$$\|g\| = \sum\{|q_\xi| : \xi \in \Xi\},$$

we get the norm $\| \cdot \|$ on G.

For every natural number $m > 0$, let

$$B_m = \{g \in G : \|g\| \le m\}$$

denote the ball in G with center at zero and with radius m. According to Lemma 5, there exists a disjoint countable family $\{P_j : j \in J\}$ of subsets of G satisfying the following relations:

(1) each set P_j $(j \in J)$ is a G-translate of some ball B_m;

(2) for any ball B_m $(m < \omega)$, there are infinitely many indices $j \in J$ such that the set P_j is a G-translate of B_m.

Let $J(m)$ denote the family of all those indices $j \in J$ for which P_j is a G-translate of B_m, i.e., $j \in J(m)$ if and only if there exists an element $g_{m,j} \in G$ such that $P_j = g_{m,j} + B_m$. Clearly, all the sets $J(m)$ $(0 < m < \omega)$ are countably infinite, pairwise disjoint, and their union coincides with J.

Consider the family of sets

$$\{X \cap (B_m + H) : 0 < m < \omega\}.$$

Obviously, we have

$$X = \cup\{X \cap (B_m + H) : 0 < m < \omega\}.$$

For any $g \in B_m$, the set $X \cap (g + H)$ is at most countable. This implies that the set $X \cap (B_m + H)$ can be represented in the form

$$X \cap (B_m + H) = \cup\{Y_{m,j} : j \in J(m)\},$$

where, for each index $j \in J(m)$ and for each element $g \in B_m$, we have

$$\mathrm{card}(Y_{m,j} \cap (g + H)) \le 1.$$

Now, we put

$$Y = \cup\{g_{m,j} + Y_{m,j} : j \in J(m), 0 < m < \omega\}.$$

It is not difficult to check that the set Y satisfies the conditions (a) and (b) of the lemma, which completes the proof.

Lemma 7. *Let $k \geq 2$ be a natural number and let $(G, ||\cdot||)$ be a vector space over $\mathbf{Q}$ representable in the form of a direct sum*

$$G = G_1 + G_2 + ... + G_k,$$

where all G_i $(i = 1, 2, ..., k)$ are vector subspaces of G of cardinality ω_1.
Then subsets $Y_1, Y_2, ..., Y_k$ of G can be found such that:
(1) for each index $i \in \{1, 2, ..., k\}$, the set

$$Y_1 \cup ... \cup Y_{i-1} \cup Y_{i+1} ... \cup Y_k$$

is G-negligible in G;
(2) there exists a family $\{g_m : m < \omega\}$ of elements from G for which we have the equality

$$\cup\{g_m + (Y_1 \cup Y_2 \cup ... \cup Y_k) : m < \omega\} = G.$$

Consequently, there is no nonzero σ-finite G-quasi-invariant measure ν on G such that all sets $Y_1, Y_2, ..., Y_k$ are ν-measurable.

Proof. The argument is based on some ideas of Sierpiński which he used in establishing the equivalence of the Continuum Hypothesis to the existence of certain decompositions of $\mathbf{R}^2$ and $\mathbf{R}^3$ (see [233], [243], Exercise 11 from Chapter 4 and Exercise 7 in this chapter).

Without loss of generality, we may suppose that the subspaces $G_1, G_2, ..., G_k$ are well-ordered by ordering relations which are isomorphic to ω_1. So let

$$\xi_i : G_i \to \omega_1 \quad (i = 1, 2, ..., k)$$

denote the corresponding isomorphisms. Consequently, if x_i belongs to G_i, where $i \in \{1, 2, ..., k\}$, then $\xi_i(x_i)$ denotes the countable ordinal corresponding to this x_i with respect to the isomorphism ξ_i between G_i and ω_1.

In addition, we may assume that every G_i $(i = 1, 2, ..., k)$ is a normed vector space over $\mathbf{Q}$ (see the proof of Lemma 6).

Now, let us consider the sets X_i $(i = 1, 2, ..., k)$ defined as follows:

$$X_i = \{x_1 + x_2 + ... + x_k \in G : \xi_i(x_i) = \max(\xi_1(x_1), \xi_2(x_2), ..., \xi_k(x_k))\}.$$

Clearly, we have the equality

$$G = X_1 \cup X_2 \cup ... \cup X_k.$$

Furthermore, each set X_i $(i \in \{1, 2, ..., k\})$ possesses the following property:

for any element $x_i \in G_i$, the set

$$X_i \cap (G_1 + ... + G_{i-1} + x_i + G_{i+1} + ... + G_k)$$

is at most countable.

Applying Lemma 6, we come to a family $\{Y_1, Y_2, ..., Y_k\}$ of subsets of G such that:

(a) each set Y_i is uniform with respect to $G_1 + ... + G_{i-1} + G_{i+1} + ... + G_k$, i.e., for any element $x_i \in G_i$, we have the inequality

$$\text{card}(Y_i \cap (G_1 + ... + G_{i-1} + x_i + G_{i+1} + ... + G_k)) \le 1;$$

(b) each set X_i can be covered by a countable family of translates of Y_i.

Notice now that relation (b) directly implies relation (2). It remains to show that relation (1) is also true. Observe that, for any integer $i \in [1, k]$ and for any element

$$x_1 + ... + x_{i-1} + x_{i+1} + ... + x_k \in G_1 + ... + G_{i-1} + G_{i+1} + ... + G_k,$$

the set

$$(Y_1 \cup ... \cup Y_{i-1} \cup Y_{i+1} \cup ... \cup Y_k) \cap (x_1 + ... + x_{i-1} + G_i + x_{i+1} + ... + x_k)$$

consists of at most $k - 1$ elements. So, by virtue of Lemma 4, we conclude that the set $Y_1 \cup ... \cup Y_{i-1} \cup Y_{i+1} \cup ... \cup Y_k$ is G-negligible in G.

Lemma 7 has thus been proved.

Lemma 8. *For any two natural numbers $n \ge 1$ and $k \ge 2$, the Euclidean space $\mathbf{R}^n$ can be represented in the form of a direct sum*

$$\mathbf{R}^n = G_1 + G_2 + ... + G_k + H,$$

where all G_i $(i = 1, 2, ..., k)$ and H are vector spaces over $\mathbf{Q}$ and the following conditions are fulfilled:

(1) $\text{card}(G_1) = \text{card}(G_2) = ... = \text{card}(G_k) = \omega_1$;
(2) $\text{card}(H) = \mathbf{c}$;
(3) H is a λ_n-thick subset of $\mathbf{R}^n$.

Proof. We use the technique of Hamel bases and the standard argument based on the method of transfinite induction. Namely, we identify $\mathbf{c}$ with the first ordinal number of cardinality continuum and denote by $\mathcal{B}$ the family of all Borel subsets of $\mathbf{R}^n$ having strictly positive λ_n-measure. Since $\text{card}(\mathcal{B}) = \mathbf{c}$, we can represent $\mathcal{B}$ in the form $\{B_\xi : \xi < \mathbf{c}\}$ where $B_\xi = B_{\xi+1}$ for all ordinals $\xi < \mathbf{c}$. Further, for any set $T \subset \mathbf{R}^n$, denote by $\text{span}_{\mathbf{Q}}(T)$ the linear span (hull) over $\mathbf{Q}$ of this T. Obviously, the relation $\text{card}(T) < \mathbf{c}$ implies the relation

$$\text{card}(\text{span}_{\mathbf{Q}}(T)) < \mathbf{c}.$$

Taking this circumstance into account and applying transfinite recursion, we are able to construct a family of points $\{x_\xi : \xi < \mathbf{c}\} \subset \mathbf{R}^n$ such that

$$x_\xi \in B_\xi \setminus \text{span}_\mathbf{Q}(\{x_\zeta : \zeta < \xi\}) \quad (\xi < \mathbf{c}).$$

Actually, here we have a certain version of the classical Bernstein construction (cf. [14], [77], [96], [128], [147], [152], [188], [190], [203]).

Let Ξ denote the set of all even ordinals strictly less than $\mathbf{c}$ and let Ξ' stand for the set of all odd ordinals strictly less than $\mathbf{c}$. Since $\text{card}(\Xi) = \mathbf{c}$, there are pairwise disjoint subsets $\Xi_1, \Xi_2, ..., \Xi_k$ of Ξ such that

$$\text{card}(\Xi_1) = \text{card}(\Xi_2) = ... = \text{card}(\Xi_k) = \omega_1.$$

Now, let us put

$$G' = \text{span}_\mathbf{Q}(\{x_\xi : \xi \in \Xi'\}),$$

$$G_i = \text{span}_\mathbf{Q}(\{x_\xi : \xi \in \Xi_i\}) \quad (i = 1, 2, ..., k).$$

Then G' and all G_i $(i = 1, ..., k)$ are vector spaces over $\mathbf{Q}$ and

$$\text{card}(G_1) = \text{card}(G_2) = ... = \text{card}(G_k) = \omega_1.$$

Moreover, keeping in mind that the family of points $\{x_\xi : \xi < \mathbf{c}\}$ is linearly independent over $\mathbf{Q}$, we infer that the sum $G' + G_1 + ... + G_k$ is direct. Further, there exists a vector space $F \subset \mathbf{R}^n$ over $\mathbf{Q}$ satisfying the relations

$$F \cap (G' + G_1 + ... + G_k) = \{0\},$$

$$F + (G' + G_1 + ... + G_k) = \mathbf{R}^n.$$

Let us denote

$$H = F + G'.$$

So we come to a certain representation of $\mathbf{R}^n$ in the form of a direct sum:

$$\mathbf{R}^n = G_1 + G_2 + ... + G_k + H.$$

Since $B_\xi = B_{\xi+1}$ for any ordinal $\xi < \mathbf{c}$, the family of points $\{x_\xi : \xi \in \Xi'\}$ is λ_n-thick in $\mathbf{R}^n$. Taking in view the relations

$$\{x_\xi : \xi \in \Xi'\} \subset G' \subset H,$$

we conclude that H is also λ_n-thick, which finishes the proof of Lemma 8.

With the aid of the above-mentioned lemmas, we are able to obtain the two main statements of this chapter.

Theorem 1. *Let $n > 0$ and $k \geq 2$ be two natural numbers. Then subsets $A_1, A_2, ..., A_k$ of the Euclidean space $\mathbf{R}^n$ can be found such that:*

(1) for each index $i \in \{1, 2, ..., k\}$, the set

$$A_1 \cup ... \cup A_{i-1} \cup A_{i+1} \cup ... \cup A_k$$

is $\mathbf{R}^n$-negligible in $\mathbf{R}^n$;

(2) for each index $i \in \{1, 2, ..., k\}$, there exists a complete translation invariant measure μ_i on $\mathbf{R}^n$ extending λ_n and satisfying the equality

$$\mu_i(A_1 \cup ... \cup A_{i-1} \cup A_{i+1} \cup ... \cup A_k) = 0;$$

consequently, all sets $A_1, ..., A_{i-1}, A_{i+1}, ..., A_k$ turn out to be measurable with respect to μ_i;

(3) there exists no nonzero σ-finite translation quasi-invariant measure μ on $\mathbf{R}^n$ for which all sets $A_1, A_2, ..., A_k$ are μ-measurable.

Proof. Consider the representation of $\mathbf{R}^n$ in the form of a direct sum

$$\mathbf{R}^n = G_1 + G_2 + ... + G_k + H,$$

where G_i $(i = 1, 2, ..., k)$ and H are as in Lemma 8. Then introduce the notation

$$G = G_1 + G_2 + ... + G_k.$$

Let $Y_1, Y_2, ..., Y_k$ be subsets of G as in Lemma 7. Further, define the sets

$$A_i = Y_i + H \qquad (i = 1, 2, ..., k).$$

By virtue of Lemmas 1 and 7, all the sets

$$B_i = A_1 \cup ... \cup A_{i-1} \cup A_{i+1} \cup ... \cup A_k \qquad (i = 1, 2, ..., k)$$

are $\mathbf{R}^n$-negligible. Moreover, keeping in mind the λ_n-thickness of H, we readily derive that, for each index $i \in \{1, 2, ..., k\}$ and for any family

$$\{g_m : m < \omega\} \subset \mathbf{R}^n,$$

the set $\cup\{g_m + B_i : m < \omega\}$ has inner λ_n-measure zero. Then the standard construction of extending invariant measures (see, e.g., [99], [123], [174], [176], [210], Exercise 18 from Chapter 5 or Lemma 2 of this chapter) enables us to conclude that there exists a translation invariant complete measure μ_i on $\mathbf{R}^n$ extending λ_n and such that $\mu_i(B_i) = 0$. So all the sets

$$A_1, \ ... \ , \ A_{i-1}, \ A_{i+1}, \ ... \ , \ A_k$$

become measurable with respect to μ_i. On the other hand, since G can be covered by countably many G-translates of the set $Y_1 \cup Y_2 \cup ... \cup Y_k$, the whole space $\mathbf{R}^n$ can also be covered by countably many G-translates of the set

$$A_1 \cup A_2 \cup ... \cup A_k = (Y_1 \cup Y_2 \cup ... \cup Y_k) + H.$$

It easily follows from the above-mentioned circumstance that there exists no nonzero σ-finite translation-quasi-invariant measure μ on $\mathbf{R}^n$ such that

$$\{A_1, A_2, ..., A_k\} \subset \text{dom}(\mu).$$

This finishes the proof of Theorem 1.

The next statement readily follows from the theorem just proved.

Theorem 2. *Let $n > 0$ and $k \geq 2$ be two natural numbers. Then subsets $C_1, C_2, ..., C_k$ of the Euclidean space $\mathbf{R}^n$ can be found such that:*

(1) for each index $i \in \{1, 2, ..., k\}$, the set C_i is $\mathbf{R}^n$-negligible in $\mathbf{R}^n$;

(2) for each index $i \in \{1, 2, ..., k\}$, there is a translation invariant complete measure μ_i on $\mathbf{R}^n$ extending λ_n, satisfying the relation $C_i \in \text{dom}(\mu_i)$ and, consequently, satisfying the equality $\mu_i(C_i) = 0$;

(3) if i and j are any two distinct indices from the set $\{1, 2, ..., k\}$, then there exists no nonzero σ-finite translation quasi-invariant measure μ on $\mathbf{R}^n$ for which both sets C_i and C_j are μ-measurable.

Proof. We preserve the notation used in the proof of Theorem 1. Let us put

$$C_i = B_i \qquad (i \in \{1, 2, ..., k\}).$$

We already know that each set C_i is $\mathbf{R}^n$-negligible in $\mathbf{R}^n$ and that there exists a translation invariant complete measure μ_i on $\mathbf{R}^n$ extending λ_n such that $\mu_i(C_i) = 0$. Further, for any two distinct indices i and j from $\{1, 2, ..., k\}$, we have

$$C_i \cup C_j = A_1 \cup A_2 \cup ... \cup A_k.$$

Since the whole space $\mathbf{R}^n$ can be covered by countably many translates of the set $A_1 \cup A_2 \cup ... \cup A_k$, we readily deduce that there is no nonzero σ-finite translation quasi-invariant measure μ on $\mathbf{R}^n$ satisfying the relation

$$\{C_i, C_j\} \subset \text{dom}(\mu).$$

Theorem 2 has thus been proved.

In connection with the obtained results, the following open combinatorial problem seems to be of some interest.

Problem. Let $n \geq 1$, $k > 2$ and $0 < l < k$ be natural numbers. Prove (or disprove) that there is a family $\{A_1, A_2, ..., A_k\}$ of subsets of $\mathbf{R}^n$ satisfying the following conditions:

(a) for any l-element subfamily of $\{A_1, A_2, ..., A_k\}$, there exists a translation invariant extension of λ_n such that all members from the subfamily are measurable with respect to this extension;

(b) for any $(l + 1)$-element subfamily of $\{A_1, A_2, ..., A_k\}$, there exists no nonzero σ-finite translation quasi-invariant measure on $\mathbf{R}^n$ whose domain contains this subfamily.

Example 3. Let us consider the Euclidean plane $\mathbf{R}^2 = \mathbf{R} \times \mathbf{R}$ and let a set $X \subset \mathbf{R}^2$ be such that

$$\operatorname{card}(X \cap (\{t\} \times \mathbf{R})) < \omega$$

for all $t \in \mathbf{R}$. Then, according to Lemma 4, X is $\mathbf{R}^2$-negligible in $\mathbf{R}^2$. At the same time, there exists a set $Z \subset \mathbf{R}^2$ which satisfies the relation

$$\operatorname{card}(Z \cap (\{t\} \times \mathbf{R})) \leq \omega$$

for any $t \in \mathbf{R}$, but which is not $\mathbf{R}^2$-negligible in $\mathbf{R}^2$ (see, for instance, [128] where a much stronger result is presented).

Remark 2. It is known that there exists a countable family

$$\{X_0, \ X_1, \ \ldots, \ X_n, \ \ldots\}$$

of subsets of $\mathbf{R}$ satisfying the following two conditions:

(1) for each natural number $k \geq 1$, for any σ-finite translation invariant (translation quasi-invariant) measure μ on $\mathbf{R}$, and for any k sets

$$X_{n_1}, \ X_{n_2}, \ \ldots, \ X_{n_k}$$

of this family, there exists a translation invariant (translation quasi-invariant) measure μ' on $\mathbf{R}$ extending μ and such that all sets $X_{n_1}, X_{n_2}, \ldots, X_{n_k}$ are measurable with respect to μ';

(2) there is no nonzero σ-finite translation quasi-invariant measure on $\mathbf{R}$ which makes all sets $X_0, X_1, \ldots, X_n, \ldots$ to be measurable with respect to it.

Actually, we may take as $\{X_n : n < \omega\}$ a countable family of those $\mathbf{R}$-absolutely negligible subsets of $\mathbf{R}$, which collectively cover $\mathbf{R}$ (for more details, see [118], [123] or Exercise 10 below).

EXERCISES

1. Demonstrate that the following two assertions are equivalent:

(a) there exists a measure extending the Lebesgue measure λ and defined on the family of all subsets of $\mathbf{R}$;

(b) there exists a nonzero σ-finite continuous measure defined on the family of all subsets of $\mathbf{R}$.

For this purpose, keep in mind the fact that λ is a non-atomic measure.

2*. Prove Ulam's result mentioned in this chapter.

Argue as follows. Let E be an uncountable set whose cardinality is strictly less than the first weakly inaccessible cardinal. By using the method of transfinite induction on $\text{card}(E)$, demonstrate that E does not admit a nonzero σ-finite diffused measure defined on the power set of E. For this purpose, consider two possible cases.

(a) $\text{card}(E) = w_{\alpha+1}$, where $\alpha \geq 0$.

In this case, utilize Ulam's matrix (see Exercise 25 of Appendix 1) and the inductive assumption that w_α is not measurable in the Ulam sense.

(b) $\text{card}(E) = w_\alpha$, where α is a limit ordinal.

In this case, take into account that w_α is a singular cardinal, i.e.,

$$E = \cup\{E_i : i \in I\},$$

where $\text{card}(I) < w_\alpha$ and $\text{card}(E_i) < w_\alpha$ for all indices $i \in I$. Apply the inductive assumption to $\text{card}(I)$ and to the cardinal numbers $\text{card}(E_i)$ $(i \in I)$.

3. Under Martin's Axiom, show that there are countably many subsets of $\mathbf{R}$ such that no nonzero σ-finite diffused measure ν on $\mathbf{R}$ can make all these sets to be measurable with respect to ν.

For this purpose, take any generalized Luzin subset L of $\mathbf{R}$ and consider a bijection of L onto $\mathbf{R}$.

4. Give a proof of Lemma 1. Also, formulate and prove the analogue of this lemma for two groups $(\Gamma_1, \cdot)$ and $(\Gamma_2, \cdot)$ which are not assumed to be commutative ones.

5. Recall that a measure μ on a ground set E satisfies the countable chain condition (briefly, ccc) if any disjoint family of μ-measurable sets of strictly positive measure is at most countable.

Prove the analogue of Lemma 3 for quasi-invariant measures on a base set E, satisfying the countable chain condition. Give examples of such measures which are not σ-finite.

6. Give a proof of Lemma 5.

7*. Demonstrate that the following two assertions are equivalent within **ZFC** set theory:

(a) the Continuum Hypothesis (**CH**);

(b) there exists a partition $\{A, B, C\}$ of the space $\mathbf{R}^3$ such that the set A meets every straight line in $\mathbf{R}^3$ parallel to Ox axis in finitely many points, the set B meets every straight line in $\mathbf{R}^3$ parallel to Oy axis in finitely many points, and the set C meets every straight line in $\mathbf{R}^3$ parallel to Oz axis in finitely many points.

This beautiful result is due to Sierpiński [243] and is usually regarded as a certain geometric equivalent of **CH**.

To obtain the equivalence of (a) and (b), argue similarly to the argument sketched in Exercise 11 from Chapter 4.

8. Let $(G, +)$ be a commutative group and let X be a subset of G. Suppose that, for some uncountable subgroup H of G, the following condition holds:

$$(\forall g \in G)(\operatorname{card}((g + H) \cap X) < \omega).$$

Can one assert that X is a G-negligible set in G?

9. Show that there exists a subset Z of $\mathbf{R}^2$ satisfying these two relations:
(a) $\operatorname{card}(Z \cap (\{t\} \times \mathbf{R})) \leq \omega$ for each $t \in \mathbf{R}$;
(b) Z is not an $\mathbf{R}^2$-negligible set in $\mathbf{R}^2$.

10*. Construct a countable covering of $\mathbf{R}$ with $\mathbf{R}$-absolutely negligible sets
For this purpose, consider $\mathbf{R}$ as a vector space E over $\mathbf{Q}$ and equip it with the norm $|| \cdot ||$ described in the proof of Lemma 6. In this manner, the nonseparable normed vector space $(E, || \cdot ||)$ is obtained, where $E = \mathbf{R}$. Further, for every natural number $n > 0$, denote

$$B_n = \{x \in E : ||x|| \leq n\}$$

and verify the validity of the following two assertions:
(a) $\cup\{B_n : 0 < n < \omega\} = E = \mathbf{R}$;
(b) each ball B_n $(0 < n < \omega)$ is an $\mathbf{R}$-absolutely negligible set in $\mathbf{R}$.
Finally, putting $X_n = B_n$ for any nonzero $n < \omega$, check that the sequence of sets

$$\{X_0, \ X_1, \ \ldots, \ X_n, \ \ldots \}$$

satisfies conditions (1) and (2) of Remark 2.

Remark 3. Many years ago, Sierpiński posed the problem of whether any translation invariant measure on $\mathbf{R}$ extending λ admits a proper translation invariant extension (see [176]). The result of Exercise 10 yields a positive solution to this problem. For more detailed information on Sierpiński's problem and further related results, see [37], [99], [123], [126], [128], [174], [210], [272], and references in [272].

18. Countable almost invariant partitions of G-spaces

In this chapter we will be dealing with some kinds of partitions of a ground (base) set E, which are almost invariant with respect to a given transformation group G acting in E. For a σ-finite G-quasi-invariant measure μ on E which is G-ergodic and has the Steinhaus property, it will be shown that every nontrivial countable μ-almost G-invariant partition has a μ-nonmeasurable member.

At present, several interesting countable partitions of the real line $\mathbf{R}$ into pairwise congruent subsets are known (see especially [232] and [266]).

Perhaps, the historically first example of such a partition was given by Vitali [266] in 1905. Recall that, by using a certain uncountable form of the Axiom of Choice, Vitali constructed a set $V \subset \mathbf{R}$ which has the following properties:

(a) $(V + p) \cap (V + q) = \emptyset$ for any two distinct rational numbers p and q;

(b) the union of the sets $V + q$, where q ranges over the field $\mathbf{Q}$ of all rational numbers, coincides with $\mathbf{R}$.

Recall also that the set V is, in fact, a selector of the quotient set $\mathbf{R}/\mathbf{Q}$ and provides the first example of a Lebesgue nonmeasurable subset of $\mathbf{R}$. This V is usually called a Vitali subset of $\mathbf{R}$. Notice that Vitali type subsets of uncountable groups were discussed in many articles, surveys and monographs (see, e.g., [32], [41], [95], [128], [139], [210], [252], [272]).

After Vitali's classical result [266], Sierpiński gave in [232] another example of a countable partition of $\mathbf{R}$ into pairwise congruent sets. Namely, he constructed a disjoint countable family $\{E_i : i \in I\}$ of subsets of $\mathbf{R}$ such that:

(1) all sets E_i $(i \in I)$ are translates of each other and collectively cover $\mathbf{R}$;

(2) all sets E_i $(i \in I)$ are thick with respect to the Lebesgue measure on $\mathbf{R}$; in particular, each E_i is nonmeasurable in the Lebesgue sense.

The main purpose of this chapter is to present some related results concerning countable almost invariant partitions of the real line (of a finite-dimensional Euclidean space equipped with an appropriate transformation group).

As in the preceding chapters of the book, we will use the following fairly standard notation.

ω = the set of all natural numbers, i.e.,

$$\omega = \mathbf{N} = \{0, 1, 2, ..., n, ...\};$$

simultaneously, ω stands for the least infinite ordinal (cardinal) number.

$\mathbf{c}$ = the cardinality of the continuum; in many cases it is convenient to identify $\mathbf{c}$ with the least ordinal number which is equinumerous with $\mathbf{R}$.

$\mathbf{R}^m$ = the Euclidean space whose dimension is m, where $m \in \mathbf{N}$.

λ_m = the standard m-dimensional Lebesgue measure on $\mathbf{R}^m$. If $m = 1$, then, for the sake of brevity, we write λ instead of λ_1.

If E is a set, then the symbol Id_E denotes the identity transformation of E.

If X and Y are any two sets, then $X \triangle Y$ stands for the symmetric difference of X and Y.

Let E be a ground (base) space equipped with a transformation group G. In this case, for the sake of brevity, we say that E is a G-space.

Let μ be a nonzero σ-finite measure on E. As usual, we denote by $\mathrm{dom}(\mu)$ the σ-algebra of all μ-measurable subsets of E. The symbol $\mathcal{I}(\mu)$ stands for the σ-ideal generated by the family of all μ-measure zero sets in E.

A subset X of E is μ-thick (or μ-massive) if the equality $\mu_*(E \setminus X) = 0$ holds true, where the symbol μ_* denotes the inner measure associated with μ.

We say that μ is a G-quasi-invariant measure on E if both $\mathrm{dom}(\mu)$ and $\mathcal{I}(\mu)$ are G-invariant classes of sets.

We recall that μ is G-ergodic (or G-metrically transitive) if, for an arbitrary μ-measurable set $X \subset E$ with $\mu(X) > 0$, there exists a countable family

$$\{g_j : j \in J\} \subset G$$

such that

$$\mu(E \setminus \cup\{g_j(X) : j \in J\}) = 0.$$

Let G be a group of transformations of E and suppose that G is endowed with some topology. In general, we will not assume in our further consideration that this topology is compatible with the algebraic structure of G.

We shall say that μ has (or possesses) the Steinhaus property if, for each μ-measurable set $X \subset E$ with $\mu(X) > 0$, there exists a neighborhood U of the identity transformation Id_E such that

$$(\forall g \in U)(\mu(g(X) \cap X) > 0).$$

In particular, the above relation immediately implies

$$(\forall g \in U)(g(X) \cap X \neq \emptyset).$$

Remark 1. The Steinhaus property is well known in the theory of locally compact topological groups. Let $(G, \cdot)$ be a σ-compact locally compact topological group, μ be the left Haar measure on G, and let μ' denote the completion of μ. Further, let X be a μ'-measurable subset of G satisfying the condition $\mu(X) < +\infty$. Then we have the equality

$$\lim_{g \to e} \mu'((g \cdot X) \triangle X) = 0,$$

where e stands for the neutral element of G (see, e.g., [41], [89], [95] or Exercise 9 from Chapter 19). Obviously, the above equality implies the Steinhaus property of μ.

Let E be a G-space equipped with some σ-finite measure μ (not necessarily G-invariant or G-quasi-invariant) and let $\{X_i : i \in I\}$ be a family of subsets of E.

We say that $\{X_i : i \in I\}$ is a μ-almost disjoint family if

$$\mu(X_i \cap X_{i'}) = 0 \quad (i \in I, i' \in I, i \neq i').$$

We say that $\{X_i : i \in I\}$ is μ-almost G-invariant if, for any $g \in G$, the family $\{g(X_i) : i \in I\}$ almost (more precisely, μ-almost) coincides with $\{X_i : i \in I\}$.

The latter phrase means that, for any index $i \in I$, there exists an index $i' = i'(i)$ such that

$$\mu(X_{i'} \triangle g(X_i)) = 0.$$

Recall that an abstract group $(G, \cdot)$ (not necessarily commutative) is divisible if, for each element $g \in G$ and for every natural number $n > 0$, the equation $x^n = g$ is solvable in G.

Remark 2. The structure of all commutative divisible groups is well known (see, for instance, $23 in a fundamental monograph [159]). On the other hand, the structure of non-commutative divisible groups is still unclear and may be very complicated. A simple example of a non-commutative divisible group is provided by the group Is_2^+ of all orientation preserving isometric transformations of the Euclidean plane $\mathbf{R}^2$. It is a widely known fact that any (commutative) group can be isomorphically embedded in a (commutative) divisible group (see $23 and $67 in [159]).

Remark 3. It should be noticed that if $(G, +)$ is an infinite commutative divisible group and X is a nonempty proper subset of G, then the family

$$\{X + g : g \in G\}$$

is necessarily infinite. Actually, this fact was first proved by Sierpiński (cf. [242]). Also, it is worth noticing that there exists a Vitali set V in the divisible commutative group $(\mathbf{R}, +)$ such that the family $\{V + t : t \in \mathbf{R}\}$ is countably infinite (see, e.g., Chapter 9).

Let E be a G-space and suppose that G is endowed with some topology.

We shall say that G is admissible if, for any neighborhood U of Id_E and for any element $g \in G$, there exists a natural number n such that the equation $x^n = g$ has a solution belonging to U.

Example 1. Clearly, the topological group T_m of all translations of $\mathbf{R}^m$ is admissible. Also, the topological group $\mathrm{O}^+(m)$ of all rotations of $\mathbf{R}^m$ about its origin is admissible. These two simple facts will be substantially exploited below.

Let E be again a G-space and suppose that G is endowed with some topology.

We shall say that G is weakly admissible if there are finitely many admissible subgroups $G_1, G_2, \ldots, G_r$ of G such that

$$G = G_1 \circ G_2 \circ \ldots \circ G_r.$$

Example 2. Consider the topological group Is_m^+ of all orientation preserving isometric transformations of $\mathbf{R}^m$. If g is any element of Is_m^+, then g can be uniquely represented in the form $g = h \circ g_0$, where $h \in \mathrm{T}_m$ and $g_0 \in \mathrm{O}_m^+$. In view of the above (see Example 1), one may conclude that the group Is_m^+ is weakly admissible. On the other hand, if $m \geq 1$, then the topological group Is_m of all isometric transformations of $\mathbf{R}^m$ is not weakly admissible (cf. Example 5 at the end of this chapter).

We will be dealing with countable μ-almost G-invariant partitions of a G-space E, where μ is a nonzero σ-finite G-quasi-invariant measure on E and G is a weakly admissible group of transformations of E.

First, we formulate and prove the main result of the chapter. Then several consequences of the main result will be discussed.

Theorem 1. *Let G be a weakly admissible group of transformations of E, let μ be a nonzero σ-finite G-ergodic measure on E having the Steinhaus property, and let $\{X_i : i \in I\}$ be a countable μ-almost disjoint μ-almost G-invariant covering of E. Then the disjunction of the following two assertions holds:*

(1) there exists at least one index $i \in I$ such that the set X_i is not μ-measurable;

(2) there exists an index $k \in I$ such that $\mu(E \setminus X_k) = 0$, i.e., the given covering is trivial in the sense of μ.

Proof. Suppose that (1) is not satisfied, i.e., all the sets X_i ($i \in I$) are μ-measurable. Then, since μ is not identically equal to zero, there exists an index $k \in I$ such that $\mu(X_k) > 0$. We assert that $\mu(E \setminus X_k) = 0$.

Suppose to the contrary that $\mu(E \setminus X_k) > 0$. Since μ is G-ergodic, we can find a countable family $\{g_j : j \in J\} \subset G$ for which the equality

$$\mu(E \setminus \cup \{g_j(X_k) : j \in J\}) = 0$$

is valid. This equality implies the existence of an index $k' \in I \setminus \{k\}$ such that

$$\mu(g_j(X_k) \cap X_{k'}) > 0$$

for some $j \in J$. Therefore, taking into account the μ-almost disjointness and μ-almost G-invariance of $\{X_i : i \in I\}$, we must have

$$\mu(X_{k'} \triangle g_j(X_k)) = 0.$$

Further, let us denote

$$G_k = \{g \in G : \mu(g(X_k) \triangle X_k) = 0\}.$$

Clearly, G_k is a subgroup of G and $g_j \notin G_k$. Keeping in mind the inequality $\mu(X_k) > 0$ and remembering that μ has the Steinhaus property, we can find a neighborhood U of Id_E such that

$$(\forall g \in U)(\mu(g(X_k) \cap X_k) > 0).$$

Further, since the group G is weakly admissible, we may write

$$G = H_1 \circ H_2 \circ ... \circ H_r$$

for some groups $H_1 \subset G$, $H_2 \subset G$, ..., $H_r \subset G$, all of which are admissible. In particular, we have the equality

$$g_j = h_1 \circ h_2 \circ ... \circ h_r,$$

where $h_1 \in H_1$, $h_2 \in H_2$, ... , $h_r \in H_r$. Since $g_j \notin G_k$, at least one of the transformations $h_1, h_2, ... , h_r$ does not belong to G_k. Let p be a natural number from the set $\{1, 2, ..., r\}$, for which $h_p \notin G_k$.

Now, there exists a natural number n and an element $h_0 \in U$ such that $h_0^n = h_p$. Since $h_p \notin G_k$, we also have $h_0 \notin G_k$. On the other hand, the relation

$$\mu(h_0(X_k) \cap X_k) > 0$$

holds true by virtue of the Steinhaus property of μ. Using once again the μ-almost disjointness and μ-almost G-invariance of $\{X_i : i \in I\}$, we obtain

$$\mu(h_0(X_k) \triangle X_k) = 0, \quad h_0 \in G_k.$$

So we come to a contradiction with the relation $h_0 \notin G_k$.
The obtained contradiction finishes the proof.

The following statement is a direct consequence of the above theorem.

Theorem 2. *Let $G = \mathrm{Is}_m^+$ be the topological group of all orientation preserving isometric transformations of the space $\mathbf{R}^m$, where $m \geq 1$, and let μ be a nonzero σ-finite G-ergodic measure on $\mathbf{R}^m$ possessing the Steinhaus property. Suppose that $\{X_i : i \in I\}$ is a countable μ-almost disjoint and μ-almost G-invariant covering of $\mathbf{R}^m$. Then either there exists at least one index $i \in I$ such that the set X_i is not μ-measurable or there exists an index $k \in I$ such that $\mu(\mathbf{R}^m \setminus X_k) = 0$.*

Proof. It suffices to apply Theorem 1, keeping in mind the fact that Is_m^+ is a weakly admissible group of transformations of $\mathbf{R}^m$ (see Example 2).

Now, we are going to give an application of Theorem 2 to the case where the role of μ is played by the standard Lebesgue measure λ_m on $\mathbf{R}^m$.

For this purpose, we need one auxiliary statement (probably, it is well known but, for the sake of completeness, we present its proof here).

Theorem 3. *Let G be a subgroup of the group Is_m. The following three assertions are equivalent:*

(1) the measure λ_m is G-metrically transitive;

(2) there exists a point $y \in \mathbf{R}^m$ whose G-orbit $G(y)$ is everywhere dense in $\mathbf{R}^m$;

(3) for any point $z \in \mathbf{R}^m$, its G-orbit $G(z)$ is everywhere dense in $\mathbf{R}^m$.

Proof. The equivalence $(2) \Leftrightarrow (3)$ is easy to show and, actually, this equivalence remains true in a much more general situation (e.g., in the case of a metric space E equipped with some group G of isometric transformations of E). So, in our further consideration, we will be focused only on the proof of the equivalence $(1) \Leftrightarrow (2)$.

Suppose that (1) holds and consider an arbitrary point $y \in \mathbf{R}^m$. Let ε be a strictly positive real number and let $U(y)$ be the open ε-neighborhood of y. Let us denote

$$V(0) = U(y) - y.$$

Then $V(0)$ is the open ε-neighborhood of 0. Since $\lambda_m(V(0)) > 0$ and λ_m is G-metrically transitive, there exists a countable family $\{g_i : i \in I\} \subset G$ such that

$$\lambda_m(\mathbf{R}^m \setminus \cup\{g_i(V(0)) : i \in I\}) = 0.$$

Further, since $\lambda_m(U(y)) > 0$, there is an index $i \in I$ such that

$$\lambda_m(g_i(V(0)) \cap U(y)) > 0$$

and, consequently,

$$g_i(V(0)) \cap U(y) \neq \emptyset.$$

So we infer that the point $g_i(0)$ belongs to the (2ε)-neighborhood of y. Since ε was taken arbitrarily small, we conclude that the orbit $G(0)$ is everywhere dense in the space $\mathbf{R}^m$, i.e., (2) holds true.

Suppose now that (2) is satisfied for a group $G \subset \mathrm{Is}_m$. Without loss of generality, we may assume that G is at most countable and the orbit $G(0)$ is everywhere dense in $\mathbf{R}^m$.

Let X be a λ_m-measurable subset of $\mathbf{R}^m$ with $\lambda_m(X) > 0$. We assert that

$$\lambda_m(\mathbf{R}^m \setminus G(X)) = 0.$$

Suppose otherwise, i.e., $\lambda_m(\mathbf{R}^m \setminus G(X)) > 0$, and denote

$$Z = \mathbf{R}^m \setminus G(X).$$

Since we have $\lambda_m(X) > 0$, there exists a density point x of X. Analogously, since $\lambda_m(Z) > 0$, there exists a density point z of Z. Let B be a ball in $\mathbf{R}^m$ whose center is 0 and whose radius is so small that

$$\lambda_m(X \cap (B + x)) \geq (2/3)\lambda_m(B), \quad \lambda_m(Z \cap (B + z)) \geq (2/3)\lambda_m(B).$$

Further, let $g \in G$ be such that

$$\lambda_m(g(B + x) \cap (B + z)) > (2/3)\lambda_m(B).$$

Then a straightforward calculation enables one to conclude that

$$\lambda_m(g(X \cap (B + x)) \cap (Z \cap (B + z))) > 0$$

and, therefore,

$$\lambda_m(g(X) \cap Z) > 0,$$

which contradicts the obvious equality

$$G(X) \cap Z = \emptyset.$$

The obtained contradiction establishes the validity of the implication $(2) \Rightarrow (1)$. This finishes the proof of Theorem 3.

Combining Theorems 2 and 3, we obtain the next result.

Theorem 4. *Let G be a subgroup of Is_m^+ with the property that at least one point of the space $\mathbf{R}^m$ has everywhere dense G-orbit, and let $\{X_i : i \in I\}$ be a countable λ_m-almost disjoint and λ_m-almost G-invariant covering of $\mathbf{R}^m$. Then either there exists an index $i \in I$ such that the set X_i is not λ_m-measurable or there exists an index $k \in I$ such that $\lambda_m(E \setminus X_k) = 0$.*

We would like to underline that the proof of Theorem 1 substantially utilizes the following two conditions:
(*) the G-ergodicity of a given measure μ;
(**) the Steinhaus property of μ.
Now, we are going to show by several relevant examples that none of the conditions (*) and (**) can be omitted.

In order to present appropriate examples, we shall consider some quasi-invariant and invariant extensions of the Lebesgue measure. Notice that various constructions of extensions of such a type are discussed in [37], [41], [95], [99], [107], [123], [137], [144], [210], [272].

Example 3. Let $m \geq 1$ be a natural number. There exists a countable partition $\{A_n : n < \omega\}$ of $\mathbf{R}^m$ such that:
(a) for each $n < \omega$ and each $g \in \mathrm{Is}_m^+$, we have

$$\mathrm{card}(g(A_n) \triangle A_n) < \mathbf{c};$$

(b) if Z is a Borel subset of $\mathbf{R}^m$ with $\lambda_m(Z) > 0$ and n is any natural index, then $\operatorname{card}(Z \cap A_n) = \mathbf{c}$; in particular, all sets A_n $(n < \omega)$ are λ_m-thick in $\mathbf{R}^m$.

The transfinite construction of $\{A_n : n < \omega\}$ is fairly standard and may be found in several works (see, e.g., [99], [123], [203], [210] or Exercise 15).

Further, take the family $\mathcal{F}$ of all those sets which admit a representation in the form

$$\cup\{A_n \cap Z_n : (\forall n < \omega)(Z_n \in \operatorname{dom}(\lambda_m))\}.$$

Notice that $\mathcal{F}$ is a σ-algebra of subsets of $\mathbf{R}^m$ containing $\operatorname{dom}(\lambda_m)$. Introduce a functional μ on $\mathcal{F}$ by the formula

$$\mu(\cup\{A_n \cap Z_n : n < \omega\}) = \sum\{(1/2^{n+1})\lambda_m(Z_n) : n < \omega\}.$$

This μ is well defined and is a measure extending λ_m. Moreover, μ can be uniquely extended to an Is_m^+-quasi-invariant measure μ' by adding to the domain of μ the family of all subsets of $\mathbf{R}^m$ whose cardinalities are strictly less than $\mathbf{c}$. For the extended measure μ', one can readily conclude that:

(i) μ' is not Is_m^+-ergodic;

(ii) μ' has the Steinhaus property;

(iii) $\{A_n : n < \omega\}$ is a nontrivial μ'-almost Is_m^+-invariant partition of $\mathbf{R}^m$ into countably many μ'-measurable subsets of $\mathbf{R}^m$, each of which is of strictly positive μ'-measure.

The next example essentially relies on the existence of a Hamel basis in $\mathbf{R}$.

Example 4. Consider the real line $\mathbf{R}$ as a vector space over the field $\mathbf{Q}$ of all rational numbers. Let $\{e_\xi : \xi < \mathbf{c}\}$ denote a Hamel basis of $\mathbf{R}$. Now, consider the vector space over $\mathbf{Q}$ generated by the partial family $\{e_\xi : 0 < \xi < \mathbf{c}\}$. Denote this vector space by H and observe that H is a hyperplane in $\mathbf{R}$ complementary to the line $\mathbf{Q}e_0$ (cf. the proof of Theorem 3 from Chapter 9). So we come to the countable partition

$$\{H_n : n < \omega\} = \{H + qe_0 : q \in \mathbf{Q}\}$$

of $\mathbf{R}$. Obviously, this partition is T_1-invariant, where T_1 denotes the group of all translations of $\mathbf{R}$. Moreover, the following relations are satisfied:

(a) for each λ-measurable set Z with $\lambda(Z) > 0$ and for each $n < \omega$, we have $Z \cap H_n \neq \emptyset$;

(b) for any two natural indices n and m, there exists $h \in \mathbf{R}$ such that $h + H_n = H_m$.

Now, we introduce the σ-algebra of sets

$$\mathcal{F} = \{\cup\{H_n \cap Z_n : n < \omega\} : (\forall n < \omega)(Z_n \in \operatorname{dom}(\lambda))\}$$

and define a functional μ on $\mathcal{F}$ by the formula

$$\mu(\cup\{H_n \cap Z_n : n < \omega\}) = \sum\{(1/2^{n+1})\lambda(Z_n) : n < \omega\}.$$

It is not difficult to check that μ is a T_1-quasi-invariant T_1-ergodic extension of λ for which $\{H_n : n < \omega\}$ is a nontrivial μ-almost T_1-invariant countable partition of $\mathbf{R}$, and all sets H_n $(n < \omega)$ are μ-measurable and have strictly positive μ-measure.

Consequently, in view of Theorem 1, μ does not possess the Steinhaus property.

Remark 4. Actually, for the measure μ of Example 4, the Steinhaus property fails in a very strong form. Namely, one can see that

$$H_n \cap (H_n + qe_0) = \emptyset$$

for any $n < \omega$ and $q \in \mathbf{Q} \setminus \{0\}$. Of course, $|q|$ and hence $|qe_0|$ may be arbitrarily small here (cf. Exercise 18 in this chapter).

Let us give one more example which shows that the assumption on a transformation group G be weakly admissible is very essential for the validity of Theorem 1.

Example 5. Let $m \geq 1$ be a natural number. There exists a partition $\{A, B, C\}$ of the space $\mathbf{R}^m$ such that:
(a) for each transformation $g \in \mathrm{Is}_m^+$, we have

$$\mathrm{card}(g(A) \triangle A) < \mathbf{c}, \quad \mathrm{card}(g(B) \triangle B) < \mathbf{c};$$

(b) for each transformation $g \in \mathrm{Is}_m \setminus \mathrm{Is}_m^+$, we have

$$\mathrm{card}(g(A) \triangle B) < \mathbf{c}, \quad \mathrm{card}(g(B) \triangle A) < \mathbf{c};$$

(c) for each transformation $g \in \mathrm{Is}_m$, we have

$$\mathrm{card}(g(C) \triangle C) < \mathbf{c};$$

(d) if Z is any Borel subset of $\mathbf{R}^m$ with $\lambda_m(Z) > 0$, then

$$\mathrm{card}(Z \cap A) = \mathrm{card}(Z \cap B) = \mathbf{c},$$

in particular, both sets A and B are λ_m-thick in $\mathbf{R}^m$.

A detailed transfinite construction of the partition $\{A, B, C\}$ is presented in [116] (see also Exercises 16 and 17 for this chapter). Notice, by the way, that condition (c) directly follows from the conjunction of the conditions (a) and (b).

Now, consider the σ-algebra of sets

$$\mathcal{F} = \{(A \cap X) \cup (B \cap Y) \cup (C \cap Z) : \{X, Y, Z\} \subset \mathrm{dom}(\lambda_m)\}$$

and define on $\mathcal{F}$ a functional μ by the formula

$$\mu((A \cap X) \cup (B \cap Y) \cup (C \cap Z)) = (1/2)(\lambda_m(X) + \lambda_m(Y)).$$

It can be checked that μ is well defined and is a measure extending λ_m. Furthermore, by adding to $\text{dom}(\mu)$ the class of all those subsets of $\mathbf{R}^m$ whose cardinalities are strictly less than $\mathbf{c}$, we obtain the measure μ' which is Is_m-invariant, Is_m-ergodic and has the Steinhaus property.

However, we see that $\{A, B, C\}$ is a μ'-almost Is_m-invariant partition of $\mathbf{R}^m$ such that all three sets A, B, C are μ'-measurable and

$$\mu'(A) = \mu'(B) = +\infty, \quad \mu'(C) = 0.$$

This circumstance can be explained by taking into account the fact that the group Is_m is not weakly admissible.

EXERCISES

1. Let E be a G-space such that the group G is endowed with some topology, and let μ be a finite G-invariant measure on E.

We say that μ has the strong Steinhaus property if, for any μ-measurable set X, the equality

$$\lim_{g \to \text{Id}_E} \mu(X \triangle g(X)) = 0$$

is satisfied.

Suppose that a G-invariant algebra $\mathcal{A}$ of subsets of E generates $\text{dom}(\mu)$ and the above equality holds true for each set $X \in \mathcal{A}$.

Show that in this case μ has the strong Steinhaus property.

Argue as follows. Take an arbitrary set $Y \in \text{dom}(\mu)$ and let ε be any strictly positive real number. For this ε, there exists a set $X \in \mathcal{A}$ such that $\mu(X \triangle Y) < \varepsilon/3$. According to the assumption, one can find a neighborhood U of Id_E satisfying the relation

$$(\forall g \in U)(\mu(g(X) \triangle X) < \varepsilon/3).$$

Further, verify the validity of the inclusion

$$g(Y) \triangle Y \subset (X \triangle Y) \cup (g(X) \triangle g(Y)) \cup (g(X) \triangle X)$$

and obtain the required result.

2. Let E_1 be a G_1-space, E_2 be a G_2-space, and suppose that the groups G_1 and G_2 are equipped with some topologies $\mathcal{T}_1$ and $\mathcal{T}_2$ respectively. Suppose also that a finite G_1-invariant measure μ_1 on E_1 possesses the strong Steinhaus property and a finite G_2-invariant measure μ_2 on E_2 possesses the strong Steinhaus property.

Demonstrate that the $(G_1 \times G_2)$-invariant product measure $\mu_1 \otimes \mu_2$ possesses the strong Steinhaus property (of course, here the product group $G_1 \times G_2$ is assumed to be equipped with the product topology $\mathcal{T}_1 \times \mathcal{T}_2$).

In order to get the required result, apply to $\mu_1 \otimes \mu_2$ the assertion formulated in Exercise 1.

Finally, generalize the above result to the case of an arbitrary family of probability invariant measures, all of which have the strong Steinhaus property.

3. Prove that the product group of any family of divisible commutative groups is a divisible commutative group.

Remark 5. The proof of this simple fact is based on the Axiom of Choice.

4*. Let p be an arbitrary prime natural number. For every natural number k, let us denote by G_k the subgroup of the circle group $(\mathbf{S}_1, \cdot)$, consisting of all those elements $z \in \mathbf{S}_1$ which satisfy the equality $z^{p^k} = 1$. In this way, a strictly increasing (by inclusion) sequence of finite commutative groups

$$G_0 \subset G_1 \subset ... \subset G_k \subset ...$$

is obtained. Let us define

$$\Gamma_p = \cup \{G_k \ : \ k < \omega\}.$$

Then Γ_p is also a subgroup of $\mathbf{S}_1$ and $\mathrm{card}(\Gamma_p) = \omega$.

Any group isomorphic to Γ_p is usually called a quasi-cyclic group of type p^∞.

Actually, the group Γ_p may be considered as the inductive limit of the family of groups $\{G_k : k < \omega\}$ with respect to their canonical embeddings

$$\phi_{k,k+1} \ : \ G_k \rightarrow G_{k+1} \qquad (k < \omega).$$

Verify that Γ_p is divisible.

For this purpose, first observe that it suffices to establish the following fact:

For any prime natural number q and for each element $s \in \Gamma_p$, there exists an element $z \in \Gamma_p$, such that $z^q = s$.

Indeed, assume that this fact is true. If n is an arbitrary nonzero natural number, then we can represent n in the form

$$n = q_1 q_2 ... q_k$$

for some finite sequence $(q_1, q_2, ..., q_k)$ of prime numbers. According to the made assumption, the equalities

$$s = z_1^{q_1}, \quad z_1 = z_2^{q_2}, \quad ...., \quad z_{k-2} = z_{k-1}^{q_{k-1}}, \quad z_{k-1} = z_k^{q_k}$$

are satisfied for certain elements $z_1, z_2, ..., z_{k-1}, z_k$ of Γ_p, from which it immediately follows that

$$s = z_k^{q_1 q_2 ... q_k} = z_k^n.$$

Now, let q be a prime natural number and let s be an arbitrary element from Γ_p. Then there exists a natural number k such that $s^{p^k} = 1$.

Only two cases are possible.

(i) $q \neq p$. In this case, the numbers p^k and q are co-prime. According to a well-known theorem from elementary number theory,

$$mp^k + lq = 1$$

for some two integers m and l. Therefore, putting $z = s^l$, one obtains

$$z^q = s^{lq} = s^{1-mp^k} = s.$$

Thus, the element z belongs to Γ_p and is a solution of the equation $x^q = s$.

(ii) $q = p$. In this case, consider an element $z \in S_1$ such that $z^p = s$. Since

$$z^{p^{k+1}} = s^{p^k} = 1,$$

one may infer that z belongs to Γ_p and is a solution of the equation $x^q = s$.

This argument establishes the divisibility of the group Γ_p.

5. Prove that each proper subgroup of Γ_p is necessarily finite.

Remark 6. A commutative group $(G, +)$ is called injective if, for any two commutative groups $(H, +)$ and $(H', +)$ such that $(H, +)$ is a subgroup of $(H', +)$, and for an arbitrary homomorphism

$$\phi : H \to G,$$

there always exists a homomorphism $\phi' : H' \to G$ extending ϕ.

It can be shown that, for a commutative group $(G, +)$, the following three assertions are equivalent:

(a) G is a divisible group;

(b) G can be represented as a direct sum of groups, each of which is isomorphic either to $\mathbf{Q}$ or to a group of type p^∞, where p is any prime number;

(c) G is an injective group.

The proof of this equivalence may be found, e.g., in [72] or [159].

6. Infer from Remark 6 that if $(G, +)$ is a divisible commutative group, all elements of which are of infinite order, then $(G, +)$ may be treated as a vector space over $\mathbf{Q}$.

7. Show that if G is a divisible subgroup of a commutative group $(H, +)$, then G is a direct summand in H. In other words, one has the representation

$$H = G + G' \qquad (G \cap G' = \{0\})$$

for some subgroup G' of H.

For this purpose, consider the identity mapping

$$\mathrm{Id}_G \; : \; G \to G$$

which obviously is an isomorphism of G onto itself. Applying the injectivity of G (see Remark 6), deduce that there exists a homomorphism

$$\phi \; : \; H \to G$$

extending Id_G. Then put

$$G' = \ker(\phi) = \phi^{-1}(0)$$

and verify that H is a direct sum of G and G'.

8. Check that the property of a commutative group to be a direct summand in any larger commutative group characterizes divisible commutative groups.

9. Verify that any direct summand in a divisible commutative group is divisible itself.

For this purpose, suppose that $(G, +)$ is a divisible commutative group and let

$$G = H + H' \qquad (H \cap H' = \{0\})$$

be a representation of G in the form of the direct sum of its two subgroups H and H'. To demonstrate that H is divisible, take any natural number $n > 0$ and an arbitrary element $h \in H$. Since G is divisible, there exists an element $x \in G$ such that $nx = h$. Clearly, one can write $x = y + z$, where $y \in H$ and $z \in H'$. Further, one has

$$nx = n(y + z) = ny + nz = h, \quad h - ny = nz, \quad h - ny \in H, \quad nz \in H'.$$

Consequently, the equalities

$$h - ny = nz = 0, \qquad h = ny$$

are valid, which shows that H is divisible.

10*. Starting with Remark 6, prove that, for any commutative group $(G, +)$, there are sufficiently many homomorphisms acting from G into $\mathbf{S}_1$. In other words, the family of all homomorphisms acting from G into $\mathbf{S}_1$ separates the elements of G.

For this purpose, take any nonzero element $g \in G$ and consider the group $[g]$ generated by g. Obviously, the circle-group $\mathbf{S}_1$ contains a subgroup H isomorphic to $[g]$. Let

$$\phi \; : \; [g] \to H$$

be an isomorphism between these two subgroups, and let

$$\phi' \ : \ G \to \mathbf{S}_1$$

be a homomorphism extending ϕ. Clearly, $\phi'(g) \neq e$ where e stands for the neutral element of $\mathbf{S}_1$.

Deduce from the latter fact that if $x \in G$, $y \in G$ and $x \neq y$, then there always exists a homomorphism

$$\phi_{x,y} : G \to \mathbf{S}_1$$

such that $\phi(x) \neq \phi(y)$.

11*. According to the result stated in Exercise 2 of Chapter 15, every commutative group $(G, +)$ can be embedded in some product group of the form $\mathbf{S}_1^\kappa$, where κ is an appropriate cardinal number.

Infer from this fact that, for any commutative group $(G, +)$, there exists a divisible commutative group $(G', +)$ containing G as a subgroup. Moreover, $(G', +)$ can be chosen satisfying the inequality

$$\mathrm{card}(G') \leq \mathrm{card}(G) + \omega.$$

Finally, conclude from the above results that every infinite commutative group $(G, +)$ can be endowed with a nondiscrete Hausdorff topology compatible with the group structure of G.

Remark 7. The last fact fails to be true for infinite non-commutative groups. In 1976, supposing **CH**, Shelah constructed a group of cardinality continuum which does not admit a nondiscrete Hausdorff group topology (see [228]). The same result was then obtained by Hesse [94] without assuming any additional set-theoretical hypotheses. Later, Ol'shanskii [201] constructed an example of a countably infinite group which does not admit a nontrivial Hausdorff group topology, and his example is also free of using any extra axioms.

12*. Give a detailed proof of Sierpiński's result mentioned in Remark 3.

For this purpose, argue by induction on $k \in \mathbf{N}$ and demonstrate that if a nonempty proper subset X of an infinite divisible commutative group $(G, +)$ has at least k distinct translates, then X also has at least $k + 1$ distinct translates.

13. Give a detailed proof of the assertion that the group $\mathrm{O}^+(m)$ of all rotations of $\mathbf{R}^m$ about the origin is admissible.

14. Check that the group Is_m^+ is weakly admissible.

15*. By using the method of transfinite recursion, construct the partition $\{A_n : n < \omega\}$ of $\mathbf{R}^m$ mentioned in Example 3.

Argue as follows. Identify, as usual, $\mathbf{c}$ with the initial ordinal number equinumerous with $\mathbf{c}$ and denote by $\{Z_\xi : \xi < \mathbf{c}\}$ the family of all Borel subsets of $\mathbf{R}^m$ having strictly positive λ_m-measure. Assume, in addition, that every Borel set $Z \subset \mathbf{R}^m$ with $\lambda_m(Z) > 0$ occurs continuum many times in this family. Further, represent the group Is_m^+ in the form

$$\mathrm{Is}_m^+ = \cup\{G_\xi : \xi < \mathbf{c}\},$$

where a $\mathbf{c}$-sequence $\{G_\xi : \xi < \mathbf{c}\}$ of subgroups of Is_m^+ satisfies these two conditions:

(a) $\{G_\xi : \xi < \mathbf{c}\}$ is increasing by inclusion;
(b) $\mathrm{card}(G_\xi) \leq \mathrm{card}(\xi) + \omega$.

Then define by transfinite recursion countably many families of sets

$$\{A_{n,\xi} : \xi < \mathbf{c}\} \qquad (n < \omega),$$

which satisfy the following relations:

(c) $A_{n,\xi} \cap A_{m,\varsigma} = \emptyset$ for any two distinct natural numbers n and m and for all ordinals $\xi < \mathbf{c}$ and $\varsigma < \mathbf{c}$;

(d) $A_{n,\xi} \cap A_{n,\varsigma} = \emptyset$ for each natural number n and for any two distinct ordinals $\xi < \mathbf{c}$ and $\varsigma < \mathbf{c}$;

(e) every set $A_{n,\xi}$ $(n < \omega, \xi < \mathbf{c})$ is the G_ξ-orbit of a point from $\mathbf{R}^m$;

(f) $A_{n,\xi} \cap Z_\xi \neq \emptyset$ for any $n < \omega$ and $\xi < \mathbf{c}$.

As soon as the above-mentioned families of sets are constructed, obtain the desired partition $\{A_n : n < \omega\}$ of $\mathbf{R}^m$.

16*. Let X be a λ_m-measurable subset of $\mathbf{R}^m$ with $\lambda_m(X) > 0$ and let G be a group of isometric transformations of $\mathbf{R}^m$ with $\mathrm{card}(G) < \mathbf{c}$.

Show that X contains a subset Y with $\mathrm{card}(Y) = \mathbf{c}$, all points of which are in general position.

For this purpose, argue by induction on m, use Fubini's theorem, and demonstrate that X cannot be covered by any family of affine hyperplanes in $\mathbf{R}^m$, whose cardinality is strictly less than $\mathbf{c}$.

Infer from the above fact that there exists a point $x \in X$ such that the group G acts freely on the orbit $G(x)$. The latter means that if $g \in G$ and $g(x) = x$, then g necessarily coincides with the identical transformation of $\mathbf{R}^m$.

17*. Let $m \geq 1$ be a natural number. Prove that there exists a partition $\{A, B, C\}$ of the space $\mathbf{R}^m$, for which the following conditions are fulfilled:

(a) for any $g \in \mathrm{Is}_m^+$, we have

$$\mathrm{card}(g(A)\triangle A) < \mathbf{c}, \quad \mathrm{card}(g(B)\triangle B) < \mathbf{c};$$

(b) for any $g \in \mathrm{Is}_m \setminus \mathrm{Is}_m^+$, we have

$$\mathrm{card}(g(A)\triangle B) < \mathbf{c}, \quad \mathrm{card}(g(B)\triangle A) < \mathbf{c};$$

(c) for any $g \in \mathrm{Is}_m$, we have

$$\mathrm{card}(g(C) \triangle C) < \mathbf{c};$$

(d) if Z is any Borel subset of $\mathbf{R}^m$ with $\lambda_m(Z) > 0$, then

$$\mathrm{card}(Z \cap A) = \mathrm{card}(Z \cap B) = \mathbf{c},$$

in particular, both sets A and B are λ_m-thick in $\mathbf{R}^m$.

For this purpose, keep in mind the result of the previous exercise and construct the required sets A, B, C by using the method of transfinite recursion (cf. Exercise 15).

18*. Let E be a G-space such that G is endowed with some topology and E is equipped with a σ-finite G-quasi-invariant measure μ.

We shall say that a μ-measurable set X with $\mu(X) > 0$ has the Steinhaus property if there exists a neighborhood U of Id_E satisfying the relation

$$(\forall g \in U)(\mu(g(X) \cap X) > 0).$$

Let now E be a G-space for which the identity transformation Id_E possesses a countable local base. Suppose also that some μ-measurable set Y with $\mu(Y) > 0$ does not have the Steinhaus property.

Show that there exist a μ-measurable set Z with $\mu(Z) > 0$ and a sequence $\{g_n : n < \omega\} \subset G$ converging to Id_E such that

$$g_n(Z) \cap Z = \emptyset \qquad (n < \omega).$$

Argue as follows. Let $\{U_n : n < \omega\}$ be a local base of G at Id_E. It may be assumed, without loss of generality, that this local base is decreasing by inclusion. Since Y does not have the Steinhaus property, for any $n < \omega$ there exists a transformation $g_n \in U_n$ satisfying the equality

$$\mu(g_n(Y) \cap Y) = 0.$$

Clearly, the sequence $\{g_n : n < \omega\}$ converges to Id_E. Now, put

$$Z = Y \setminus (\cup\{g_n^{-1}(Y) : n < \omega\})$$

and verify that the set Z is as required.

19. Let I be an arbitrary set of indices and let, for each index $i \in I$, a set E_i be given endowed with a group G_i of transformations of E_i and equipped with a σ-finite G_i-invariant measure μ_i. Suppose also that there exists a subset J of I such that $\mathrm{card}(I \setminus J) < \omega$ and all μ_i $(i \in J)$ are probability measures.

Consider the product group $G = \prod\{G_i : i \in I\}$ and the product measure $\mu = \otimes\{\mu_i : i \in I\}$ on the product set $E = \prod\{E_i : i \in I\}$.

Prove that μ is a σ-finite G-invariant measure on E.

20*. Let I be again an arbitrary set of indices and let, for each index i from I, a set E_i be given endowed with a group G_i of transformations of E_i and equipped with a σ-finite G_i-quasi-invariant measure μ_i. As in Exercise 19, suppose that there exists a subset J of I such that $\mathrm{card}(I \setminus J) < \omega$ and all μ_i ($i \in J$) are probability measures. Once again consider the product group $G = \prod\{G_i : i \in I\}$ and the product measure $\mu = \otimes\{\mu_i : i \in I\}$ on the product set $E = \prod\{E_i : i \in I\}$.

Demonstrate that μ is a σ-finite G^*-quasi-invariant measure, where G^* stands for the weak product of the family $\{G_i : i \in I\}$ of groups, i.e.,

$$G^* = \{\{g_i : i \in I\} \in G : \mathrm{card}(\{i \in I : g_i \neq \mathrm{Id}_{E_i}\}) < \omega\}.$$

For this purpose, apply to μ the Fubini theorem.

Also, give an example which shows that, in general, the measure μ does not need to be G-quasi-invariant.

21*. Let I be an uncountable set of indices. Consider the topological vector space $\mathbf{R}^I$ and its everywhere dense vector subspace

$$\mathbf{R}^{(I)} = \{\{x_i : i \in I\} \in \mathbf{R}^I : \mathrm{card}(\{i \in I : x_i \neq 0\}) < \omega\}.$$

Actually, $\mathbf{R}^{(I)}$ is a direct sum of $\mathrm{card}(I)$ many copies of the additive group $(\mathbf{R}, +)$ or, in other words, $\mathbf{R}^{(I)}$ is a weak product of $\mathrm{card}(I)$ many copies of $(\mathbf{R}, +)$ (see the previous exercise).

Let μ be a nonzero σ-finite $\mathbf{R}^{(I)}$-quasi-invariant measure on $\mathbf{R}^I$.

Prove that μ is not a Radon measure on $\mathbf{R}^I$.

For this purpose, first check that if K is an arbitrary compact subset of $\mathbf{R}^I$, then there are uncountably many pairwise disjoint $\mathbf{R}^{(I)}$-translates of K in $\mathbf{R}^I$. Consequently, the relation $K \in \mathrm{dom}(\mu)$ necessarily implies the equality $\mu(K) = 0$, from which the required result follows.

Remark 8. It can be proved that, for any infinite set I of indices, there exists a Borel probability measure μ on $\mathbf{R}^I$ which is $\mathbf{R}^{(I)}$-quasi-invariant (see, e.g., [126]). By virtue of Exercise 21, if I is uncountable, then such a μ cannot be Radon.

22*. Let I be again an uncountable set of indices and let, for each index $i \in I$, the symbol μ_i denote a probability measure on $\mathbf{R}$ equivalent to the standard Lebesgue measure λ (thus, μ_i is an $\mathbf{R}$-quasi-invariant Radon probability measure on $\mathbf{R}$).

Verify that the product measure $\mu = \prod\{\mu_i : i \in I\}$ on $\mathbf{R}^I$ is not Radon. Moreover, check that if μ' is any $\mathbf{R}^{(I)}$-quasi-invariant measure on $\mathbf{R}^I$ extending μ, then μ' cannot be a Radon measure.

In order to establish this fact, first observe that according to Exercise 20, μ is an $\mathbf{R}^{(I)}$-quasi-invariant measure, and then apply to μ (respectively, to μ') the result of Exercise 21.

Remark 9. The previous exercise shows, in particular, that the product of an uncountable family of Radon probability measures does not need to be a Radon measure.

23^*. Let E be a topological space and let Γ be a group of homeomorphisms of E.

We say that Γ acts topologically transitively in E if, for any two nonempty open sets $U \subset E$ and $V \subset E$, there exists a transformation $g \in \Gamma$ such that

$$g(U) \cap V \neq \emptyset.$$

Suppose that Γ acts topologically transitively in E and let $\phi : E \to \mathbf{R}$ be a function having the Baire property and almost invariant with respect to Γ, i.e. for each $g \in \Gamma$, the set $\{x \in E : \phi(x) \neq \phi(g(x))\}$ is of first category in E.

Demonstrate that the function ϕ is constant on a co-meager subset of E.

For this purpose, assume (without loss of generality) that E is a Baire space and then consider the sets of the form $\{x \in E : a \leq \phi(x) \leq b\}$, where a and b are any two real numbers satisfying the inequality $a < b$ (cf. Exercise 2 from Chapter 3).

24^*. Work in $\mathbf{ZF}$ set theory and define some linear functional

$$f : \mathbf{R}^{(\omega)} \to \mathbf{R}$$

which cannot be extended, within $\mathbf{ZF}$ & $\mathbf{DC}$ theory, to an additive (not necessarily continuous) functional on $\mathbf{R}^\omega$.

For this purpose, take any $x = \{x_n : n < \omega\} \in \mathbf{R}^{(\omega)}$ and put

$$f(x) = \sum \{x_n : n < \omega\}.$$

Keeping in mind the result of Solovay [253], check that this f is as required.

19. Nonmeasurable unions of measure zero sections of plane sets

Let λ denote, as usual, the standard Lebesgue measure on the real line $\mathbf{R}$. In preceding sections of this book, we have repeatedly mentioned that there are subsets of $\mathbf{R}$ nonmeasurable with respect to λ and that many works were devoted to various constructions of such pathological sets in $\mathbf{R}$ (see especially [24], [27], [30], [31], [32], [63], [71], [98], [128], [168], [184], [216], [232], [239], [252], [262]). Of course, the list of references to constructions of this kind can be significantly continued and expanded. Here we would like to begin with making some remarks in connection with the relatively recent paper by Reclaw [216], in which the following statement is established.

Theorem 1. *Assume Martin's Axiom and let B be a Borel subset of the Euclidean plane $\mathbf{R}^2$ such that:*

(1) for each point $y \in \mathbf{R}$, the horizontal section

$$B^{-1}(y) = \{x \in \mathbf{R} : (x, y) \in B\}$$

is of λ-measure zero;

(2) $\lambda(\mathrm{pr}_1(B)) = \lambda(\cup\{B^{-1}(y) : y \in \mathbf{R}\}) > 0$.

Then there exists a set $Y \subset \mathbf{R}$ for which the set

$$B^{-1}(Y) = \cup\{B^{-1}(y) : y \in Y\}$$

is not measurable with respect to λ.

In particular, under Martin's Axiom this result yields a positive solution to one problem formulated by Cichoń (for more details about this problem, see [216]).

The proof of Theorem 1 is based on the famous Luzin–Jankov–von Neumann theorem concerning the existence of measurable selectors (see, e.g., [10], [115], [167], [191], [264], or Appendix 5) and on the following fact which probably is also well-known.

Lemma 1. *Assume Martin's Axiom. Let λ_2 denote the standard two-dimensional Lebesgue measure on the plane $\mathbf{R}^2$ and let Z be a λ_2-measure zero subset of $\mathbf{R}^2$. Then there exist two sets $X \subset \mathbf{R}$ and $Y \subset \mathbf{R}$ such that:*

(1) both X and Y are λ-thick in $\mathbf{R}$, i.e., we have the equality

$$\lambda_*(\mathbf{R} \setminus X) = \lambda_*(\mathbf{R} \setminus Y) = 0;$$

(2) $(X \times Y) \cap Z = \emptyset$.

Proof. The required sets X and Y will be constructed by using the method of transfinite recursion and utilizing the classical Fubini theorem at each step of the recursion.

As usual, we identify the cardinal $\mathbf{c}$ with the equinumerous initial ordinal number, so the inequality

$$\operatorname{card}(\alpha) < \mathbf{c}$$

holds true for each ordinal $\alpha < \mathbf{c}$. Then we denote by $\{K_\alpha : \alpha < \mathbf{c}\}$ the family of all those compact subsets of $\mathbf{R}^2$ which have strictly positive λ_2-measure. According to the assumption, $\lambda_2(Z) = 0$. Consequently, by virtue of Fubini's theorem, both sets

$$U = \{x \in \mathbf{R} : \lambda^*(\{y \in \mathbf{R} : (x, y) \in Z\}) > 0\},$$

$$V = \{y \in \mathbf{R} : \lambda^*(\{x \in \mathbf{R} : (x, y) \in Z\}) > 0\}$$

are of λ-measure zero. Notice that if $x \in \mathbf{R} \setminus U$ (respectively, if $y \in \mathbf{R} \setminus V$), then again by Fubini's theorem,

$$\lambda(\{y \in \mathbf{R} : (x, y) \in Z\}) = 0$$

and, respectively,

$$\lambda(\{x \in \mathbf{R} : (x, y) \in Z\}) = 0.$$

Now, we are going to construct by transfinite recursion two injective $\mathbf{c}$-sequences

$$\{x_0, x_1, ..., x_\alpha, ...\}, \quad \{y_0, y_1, ..., y_\alpha, ...\} \quad (\alpha < \mathbf{c})$$

of points in $\mathbf{R} \setminus (U \cup V)$. Suppose that, for an ordinal $\alpha < \mathbf{c}$, the partial α-sequences

$$\{x_0, x_1, ..., x_\xi, ...\}, \quad \{y_0, y_1, ..., y_\xi, ...\} \quad (\xi < \alpha)$$

of points in $\mathbf{R} \setminus (U \cup V)$ have already been defined.

Consider the following subsets of $\mathbf{R}$:

$$Z_{1,\xi} = \{y \in \mathbf{R} : (x_\xi, y) \in Z\} \quad (\xi < \alpha),$$

$$Z_{2,\xi} = \{x \in \mathbf{R} : (x, y_\xi) \in Z\} \quad (\xi < \alpha).$$

Clearly, we have

$$\lambda(Z_{1,\xi}) = \lambda(Z_{2,\xi}) = 0 \quad (\xi < \alpha).$$

Applying Martin's Axiom to the set

$$Z_\alpha = \cup\{Z_{1,\xi} \cup Z_{2,\xi} : \xi < \alpha\},$$

we obtain the equality $\lambda(Z_\alpha) = 0$. Further, consider the set

$$K_{1,\alpha} = \{x \in \mathbf{R} : \lambda(\{y \in \mathbf{R} : (x,y) \in K_\alpha\}) > 0\}.$$

Since $\lambda_2(K_\alpha) > 0$, we have $\lambda(K_{1,\alpha}) > 0$ by Fubini's theorem. Consequently, there exists a point

$$x \in K_{1,\alpha} \setminus (U \cup V \cup Z_\alpha \cup \{x_\xi : \xi < \alpha\}).$$

So we may put $x_\alpha = x$.

Now, let us introduce the set

$$K_{2,\alpha} = \{y \in \mathbf{R} : (x_\alpha, y) \in K_\alpha\}.$$

In view of the definition of x_α, the inequality $\lambda(K_{2,\alpha}) > 0$ holds true. Besides, we have

$$\lambda(\{y \in \mathbf{R} : (x_\alpha, y) \in Z\}) = 0.$$

Consequently, there exists a point

$$y \in K_{2,\alpha} \setminus (U \cup V \cup Z_\alpha \cup \{y \in \mathbf{R} : (x_\alpha, y) \in Z\} \cup \{y_\xi : \xi < \alpha\}).$$

So we may put $y_\alpha = y$.

Proceeding in this manner, we come to the two injective **c**-sequences

$$\{x_0, x_1, ..., x_\alpha, ...\}, \quad \{y_0, y_1, ..., y_\alpha, ...\} \quad (\alpha < \mathbf{c})$$

of points in $\mathbf{R}$. Finally, putting

$$X = \{x_\alpha : \alpha < \mathbf{c}\}, \quad Y = \{y_\alpha : \alpha < \mathbf{c}\},$$

and analyzing the above construction, we readily conclude that the sets X and Y are as required.

This completes the proof of Lemma 1.

Remark 1. Actually, Lemma 1 does not need the full power of Martin's Axiom. It suffices to suppose that the covering number of the σ-ideal $\mathcal{I}(\lambda)$ of all λ-measure zero sets is equal to **c**. In other words, it suffices to assume that **c** coincides with the smallest cardinality of a covering of $\mathbf{R}$ by λ-measure zero sets. Also, Lemma 1 admits a straightforward generalization to the case of an uncountable Polish space E (instead of $\mathbf{R}$). In this case the role of λ is played by the completion of any nonzero σ-finite diffused Borel measure on E.

It should be emphasized that an abstract analogue of Lemma 1 holds under the Continuum Hypothesis (which is much stronger than Martin's Axiom).

In order to formulate this analogue, let us recall that a pseudo-base for a given measure μ is usually defined as any family $\mathcal{F} \subset \mathrm{dom}(\mu)$ satisfying the following two conditions:

(i) every set from $\mathcal{F}$ is of strictly positive μ-measure;

(ii) for any set $X \in \mathrm{dom}(\mu)$ with $\mu(X) > 0$, there exists a set $Y \in \mathcal{F}$ such that $Y \subset X$.

Lemma 2. *Assume the Continuum Hypothesis. Let μ be a σ-finite diffused measure given on a ground set E and having a pseudo-base whose cardinality does not exceed* $\mathbf{c}$. *Further, let Z be a subset of $E \times E$ such that μ-almost all horizontal sections and μ-almost all vertical sections of Z are of μ-measure zero. Then there exist two μ-thick subsets X and Y of E satisfying the equality* $(X \times Y) \cap Z = \emptyset$.

The proof of Lemma 2 is very similar to the proof of Lemma 1, so we leave it to the reader as an exercise.

Remark 2. In general, a set Z of Lemma 2 is not measurable with respect to the completion of the product measure $\mu \otimes \mu$. Moreover, the classical example due to Sierpiński shows that Z even may be $(\mu \otimes \mu)$-thick in the product space $E \times E$ (cf. Exercise 14 from Chapter 5).

Remark 3. Notice that a measure μ of Lemma 2 can be nonseparable (or, equivalently, the Hilbert space $L_2(\mu)$ of all μ-square-integrable real-valued functions on E can be nonseparable).

Theorem 1 admits an extension to the case of an analytic (i.e., Suslin) subset A of $\mathbf{R}^2$. Namely, the following statement is valid.

Theorem 2. *Suppose that the covering number of the σ-ideal of all λ-measure zero sets is equal to* $\mathbf{c}$. *Let A be an analytic subset of the Euclidean plane $\mathbf{R}^2$ such that:*

(1) for each point $y \in \mathbf{R}$, the horizontal section

$$A^{-1}(y) = \{x \in \mathbf{R} : (x, y) \in A\}$$

is of λ-measure zero;

(2) $\lambda(\mathrm{pr}_1(A)) = \lambda(\cup\{A^{-1}(y) : y \in \mathbf{R}\}) > 0$.

Then there exists a set $Y \subset \mathbf{R}$ for which the set

$$A^{-1}(Y) = \cup\{A^{-1}(y) : y \in Y\}$$

is not measurable with respect to λ.

Proof. The argument is quite similar to that of [216]. Only a few technical details occur. According to the Luzin–Jankov–von Neumann theorem (see Appendix 5), there exists a λ-measurable function

$$f : \mathrm{pr}_1(A) \to \mathbf{R}$$

whose graph is entirely contained in A. Further, there is a Borel subset T of $\mathrm{pr}_1(A)$ such that:

(a) $\lambda(\mathrm{pr}_1(A) \setminus T) = 0$;

(b) the restriction $f|T$ is a Borel function.

Let us consider the product set $T \times \mathbf{R}$ and define a mapping

$$\Phi : T \times \mathbf{R} \to \mathbf{R} \times \mathbf{R}$$

by the formula

$$\Phi(x, y) = (y, f(x)) \qquad (x \in T, \ y \in \mathbf{R}).$$

Also, let us put $Z = \Phi^{-1}(A)$. Since Φ is a Borel mapping and A is an analytic set, Z is analytic, too. Consequently, Z is λ_2-measurable (and, more generally, Z is absolutely measurable with respect to the class of completions of all σ-finite Borel measures on $\mathbf{R}^2$). For any $x \in T$, we have

$$Z(x) = \{y : (x, y) \in Z\} = \{y : (y, f(x)) \in A\} = A^{-1}(f(x)).$$

This relation shows that all x-sections of Z are of λ-measure zero, from which it follows (in view of the λ_2-measurability of Z) that $\lambda_2(Z) = 0$.

Applying Lemma 1, we can find two sets $X_1 \subset \mathbf{R}$ and $X_2 \subset \mathbf{R}$ such that

$$\lambda_*(\mathbf{R} \setminus X_1) = \lambda_*(\mathbf{R} \setminus X_2) = 0,$$

$$(X_1 \times X_2) \cap Z = \emptyset.$$

Now, we are going to verify that the set

$$Y = f(X_1 \cap T)$$

is as required, i.e., the corresponding set $\cup\{A^{-1}(y) : y \in Y\}$ is nonmeasurable with respect to λ.

First, let us check the validity of the inclusion

$$X_1 \cap T \subset \cup\{A^{-1}(y) : y \in Y\}.$$

Indeed, take an arbitrary point $x_1 \in X_1 \cap T$ and denote $y = f(x_1)$. Then

$$y \in Y, \quad (x_1, y) = (x_1, f(x_1)) \in A.$$

Therefore, $x_1 \in A^{-1}(y)$, which yields the desired result. On the other hand, let us verify that

$$(X_2 \cap T) \cap (\cup\{A^{-1}(y) : y \in Y\}) = \emptyset.$$

Indeed, take an arbitrary point $x_2 \in X_2 \cap T$ and suppose to the contrary that $x_2 \in \cup\{A^{-1}(y) : y \in Y\}$. This means that there exists $x_1 \in X_1 \cap T$ for which

$$x_2 \in A^{-1}(f(x_1)), \quad (x_2, f(x_1)) \in A, \quad (x_1, x_2) \in Z,$$

and we come to a contradiction with the equality $(X_1 \times X_2) \cap Z = \emptyset$.

Thus, the union $\cup\{A^{-1}(y) : y \in Y\}$ is almost contained in T, contains the set $X_1 \cap T$, and does not intersect the set $X_2 \cap T$. By virtue of the relation

$$\lambda^*(X_1 \cap T) = \lambda^*(X_2 \cap T) = \lambda(T) > 0,$$

we are able to conclude that the set

$$A^{-1}(Y) = \cup\{A^{-1}(y) : y \in Y\}$$

is not λ-measurable.

This completes the proof of the theorem.

From Theorem 2 one can readily infer the next statement.

Theorem 3. *Suppose again that the covering number of the σ-ideal of all λ-measure zero sets is equal to $\mathbf{c}$. Let X and Y be any two analytic sets in $\mathbf{R}$, let X be of λ-measure zero, and let the algebraic sum*

$$X + Y = \{x + y : x \in X, \ y \in Y\}$$

have strictly positive λ-measure. Then there exists a subset Y' of Y such that the algebraic sum

$$X + Y' = \{x + y' : x \in X, \ y' \in Y'\}$$

is not measurable with respect to λ.

Proof. It suffices to consider the analytic set

$$A = \{(x, y) : x - y \in X, \ y \in Y\}$$

in the plane $\mathbf{R}^2$ and to apply Theorem 2 to this A.

Theorem 4. *Suppose that Martin's Axiom and the negation of the Continuum Hypothesis hold. Let A be a Σ_2^1-subset of $\mathbf{R}^2$ satisfying the relations:*
(1) for each point $y \in \mathbf{R}^2$, the horizontal section

$$A^{-1}(y) = \{x \in \mathbf{R} : (x, y) \in A\}$$

is of λ-measure zero;

(2) $\lambda(\mathrm{pr}_1(A)) = \lambda(\cup\{A^{-1}(y) : y \in \mathbf{R}\}) > 0.$

Then there exists a set $Y \subset \mathbf{R}$ for which the set

$$A^{-1}(Y) = \cup\{A^{-1}(y) : y \in Y\}$$

is not measurable with respect to λ.

The proof of Theorem 4 can be carried out similarly to the argument used in the proof of Theorem 2. We only must take into account the following two well-known facts of descriptive set theory:

(*) every Σ_2^1-subset of the plane admits a Σ_2^1-uniformization (a consequence of Kondo's classical theorem);

(**) under $\mathbf{MA}$ & ($\neg\mathbf{CH}$), every Σ_2^1-subset of $\mathbf{R}$ (of $\mathbf{R}^2$) is Lebesgue measurable and, moreover, is absolutely measurable with respect to the class of completions of all σ-finite Borel measures on $\mathbf{R}$ (on $\mathbf{R}^2$).

For more details about (*) and (**), see [10], [103], [115], [152], [191].

On the other hand, in Gödel's Constructible Universe there are Σ_2^1-subsets of the plane, for which the assertion of Theorem 4 fails to be true. In particular, if A is a well-ordering of $\mathbf{R}$ isomorphic to w_1 and, simultaneously, A is a Σ_2^1-subset of $\mathbf{R}^2$, then both relations (1) and (2) of Theorem 4 are fulfilled but there exists no set $Y \subset \mathbf{R}$ for which $\cup\{A^{-1}(y) : y \in Y\}$ is λ-nonmeasurable.

In this context, the following three examples are relevant and should also be mentioned.

Example 1. Suppose that all those subsets of $\mathbf{R}$ which have cardinality strictly less than $\mathbf{c}$ are of λ-measure zero. Let $\preceq$ be an arbitrary well-ordering of $\mathbf{R}$ isomorphic to the smallest ordinal number of cardinality $\mathbf{c}$. Denote

$$S = \{(x, y) : x \preceq y\}.$$

By virtue of the Fubini theorem, S is not λ_2-measurable (cf. the proof of Theorem 4 from Chapter 4). Further, for any point $y \in \mathbf{R}$, the horizontal section

$$S^{-1}(y) = \{x : x \preceq y\}$$

is of λ-measure zero. At the same time, it can be easily verified that, for each set $Y \subset \mathbf{R}$, the corresponding union $\cup\{S^{-1}(y) : y \in Y\}$ is either of λ-measure zero or coincides with the whole real line $\mathbf{R}$.

Actually, Example 1 copies Sierpiński's construction [233] in which the Continuum Hypothesis is used instead of the assumption formulated above.

Example 2. Suppose again that all those subsets of $\mathbf{R}$ which have cardinality strictly less than $\mathbf{c}$ are of λ-measure zero. Let $C \subset \mathbf{R}$ be a set of cardinality continuum such that $\lambda(C) = 0$ (e.g., the role of C can be played by the classical Cantor set in $\mathbf{R}$). Let α denote the least ordinal number of cardinality continuum. Fix two bijective enumerations

$$\mathbf{R} = \{x_\xi : \xi < \alpha\}, \quad C = \{y_\zeta : \zeta < \alpha\}.$$

Further, define the set

$$G = \{(x_\xi, y_\zeta) : \xi < \zeta\}.$$

Again, all y-sections of G are of λ-measure zero and $\mathrm{pr}_1(G) = \mathbf{R}$. Furthermore, since $G \subset \mathbf{R} \times C$, we deduce that G is of λ_2-measure zero. At the same time, as in Example 1, for each set $Y \subset \mathbf{R}$, the union $\cup\{G^{-1}(y) : y \in Y\}$ is either of λ-measure zero or coincides with the whole $\mathbf{R}$.

Example 3. Assuming Martin's Axiom, there exists a subset D of $\mathbf{R}$ satisfying the following three conditions:

(a) $\mathrm{card}(D) = \mathbf{c}$ and D is of λ-measure zero;

(b) D is almost translation-invariant, i.e., the relation

$$(\forall h \in \mathbf{R})(\mathrm{card}((h + D)\triangle D) < \mathbf{c})$$

holds true;

(c) D is almost symmetric with respect to the origin of $\mathbf{R}$, i.e.,

$$\mathrm{card}(D\triangle(-D)) < \mathbf{c}.$$

Actually, the construction of such a set D also goes back to Sierpiński (cf. [99], [123], [176], [190], [203], [210] or Chapter 18).

It is not difficult to check that $D + H = \mathbf{R}$ for every set $H \subset D$ with $\mathrm{card}(H) = \mathbf{c}$. This circumstance directly implies that all algebraic sums of the form

$$D + H \quad (H \subset D)$$

are either of λ-measure zero or coincide with the whole real line $\mathbf{R}$; hence all of these algebraic sums are λ-measurable.

The three examples presented above show that some regular descriptive properties of a set $A \subset \mathbf{R}^2$ are necessary for the validity of appropriate analogues of Theorem 2.

We shall say that a class $\mathcal{K}$ of subsets of $\mathbf{R}^2$ is admissible if the following two conditions are satisfied:

(i) for any set from $\mathcal{K}$, there exists its uniformization by the graph of a partial λ-measurable function;

(ii) if P is an arbitrary member of $\mathcal{K}$ and $\Phi : \mathbf{R}^2 \to \mathbf{R}^2$ is an arbitrary Borel mapping, then the pre-image $\Phi^{-1}(P)$ is λ_2-measurable.

For admissible classes of sets, we have a suitable analogue of Theorem 2.

Theorem 5. *Suppose that the covering number of the σ-ideal of all λ-measure zero sets is equal to $\mathbf{c}$. Let $\mathcal{K}$ be an admissible class of subsets of $\mathbf{R}^2$ and let a set $P \in \mathcal{K}$ satisfy the relations:*
(1) for each point $y \in \mathbf{R}$, the horizontal section

$$P^{-1}(y) = \{x \in \mathbf{R} : (x, y) \in P\}$$

is of λ-measure zero;
(2) $\lambda(\mathrm{pr}_1(P)) = \lambda(\cup\{P^{-1}(y) : y \in \mathbf{R}\}) > 0$.
Then there exists a set $Y \subset \mathbf{R}$ for which the set

$$P^{-1}(Y) = \cup\{P^{-1}(y) : y \in Y\}$$

is not measurable with respect to λ.

The proof is similar to the proof of Theorem 2, so is left to the reader.

Obviously, under certain set-theoretical assumptions, Theorem 5 can be applied to projective plane sets of higher levels. Moreover, by utilizing Lemma 2, natural analogues of Theorem 2 can be obtained for a wide class of extensions of the Lebesgue measure λ. Notice that among those extensions there is some nonseparable measure μ having the property that the σ-ideal of all μ-measure zero sets coincides with the σ-ideal of all λ-measure zero sets (see [134]).

Theorem 6. *Assume Martin's Axiom. Let $(G, \cdot)$ be an uncountable locally compact Polish topological group and let μ' denote the completion of the left Haar measure μ on G. Then, for any nonempty μ'-measure zero set Y, there exist two μ'-thick sets $X_1 \subset G$ and $X_2 \subset G$ such that*

$$(Y \cdot X_1) \cap X_2 = \emptyset.$$

In particular, the set $Y \cdot X_1$ is nonmeasurable with respect to μ'.

Proof. The argument is similar to the proof of Theorem 3. We may suppose, without loss of generality, that Y is a Borel subset of G. In the product group $G \times G$ consider the set

$$Z = \{(x_1, x_2) \in G \times G : x_2 \in Y \cdot x_1\}.$$

Obviously, all horizontal and all vertical sections of Z are μ'-measure zero subsets of G, so we may apply to Z the generalized version of Lemma 1 formulated in Remark 1. Consequently, there exist two μ'-thick subsets X_1 and X_2 of G such that

$$(X_1 \times X_2) \cap Z = \emptyset.$$

This equality readily implies that

$$(Y \cdot X_1) \cap X_2 = \emptyset.$$

Since $Y \neq \emptyset$, we infer that the set $Y \cdot X_1$ is μ'-thick and its complement is μ'-thick, too. So we finally come to the conclusion that the set $Y \cdot X_1$ is non-measurable with respect to μ'.

Theorem 6 has thus been proved.

It is useful to compare the above theorem with the situation described in Example 3.

Remark 4. Theorem 6 does not need the full power of Martin's Axiom. Indeed, since G and $\mathbf{R}$ are Borel isomorphic, it suffices to suppose that the covering number of the σ-ideal $\mathcal{I}(\lambda)$ of all λ-measure zero sets is equal to $\mathbf{c}$.

EXERCISES

1. Give a proof of Lemma 1 under the assumption which is formulated in Remark 1. Also, generalize this lemma to the case of an uncountable Polish topological space E (instead of $\mathbf{R}$). As indicated in Remark 1, in this more general case the role of λ can be played by the completion of an arbitrary nonzero σ-finite diffused Borel measure on E.

2. Give a detailed proof of Lemma 2.

3*. Construct a measure μ on $\mathbf{R}$ satisfying the following three conditions:
(a) μ is a translation invariant extension of the Lebesgue measure λ;
(b) the topological weight of the space $L_2(\mu)$ is equal to $\mathbf{c}$;
(c) μ possesses a pseudo-base whose cardinality does not exceed $\mathbf{c}$.
For this purpose, argue as follows (cf. Exercise 3 from Chapter 15).
Let I be a set of indices with $\mathrm{card}(I) = \mathbf{c}$ and let, for each index $i \in I$, the symbol ν_i denote the probability Haar measure on the circle group $(\mathbf{S}_1, \cdot)$. Consider the product measure

$$\nu = \otimes\{\nu_i : i \in I\}.$$

Check that the cardinality of $\mathrm{dom}(\nu)$ is equal to $\mathbf{c}$. By using the method of transfinite recursion, define a group homomorphism

$$\phi : (\mathbf{R}, +) \to (\mathbf{S}_1^{\mathbf{c}}, \cdot)$$

whose graph is $(\lambda \otimes \nu)$-thick in the product space $\mathbf{R} \times \mathbf{S}_1^{\mathbf{c}}$. Further, for any $(\lambda \otimes \nu)$-measurable set Z, put

$$Z' = \{x \in \mathbf{R} : (x, \phi(x)) \in Z\}.$$

Verify that the family of sets

$$\mathcal{S} = \{Z' : Z \in \mathrm{dom}(\lambda \otimes \nu)\}$$

is a translation invariant σ-algebra in $\mathbf{R}$ and that the formula

$$\mu(Z') = (\lambda \otimes \nu)(Z) \qquad (Z \in \mathrm{dom}(\lambda \otimes \nu))$$

yields the required measure μ on $\mathbf{R}$.

4. Give a more detailed proof of Theorem 3.

5. Taking into account the statements (*) and (**), prove Theorem 4.

6. Verify the validity of the statement formulated in Example 1.

7. Formulate and prove an appropriate topological analogue of Example 2 (in terms of sets of first category in $\mathbf{R}$).

8. Give a detailed proof of Theorem 5.

9*. Let $(G, \cdot)$ be a σ-compact locally compact topological group with the neutral element e and let μ be the left Haar measure on G.
Prove that μ possesses the strong Steinhaus property, i.e., the equality

$$\lim_{g \to e} \mu((g \cdot X) \triangle X) = 0$$

holds true for any μ-measurable set X with $\mu(X) < +\infty$.
Argue in the following manner. First observe that the strong Steinhaus property is equivalent to the equality $\lim_{g \to e} \mu((g \cdot X) \cap X) = \mu(X)$ for any μ-measurable set X with $\mu(X) < +\infty$.
Let ε be an arbitrary strictly positive real number and let K be any compact subset of G. Verify the validity of these three relations:
(a) there exists an open set $U \subset G$ such that $K \subset U$ and $\mu(U \setminus K) < \varepsilon/2$;
(b) there exists an open neighborhood V of e such that $V \cdot K \subset U$;
(c) $\mu((g \cdot K) \cap K) > \mu(K) - \varepsilon$ for each element $g \in V$.
Infer from the above relations that $\lim_{g \to e} \mu((g \cdot K) \cap K) = \mu(K)$ for every compact subset K of G.
Now, let X be any μ-measurable set with $\mu(X) < +\infty$ and let ε be again an arbitrary strictly positive real number. Keeping in mind the fact that μ is a Radon measure, find a compact set $C \subset X$ such that $\mu(X \setminus C) < \varepsilon/2$. Further, find an open neighborhood W of e satisfying the relation

$$(\forall g \in W)(\mu((g \cdot C) \cap C) > \mu(C) - \varepsilon/2)$$

and, taking into account the trivial inclusion

$$(g \cdot C) \cap C \subset (g \cdot X) \cap X \qquad (g \in G),$$

deduce that if $g \in W$, then

$$\mu(X) - \varepsilon < \mu(C) - \varepsilon/2 < \mu((g \cdot C) \cap C) \leq \mu((g \cdot X) \cap X).$$

The latter yields the strong Steinhaus property of μ.

Deduce from the above that if $\mu(X) > 0$ and $\mu(Y) > 0$, then the set $X \cdot Y$ has nonempty interior.

10*. Let $(G, \cdot)$ be a locally compact Polish topological group equipped with the left Haar measure μ and let $(H, \cdot)$ be an arbitrary topological group. Suppose that $f : (G, \cdot) \to (H, \cdot)$ is a group homomorphism measurable with respect to the completion μ' of μ.

Assuming Martin's Axiom, demonstrate that f is continuous, so f is a homomorphism of the topological group $(G, \cdot)$ into the topological group $(H, \cdot)$.

Argue in the following manner. It suffices to show the continuity of f at the neutral element e_G of G. Take any open neighborhood V of the neutral element e_H of H. There exists an open neighborhood W of e_H such that $W = W^{-1}$ and $W \cdot W \subset V$. Consider the set $A = f^{-1}(W)$. Since f is μ'-measurable, the set A must be μ'-measurable, too. Only two cases are possible.

(a) $\mu'(A) > 0$.

In this case, utilize the strong Steinhaus property of μ (see Exercise 9) and infer that there exists a neighborhood U of e_G satisfying the inclusion

$$U \subset A \cdot A^{-1}.$$

Then show for this neighborhood U that

$$f(U) \subset f(A \cdot A^{-1}) \subset f(A) \cdot f(A^{-1}) \subset W \cdot W^{-1} \subset V.$$

(b) $\mu'(A) = 0$.

Verify that this case is impossible. Indeed, taking into account that A is nonempty and applying Theorem 6, find a set $B \subset G$ such that $A \cdot B$ is nonmeasurable with respect to μ'. Then check the equality

$$A \cdot B = f^{-1}(W \cdot f(B)).$$

Finally, observe that the set $W \cdot f(B)$ is open in H, so its f-pre-image must be μ'-measurable, which yields a contradiction.

Conclude from the stated above that only case (a) is realizable. Thus f turns out to be continuous at the point e_G and, consequently, at all points of G.

Remark 5. A further extension of the result presented in Exercise 10 is discussed in [160].

20. Measurability properties of well-orderings

In the present (last) chapter our main goal is to show that well-orderings are either negligible or nonmeasurable with respect to the completions of σ-finite diffused product measures. In addition, several consequences of this fact are discussed in the light of some classical problems formulated by Hilbert, Lebesgue and Luzin. In preceding sections of the book we were partially concerned with questions of similar type (cf., for instance, Chapter 4). Here we would like to envisage this topic in more details.

Let us consider an arbitrary base (ground) set E and suppose that some subset of E is well-ordered by a certain binary relation $G \subset E \times E$. In other words, we assume below that G is the graph of a well-ordering of $\mathrm{pr}_1(G)$ and $\mathrm{pr}_1(G) \subset E$.

We also suppose that some nonzero σ-finite measure μ is given on E and, in addition, this μ is continuous (i.e., diffused), which means that the equality $\mu(\{x\}) = 0$ holds true for every element $x \in E$.

In our further considerations we are going to analyze some relationships between G and μ from the viewpoint of the compatibility of these two very important mathematical structures. Actually, we will see that the compatibility of G and μ is not a typical phenomenon.

Throughout this chapter we use the following standard notation.

Let Γ be an arbitrary binary relation on E, i.e., $\Gamma \subset E \times E$.

For any $x \in E$, the symbol $\Gamma(x)$ stands for the vertical section of Γ at x, i.e., we have $\Gamma(x) = \{y \in E : (x, y) \in \Gamma\}$.

For any $y \in E$, the symbol $\Gamma^{-1}(y)$ stands for the horizontal section of Γ at y, i.e., we have $\Gamma^{-1}(y) = \{x \in E : (x, y) \in \Gamma\}$.

If μ is a σ-finite measure on E, then the symbol μ^* (respectively, μ_*) denotes the outer measure (respectively, the inner measure) canonically associated with a given measure μ.

The symbol $\mathrm{dom}(\mu)$ denotes, as usual, the domain of μ, i.e., the σ-algebra of all μ-measurable sets.

The symbol μ' denotes the completion of μ.

Suppose that μ and ν are measures on E and ν extends μ. We shall say that ν is a normal extension of μ if, for any ν-measurable set X, there exists a μ-measurable set Y such that $\nu(X \triangle Y) = 0$, where $X \triangle Y$ is the symmetric difference of X and Y, i.e.,

$$X \triangle Y = (X \setminus Y) \cup (Y \setminus X).$$

The symbol $\mu \otimes \mu$ stands, as usual, for the product measure whose both multipliers (factors) coincide with a given σ-finite measure μ.

Theorem 1. *Let E be a ground set, μ be a σ-finite continuous measure on E, and let G be a well-ordering of some subset of E. The following two assertions are equivalent:*
   (1) $\mu^*(\mathrm{pr}_1(G)) = 0$;
   (2) $(\mu \otimes \mu)^*(G) = 0$.

Proof. The implication $(1) \Rightarrow (2)$ is trivial because of the relations

$$\mathrm{pr}_1(G) = \mathrm{pr}_2(G), \quad G \subset \mathrm{pr}_1(G) \times \mathrm{pr}_2(G).$$

Let us verify the validity of the converse implication. Assume that (2) holds and suppose to the contrary that

$$\mu^*(\mathrm{pr}_1(G)) = \mu^*(\mathrm{pr}_2(G)) > 0.$$

Since G has $(\mu \otimes \mu)^*$-measure zero, μ'-almost all of its horizontal sections must be of μ'-measure zero (by virtue of Fubini's theorem). Consequently, there are elements $y \in \mathrm{pr}_2(G) = \mathrm{pr}_1(G)$ such that $\mu^*(G^{-1}(y)) = 0$. Only two cases are possible.
   (a) For every $y \in \mathrm{pr}_2(G)$, we have $\mu^*(G^{-1}(y)) = 0$.
   In this case, for any $y \in \mathrm{pr}_2(G) = \mathrm{pr}_1(G)$, we may write

$$\mu^*(G(y)) = \mu^*(\mathrm{pr}_1(G) \setminus G^{-1}(y)) > 0.$$

The above relation shows that all vertical sections of G corresponding to the elements of $\mathrm{pr}_1(G)$ $(= \mathrm{pr}_2(G))$ are of strictly positive μ^*-measure. On the other hand, according to Fubini's theorem, μ'-almost all vertical sections of G must be of μ^*-measure zero, so we come to a contradiction.
   (b) There exist elements $y \in \mathrm{pr}_2(G)$ such that $\mu^*(G^{-1}(y)) > 0$.
   In this case, let us put

$$y_0 = \inf\{y \in \mathrm{pr}_2(G) : \mu^*(G^{-1}(y)) > 0\}.$$

Consequently, $\mu^*(G^{-1}(y_0)) > 0$ and $\mu^*(G^{-1}(y)) = 0$ for an arbitrary element $y \in G^{-1}(y_0) \setminus \{y_0\}$. Consider the set

$$Z = G \cap ((G^{-1}(y_0) \setminus \{y_0\}) \times (G^{-1}(y_0) \setminus \{y_0\})).$$

According to the definition of Z, we have

$$Z \subset G, \quad (\mu \otimes \mu)^*(Z) = 0, \quad \mu^*(\mathrm{pr}_1(Z)) > 0$$

and $\mu^*(Z^{-1}(y)) = 0$ for all elements $y \in \mathrm{pr}_2(Z)$. So we again come to the case (a) which, as has already been shown, leads to a contradiction.

The obtained contradiction finishes the proof.

In a similar manner, the next statement can be established.

Theorem 2. *If μ is a σ-finite continuous measure on E and G is a well-ordering of a subset of E, then the disjunction of the following two assertions holds true:*

(1) G is of $(\mu \otimes \mu)'$-measure zero;

(2) G is nonmeasurable with respect to $(\mu \otimes \mu)'$.

Proof. Actually, it should be demonstrated that if G is measurable with respect to $(\mu \otimes \mu)'$, then $(\mu \otimes \mu)'(G) = 0$. Suppose to the contrary that G is measurable with respect to $(\mu \otimes \mu)'$ but $(\mu \otimes \mu)'(G) > 0$. In view of Fubini's theorem, there exists an element $y \in \mathrm{pr}_2(G)$ such that $\mu^*(G^{-1}(y)) > 0$. Again, let us denote

$$y_0 = \inf\{y \in \mathrm{pr}_2(G) : \mu^*(G^{-1}(y)) > 0\}$$

and let Y_0 be a μ'-measurable hull of the set $G^{-1}(y_0)$. It directly follows from the definition of Y_0 that $\mu'(Y_0) > 0$. Also, according to the definition of y_0, we have $\mu^*(G^{-1}(y)) = 0$ for all elements y belonging to $G^{-1}(y_0) \setminus \{y_0\}$. Further, consider the set

$$Z = (Y_0 \times Y_0) \cap G.$$

Obviously, Z is measurable with respect to $(\mu \otimes \mu)'$. Notice now that

$$\mu^*(Z^{-1}(y)) = \mu^*(G^{-1}(y)) = 0$$

for each element $y \in G^{-1}(y_0) \setminus \{y_0\}$. Taking into account the equality

$$\mu_*(Y_0 \setminus G^{-1}(y_0)) = 0$$

and using again Fubini's theorem, we get $(\mu \otimes \mu)'(Z) = 0$.

On the other hand, let us show that $\mu^*(Z(x)) > 0$ for all elements x from $G^{-1}(y_0) \setminus \{y_0\}$. Indeed, we may write

$$Z(x) = Y_0 \cap G(x) \quad (x \in G^{-1}(y_0) \setminus \{y_0\}).$$

Suppose for a while that $\mu'(Z(x)) = 0$ for some $x \in G^{-1}(y_0) \setminus \{y_0\}$. Then

$$\mu'(G^{-1}(y_0) \cap G(x)) = 0.$$

Keeping in mind the trivial inclusion

$$G^{-1}(y_0) \subset G(x) \cup G^{-1}(x) = E,$$

we infer that the set $G^{-1}(y_0)$ is μ'-almost contained in the set $G^{-1}(x)$. Consequently, in view of $\mu^*(G^{-1}(y_0)) > 0$, we must have $\mu^*(G^{-1}(x)) > 0$, which contradicts the definition of y_0. Thus, the inequality $\mu^*(Z(x)) > 0$ is valid for all elements x belonging to the thick subset $G^{-1}(y_0) \setminus \{y_0\}$ of Y_0. Applying once more Fubini's theorem, we conclude that $(\mu \otimes \mu)'(Z) > 0$, which yields a contradiction.

The obtained contradiction completes the proof.

To continue our presentation, let us recall several measure-theoretical concepts which were introduced in Chapter 5.

Let E be a base (ground) set and let $\mathcal{M}$ be some class of measures on E.

We say that a set $X \subset E$ is absolutely (or universally) measurable with respect to $\mathcal{M}$ if, for any measure $\mu \in \mathcal{M}$, we have $X \in \mathrm{dom}(\mu)$.

It is easy to see that the family of all absolutely measurable sets with respect to $\mathcal{M}$ forms a σ-subalgebra of the power set of E.

We say that a set $X \subset E$ is relatively measurable with respect to $\mathcal{M}$ if there exists at least one measure $\mu \in \mathcal{M}$ such that $X \in \mathrm{dom}(\mu)$.

We say that a set $X \subset E$ is absolutely nonmeasurable with respect to $\mathcal{M}$ if, for every measure $\mu \in \mathcal{M}$, we have $X \notin \mathrm{dom}(\mu)$.

Further, we shall say that a set $X \subset E$ is an absolute null set with respect to $\mathcal{M}$ if, for any measure $\mu \in \mathcal{M}$, we have $\mu(X) = 0$.

Clearly, any absolute null set with respect to $\mathcal{M}$ is absolutely measurable with respect to $\mathcal{M}$. The converse assertion is not true in general.

In connection with the above notions and their applications, see Chapter 5 and subsequent chapters of this book. Let us recall two typical examples illustrating these notions.

Example 1. Let E be a topological space and let $\mathcal{CBM}_0(E)$ denote the class of completions of all nonzero σ-finite continuous Borel measures on E. According to a widely known result of descriptive set theory, every analytic (co-analytic) subset of E is absolutely measurable with respect to $\mathcal{CBM}_0(E)$ (see, for instance, [10], [33], [103], [115], [152], [191], [199] or Appendix 5). On the other hand, if E is an uncountable Polish topological space, then every Bernstein subset of E is absolutely nonmeasurable with respect to the same class $\mathcal{CBM}_0(E)$. More detailed information on Bernstein sets may be found, e.g., in [33], [77], [96], [128], [147], [152], [190], [203]; see also Chapters 3 and 5 of the present book.

Remark 1. For an uncountable Polish topological space E, several beautiful constructions of uncountable absolute null sets with respect to the class $\mathcal{CBM}_0(E)$ are known (within **ZFC** theory). One of the earliest constructions is due to Luzin. It exploits some profound facts of classical descriptive set theory (see Appendix 5). If one assumes the Continuum Hypothesis (respectively, Martin's Axiom), then any Luzin set (respectively, any generalized Luzin set) in E can be regarded as an absolute null set. Various properties of Luzin sets (generalized Luzin sets) and their applications to point set topology, measure theory and real analysis are discussed in [133], [147], [152], [188], [190], [203] (see also Chapters 4, 10, 12, and 13).

Example 2. As we know, the family of all absolutely measurable subsets of $\mathbf{R}$ (with respect to the class $\mathcal{CBM}_0(\mathbf{R})$) is a σ-algebra containing all analytic subsets of $\mathbf{R}$. Denote this σ-algebra by $\mathcal{S}$. Suppose that every uncountable co-analytic set in $\mathbf{R}$ contains a nonempty perfect subset (this assumption does not contradict **ZFC** theory; see [103]). In this case, it can be shown that $\mathcal{S}$ is not generated by the combinations of all Borel sets in $\mathbf{R}$ with all absolute null sets with respect to $\mathcal{CBM}_0(\mathbf{R})$, i.e., we have the relation

$$\mathcal{S} \neq \{B \triangle U : B \in \mathcal{B}(\mathbf{R}), \ U \in \mathcal{N}(\mathbf{R})\},$$

where $\mathcal{B}(\mathbf{R})$ stands for the Borel σ-algebra of $\mathbf{R}$ and $\mathcal{N}(\mathbf{R})$ denotes the σ-ideal of all absolute null sets with respect to $\mathcal{CBM}_0(\mathbf{R})$. To demonstrate the validity of the above-mentioned relation, it suffices to utilize Suslin's theorem stating the existence of an analytic subset of $\mathbf{R}$ which does not belong to $\mathcal{B}(\mathbf{R})$ (see [10], [115], [152], [154], [167], [191] or Exercise 12 in this chapter).

Theorem 3. *Let E be a ground set and let $\mathcal{M}$ be a class of measures on $E \times E$ containing the family of completions of all product measures $\mu \otimes \mu$, where μ ranges over a class $\mathcal{M}_0$ of σ-finite continuous measures on E. Suppose that $G \subset E \times E$ is the graph of some well-ordering of a subset of E, and let G be absolutely measurable with respect to $\mathcal{M}$. Then $\mathrm{pr}_1(G)$ is an absolute null set with respect to the class of completions of all measures from $\mathcal{M}_0$.*

Proof. Let μ be an arbitrary measure from $\mathcal{M}_0$. We have to show that $\mu'(\mathrm{pr}_1(G)) = 0$. Suppose to the contrary that $\mu^*(\mathrm{pr}_1(G)) > 0$. Then, by virtue of Theorem 1, we get $(\mu \otimes \mu)^*(G) > 0$. According to the assumption, G is absolutely measurable with respect to $\mathcal{M}$, so G is measurable with respect to $(\mu \otimes \mu)'$ and, consequently, $(\mu \otimes \mu)'(G) > 0$, which contradicts Theorem 2. The obtained contradiction finishes the proof.

Example 3. Let E be a Polish space and let G be the graph of a well-ordering of a subset of E. Suppose that G itself is an analytic (or co-analytic) subset of the product space $E^2 = E \times E$. It easily follows from Theorem 3 that G is at most countable (this is a well-known result of descriptive set theory; see, e.g., [103], [115], [191]). Indeed, consider the case when G is analytic. According to Example 1 and Theorem 3, $\mathrm{pr}_1(G)$ is an absolute null set with respect to the class $\mathcal{CBM}_0(E)$. At the same time, $\mathrm{pr}_1(G)$ is also an analytic set in E. Suppose for a moment that $\mathrm{pr}_1(G)$ is uncountable. Then, by virtue of the theorem of Alexandrov and Hausdorff, $\mathrm{pr}_1(G)$ contains a subset homeomorphic to the Cantor space $\{0,1\}^\omega$. So $\mathrm{pr}_1(G)$ cannot be an absolute null set with respect to $\mathcal{CBM}_0(E)$ (since the Cantor space carries a canonical continuous Borel probability product measure). This contradiction shows that $\mathrm{pr}_1(G)$ ($= \mathrm{pr}_2(G)$) is at most countable, and so is G. The case of a co-analytic set G can readily be reduced to the previous case. Notice also that an analogous result holds true for all projective subsets of E assuming that all of them have the following two regularity properties:

(a) every uncountable projective set contains a nonempty perfect subset;

(b) every projective set is absolutely measurable with respect to the class $CBM_0(E)$.

Under these assumptions, which do not contradict **ZFC** theory (supposing the existence of some large cardinals), we obtain with the aid of Theorem 3 that any projective well-ordering of a subset of a Polish space E is at most countable. On the other hand, in a certain model of **ZFC** (constructed by Harrington) there is a projective well-ordering G of a subset of $\mathbf{R}$ such that the order type of $\mathrm{pr}_1(G)$ is as large as possible (for more details, see [103]).

Remark 2. As mentioned above, for an analytic set Z in a Polish space E, the following two assertions are equivalent:

(i) Z is at most countable;

(ii) Z is an absolute null set with respect to the class $CBM_0(E)$.

Recall that the equivalence (i) $\Leftrightarrow$ (ii) does not any longer hold if Z is a co-analytic subset of $\mathbf{R}$. There is a well-known model of **ZFC** theory, the Constructible Universe $\mathbf{L}$ of Gödel (see [10], [42], [103], [148]), in which there exists an uncountable co-analytic set $Z \subset \mathbf{R}$ not containing any nonempty perfect subset of $\mathbf{R}$. One can easily check that this Z is an absolute null set with respect to the class $CBM_0(\mathbf{R})$.

Remark 3. For continuous images of co-analytic sets (i.e., for Σ_2^1-sets), the assumption (a) of Example 3 is not provable within **ZFC** theory. Indeed, in the Constructible Universe $\mathbf{L}$ some Σ_2^1-subset of $\mathbf{R}$ is the graph of a well-ordering of $\mathbf{R}$ (see [103]). It follows from this fact that in $\mathbf{L}$ there are Σ_2^1-subsets of $\mathbf{R}$ which are not measurable in the Lebesgue sense (actually, one of such sets coincides with a certain Vitali set).

Example 4. Let E be a ground set and let $\mathcal{M}_0$ denote the class of all nonzero σ-finite continuous measures on E. Let us introduce the class of measures

$$\mathcal{M} = \{(\mu \otimes \mu)' : \mu \in \mathcal{M}_0\}.$$

Further, consider a well-ordering $\preceq$ of E and let G be the graph of $\preceq$. We may assert that G is absolutely nonmeasurable with respect to $\mathcal{M}$. Indeed, suppose to the contrary that G is $(\mu \otimes \mu)'$-measurable for some $\mu \in \mathcal{M}_0$. Then, by Theorem 2, we must have $(\mu \otimes \mu)'(G) = 0$. Consequently, $(\mu \otimes \mu)'(G^{-1}) = 0$, too. Now, since

$$G \cup G^{-1} = E \times E,$$

we easily get

$$(\mu \otimes \mu)'(E \times E) \le (\mu \otimes \mu)'(G) + (\mu \otimes \mu)'(G^{-1}) = 0,$$

which contradicts the assumption that μ is not identically equal to zero. The obtained contradiction shows that G is absolutely nonmeasurable with respect to the class $\mathcal{M}$.

Example 5. Let λ be the standard Lebesgue measure on $\mathbf{R}$ and let

$$\lambda_2 = (\lambda \otimes \lambda)'$$

denote the standard two-dimensional Lebesgue measure on the plane $\mathbf{R}^2$. Consider any well-ordering $\preceq$ of $\mathbf{R}$ isomorphic to the least ordinal number of cardinality $\mathbf{c}$. Let G denote the graph of $\preceq$. As has already been mentioned, G is not λ_2-measurable. Now, let $\mathcal{M}$ be the class of all those measures on $\mathbf{R}^2$ which are normal extensions of λ_2 and are translation invariant. Then it can be shown that there exist two measures $\mu_1 \in \mathcal{M}$ and $\mu_2 \in \mathcal{M}$ such that

$$G \in \mathrm{dom}(\mu_1), \quad \mu_1(\mathbf{R}^2 \setminus G) = 0, \quad G \in \mathrm{dom}(\mu_2), \quad \mu_2(G) = 0.$$

The constructions of μ_1 and μ_2 can be carried out by Marczewski's method, with the aid of appropriate translation invariant σ-ideals of subsets of $\mathbf{R}^2$ (cf. [99], [123], [174], [176], [210] or Exercise 18 from Chapter 5).

Example 6. The previous example admits a further generalization. Let E be a ground set such that $\mathrm{card}(E)$ is not cofinal with ω. Let ν be a nonzero σ-finite continuous measure on E satisfying the following condition:

$$(*) \quad (\forall X \in \mathrm{dom}(\nu))(\mathrm{card}(X) < \mathrm{card}(E) \Rightarrow \nu(X) = 0).$$

Let $\preceq$ be a well-ordering of E isomorphic to the least ordinal number whose cardinality equals $\mathrm{card}(E)$. Let G denote the graph of $\preceq$. We know that G is not measurable with respect to $(\nu \otimes \nu)'$. At the same time, there exist two normal extensions μ_1 and μ_2 of $(\nu \otimes \nu)'$ such that

$$G \in \mathrm{dom}(\mu_1), \quad \mu_1(E^2 \setminus G) = 0, \quad G \in \mathrm{dom}(\mu_2), \quad \mu_2(G) = 0.$$

Moreover, if E itself is a commutative group and the original measure ν on E is translation invariant (or translation quasi-invariant), then both normal extensions μ_1 and μ_2 can be taken as translation invariant (translation quasi-invariant) measures on the product group $E \times E$. Notice that, in this case, the condition $(*)$ is satisfied automatically.

The two preceding examples show that the graphs of certain well-orderings of E can be relatively measurable with respect to more or less natural classes of measures. We thus see that, sometimes, well-orderings and measures behave as mutually compatible structures.

When speaking of Hilbert's first problem, one always means Cantor's famous Continuum Hypothesis which asserts that each subset X of the real line $\mathbf{R}$ either is at most countable (i.e., $\text{card}(X) \leq \omega$) or is of cardinality continuum (i.e., $\text{card}(X) = \mathbf{c} = 2^\omega$). But having formulated this hypothesis as Problem 1 in his celebrated lecture of 1900, Hilbert especially underlined another problem concerning the existence of concrete well-orderings of the real line (see [97]). Saying more precisely, Hilbert was interested to know whether it is possible to actually give (define or describe by an appropriate constructive method) a well-ordering of the set $\mathbf{R}$ of all real numbers. Obviously, this question can be reformulated in the following manner:

Does there exist a concrete subset G of the plane $\mathbf{R}^2 = \mathbf{R} \times \mathbf{R}$ such that G is the graph of a well-ordering of $\mathbf{R}$?

In this context, it seems reasonable to notice here that the graph of the standard linear order $\leq$ in $\mathbf{R}$ is a very simple subset of $\mathbf{R}^2$, namely, it coincides with the closed upper half-plane determined by the straight line

$$l = \{(x, y) \in \mathbf{R}^2 : y = x\}.$$

Of course, if the Axiom of Choice ($\mathbf{AC}$) is adopted, then the posed question becomes trivial. In 1904, i.e., shortly after Hilbert's report was published, Zermelo [275] proved by using $\mathbf{AC}$ that every set can be made well-ordered. From the purely logical point of view, this fundamental result of Zermelo as well as the concepts and methods introduced by him may be regarded as an evident border line separating classical and contemporary mathematics.

It immediately follows from the above-mentioned result of Zermelo that the set $\mathbf{R}$ can be equipped with some well-ordering. But this yields nothing in the direction of solving Hilbert's problem, because Hilbert himself was interested in the existence of those well-orderings of $\mathbf{R}$ which are definable without appealing to nonconstructive or noneffective methods.

In 1930, Luzin posed a question very similar to Hilbert's one (see [167], [264]). More precisely, Luzin was interested in whether a subset of $\mathbf{R}$ with cardinality ω_1 could be effectively defined. Observe that Hilbert's and Luzin's questions are closely related to each other. Namely, it is easy to demonstrate within $\mathbf{ZF}$ set theory that if there exists a well-ordering of $\mathbf{R}$, then there exists a subset of $\mathbf{R}$ whose cardinality is ω_1. Indeed, it is widely known that the inequality $\omega_1 > \mathbf{c}$ is false, and this almost trivial fact does not need any form of the Axiom of Choice, i.e., the fact is a true sentence of $\mathbf{ZF}$ theory. Now, suppose that a binary relation $\preceq$ is some well-ordering of $\mathbf{R}$. Since any two well-ordered sets are comparable (again, without the aid of the Axiom of Choice), we must have the inequality $\omega_1 \leq \mathbf{c}$. Obviously, this inequality gives a concrete subset of $\mathbf{R}$ with cardinality ω_1.

At the beginning of the 20th century, Lebesgue introduced his measure λ on $\mathbf{R}$ (and, some time later, the analogous measure λ_n on the n-dimensional Euclidean space $\mathbf{R}^n$). He first supposed that all subsets of $\mathbf{R}$ are λ-measurable. But, in 1905, Vitali [266] gave a noneffective construction of a subset of $\mathbf{R}$ which is out of the domain of λ. Vitali's classical construction is widely known and is presented in almost all textbooks of real analysis and measure theory (see, e.g., [17], [89], [96], [128], [133], [197], [203]; cf. also Chapter 9). In the same 1905 year, Lebesgue formulated the fundamental question of whether it is possible to indicate an effective example of a λ-nonmeasurable subset of $\mathbf{R}$ (see [163]). Luzin reacted to this question very actively. In particular, he underlined several times in his works that the existence of Lebesgue nonmeasurable sets in $\mathbf{R}$ is one of the deepest phenomena in the theory of real functions (see [167], cf. also [264]). Certainly, all of the great mathematicians of that time, who were concerned with difficult problems of set theory, topology and analysis (e.g., Lebesgue, Hausdorff, Luzin, Sierpiński, Gödel, Kolmogorov, Novikov) believed that no effective example of a Lebesgue nonmeasurable subset of $\mathbf{R}$ can be presented. Only after a long-term period, Solovay was able to resolve negatively the above-mentioned Lebesgue problem. Namely, he proved in his seminal article [253] that under the assumption of the existence of a strongly inaccessible cardinal, there is a model of $\mathbf{ZF}$ & $\mathbf{DC}$ theory, in which all subsets of $\mathbf{R}$ are λ-measurable and possess the Baire property, and every uncountable subset of $\mathbf{R}$ contains a nonempty perfect subset (it directly follows from the latter fact that the inequality $\omega_1 \leq \mathbf{c}$ cannot be established within $\mathbf{ZF}$ & $\mathbf{DC}$ theory, which yields a negative solution of Luzin's above-mentioned problem). Some years later, Shelah [229] discovered that the assumption concerning the existence of a large cardinal is necessary for Solovay's result. In this context, we would like to point out the extensive monograph [110] where the theory of large cardinals is presented in all of its aspects (see also [47], [154]).

Theorem 4. *Let E be a Polish topological space and let G be the graph of a well-ordering of some subset of E. Suppose that $\mathrm{pr}_1(G)$ is not an absolute null set with respect to the class $\mathcal{CBM}_0(E)$. Then, in $\mathbf{ZF}$ & $\mathbf{DC}$ theory, for any measure $\mu \in \mathcal{CBM}_0(E)$, there exists a μ-nonmeasurable subset of E.*

Proof. Since $\mathrm{pr}_1(G)$ is not an absolute null set with respect to $\mathcal{CBM}_0(E)$, there exists a measure $\nu \in \mathcal{CBM}_0(E)$ such that $\nu^*(\mathrm{pr}_1(G)) > 0$. So, according to Theorem 3, G cannot be absolutely measurable with respect to the following class of measures:

$$\{(\theta \otimes \theta)' : \theta \in \mathcal{CBM}_0(E)\}.$$

Hence there exists a measure $\theta \in \mathcal{CBM}_0(E)$ such that $G \notin \mathrm{dom}((\theta \otimes \theta)')$. Notice now that the measures

$$\mu, \quad \theta, \quad (\theta \otimes \theta)'$$

are mutually equivalent via appropriate Borel isomorphisms, and this is a fact of $\mathbf{ZF}$ & $\mathbf{DC}$ theory. In particular, there exists a Borel isomorphism from E onto $E \times E$ which maps all μ-measurable sets onto all $(\theta \otimes \theta)'$-measurable sets. The latter circumstance implies at once the existence of a subset of E nonmeasurable with respect to μ.

Remark 4. Another proof of Theorem 4 can be derived from one deep result of Shelah [229] and Raisonnier [214].

Remark 5. Suppose that Martin's Axiom and the negation of the Continuum Hypothesis are valid. Let G be the graph of a well-ordering of some subset of a Polish space E and, in addition, let G be a Σ_2^1-subset of the product space $E \times E$. Then, since G turns out to be absolutely measurable with respect to $\mathcal{CBM}_0(E \times E)$, it follows from Theorem 3 that $\mathrm{pr}_1(G)$ is an absolute null set with respect to $\mathcal{CBM}_0(E)$. Consequently, the cardinality of $\mathrm{pr}_1(G)$ does not exceed ω_1. Notice that the latter fact also holds in **ZFC** theory (see, for instance, [103] where a much stronger result is established for the well-founded Σ_2^1-relations on E).

In connection with this theme, many old works of Sierpiński can be mentioned, in which he presented a profound logical analysis of various important statements in mathematics from the point of view of the existence of λ-nonmeasurable sets on the real line **R**. Let us give a short list of some of his results which are theorems of **ZF** & **DC** theory and which were touched upon in preceding sections of this book (cf. also [93]).

(1) If the space $\{0,1\}^{\mathbf{c}}$ is countably compact, then there exists a subset of **R** nonmeasurable with respect to λ.

(2) If the family of all countable subsets of **R** has cardinality **c** (more generally, if this family can be linearly ordered), then there exists a λ-nonmeasurable subset of **R**.

(3) If an arbitrary $(2-2)$-correspondence between any two sets admits an injective selector (a special case of Hall's combinatorial theorem), then there exists a λ-nonmeasurable subset of **R**.

(4) If there exists a nontrivial ultrafilter in the power set of ω, then there exists a λ-nonmeasurable subset of **R**.

In this direction, the most impressive and beautiful result is due to Shelah [229] and Raisonnier [214], which yields a strongly negative answer to the question posed by Luzin. As was already mentioned, Shelah and Raisonnier proved that in the theory **ZF** & **DC** the existence of a subset of **R** with cardinality ω_1 implies the existence of a λ-nonmeasurable subset of **R**. All the known proofs of this result are rather difficult.

EXERCISES

1^*. A class $\mathcal{K}$ of subsets of $\mathbf{R}^2$ is called admissible if the following five conditions are satisfied:

(a) the relations $X \in \mathcal{K}$ and $Y \in \mathcal{K}$ imply $X \cap Y \in \mathcal{K}$;

(b) the relations $X \subset \mathbf{R}$, $Y \subset \mathbf{R}$, $X \in \mathcal{K}$, $Y \in \mathcal{K}$ imply $X \times Y \in \mathcal{K}$;

(c) the relation $Z \in \mathcal{K}$ implies $\mathrm{pr}_1(Z) \in \mathcal{K}$;

(d) the relations $Z \in \mathcal{K}$ and $x \in \mathbf{R}$ imply $(Z \cap (\{x\} \times \mathbf{R})) \in \mathcal{K}$;

(e) every set $Z \in \mathcal{K}$ is absolutely measurable with respect to the class of completions of all σ-finite Borel measures on $\mathbf{R}^2$.

Verify that the conditions (a)–(e) for a class $\mathcal{K}$ are mutually independent. In other words, give five relevant examples of classes $\mathcal{K}$ of subsets of $\mathbf{R}^2$, which satisfy all these conditions except for only one of them.

For this purpose, argue as follows.

(a') Consider the class $\mathcal{K}$ whose members are the line segments on $\mathbf{R} \times \{0\}$, the discs in $\mathbf{R}^2$, the rectangles in $\mathbf{R}^2$ whose sides are parallel to $\mathbf{R} \times \{0\}$ and $\{0\} \times \mathbf{R}$ respectively, and the empty set.

Check that this $\mathcal{K}$ satisfies conditions (b)–(e) but does not satisfy (a).

(b') Consider the class $\mathcal{K}$ whose members are the subsets of $\mathbf{R} \times \{0\}$ absolutely measurable with respect to the class of completions of all σ-finite Borel measures on $\mathbf{R}^2$.

Check that this $\mathcal{K}$ satisfies conditions (a) and (c)–(e) but does not satisfy condition (b).

(c') Consider the class $\mathcal{K}$ whose members are all Borel subsets of $\mathbf{R}^2$.

Check that this $\mathcal{K}$ satisfies conditions (a), (b), (d), (e) but does not satisfy condition (c).

(d') Define $\mathcal{K}$ as follows:

$$\mathcal{K} = \{[r, +\infty[\ : r \in \mathbf{R}\} \cup \{[r, +\infty[\ \times \ [t, +\infty[\ : \ r \in \mathbf{R}, t \in \mathbf{R}\}.$$

Check that this $\mathcal{K}$ satisfies conditions (a)–(c) and (e) but does not satisfy condition (d).

(e') Let $\mathcal{K}$ be the family of all subsets of $\mathbf{R}^2$.

Check that this $\mathcal{K}$ satisfies conditions (a)–(d) but does not satisfy condition (e).

Remark 6. In some models of **ZFC** the role of $\mathcal{K}$ can be played by the class of all projective subsets of $\mathbf{R}$ (see [10], [115], [152], [154], [191]). All of such sets are defined effectively. The initial and most simple representatives of the projective hierarchy are the Borel sets and analytic sets. Recall that a subset Z of the Euclidean space $\mathbf{R}^n$ is analytic if Z is either empty or Z is a continuous image of the set of all irrational numbers. Another (of course, equivalent) definition of analytic sets in $\mathbf{R}^n$ is based on the notion of A-operation (see again [10], [115], [152], [154], [191] or Appendix 5). The theory of analytic sets was created by Suslin and Luzin and was then developed in numerous works by many authors. It also found nontrivial applications in various fields of mathematics. The theory of projective sets was created by Luzin and Sierpiński and also was intensively studied. Recall that Borel and analytic subsets of $\mathbf{R}^n$ have certain regularity properties provable within **ZF** & **DC** theory. In particular, the following assertions are valid:

(1) every uncountable analytic set $Z \subset \mathbf{R}^n$ contains a subset homeomorphic to the Cantor space $\{0, 1\}^\omega$;

(2) every analytic set $Z \subset \mathbf{R}^n$ is absolutely measurable with respect to the class of completions of all σ-finite Borel measures on $\mathbf{R}^n$;

(3) every analytic set $Z \subset \mathbf{R}^n$ admits a canonical representation in the form of the union of ω_1 many Borel sets (which are called the constituents of Z).

It should be noticed that in (c′) of Exercise 1 we used the deep fact due to Suslin and stating that the canonical projection of a Borel subset of $\mathbf{R}^2$ to $\mathbf{R} = \mathbf{R} \times \{0\}$ is not, in general, a Borel subset of $\mathbf{R}$ (but is always an analytic set in $\mathbf{R}$). As is well known, this fact inspired the emergence of classical descriptive set theory (see especially [108], [152], [167], and [264]).

2. Let $\mathcal{K}$ denote the class of all analytic subsets of $\mathbf{R}^2$. Verify that the conditions (a)–(e) of Exercise 1 are fulfilled for $\mathcal{K}$.

3. Let $\mathcal{K}$ denote the family of all absolutely measurable subsets of $\mathbf{R}^2$ with respect to the class of completions of all σ-finite Borel measures on $\mathbf{R}^2$.

Verify that $\mathcal{K}$ satisfies the conditions (a), (b), (d), and (e).

Remark 7. If Gödel's Constructibility Axiom holds, then there exists a set $X \subset \mathbf{R}$ which is not λ-measurable and which coincides with the projection of some co-analytic set $Z \subset \mathbf{R}^2$. The set Z is absolutely measurable with respect to the class of completions of all σ-finite Borel measures on $\mathbf{R}^2$, while its projection $\mathrm{pr}_1(Z) = X$ is not λ-measurable. This circumstance indicates that, within **ZFC** theory, condition (c) of Exercise 1 does not hold for the family of absolutely measurable sets with respect to the class of completions of all σ-finite Borel measures on $\mathbf{R}^2$.

4*. In the theory **ZF** & **DC**, let $\mathcal{K}$ be a fixed admissible class of subsets of $\mathbf{R}^2$ (see Exercise 1) and let G denote the graph of a well-ordering of some subset of $\mathbf{R}$. Suppose also that μ is a σ-finite diffused Borel measure on $\mathbf{R}$ and μ' stands for the completion of μ.

Prove that if $G \in \mathcal{K}$, then the equality $\mu'(\mathrm{pr}_1(G)) = 0$ holds true (consequently, $\mathrm{pr}_1(G) \neq \mathbf{R}$).

Argue in the following manner. Since $G \in \mathcal{K}$, we have $\mathrm{pr}_1(G) \in \mathcal{K}$ and $\mathrm{pr}_1(G)$ is a μ'-measurable subset of $\mathbf{R}$. Suppose to the contrary that $\mu'(\mathrm{pr}_1(G)) > 0$. Only two cases are possible.

(i) For each element $x \in \mathrm{pr}_1(G)$, the set $\{t : (t, x) \in G\}$ has μ'-measure zero. In this case, put $X = \mathrm{pr}_1(G)$.

(ii) There are elements $x \in \mathrm{pr}_1(G)$ for which $\mu'(\{t : (t, x) \in G\}) > 0$.

In this case, let x_0 denote the least element (with respect to G) such that $\mu'(\{t : (t, x_0) \in G\}) > 0$ and put $X = \{t : (t, x_0) \in G\}$.

Further, consider the set

$$Z = (X \times X) \cap G.$$

Under conditions (a)–(e), this set belongs to the class $\mathcal{K}$ and hence is measurable with respect to the completion of $\mu \otimes \mu$. Notice that if $y \in X \setminus \{x_0\}$, then the section $Z \cap (\mathbf{R} \times \{y\})$ is of μ'-measure zero. So, by virtue of Fubini's theorem (which is a true sentence within **ZF** & **DC** theory), one must have $(\mu \otimes \mu)'(Z) = 0$. On the other hand, if $x \in X \setminus \{x_0\}$, then the section $Z \cap (\{x\} \times \mathbf{R})$ is of strictly positive μ'-measure and, applying again Fubini's theorem, one has $(\mu \otimes \mu)'(Z) > 0$, which is a contradiction. The obtained contradiction yields the desired result.

5*. Work in **ZF** & **DC** theory and prove that the so-called perfect subset property is valid within the class of all analytic sets in $\mathbf{R}^n$ (this old result is due to Alexandrov and Hausdorff). More generally, let E be a Polish topological space, Y be an uncountable Suslin subset of E and let $g : E \to Y$ be a continuous surjection.

Show that there exists a set $X \subset E$ homeomorphic to the Cantor space $\{0, 1\}^\omega$ and such that the restriction

$$g|X : X \to g(X)$$

is a bijection (hence, in view of the compactness of X, this restriction is also a homeomorphism between X and $g(X)$).

Argue in the following manner. First, use ordinary recursion and construct a dyadic system

$$\{F_s : s \in \{0, 1\}^{<\omega}\}$$

of closed balls in E such that, for any sequence $s \in \{0, 1\}^{<\omega}$, the following four relations would be satisfied:

(1) $F_{s,0} \cup F_{s,1} \subset F_s$;
(2) $g(F_{s,0}) \cap g(F_{s,1}) = \emptyset$;
(3) $g(F_s)$ is an uncountable set;
(4) the radius of F_s does not exceed $1/(\mathrm{lh}(s) + 1)$, where $\mathrm{lh}(s)$ stands for the length of s.

Then, having this system of balls, put

$$X(k) = \cup\{F_s : s \in \{0, 1\}^{<\omega}, \ \mathrm{lh}(s) = k\} \qquad (k < \omega),$$

$$X = \cap\{X(k) : k < \omega\}.$$

A direct verification yields that the set X is homeomorphic to the Cantor space by virtue of the relations (1), (2), and (4). In addition to this circumstance, relation (2) implies that $g|X$ is an injection. Keeping in mind the compactness of X, conclude that $g|X$ is a homeomorphism between the sets X and $g(X)$.

Remark 8. As an immediate consequence of Exercise 5, we obtain that the cardinality of any uncountable analytic (in particular, Borel) subset of a Polish space is equal to **c**. We thus see that analytic sets in some sense realize the Continuum Hypothesis: they are either countable or they are of cardinality continuum. Unfortunately, the same statement fails to be true for co-analytic sets (in this connection, see [10], [103], [115], [152], [154], [191]; cf. also Exercise 13 from Appendix 5).

6. Prove the analogue of Example 3 for co-analytic sets; more precisely, demonstrate that if a co-analytic set $G \subset \mathbf{R}^2$ is a well-ordering of some part of $\mathbf{R}$, then $\mathrm{card}(G) \leq \omega$.

For this purpose, consider the analytic set $\mathbf{R}^2 \setminus G$ and apply to it the argument similar to that of Example 3.

7. Let E be a ground set equipped with a σ-finite measure μ and let $\mu \otimes \mu$ denote the corresponding product measure on $E \times E$.

Show that, for a set $G \subset E \times E$, the following two assertions are equivalent:
(a) G is measurable with respect to $\mu \otimes \mu$;
(b) G^{-1} is measurable with respect to $\mu \otimes \mu$.
Moreover, if one of these two assertions is valid, then the equality

$$(\mu \otimes \mu)(G) = (\mu \otimes \mu)(G^{-1})$$

holds true.

In order to obtain the required result, consider the class $\mathcal{F}$ of subsets G of $E \times E$ for which (a) (or, equivalently, (b)) is fulfilled and the above equality is valid. Verify that $\mathcal{F}$ contains the domain of $\mu \otimes \mu$.

8. Give a proof of the assertion formulated in Example 5.

9. Give a proof of the assertion formulated in Example 6.

10. Let Γ be a subgroup of the additive group $(\mathbf{R}, +)$.
Show that Γ satisfies the disjunction of the following three assertions:
(a) Γ coincides with $\mathbf{R}$;
(b) Γ is of λ-measure zero;
(c) Γ is a λ-nonmeasurable subset of $\mathbf{R}$.
For this purpose, utilize the Steinhaus property of λ which says that if a λ-measurable set $X \subset \mathbf{R}$ has strictly positive λ-measure, then the set

$$X - X = \{x - y : x \in X, \ y \in X\}$$

is a neighborhood of zero (see [257] or Exercise 1 from Chapter 3).

Remark 9. The result of Exercise 10 can be applied to some questions concerning the descriptive structure of $\mathbf{R}$. Namely, suppose that there exists a nonconstructible real, i.e., $\mathbf{R} \setminus (\mathbf{L} \cap \mathbf{R}) \neq \emptyset$. Then the set $\mathbf{L} \cap \mathbf{R}$ either is of λ-measure zero or is λ-nonmeasurable. Notice that each of these possibilities can be realized in suitable models of **ZFC** (for more details, see [103]).

11. Recall the result of Shelah and Raisonnier which states that, within **ZF** & **DC** theory, the inequality $\omega_1 \leq \mathbf{c}$ implies the existence of a Lebesgue nonmeasurable subset of $\mathbf{R}$.

Derive Theorem 4 from the above-mentioned result.

12*. Let E be an arbitrary uncountable Polish topological space.

Work in **ZF** & **DC** theory and prove the classical theorem of Suslin stating that there exists an analytic subset of E which is not Borel.

Argue in the following manner. It suffices to establish this fact for some concrete Polish space E, e.g., in the case where E coincides with Cantor's discontinuum $\{0,1\}^\omega$. Consider the product space

$$K = \{0,1\}^\omega \times [0,1]$$

and denote by $Comp(K)$ the family of all compact subsets of K. Endowed with the standard Hausdorff metric, this family becomes a compact metric space (see, e.g., [49], [153]). Consequently, there exists a Peano type mapping

$$g : \{0,1\}^\omega \to Comp(K),$$

i.e., g is a continuous surjection. Let I be the set of all irrational numbers from $[0,1]$, which is homeomorphic to the Baire space ω^ω. Denoting, as usual, by $\mathcal{P}(\{0,1\}^\omega)$ the family of all subsets of $\{0,1\}^\omega$, define a set-valued mapping

$$G : \{0,1\}^\omega \to \mathcal{P}(\{0,1\}^\omega)$$

by the formula

$$G(t) = \mathrm{pr}_1(((\{0,1\}^\omega \times I) \cap g(t)) \qquad (t \in \{0,1\}^\omega).$$

Verify the validity of the relation $\mathrm{ran}(G) = \mathcal{A}(\{0,1\}^\omega)$, where $\mathcal{A}(\{0,1\}^\omega)$ stands for the family of all analytic subsets of $\{0,1\}^\omega$. Further, in $\{0,1\}^\omega$ consider the set

$$X = \{t \in \{0,1\}^\omega : t \notin G(t)\}.$$

Suppose for a while that X is an analytic set in $\{0,1\}^\omega$. Then, for some point $t_0 \in \{0,1\}^\omega$, the equality $G(t_0) = X$ must be valid. It easily follows from this equality that

$$t_0 \in X \Leftrightarrow t_0 \notin X,$$

which yields a contradiction. Therefore, X cannot be an analytic subset of $\{0,1\}^\omega$. On the other hand, show that the set

$$\{0,1\}^\omega \setminus X = \{t \in \{0,1\}^\omega : t \in G(t)\}$$

is analytic. For this purpose, check the validity of the relation

$$\{0,1\}^\omega \setminus X = \mathrm{pr}_1(D),$$

where $D \subset K$ is a Borel set defined as follows:

$$D = \{(t,r) \in K : (t,r) \in g(t) \ \& \ r \in I\}.$$

Summarize all the facts stated above and infer the required result.

13*. Let E be a topological space and let A and B be two subsets of E. These A and B are called separated by two disjoint Borel sets if there exist Borel subsets A' and B' of E such that

$$A \subset A', \quad B \subset B', \quad A' \cap B' = \emptyset.$$

Let now X and Y be two subsets of E and let

$$X = \cup\{X_n : n < \omega\}, \quad Y = \cup\{Y_n : n < \omega\}.$$

Work in **ZF** & **CC** theory and prove that if X and Y are not separated by any two disjoint Borel sets, then there exists a pair (m, n) of natural numbers such that the sets X_m and Y_n also are not separated by any two disjoint Borel sets.

Argue as follows. Suppose to the contrary that for any pair (m, n) of natural numbers, the sets X_m and Y_n are separated by some two disjoint Borel sets (of course, depending on (m, n)). This assumption means that there exists a Borel set $P_{mn} \subset E$ satisfying the relation

$$X_m \subset P_{mn} \subset E \setminus Y_n.$$

Now, putting

$$B_m = \cap\{P_{mn} : n < \omega\} \quad (m < \omega),$$

deduce that

$$X \subset \cup\{B_m : m < \omega\}, \quad Y \cap (\cup\{B_m : m < \omega\}) = \emptyset,$$

which contradicts the condition that X and Y are not separated by any two disjoint Borel sets.

The obtained contradiction yields the desired result.

14*. Let E be a Polish topological space and let X and Y be two disjoint analytic subsets of E.

Work in **ZF** & **DC** theory and prove Luzin's separation principle stating that X and Y are separated by two disjoint Borel sets, i.e., there exist Borel sets X' and Y' for which the relations

$$X \subset X', \quad Y \subset Y', \quad X' \cap Y' = \emptyset$$

hold true.

Argue as follows. Suppose to the contrary that X and Y are not separated by any two disjoint Borel sets. Then, obviously, $X \neq \emptyset$ and $Y \neq \emptyset$. According to the definition of analytic sets (cf. Appendix 5), there are continuous surjections

$$f : \omega^\omega \to X, \quad g : \omega^\omega \to Y.$$

For any nonempty finite sequence $(n_0, n_1, ..., n_k)$ of natural numbers, denote

$$U(n_0, n_1, ..., n_k) = \{z \in \omega^\omega : (z_0, z_1, ..., z_k) = (n_0, n_1, ..., n_k)\}.$$

Using this notation, one may write

$$X = \cup\{f(U(n)) : n < \omega\}, \quad Y = \cup\{g(U(m)) : m < \omega\}.$$

By virtue of Exercise 13, there exist natural numbers n_0 and m_0 such that the sets $f(U(n_0))$ and $g(U(m_0))$ are not separated by any two disjoint Borel sets. Proceeding by ordinary recursion, construct two infinite sequences

$$a = (n_0, n_1, ..., n_k, ...) \in \omega^\omega, \quad b = (m_0, m_1, ..., m_k, ...) \in \omega^\omega$$

such that, for every $k < \omega$, the corresponding sets

$$X(k) = f(U(n_0, n_1, ..., n_k)), \quad Y(k) = g(U(m_0, m_1, ..., m_k))$$

are not separated by any two disjoint Borel sets. On the other hand, keeping in mind the continuity of both mappings f and g, infer that, for a sufficiently large natural index k, the sets $X(k)$ and $Y(k)$ are contained in some disjoint open neighborhoods of the two distinct points $f(a) \in X$ and $g(b) \in Y$. Thus, a contradiction is obtained which yields the required result.

Remark 10. For any uncountable Polish space E, it was demonstrated by Luzin and Novikov that there exist two disjoint co-analytic sets in E, which are not separated by any two disjoint Borel subsets of E (see [10], [115], [152], [167], [191]). This profound and important result immediately yields that there exists an analytic subset of E which is not Borel (cf. Exercise 12).

15. Suppose that E and F are any two Polish topological spaces and h is a mapping acting from E into F such that the graph $\mathrm{Gr}(h)$ of h is an analytic subset of the product space $E \times F$.

Verify that h is a Borel mapping and, therefore, $\mathrm{Gr}(h)$ is a Borel subset of $E \times F$.

To see this fact, take any Borel set $B \subset F$. Clearly, the equalities

$$h^{-1}(B) = \mathrm{pr}_1(\mathrm{Gr}(h) \cap (E \times B)),$$

$$h^{-1}(F \setminus B) = \mathrm{pr}_1(\mathrm{Gr}(h) \cap (E \times (F \setminus B)))$$

hold true, whence it follows that $h^{-1}(B)$ and $h^{-1}(F \setminus B)$ are analytic sets in E. In addition,

$$h^{-1}(B) \cap h^{-1}(F \setminus B) = \emptyset, \quad h^{-1}(B) \cup h^{-1}(F \setminus B) = E.$$

By virtue of the separation principle (see Exercise 14), both sets $h^{-1}(B)$ and $h^{-1}(F \setminus B)$ must be Borel, which shows that h is a Borel mapping.

16*. Let E be a Polish topological space and let $\{X_n : n < \omega\}$ be a countable disjoint family of analytic sets in E.

Work in **ZF** & **DC** theory and establish a generalized version of Luzin' separation principle, namely, demonstrate that there exists a countable disjoint family $\{Y_n : n < \omega\}$ of Borel sets in E such that $X_n \subset Y_n$ for all $n < \omega$.

Argue in the following manner. According to Exercise 14, for any pair (n, m) of distinct natural numbers, there exists a Borel set B_{nm} such that

$$X_n \subset B_{nm} \subset E \setminus X_m.$$

Further, put

$$Y_0 = \cap\{B_{0m} : 1 \leq m < \omega\}$$

and, for each $n \in \omega \setminus \{0\}$, define by recursion

$$Y_n = (\cap\{B_{nm} : m < \omega, m \neq n\}) \setminus (Y_0 \cup Y_1 \cup ... \cup Y_{n-1}).$$

Finally, check that $\{Y_n : n < \omega\}$ is the required family of pairwise disjoint Borel subsets of E.

17. Luzin conjectured that every subset of $\mathbf{R}$ of cardinality ω_1 is co-analytic. It was shown later that this conjecture is consistent with **ZFC** set theory (see e.g., [264]).

Deduce from Luzin's conjecture that $2^\omega = 2^{\omega_1}$.

Luzin called the above equality the Second Continuum Hypothesis (obviously, it contradicts Cantor's Continuum Hypothesis).

Appendix 1: The axioms of set theory

In this appendix we would like to provide the reader with several more or less simple facts from Zermelo–Fraenkel set theory which nowadays is regarded as a logical base of all contemporary mathematics. Our presentation of the material is informal and should be treated at an intuitive or somewhat naive level. However, one might readily translate all considerations presented below into a formal language of mathematical logic (more precisely, first-order logic). Of course, in a book devoted primarily to topics of classical analysis and measure theory, routine details of such a translation do not seem relevant. So, below we only speak of semantical aspects of set theory and leave aside its syntactic aspects (cf. [18], [154]).

As is well known, Euclid begins his monumental manuscript of geometry with a list of axioms and postulates. Following him, it is natural and reasonable to formulate here the axioms of set theory. As we have already mentioned, they turned out to be the axioms of all modern mathematics (a detailed logical analysis of these axioms is given in a number of important works; see especially [10], [38], [40], [42], [60], [87], [93], [102], [103], [109], [111], [148], [154], [164], [172], [173], [187], [231], [276], [278]).

The only objects considered in this theory are sets. An initial relation between them is the binary membership relation $\in$. The formula $x \in y$ is usually expressed by the phrases:

$$x \text{ is an element of } y, \quad x \text{ is a member of } y.$$

Remark 1. Of course, set theory relies on general logical concepts and exploits the standard equality relation $=$ with widely known properties of $=$. Roughly speaking, set theory being a substantial extension of classical propositional calculus, contains in its alphabet the two symbols of quantifiers $\exists$ and $\forall$, the two binary relation symbols $=$ and $\in$, and uses some axioms reflecting the most general logical laws. Here we do not intend to specify those axioms (for further details, see [18], [87], [154], [187], [231]).

A complete list of proper axioms of contemporary set theory looks at follows.

Axiom 0. There exists a set without elements, i.e.,

$$(\exists x)(\forall y)(y \notin x).$$

It immediately follows from Axiom 0 that there are some sets in the mathematical universe which is under our consideration.

Axiom 1. If two sets A and B are such that the equivalence

$$x \in A \Leftrightarrow x \in B$$

holds true for all x, then $A = B$ (the so-called Extensionality Axiom).

Actually, Axiom 1 states that any set A is completely determined by all elements of A.

In particular, Axioms 0 and 1 yield immediately that there exists only one set without elements. It is called the empty set and is denoted by $\emptyset$. Since $x = x$ is a theorem of any reasonable theory, we may say that $\emptyset$ consists of all those elements x which satisfy $x \neq x$.

More generally, let $S(x)$ be a property written in the language of set theory and containing x as a free variable. In many cases it may happen that there exists a set X consisting precisely of all those elements x for which $S(x)$ is valid. Formally, we have

$$(\forall x)(x \in X \Leftrightarrow S(x)).$$

Then Axiom 1 implies that such a set X is unique, and it is convenient to denote this X by the symbol $\{x : S(x)\}$.

Remark 2. There are many properties $S(x)$ which do not determine any set. A very simple example of such a property was suggested by Russell in 1901. Namely, he considered the relation $x \notin x$ and supposed that there exists a set X satisfying

$$(\forall x)(x \in X \Leftrightarrow x \notin x).$$

Then, putting $x = X$, he obtained $X \in X \Leftrightarrow X \notin X$ which is an obvious contradiction. However, it is sometimes convenient to assume that an arbitrary relation $S(x)$ in set theory determines a certain collection (class) of all those sets x for which $S(x)$ is valid. Clearly, every set Y can be regarded as a class determined by the relation $x \in Y$. So we see that a class of sets may coincide with some set or may be a proper class distinct from any set. For example, the quite simple binary relations $x \notin x$ and $x = x$ determine two corresponding proper classes (cf. also Exercises 6 and 10 of this appendix).

If A and B are two sets and

$$(\forall x)(x \in B \Rightarrow x \in A),$$

then we say that B is a subset (or a part) of A and write $B \subset A$. Clearly, $\emptyset$ is a unique set which is a part of any set.

Axiom 2. For every set A, there exists its power set (or its Boolean), i.e., there exists a set

$$\mathcal{P}(A) = \{B : B \subset A\}.$$

In view of Axiom 1, $\mathcal{P}(A)$ is uniquely determined by a given set A.

Applying Axiom 2 to $A = \emptyset$, we obtain the set $\{\emptyset\}$ whose unique element is $\emptyset$. In view of Axiom 1, we have $\emptyset \neq \{\emptyset\}$, so there are at least two different sets in our mathematical universe (cf. Axiom 6 presented below).

Axiom 3. For any set A, there exists a union of all members of A, i.e., there exists a set

$$\cup A = \{x : (\exists X \in A)(x \in X)\}.$$

Again, by virtue of Axiom 1, $\cup A$ is uniquely determined by a given set A. Notice also that $\cup \emptyset = \emptyset$.

Axiom 4. Let $S(x, y)$ be a binary relation between free variables x and y and let this relation be functional with respect to y, i.e., for every x there is exactly one y such that $S(x, y)$. Then, for any set A, there exists a set

$$S(A) = \{y : (\exists x)(x \in A \ \& \ S(x, y))\}.$$

This is the so-called Replacement Axiom. Its intuitive meaning is obvious: if a relation $S(x, y)$ is functional with respect to y, then the S-image of any set is also a set. In addition, Axiom 1 yields that $S(A)$ is uniquely determined by a set A.

In connection with the Replacement Axiom, it should also be mentioned that it provides us with a lot of axioms corresponding to various functional relations $S(x, y)$. Actually, we have here a certain scheme of axioms.

Let X and Y be any two sets. We say that z is a common element of X and Y if $z \in X$ and $z \in Y$.

Accordingly, two sets X and Y are called having no common elements (briefly, disjoint) if there exists no z such that $z \in X$ and $z \in Y$.

Axiom 5. For an arbitrary set A consisting of nonempty pairwise disjoint sets, there exists a selector of A, i.e., there exists a set $Z \subset \cup A$ such that, for every $Y \in A$, the sets Z and Y have exactly one common element.

This is the famous Axiom of Choice denoted usually by the symbol **AC**. Clearly, this axiom is of extremely nonconstructive character. Because of its nature, **AC** inspired and stimulated many debates and discussions among various groups of mathematicians (see, e.g., [60], [93], [102], [109], [111], [154], [167], [189], [222], [232], [238], [243]).

Axiom 6. There exists an infinite set X_∞ (the so-called Infinity Axiom).

In more details, this axiom means that there exists a set X_∞ possessing the following two properties:

(a) $\emptyset \in X_\infty$;

(b) for each element $x \in X_\infty$, there exists an element $y \in X_\infty$ such that

$$(\forall z \in y)(z \in x \ \lor \ z = x).$$

It should be mentioned that the Axioms 0, 1, 2, 3, 5, 6 were introduced by Zermelo and only one Axiom 4 was suggested by Fraenkel (and, independently, by Skolem). So the obtained theory is usually denoted by **ZFC**:

Zermelo & Fraenkel & Choice.

The abbreviation **ZF** is reserved for denoting the same theory from which the Axiom of Choice is withdrawn.

It is a great mystery that the concise list of the above-mentioned axioms of **ZFC** theory turned out to be enough to develop all fields and branches of contemporary mathematics.

As was mentioned in Chapter 1, the theory **ZF** is rather weak even for purposes of classical mathematical analysis on the real line **R** (cf. Chapters 1 and 2).

On the other hand, **ZF** theory allows to establish the uncountability of **R** and, moreover, it is possible to prove within **ZF** that

$$\operatorname{card}(E) < \operatorname{card}(\mathcal{P}(E))$$

for any set E (see Theorem 1 below). This inequality means that there exists a canonical injection from E into its power set $\mathcal{P}(E)$, but there exists no injection from $\mathcal{P}(E)$ into E. The above-mentioned inequality is one of the historically first achievements of Cantor in general naive set theory (see [28]).

Let us notice that, dealing with the foundations of mathematics, a less natural axiom is usually added to the axioms listed above. It is called the Foundation Axiom (sometimes, Regularity Axiom) and is denoted by **FA**. It looks as follows.

Axiom 7. For any nonempty set A, there exists a set $X \in A$ such that

$$X \cap A = \emptyset.$$

This axiom is not used in ordinary mathematical arguments, so most of the working mathematicians may absolutely ignore it. One of the main reasons for introducing **FA** is to avoid various kinds of pathological sets, e.g., those sets X for which the relation $X \in X$ may be satisfied. Such sets cannot be met in everyday mathematical practice but, surprisingly, their nonexistence is not provable in **ZFC** theory without **FA** (see Exercise 16 from this appendix). Actually, **FA** was first introduced by Mirimanoff, but its role was recognized after von Neumann's construction of his universe **V** which serves as a natural model of **ZFC** set theory. In fact, **V** consists of all the so-called well-founded sets (for more details, see [103], [148] or some exercises of this appendix).

Thus, speaking of **ZF** theory, logicians and set theorists mean, as a rule, the Axioms 0, 1, 2, 3, 4, 6, and 7.

Let us indicate several immediate consequences of the above-mentioned Axioms 0–6 (some material connected with Axiom 7 is given in exercises).

Our first remark concerns Axiom 4. Quite frequently, it is convenient to utilize it in a slightly different form. Namely, suppose that a set A and a binary relation $S(x, y)$ are given such that, for any $a \in A$, there exists a unique b satisfying $S(a, b)$. Then there exists a set

$$B = \{b : (\exists a)(a \in A \; \& \; S(a, b))\}.$$

To get this form of Axiom 4, let us introduce the relation $S'(x, y)$:

$$(x \in A \; \& \; S(x, y)) \; \vee \; (x \notin A \; \& \; y = x).$$

Applying Axiom 4 to $S'(x, y)$ and A, we readily obtain the required set B.

Further, Axiom 4 allows to formulate the so-called Separation Principle (or Comprehension Principle) which is of paramount importance in general set theory and states that if a property $R(x)$ of a variable x is given, then for any set X, there exists a set

$$Y = \{x : x \in X \; \& \; R(x)\}.$$

To show the validity of this principle, consider two possible cases.

Case 1. There exists no element $x \in X$ such that $R(x)$. Obviously, in this case we have $Y = \emptyset$.

Case 2. There exists some $e \in X$ such that $R(e)$. In this case, consider the following binary relation $S(x, y)$:

$$(R(x) \; \& \; y = x) \; \vee \; (\neg R(x) \; \& \; y = e).$$

It can easily be checked that $S(x, y)$ is functional with respect to y. Applying Axiom 4 to $S(x, y)$ and taking $A = X$, we get the set

$$Y = S(X) = \{x : x \in X \; \& \; R(x)\}$$

as required.

The Separation Principle allows us also to introduce the notion of intersection of sets. Namely, if Z is an arbitrary nonempty set, then by definition

$$\cap Z = \{x : (\forall X)(X \in Z \Rightarrow x \in X)\}.$$

Here the assumption $Z \neq \emptyset$ is essential for applying the Separation Principle to Z (observe that $\cap Z$ is uniquely determined by Z).

Let us give another application of Axiom 4. First of all, notice that Axiom 6 directly implies that an infinite set X_∞ contains both $\emptyset$ and $\{\emptyset\}$ as its elements. Furthermore, the Separation Principle yields that there exists a subset D of X_∞ defined by the equality

$$D = \{\emptyset, \{\emptyset\}\}.$$

Now, let us take any two sets a and b and consider the binary relation $S(x, y)$:

$$(x = \emptyset \ \& \ y = a) \ \vee \ (x \neq \emptyset \ \& \ y = b).$$

It can readily be verified that $S(x, y)$ is functional with respect to y. Then applying Axiom 4 to $S(x, y)$ and $A = D$, we obtain the two-element set

$$Y = S(D) = \{a, b\},$$

whose members are exactly a and b (and Y has no other members). In particular, if $a = b$, then we come to the singleton $\{a\}$. We thus conclude that any set may be regarded as an element of some set. So the two basic terms – set and element – may replace one another in mathematical reasonings.

The above construction enables us to define the ordered pair generated (produced) by any two sets a and b. Indeed, following Kuratowski, let us put

$$(a, b) = \{\{a\}, \{a, b\}\}.$$

It is not difficult to check that the implication

$$((a, b) = (a', b')) \Rightarrow (a = a' \ \& \ b = b')$$

holds true (the converse implication is trivial).

Starting with the notion of an ordered pair, we may introduce the Cartesian product of any two sets A and B. Namely, we put

$$A \times B = \{z : (\exists a)(\exists b)(a \in A \ \& \ b \in B \ \& \ z = (a, b))\}.$$

Of course, it must be verified that this definition makes sense. To establish the existence of $A \times B$ for two given sets A and B, we again appeal to Axiom 4. Our argument will consist of two steps. First, assuming without loss of generality that $A \neq \emptyset$, we take any element $a \in A$ and consider the relation $S_a(x, y)$:

$$y = (a, x).$$

Clearly, $S_a(x, y)$ is functional with respect to y. Therefore, Axiom 4 allows us to assert the existence of

$$Z(a) = \{z : (\exists b)(b \in B \ \& \ z = (a, b))\}.$$

Then the same axiom implies the existence of $\{Z(a) : a \in A\}$. Finally, Axiom 3 directly yields the existence of $\cup\{Z(a) : a \in A\}$, and one can readily verify that

$$A \times B = \cup\{Z(a) : a \in A\}.$$

Another way to get the Cartesian product $A \times B$, for any two initial sets A and B, is sketched in Exercise 2 of this appendix.

Remark 3. Having Cartesian products, one successively may introduce various kinds of binary relations such as: equivalence relations, partial orderings, linear orderings, well-orderings, functions, injections, surjections, bijections, etc. Then various mathematical structures, defined on corresponding base (ground) sets, enter the scene: algebraic structures, topologies, differential structures, and so on (cf. Chapter 4 in [18]).

It follows from the Infinity Axiom that all the sets

$$0 = \emptyset, \quad 1 = \{\emptyset\}, \quad 2 = \{\emptyset, \{\emptyset\}\}, \quad 3 = \{\emptyset, \{\emptyset\}, \{\emptyset, \{\emptyset\}\}\}, \ \ldots$$

belong to X_∞.

Let X be a subset of X_∞. We shall say that X is inductive if the conditions

$$\emptyset \in X,$$

$$(\forall x)(x \in X \Rightarrow (x \cup \{x\}) \in X)$$

are satisfied.

The intersection of all inductive subsets of X_∞ is the smallest inductive set denoted by

$$\mathbf{N} = \{0,\ 1,\ 2,\ 3,\ \ldots,\ n,\ \ldots\}.$$

Since every inductive set X has the properties analogous to (a) and (b) of Axiom 6, we see that $\mathbf{N}$ is infinite in the sense of this axiom. $\mathbf{N}$ is called the set of all natural numbers and quite often is also denoted by ω.

A set Z is called finite if there exists a bijection of Z onto some natural number. Otherwise, Z is called infinite.

Remark 4. As is widely known, Bolzano and Dedekind introduced another notion of a finite (respectively, infinite) set. According to their definition, a set Z is finite if there exists no bijection of Z onto a proper subset of Z. Consequently, Z is infinite if such a bijection does exist. Finite and infinite sets in the sense of Bolzano and Dedekind are usually called D-finite and D-infinite. The equivalence between the standard definition of a finite (infinite) set and the definition of a D-finite (D-infinite) set can be established with the aid of a weak form of **AC**. Moreover, it was shown that this equivalence is not provable within **ZF** theory (see [93], [102]). Notice, by the way, that Bolzano's and Dedekind's definition of a D-infinite set may be regarded as a prototype of elementary embeddings which were explicitly defined much later in model theory (see [29]).

All members of the set $\mathbf{N}$ and $\mathbf{N}$ itself are examples of the so-called transitive sets (see Exercise 4). Actually, all $n \in \mathbf{N}$ are finite ordinal numbers and ω is the least infinite ordinal number (see Exercise 5). In modern set theory, the ordinal numbers (briefly, ordinals) are defined as transitive sets, all elements of which are transitive, too. This definition of ordinals was introduced by von Neumann and is convenient in many respects.

The standard principle of mathematical induction for $\mathbf{N}$ turns out to be a very particular case of the principle of transfinite induction. Let us recall the formulation of this general and extremely useful principle (which was exploited many times in the present book).

Let $(X, \preceq)$ be a well-ordered set, which means that every nonempty subset of X has a least element (with respect to $\preceq$), and let $R(x)$ be a property of an element of X. Suppose that

$$(\forall x \in X)(((\forall y \prec x)R(y)) \Rightarrow R(x)).$$

Then one can assert that

$$(\forall x \in X)R(x).$$

The principle of transfinite induction allows one to construct, by the method of transfinite recursion, various mathematical objects with the desired properties. This general method can be formulated as follows.

Let $F(x)$ be a term function in $\mathbf{ZFC}$ theory (intuitively, $F(x)$ is a certain object depending on a free variable x and, for every set t, the object $F(t)$ is a set uniquely determined by t; in another interpretation, $F(x)$ can be associated with some binary relation $S(x, y)$ which is functional with respect to y). Let $(X, \preceq)$ be again a well-ordered set. Then there exists a unique surjective function f whose domain coincides with X and for which we have

$$f(x) = F(f|X_x) \qquad (x \in X),$$

where $X_x = \{z \in X : z \prec x\}$ denotes, as usual, the initial interval of X determined by x, and $f|X_x$ stands for the restriction of f to X_x.

As is well known, in $\mathbf{ZF}$ set theory the Axiom of Choice is equivalent to the fundamental theorem of Zermelo stating that every set can be endowed with some well-ordering (cf. Exercise 9). Consequently, the principle of transfinite induction and method of transfinite recursion are applicable to all sets of the mathematical universe.

By using $\mathbf{AC}$, the most important statements of contemporary mathematics have been established. Among them there are:

(1) the theorem that, for any two sets X and Y, either there exists an injection of X into Y or there exists an injection of Y into X;

(2) the theorem that, for any infinite set Z, there exists a bijection of Z onto the Cartesian square $Z \times Z$;

(3) Kuratowski-Zorn lemma stating that in every partially ordered set there exists a maximal (with respect to the inclusion relation) subchain;

(4) Tychonoff's theorem stating that the topological product of any family of quasi-compact spaces is also quasi-compact;

(5) the theorem on the existence of a spanning subtree in any nonempty connected graph;

(6) the theorem on the existence of a basis in an arbitrary vector space.

Actually, all statements (1)–(6) turn out to be equivalent to the Axiom of Choice (within **ZF** set theory). Notice, by the way, that the proof in **ZF** theory of the equivalence between (6) and **AC** essentially relies on Axiom 7 (see [16]).

A nice geometric form of the Axiom of Choice was found by Bell and Fremlin in their joint work [12]. Moreover, an interesting purely logical effect of choice was discovered in [43] and [81].

In addition to the facts mentioned above, there are many statements which are not equivalent to **AC** but cannot be proved without the aid of it, i.e., cannot be proved within **ZF** set theory. Let us indicate only several of them:

(i) the theorem that there exists a subset of **R** which is not Lebesgue measurable;

(ii) the theorem that there exists a subset of **R** which does not possess the Baire property;

(iii) the statement on the existence of a Hamel basis in **R** (when **R** is considered as a vector space over the field **Q** of all rational numbers);

(iv) the statement that there exists a nontrivial ultrafilter of subsets of ω;

(v) the Hahn–Banach theorem on extensions of continuous linear functionals;

(vi) the statement that the least uncountable cardinal ω_1 and the cardinality of the continuum $\mathbf{c}$ (= $\mathrm{card}(\mathbf{R})$ = 2^ω) are comparable as cardinal numbers.

Moreover, it should be pointed out that all (i)–(vi) need some uncountable forms of the Axiom of Choice (cf. Chapter 3 of the present book).

Remark 5. Let us especially stress that it is impossible completely avoid **AC**. Certain forms of it are absolutely indispensable. Indeed, omitting **AC** we can come to very surprising results. For example, it may happen that the real line **R** is representable as a union of a countable family of countable sets. More precisely, there are models of **ZF** theory in which this fact is valid (see [57], [93], [102]). Some versions of **AC** with their consequences are discussed in Chapters 1, 2 and 3 of this book (see also [10], [58], [59], [60], [87], [93], [102], [109], [111], [119], [154], [167], [189], [222], [232], [238], [243]).

Now, we would like to recall the two classical and fundamental results of **ZF** set theory. First of them is due to Cantor [28]. In his proof the famous diagonal method was exploited for the first time.

Theorem 1. *For any set X, the inequality* $\mathrm{card}(X) < \mathrm{card}(\mathcal{P}(X))$ *is valid.*

Proof. Clearly, we have a canonical embedding of X into the power set $\mathcal{P}(X)$, namely, we can associate to any element $x \in X$ the corresponding singleton $\{x\}$. So it remains to establish that there exists no injection from $\mathcal{P}(X)$ into X. By virtue of Exercise 19 from this appendix, it suffices to demonstrate that there exists no surjection acting from X onto $\mathcal{P}(X)$. Suppose to the contrary that

$$g : X \to \mathcal{P}(X)$$

is such a surjection and introduce the set

$$X_0 = \{x \in X : x \notin g(x)\}.$$

In view of the surjectivity of g, there is $x_0 \in X$ such that $g(x_0) = X_0$. Then an easy verification shows that the relation $x_0 \in g(x_0)$ leads to a contradiction and the relation $x_0 \notin g(x_0)$ yields a contradiction, too.

This finishes the proof of Cantor's theorem.

Remark 6. For infinite sets, Cantor's classical inequality admits a far-going generalization. More precisely, for any set X, let $\mathcal{F}in(X) = [X]^{<\omega}$ denote the family of all finite subsets of X. Halbeisen [87] was able to demonstrate within **ZF** theory that if X is an infinite set, then the cardinality of $\mathcal{P}(X)$ is strictly greater than the cardinality of $\mathcal{F}in(X)$.

The proof of Theorem 1 does not rely on the Axiom of Infinity. The same can be said about the important theorem of Cantor and Bernstein which is presented below.

Theorem 2. *Let A and B be any two sets and let*

$$f : A \to B, \quad g : B \to A$$

be two injections. Then there exist four sets A', A'', B', B'' satisfying the following four conditions:

(1) $A = A' \cup A''$ and $A' \cap A'' = \emptyset$;
(2) $B = B' \cup B''$ and $B' \cap B'' = \emptyset$;
(3) the restriction $f|A'$ is a bijection between A' and B';
(4) the restriction $g|B''$ is a bijection between B'' and A''.
Consequently, there is a bijection

$$h : A \to B$$

such that $h|A' = f|A'$ and $h^{-1}|B'' = g|B''$.

Proof. Let us introduce the notation:

$$Y = B \setminus f(A),$$

$$j = f \circ g,$$

$$B'' = \cap \{Z \subset B : Y \cup j(Z) \subset Z\}.$$

By virtue of the definition of B'', it can easily be checked that

$$B'' = Y \cup j(B'').$$

Now, let us define

$$A'' = g(B''), \quad A' = A \setminus A'', \quad B' = B \setminus B''.$$

It suffices to show that $f(A') = B'$. Indeed, we may write

$$f(A') = f(A \setminus A'') = f(A) \setminus f(A'') = (B \setminus Y) \setminus f(A'') =$$

$$(B \setminus Y) \setminus f(g(B'')) = B \setminus (Y \cup j(B'')) = B \setminus B'' = B'.$$

The Cantor–Bernstein theorem has thus been proved.

Although there are some nontrivial and deep results within **ZF** set theory (such as Theorems 1 and 2 presented above; see also [87]), this theory is not suitable for contemporary mathematics. For example, as was demonstrated in Chapters 1 and 2, even classical mathematical analysis needs certain weak forms of the Axiom of Choice. Further, in more or less advanced topics of real analysis and measure theory, extra set-theoretical assumptions turn out to be helpful (such as the Continuum Hypothesis or Martin's Axiom). In this context, an important circumstance should be mentioned. Namely, according to Gödel's program for resolving various difficult problems and questions of set theory, which are not decidable within **ZFC**, some strengthened and far-going variants of Axiom 6 may be useful and may lead to solutions of the posed problems and questions. Actually, reasonable hypotheses pertaining to the existence of so-called large cardinal numbers turned out to be crucial for further progress in set theory (see [10], [47], [103], [110], [154]).

There are many types of large cardinal numbers. The historically first of them are weakly inaccessible and strongly inaccessible cardinals.

Recall that a cardinal w_α is said to be weakly inaccessible if w_α is regular and its index α is a nonzero limit ordinal.

A cardinal w_α is said to be strongly inaccessible if w_α is weakly inaccessible and $2^{w_\beta} < w_\alpha$ for any $\beta < \alpha$.

It can readily be checked that, under the Generalized Continuum Hypothesis (see Appendix 2), these two notions are the same. The existence of a weakly inaccessible cardinal (hence the existence of a strongly inaccessible cardinal) is not provable within **ZFC** set theory (cf. Remark 8 below).

There are two concepts of large cardinals which found numerous applications in diverse fields of modern mathematics. They are a cardinal measurable in the Ulam sense (also called a real-valued measurable cardinal) and a two-valued measurable cardinal.

Let us briefly review their definitions (cf. [10], [47], [76], [103], [110], [154], [263]).

A cardinal **a** is real-valued measurable (respectively, two-valued measurable) if there exists a probability (respectively, a two-valued probability) diffused measure μ whose domain $\text{dom}(\mu)$ coincides with the power set $\mathcal{P}(\mathbf{a})$.

Clearly, every two-valued measurable cardinal is real-valued measurable.

According to the classical result of Ulam [263], the least real-valued measurable cardinal is weakly inaccessible and the least two-valued measurable cardinal is strongly inaccessible (consequently, the existence of these cardinals cannot be proved within **ZFC** set theory). In connection with this fact, Ulam conjectured that there exists some model of set theory in which the first weakly inaccessible cardinal is real-valued measurable. However, his conjecture turned out to be false. Indeed, Solovay proved that there are many weakly inaccessible cardinals preceding the first real-valued measurable cardinal number. Analogously, as follows from some results of Hanf and Tarski, there are many strongly inaccessible cardinals preceding the first two-valued measurable cardinal number.

Moreover, it was shown by Solovay (see, e.g., [103]) that the following two theories are equiconsistent:

(*) **ZFC** & (there exists a two-valued measurable cardinal);

(**) **ZFC** & (there exists a measure on **R** extending the Lebesgue measure λ and defined on the family of all subsets of **R**).

The existence of a two-valued measurable cardinal yields some unexpected effects in concrete questions of infinite combinatorics, descriptive set theory, general topology, and real analysis. For example, as demonstrated by Solovay, if such a cardinal does exist, then every Σ_2^1-subset of **R** turns out to be absolutely measurable with respect to the class of completions of all σ-finite Borel measures on **R** (see [103]).

Briefly speaking, additional axioms concerning the existence of large cardinals of certain types may detect and highlight new interesting aspects of concrete mathematical theories. For more detailed discussions of large cardinals and their profound properties, we refer the reader to [10], [47], [103], [110] and [154]. Also, in connection with this topic it makes sense to point out the old article by Tarski [259], in which it is shown that the existence of sufficiently many strongly inaccessible cardinal numbers (defined in an appropriate manner) implies, within the framework of **ZF** set theory, the validity of **AC**.

EXERCISES

1. Give a detailed proof of the implication

$$(a, b) = (a', b') \Rightarrow (a = a' \ \& \ b = b').$$

2. Let X and Y be two sets. Check that, for any two elements $x \in X$ and $y \in Y$, we have

$$(x, y) = \{\{x\}, \{x, y\}\} \in \mathcal{P}(\mathcal{P}(X \cup Y)).$$

Deduce from this fact another proof of the existence of the product set $X \times Y$. For this purpose, utilize the Separation Principle.

3. Let I and E be two sets and let $G \subset I \times E$. Such a G is usually called a graph in $I \times E$. This G is said to be a functional graph if, for any index i from I, there exists at most one $e \in E$ such that $(i, e) \in G$. Actually, G determines a partial function $g : I \to E$ acting from I into E and defined by the formula

$$g(i) = e \qquad ((i, e) \in G).$$

If the domain of g coincides with I, then g is an ordinary function acting from the whole I into E. If, in addition, $E = \mathcal{P}(X)$ for some set X, then we have the family

$$\{X_i : i \in I\} = \{g(i) : i \in I\}$$

of sets, all of which are contained in X.

Work in **ZF** theory and show that, for any family $\{Y_j : j \in J\}$ of sets, there exists its product

$$\prod \{Y_j : j \in J\} = \{\{y_j : j \in J\} : (\forall j \in J)(y_j \in Y_j)\}.$$

The elements of $\prod\{Y_j : j \in J\}$ are called selectors of the family $\{Y_j : j \in J\}$.

So Axiom 5 can be formulated as follows: if $Y_j \neq \emptyset$ for all indices $j \in J$, then $\prod\{Y_j : j \in J\} \neq \emptyset$.

4. A set X is said to be transitive if the conjunction of the relations $x \in X$ and $y \in x$ implies $y \in X$ (equivalently, $x \in X$ implies $x \subset X$).

Demonstrate that:

(a) the intersection of any nonempty family of transitive sets is transitive;

(b) the union of any family of transitive sets is transitive;

(c) if X is a transitive set, then $X \cup \{X\}$ is transitive, too;

(d) for an arbitrary set X, there exists its transitive closure, i.e., there exists a unique transitive set $\mathrm{tr}(X)$ containing X and such that every transitive set Y containing X also contains $\mathrm{tr}(X)$.

Give a direct description of $\mathrm{tr}(X)$ in the following manner. Put by recursion

$$X_0 = X, \quad X_{n+1} = \cup X_n \qquad (n < \omega),$$

and show that $\mathrm{tr}(X) = \cup\{X_n : n < \omega\}$.

5*. Let $(X, \preceq)$ be an arbitrary well-ordered set. For each element $x \in X$, define by transfinite recursion the set $o(x)$ as follows:

$$o(x) = \{o(z) : z \prec x\}.$$

Verify that this definition does not depend on the original set $(X, \preceq)$. In other words, if $(Y, \preceq)$ is another well-ordered set and elements $x \in X$ and $y \in Y$ are such that the two associated initial intervals

$$X_x = \{z \in X : z \prec x\}, \quad Y_y = \{z \in Y : z \prec y\}$$

are isomorphic, then one has the equality $o(x) = o(y)$.

The sets of the form $o(x)$ are called von Neumann's ordinal numbers (briefly, von Neumann's ordinals) and are usually denoted by Greek letters α, β, γ, ξ, η, ζ, ... Below only these ordinals are considered.

Demonstrate that:

(a) any ordinal α is a transitive set;

(b) if α is an ordinal, then $\alpha \cup \{\alpha\}$ is also an ordinal (called the successor of α and usually denoted by the symbol $\alpha + 1$);

(c) if $\{\alpha_i : i \in I\}$ is a nonempty family of ordinals, then

$$\cup\{\alpha_i : i \in I\}, \quad \cap\{\alpha_i : i \in I\}$$

are ordinals, too;

(d) any ordinal α is a well-ordered set with respect to $\subset$; in other words, the relation

$$x \in \alpha \ \& \ y \in \alpha \ \& \ x \subset y$$

is a well-ordering of α; analogously, the relation

$$x \in \alpha \ \& \ y \in \alpha \ \& \ (x \in y \ \vee \ x = y)$$

is a well-ordering of α which coincides with the previous one;

(e) if, for a transitive set α, the above relation is a well-ordering of α, then α is an ordinal;

(f) for an arbitrary ordinal α, one has $\alpha \notin \alpha$.

In the process of establishing (f) utilize the principle of transfinite induction.

6*. Check that the class **On** of all ordinal numbers is not a set (the Burali-Forti theorem which is usually called the Burali-Forti paradox).

For this purpose, use the Replacement Axiom and the fact that if $\{\alpha_i : i \in I\}$ is an arbitrary family of ordinals, then

$$\alpha = \cup\{\alpha_i : i \in I\}$$

is also an ordinal (see the previous exercise). Then consider the ordinal number $\alpha + 1$ which is strictly greater than all α_i $(i \in I)$.

7. Show that, for a set α, the following two assertions are equivalent:

(a) α is an ordinal;

(b) α itself is transitive and all elements of α are also transitive.
For this purpose, keep in mind Axiom 7.

8. Work in **ZF** theory and verify that any well-ordered set $(X, \preceq)$ is isomorphic to a unique ordinal (which is usually called the order type of $(X, \preceq)$ and is denoted by $\mathrm{ot}(X)$).

9*. Prove within **ZF** theory that, for every set X, there exists a well-ordered set (equivalently, there exists an ordinal) which cannot be injectively mapped into X (the least of such ordinals is called the Hartogs ordinal number of X and is denoted by $h(X)$).
For this purpose, consider the family of all those ordinals which can be injectively mapped into X and check that this family is an ordinal equal to $h(X)$.
By starting with the above-mentioned property of $h(X)$ and applying the principle of transfinite induction, demonstrate within **ZFC** theory that X can be made well-ordered.
To do this, first introduce some choice function

$$f : \mathcal{P}(X) \setminus \{\emptyset\} \to X,$$

i.e., $f(Y) \in Y$ for any nonempty set $Y \subset X$. Then fix an element $z \notin X$ and define by transfinite recursion a family $\{x_\xi : \xi < h(X)\}$ so that

$$x_\xi = f(X \setminus \{x_\zeta : \zeta < \xi\})$$

if $X \setminus \{x_\zeta : \zeta < \xi\} \neq \emptyset$, and $x_\xi = z$ if $X \setminus \{x_\zeta : \zeta < \xi\} = \emptyset$. Finally, deduce the existence of a least $\xi_0 < h(X)$ such that $x_{\xi_0} = z$ and conclude that the set $X = \{x_\zeta : \zeta < \xi_0\}$ can be made well-ordered.

10. The following axiom was introduced by von Neumann:
A class X is a set if and only if there exists no term function which surjectively maps X onto the universal class determined by the equality relation $x = x$.
Show that the above axiom implies the existence of a global choice function, i.e., there exists a term function $F(X)$ such that, for any nonempty set X, the relation $F(X) \in X$ holds true.
In order to check this implication, utilize the Burali-Forti theorem (see Exercise 6).

11*. For every ordinal α, define by transfinite recursion the set V_α as follows:

$$V_\alpha = \cup \{\mathcal{P}(V_\beta) : \beta < \alpha\}.$$

Finally, denote by $\mathbf{V}$ the class of all those sets x for which there exists at least one ordinal α such that $x \in V_\alpha$. The class $\mathbf{V}$ is usually called von Neumann's Universe.

Use the principle of transfinite induction for checking that:

(a) all sets V_α are transitive;

(b) for any two ordinals α and β such that $\alpha < \beta$, one has $V_\alpha \in V_\beta$ and hence $V_\alpha \subset V_\beta$;

(c) for every ordinal α, one has $\alpha \notin V_\alpha$ and $\alpha \subset V_\alpha$, so $\alpha \in V_{\alpha+1}$.

Infer from (c) that $\mathbf{V}$ is not a set, hence $\mathbf{V}$ is a proper class.

For any set $x \in \mathbf{V}$, denote by $\mathrm{rk}(x)$ the least ordinal α for which $x \in V_{\alpha+1}$. This $\mathrm{rk}(x)$ is called the rank of x. Verify that if $x \in \mathbf{V}$, then

$$\mathrm{rk}(x) = \sup\{\mathrm{rk}(y) + 1 : y \in x\}.$$

In particular, if $y \in x \in \mathbf{V}$, then $\mathrm{rk}(y) < \mathrm{rk}(x)$.

12. Consider the following binary relation $S(x, y)$ between two sets x and y: there exists a bijection of x onto y.

Show that, for any set x, it is possible to define the set

$$\mathrm{card}(x) = \{y \in \mathbf{V} : S(x, y) \ \& \ \mathrm{rk}(y) \text{ is minimal}\}.$$

Thus, one can introduce the concept of cardinal number of x (by the above-mentioned scheme) so that $\mathrm{card}(x) \in \mathbf{V}$.

Proceeding in an analogous manner, introduce the notions of relation types and, more generally, of isomorphism types for various mathematical structures. In particular, apply this approach and define the combinatorial type of a polyhedron in the Euclidean space $\mathbf{R}^n$, where $n \in \mathbf{N}$.

13. Check that $(V_\omega, \in)$ is a model of the theory which is obtained from $\mathbf{ZFC}$ after withdrawing the Infinity Axiom.

In the set $\mathbf{N}$ of all natural numbers consider a binary relation $\in'$ defined as follows: $m \in' n$ if and only if the m-th digit (counting from right to left) in the standard binary expansion of n is equal to 1.

Show that the two structures $(V_\omega, \in)$ and $(\mathbf{N}, \in')$ are isomorphic to each other.

Remark 7. The elements of V_ω are usually called hereditarily finite sets. The reader can easily verify that this term is indeed justified.

14. Demonstrate that the Foundation Axiom (i.e., Axiom 7) is equivalent to the statement that all sets belong to $\mathbf{V}$.

15. Prove that $\mathbf{V}$ is a model of $\mathbf{ZFC}$, i.e., all axioms of $\mathbf{ZFC}$ (including the Foundation Axiom) are valid in $\mathbf{V}$.

This fact indicates, in particular, that Axiom 7 does not contradict the preceding Axioms 0–6.

Remark 8. The class $\mathbf{V}$ with the restricted membership relation $\in$ between elements of $\mathbf{V}$ serves as a canonical model of $\mathbf{ZFC}$ set theory. Its fragments

$(V_\alpha, \in)$, where α is an ordinal, may be regarded as certain approximations of $\mathbf{V}$. Some of these approximations also can be treated as models of $\mathbf{ZFC}$. For instance, if α is a strongly inaccessible cardinal, then $(V_\alpha, \in)$ satisfies all the axioms of $\mathbf{ZFC}$ and von Neumann's axiom (see Exercise 10) is also valid in $(V_\alpha, \in)$. In particular, if α coincides with the least strongly inaccessible cardinal, then the model $(V_\alpha, \in)$ of $\mathbf{ZFC}$ does not contain strongly inaccessible cardinals, which shows that the existence of such cardinals cannot be proved within $\mathbf{ZFC}$ theory. Nevertheless, it is natural to assume that the existence of strongly inaccessible cardinal numbers does not contradict this theory. A very important model of $\mathbf{ZFC}$ is the Constructible Universe $\mathbf{L} \subset \mathbf{V}$ of Gödel. Actually, $\mathbf{L}$ is the least transitive model of $\mathbf{ZFC}$ containing the class $\mathbf{On}$ of all ordinals (this is a descriptive definition of $\mathbf{L}$). A much more helpful constructive definition of $\mathbf{L}$ is usually introduced by means of transfinite recursion (see [10], [42], [103], [148]). Recall that in $\mathbf{L}$ the Generalized Continuum Hypothesis and the global version of the Axiom of Choice are valid.

16. Introduce the class-bijection F of $\mathbf{V}$ onto itself by putting

$$F(0) = 1, \quad F(1) = 0,$$

$$F(x) = x \quad (x \neq 0, 1).$$

Then define a new membership relation $\in'$ by the formula

$$x \in' y \Leftrightarrow x \in F(y).$$

For this $\in'$, verify that:

(a) $\mathbf{V}$ with $\in'$ is a model of the theory obtained from $\mathbf{ZFC}$ after removing the Foundation Axiom;

(b) in this model there exists a set x satisfying the equality $x = \{x\}$.

In particular, one has $x \in' x$, so the Foundation Axiom is not deducible from the other axioms of $\mathbf{ZFC}$.

17. Consider the following set-theoretical equation: $X \times X = X$. One of its solutions is $X = \emptyset$ (called a trivial solution).

Show that:

(a) under the Foundation Axiom, $X = \emptyset$ is a unique solution of the above-mentioned equation;

(b) if the Foundation Axiom is not assumed, then the same equation may have nontrivial solutions.

Conclude that the equation $X \times X = X$ is not decidable within $\mathbf{ZFC}$ theory, which means that it is impossible to specify all solutions of this equation.

18. Verify that:

(a) $(V_{\omega+\omega}, \in)$ is a model of the theory which is obtained from $\mathbf{ZFC}$ after withdrawing the Replacement Axiom;

(b) in this model the Replacement Axiom fails to be true.

Conclude that the Replacement Axiom is not deducible from the other axioms of **ZFC** set theory.

19. Work in **ZF** theory and suppose that some injection $f : A \to B$ is given where $A \neq \emptyset$.

Show that there exists a surjection $g : B \to A$ such that the composition $g \circ f$ coincides with the identity transformation of A.

Argue as follows. Since $A \neq \emptyset$, one may pick an element $a_0 \in A$. Define a function $g : B \to A$ by the formula:

If $b \in f(A)$, then $g(b) = a$ where a is a unique element of A for which $f(a) = b$, and if $b \in B \setminus f(A)$, then $g(b) = a_0$.

Finally, check that g is as required.

20*. Let $(T, \preceq)$ be a partially ordered set. We say that $(T, \preceq)$ is a tree (in the set-theoretical sense) if the following two conditions are fulfilled:

(i) in $(T, \preceq)$ there exists a least element t_0;

(ii) for every element $t \in T$, the set $T(t) = \{u \in T : u \prec t\}$ is well-ordered by the induced order.

The least element t_0 of $(T, \preceq)$ is called the root of this tree. For any ordinal number α, the set

$$T_\alpha = \{t \in T : \mathrm{ot}(T(t)) = \alpha\}$$

is called the α-th level of $(T, \preceq)$ (here the symbol $\mathrm{ot}(T(t))$ denotes the ordinal type of $T(t)$).

Check that all elements in any level of $(T, \preceq)$ are pairwise incomparable with respect to $\preceq$.

The height of a tree $(T, \preceq)$ is the smallest ordinal β such that $T_\beta = \emptyset$.

Observe that, for all ordinals α strictly greater than the height of $(T, \preceq)$, the levels T_α are empty.

A branch in a tree $(T, \preceq)$ is a linearly ordered subset of $(T, \preceq)$ which has a common element with every nonempty level of $(T, \preceq)$.

Check that any branch is a maximal (with respect to inclusion) linearly ordered subset of $(T, \preceq)$.

Prove the following version of König's lemma:

If a tree $(T, \preceq)$ has height ω and all levels of $(T, \preceq)$ are finite, then there exists at least one branch in this tree.

Moreover, demonstrate that the following two assertions are equivalent in **ZF** set theory:

(a) the above-mentioned version of König's lemma;

(b) for any sequence $\{X_n : n < \omega\}$ of nonempty finite sets, there exists a selector of $\{X_n : n < \omega\}$.

Remark 9. König's lemma has nontrivial applications in many questions and topics of discrete mathematics. Unfortunately, this lemma cannot be directly generalized to trees with uncountable heights. Indeed there exists a so-called Aronszajn tree $(T, \preceq)$ which possesses the following three properties:

(a) the height of $(T, \preceq)$ is equal to ω_1;

(b) all levels of $(T, \preceq)$ are at most countable;

(c) there is no branch in $(T, \preceq)$.

In this connection, see [10], [103], [148] or Exercise 8 from Chapter 13.

21*. Let I be a set of indices and let $\{X_i : i \in I\}$ be a family of finite sets such that, for every finite subset J of I, the inequality

$$\operatorname{card}(J) \le \operatorname{card}(\cup\{X_j : j \in J\})$$

is satisfied.

Prove Hall's combinatorial theorem stating that there exists an injection

$$g : I \to \cup\{X_i : i \in I\}$$

having the property that $g(i) \in X_i$ for each index $i \in I$.

Argue as follows. First, consider the case when I is a finite set and in this situation use induction on $\operatorname{card}(I)$. Further, if I is infinite, then equip all sets X_i ($i \in I$) with the corresponding discrete topologies and take the topological product space

$$X = \prod\{X_i : i \in I\},$$

which is compact by virtue of Tychonoff's theorem. Then, for each finite subset J of I, denote

$$F_J = \{f \in X : f|J \text{ is an injection}\}.$$

Check that

$$\{F_J : J \subset I \ \& \ \operatorname{card}(J) < \omega\}$$

is a centered family of nonempty closed sets in X, so every function g belonging to the intersection of this family is as required.

22. Let X and Y be two sets, $n \ge 1$ be a natural number, and let Γ be an $(n-n)$-correspondence between X and Y.

Demonstrate that there exists a bijection $h : X \to Y$ such that:

(a) for any $x \in X$, one has $h(x) \in \Gamma(x)$;

(b) for any $y \in Y$, one has $h^{-1}(y) \in \Gamma^{-1}(y)$.

To show the existence of a bijection h satisfying (a) and (b), utilize Hall's combinatorial theorem (see the previous exercise) and the Cantor–Bernstein theorem.

23*. Let E be an arbitrary set and let $\{E_i : i \in I\}$ be a family of subsets of E. This family is called independent if any intersection of the form

$$E'_{i_1} \cap E'_{i_2} \cap \ldots \cap E'_{i_k}$$

differs from $\emptyset$, where k is an arbitrary natural number, $(i_1, i_2, ..., i_k)$ is an injective sequence of indices from I, and for each $r \in \{1, 2, ..., k\}$, the set E'_{i_r} either coincides with E_{i_r} or coincides with $E \setminus E_{i_r}$.

Prove the Hausdorff–Tarski theorem stating that if E is infinite, then there exists a family $\{E_i : i \in I\}$ of subsets of E, which is independent and satisfies the equality $\operatorname{card}(I) = 2^{\operatorname{card}(E)}$.

For this purpose, argue step by step as follows.

(a) Firstly, show that there exists a family $\{X_i : i \in I\}$ of subsets of E such that $\operatorname{card}(I) = 2^{\operatorname{card}(E)}$ and $X_i \setminus X_j \neq \emptyset$ for any two distinct indices i and j from I.

(b) Secondly, show that there exists a family $\{Y_i : i \in I\}$ of subsets of E such that $\operatorname{card}(I) = 2^{\operatorname{card}(E)}$ and, for any finite injective sequence $(i_0, i_1, i_2, ..., i_k)$ of indices from I, the relation

$$Y_{i_0} \setminus (Y_{i_1} \cup Y_{i_2} \cup ... \cup Y_{i_k}) \neq \emptyset$$

is valid (here k ranges over the whole set ω).

(c) Finally, show the existence of an independent family $\{E_i : i \in I\}$ of subsets of E such that $\operatorname{card}(I) = 2^{\operatorname{card}(E)}$.

24. Let E be an infinite set and let $\mathcal{F}(E)$ denote the family of all ultrafilters in the power set $\mathcal{P}(E)$.

Deduce from the result of Exercise 23 that the relation

$$\operatorname{card}(\mathcal{F}(E)) = \operatorname{card}(\mathcal{P}(\mathcal{P}(E))) = 2^{2^{\operatorname{card}(E)}}$$

holds true.

25*. Let α be any ordinal and let E be a set of cardinality $\omega_{\alpha+1}$.

Prove that there exists a family

$$\{E_{\xi,\varsigma} : \xi < \omega_\alpha, \ \varsigma < \omega_{\alpha+1}\}$$

of subsets of E, satisfying the following two conditions:

(a) for each $\xi < \omega_\alpha$, the partial family $\{E_{\xi,\varsigma} : \varsigma < \omega_{\alpha+1}\}$ is disjoint;

(b) for each $\varsigma < \omega_{\alpha+1}$, the set $E \setminus \cup\{E_{\xi,\varsigma} : \xi < \omega_\alpha\}$ has cardinality less than or equal to ω_α.

Argue as follows. First, identify E with $\omega_{\alpha+1}$. Then, for every $\varsigma < \omega_{\alpha+1}$, take some injective mapping

$$f_\varsigma : [0, \varsigma] \to \omega_\alpha.$$

Define the sets $E_{\xi,\varsigma}$ ($\xi < \omega_\alpha$, $\varsigma < \omega_{\alpha+1}$) by the formula

$$E_{\xi,\varsigma} = \{\eta < \omega_{\alpha+1} : f_\eta(\varsigma) = \xi\}$$

and check that $\{E_{\xi,\zeta} : \xi < w_\alpha,\ \zeta < w_{\alpha+1}\}$ is as required.

Remark 10. The family of subsets of E indicated above is usually called Ulam's transfinite matrix (corresponding to E). Taking into account the result of Exercise 25, it is not difficult to prove Ulam's classical theorem stating that there is no real-valued measurable cardinal strictly less than the first weakly inaccessible cardinal (see Exercise 2 from Chapter 17). An analogous theorem holds for two-valued measurable cardinals in terms of the first strongly inaccessible cardinal. At present, a number of results are known which substantially generalize the above-mentioned theorems (cf. [10], [47], [103], [110], [154]).

26. Consider the interval $[0, w_1[$ of all at most countable ordinal numbers and equip this interval with its order topology. A set $X \subset [0, w_1[$ is called stationary if X has common elements with every closed unbounded subset of $[0, w_1[$ (otherwise, X is called nonstationary).

Check that the family of all nonstationary subsets of $[0, w_1[$ forms a σ-ideal in the power set of $[0, w_1[$.

Demonstrate that if a set $X \subset [0, w_1[$ is stationary, then there exists a partition $\{X_i : i \in I\}$ of X such that $\mathrm{card}(I) = w_1$ and all sets X_i $(i \in I)$ are also stationary.

For this purpose, use Ulam's matrix corresponding to w_1.

27*. For any natural number m, let the symbol $\mathcal{F}_m(E)$ denote the family of all m-element subsets of E (quite often, the symbol $[E]^m$ is also used for denoting the same family).

Let m and n be two nonzero natural numbers, E be an infinite set, and let

$$\mathcal{F}_m(E) = \mathcal{A}_1 \cup \mathcal{A}_2 \cup \ldots \cup \mathcal{A}_n.$$

Prove the infinite version of Ramsey's theorem stating that there exist a natural number $k \in [1, n]$ and an infinite subset E' of E such that $\mathcal{F}_m(E') \subset \mathcal{A}_k$.

For this purpose, argue by induction on m. The case $m = 1$ is trivial, because $\mathrm{card}(E) \geq w$. Suppose that the statement is true for m and let

$$\mathcal{F}_{m+1}(E) = \mathcal{A}_1 \cup \mathcal{A}_2 \cup \ldots \cup \mathcal{A}_n.$$

Define by recursion a sequence of pairs $\{(E_i, e_i) : i < w\}$, where E_i is an infinite subset of E and $e_i \in E_i$. First of all, put $E_0 = E$ and pick an element $e_0 \in E$. Now, assume that the pair (E_i, e_i) has already been defined for a natural index i, and consider the family $\mathcal{F}_m(E_i \setminus \{e_i\})$. This family admits the following representation:

$$\mathcal{F}_m(E_i \setminus \{e_i\}) = \mathcal{A}'_1 \cup \mathcal{A}'_2 \cup \ldots \cup \mathcal{A}'_n,$$

where each $\mathcal{A}'_r$ $(1 \leq r \leq n)$ is introduced by the formula

$$Z \in \mathcal{A}'_r \Leftrightarrow Z \cup \{e_i\} \in \mathcal{A}_r.$$

By the inductive assumption, there exists an infinite set $T \subset E_i \setminus \{e_i\}$ such that all m-element subsets of T belong to some $\mathcal{A}'_{k(i)}$, where $1 \leq k(i) \leq n$. Put $E_{i+1} = T$ and pick an element $e_{i+1} \in T$. Proceeding in this manner, obtain the desired sequence of pairs $\{(E_i, e_i) : i < \omega\}$ and consider the corresponding sequence $\{k(i) : i < \omega\}$ of natural numbers, all of which belong to the interval $[1, n]$. Finally, observe that there are a natural index k and an infinite set $I \subset \omega$ satisfying the relation

$$(\forall i \in I)(k_i = k),$$

and deduce from this fact that the set

$$E' = \{e_i : i \in I\}$$

is as required, i.e., all $(m+1)$-element subsets of E' belong to $\mathcal{A}_k$.

28*. Infer from the infinite version of Ramsey's theorem (see Exercise 27) the following finite version:

For any three nonzero natural numbers m, n and l, there exists a natural number $r = r(m, n, l)$ possessing the property that if a finite set E has cardinality greater than or equal to r and

$$\mathcal{F}_m(E) = \mathcal{A}_1 \cup \mathcal{A}_2 \cup \ldots \cup \mathcal{A}_n,$$

then there are a natural number $k \in [1, n]$ and a subset E' of E such that $\mathrm{card}(E') \geq l$ and $\mathcal{F}_m(E') \subset \mathcal{A}_k$.

For deducing the above-mentioned finite version of Ramsey's theorem, utilize either König's lemma (see Exercise 20) or the existence of a nontrivial ultrafilter in the power set of ω.

Remark 11. It is needless to say here that both finite and infinite versions of Ramsey's theorem have numerous applications in classical number theory, finite and infinite combinatorics, graph theory, etc. The finite version of Ramsey's theorem is provable within **ZF** set theory and, moreover, is even provable within the framework of formal arithmetic. However, some statements of finite combinatorics which strengthen this version turn out to be unprovable in formal arithmetic (for details, see e.g. [10]). Also, it should be stressed that the infinite version of Ramsey's theorem needs some weak variant of the Axiom of Choice (cf. [87]).

29*. Give a proof of the following special case of the Erdős–Rado combinatorial theorem:

Let $\mathbf{a}$ be an infinite cardinal number, let E be a set with $\mathrm{card}(E) > 2^{\mathbf{a}}$, and let

$$\mathcal{F}_2(E) = \cup\{\mathcal{A}_j : j \in J\},$$

where $\mathrm{card}(J) \leq \mathbf{a}$. Then there exist a subset E' of E with $\mathrm{card}(E') > \mathbf{a}$ and an index $r \in J$ such that $\mathcal{F}_2(E') \subset \mathcal{A}_r$.

For this purpose, argue as follows. At the beginning fix some choice function

$$f : \mathcal{P}(E) \setminus \{\emptyset\} \to E.$$

Further, identify **a** with the corresponding initial ordinal number ω_α and construct by transfinite recursion a family $\{\mathcal{Q}_\xi : \xi < \omega_{\alpha+1}\}$ of partitions of E which satisfy the inequalities

$$\mathrm{card}(\mathcal{Q}_\xi) \leq 2^{\mathbf{a}} \qquad (\xi < \omega_{\alpha+1}).$$

First of all, put $\mathcal{Q}_0 = \{E\}$. Suppose that, for a nonzero $\xi < \omega_{\alpha+1}$, the family $\{\mathcal{Q}_\zeta : \zeta < \xi\}$ of partitions of E has already been constructed and consider the partition $\mathcal{Q}'_\xi$ of E which is the infimum of this family (with respect to the natural partial ordering). In other words, the elements of $\mathcal{Q}'_\xi$ are all nonempty sets of the form

$$\cap\{Z_\zeta : \zeta < \xi\} \qquad (Z_\zeta \in \mathcal{Q}_\zeta).$$

Check the validity of the inequality $\mathrm{card}(\mathcal{Q}'_\xi) \leq 2^{\mathbf{a}}$. Then, for any set $Z \in \mathcal{Q}'_\xi$ with $\mathrm{card}(Z) > 2^{\mathbf{a}}$, define an equivalence relation $R_Z(x, y)$ in Z by the formula:

$$(x = y = f(Z)) \vee (f(Z) \notin \{x, y\} \ \& \ (\exists j \in J)(\{\{x, f(Z)\}, \{y, f(Z)\}\} \subset A_j)).$$

Obviously, $R_Z(x, y)$ produces some partition of Z. Replace in $\mathcal{Q}'_\xi$ every member Z with $\mathrm{card}(Z) > 2^{\mathbf{a}}$ by the family of all equivalence classes associated with $R_Z(x, y)$ and denote the obtained partition of E by $\mathcal{Q}_\xi$. Verify that

$$\mathrm{card}(\mathcal{Q}_\xi) \leq 2^{\mathbf{a}},$$

so the described transfinite process can be continued up to $\omega_{\alpha+1}$. As a result, get the desired family $\{\mathcal{Q}_\xi : \xi < \omega_{\alpha+1}\}$ of partitions of E.

Now, establish the validity of these two relations:

(i) if $\zeta < \xi < \omega_{\alpha+1}$, then the partition $\mathcal{Q}_\xi$ is inscribed into the partition $\mathcal{Q}_\zeta$, i.e.,

$$(\forall X \in \mathcal{Q}_\xi)(\exists Y \in \mathcal{Q}_\zeta)(X \subset Y);$$

(ii) for each ordinal $\xi < \omega_{\alpha+1}$, there exists a function g_ξ which maps every set $Z \in \mathcal{Q}_\xi$ with $\mathrm{card}(Z) > 2^{\mathbf{a}}$ to some element $g_\xi(Z) \in E$ such that

$$g_\xi(Z) \notin Z, \qquad (\exists j \in J)(\forall z \in Z)(\{z, g_\xi(Z)\} \in A_j);$$

in addition, it can be assumed that if $\zeta < \xi < \omega_{\alpha+1}$, then both Z and $g_\xi(Z)$ are contained in a member of $\mathcal{Q}_\zeta$.

Further, introduce the set

$$F = \cup\{Z : (\exists \xi < \omega_{\alpha+1})(Z \in \mathcal{Q}_\xi \ \& \ \mathrm{card}(Z) \leq 2^{\mathbf{a}})\}$$

and observe that $\operatorname{card}(F) \leq 2^{\mathbf{a}}$. So there exists an element $z \in E \setminus F$. For each ordinal $\xi < \omega_{\alpha+1}$, this z determines a unique set $Z_\xi \in \mathcal{Q}_\xi$ such that

$$z \in Z_\xi, \quad \operatorname{card}(Z_\xi) > 2^{\mathbf{a}}.$$

Consequently, one may define

$$z_\xi = g_\xi(Z_\xi) \qquad (\xi < \omega_{\alpha+1}).$$

Also, for any $\xi < \omega_{\alpha+1}$, there is an index $j_\xi \in J$ such that $\{z, z_\xi\} \in A_{j_\xi}$. Since $\operatorname{card}(J) \leq \mathbf{a}$ and $\mathbf{a} = \omega_\alpha$ is not cofinal with $\omega_{\alpha+1}$, there exists an index $r \in J$ for which

$$\operatorname{card}(\{\xi < \omega_{\alpha+1} : j_\xi = r\}) = \omega_{\alpha+1}.$$

Finally, denoting

$$\Xi = \{\xi < \omega_{\alpha+1} : j_\xi = r\},$$
$$E' = \{z_\xi : \xi \in \Xi\},$$

show that E' and r are as required, i.e., $\mathcal{F}_2(E') \subset \mathcal{A}_r$.

Remark 12. The full version of the Erdös–Rado combinatorial theorem can be found, e.g., in [10] and [54]. This theorem has a lot of nontrivial applications in set theory, general topology, model theory, measure theory, and other fields of contemporary mathematics (see, for instance, [29], [50], [106], [223]).

Appendix 2: The Axiom of Choice and Generalized Continuum Hypothesis

After the Axiom of Choice (**AC**), the most intriguing set-theoretical statement is the Continuum Hypothesis (**CH**) which goes back to Cantor and was especially distinguished by Hilbert [97] in his famous lecture given in Paris, 1900.

It is known that there are two standard but essentially different formulations of **CH**. We present them as assertions (*) and (**) stated below.

(*) the equality $2^\omega = \omega_1$ is valid;

(**) there exists no cardinal number **a** such that $\omega < \mathbf{a} < 2^\omega$.

Notice that the implication (*) $\Rightarrow$ (**) trivially holds within **ZF** set theory, but the converse implication (**) $\Rightarrow$ (*) is fulfilled only under assuming some versions of the Axiom of Choice (see [93], [102]).

There are many works devoted to various aspects of the Continuum Hypothesis and of its natural extension which is called the Generalized Continuum Hypothesis (abbreviation: **GCH**). Also, it is well known that both **CH** and **GCH** are independent of **ZFC** set theory (see [103], [148]).

Further, according to Sierpiński's classical result (see, for instance, [243]), the implication

$$\mathbf{GCH} \Rightarrow \mathbf{AC}$$

holds true even within **ZF** set theory. Here **GCH** is stated in the form completely analogous to (**), namely:

If **b** is an arbitrary infinite cardinal number, then there is no cardinal **a** such that $\mathbf{b} < \mathbf{a} < 2^\mathbf{b}$.

We are going to prove the implication **GCH** $\Rightarrow$ **AC** within **ZF** set theory. Actually, a much stronger result will be established below (see [112]).

For this purpose, we need several auxiliary propositions. Of course, all of them are effective, i.e., are provable within **ZF** theory.

The first of these propositions is due to Cantor (Theorem 1 from Appendix 1), but the proof given here was suggested by Zermelo.

Lemma 1. *Let X be a set and let $\Phi : \mathcal{P}(X) \to X$ be a function. Then this Φ is not injective.*

Proof. Let $\alpha = h(X)$ denote the Hartogs ordinal (cardinal) number associated with X; in other words, α is the smallest ordinal which cannot be injectively mapped into X (see Exercise 9 for Appendix 1). By using the method of transfinite recursion, we can define an α-sequence $\{x_\beta : \beta < \alpha\}$ of elements of X so that

$$x_\xi = \Phi(\{x_\zeta : \zeta < \xi\}) \qquad (\xi < \alpha).$$

As mentioned above, this α-sequence cannot be injective. Hence there exists a least ordinal number $\beta < \alpha$ for which

$$x_\beta \in \{x_\zeta : \zeta < \beta\}.$$

Consequently, for some $\xi < \beta$, we may write $x_\beta = x_\xi$ and

$$\Phi(\{x_\zeta : \zeta < \xi\}) = x_\xi = x_\beta = \Phi(\{x_\zeta : \zeta < \beta\}).$$

Observe now that both families $\{x_\zeta : \zeta < \xi\}$ and $\{x_\zeta : \zeta < \beta\}$ are injective and their ranges differ from each other. This yields a contradiction and completes the proof.

Lemma 2. *The inequality* $\mathrm{card}(X) + 1 < 2^{\mathrm{card}(X)}$ *holds true whenever a set* X *contains at least two elements.*

Proof. Let x and y be any two distinct elements of X and let $z \notin X$. Define a mapping

$$\Psi : X \cup \{z\} \to \mathcal{P}(X)$$

as follows: for any element t of X put $\Psi(t) = \{t\}$, and for z put $\Psi(z) = \{x, y\}$. Clearly, Ψ is an injection or, equivalently,

$$\mathrm{card}(X) + 1 \leq 2^{\mathrm{card}(X)}.$$

Now, we have to demonstrate that there exists no injective mapping

$$\Phi : \mathcal{P}(X) \to X \cup \{z\}.$$

Suppose to the contrary that the above-mentioned injection Φ does exist. Without loss of generality, we may assume that X is infinite and $\Phi(X) = z$. Denote $\alpha = h(X)$. Again, by using the method of transfinite recursion, we can define an α-sequence $\{x_\beta : \beta < \alpha\}$ of elements of $X \cup \{z\}$ such that

$$x_\xi = \Phi(\{x_\zeta : \zeta < \xi\} \setminus \{z\}) \qquad (\xi < \alpha).$$

Only two cases are possible for this α-sequence.
(1) The element z does not belong to $\{x_\beta : \beta < \alpha\}$.

In this case, we apply to Φ the argument presented in the proof of Lemma 1. Indeed, as has already been demonstrated, there exists a least $\beta < \alpha$ such that $x_\beta \in \{x_\zeta : \zeta < \beta\}$. So, for some ordinal $\xi < \beta$, we come to the relation

$$\Phi(\{x_\zeta : \zeta < \xi\}) = \Phi(\{x_\zeta : \zeta < \beta\}),$$

while the sets $\{x_\zeta : \zeta < \xi\}$ and $\{x_\zeta : \zeta < \beta\}$ are distinct. But this circumstance contradicts the injectivity of Φ.

(2) The element z belongs to $\{x_\beta : \beta < \alpha\}$.

In this case, let ξ denote the least ordinal number for which

$$z = x_\xi = \Phi(\{x_\zeta : \zeta < \xi\}).$$

Since Φ is injective and $\Phi(X) = z$, the above formula implies $X = \{x_\zeta : \zeta < \xi\}$. Therefore, X can be made well-ordered and, as is known, for such an X the equality

$$\mathrm{card}(X) + 1 = \mathrm{card}(X)$$

holds true. Consequently,

$$\mathrm{card}(X) + 1 = \mathrm{card}(X) < 2^{\mathrm{card}(X)}$$

by virtue of Cantor's theorem and we again come to a contradiction.

Thus, in both cases a contradiction is obtained, which finishes the proof.

The next statement generalizes Theorem 2 and Remark 2 from Chapter 1.

Lemma 3. Let X be an arbitrary set, let $\mathbf{m}$ denote the cardinality of X, and let $\mathrm{h}(X)$ be the Hartogs ordinal number of X. Then:

(1) there exists a partition $\{\mathcal{W}_\alpha : \alpha \leq \mathrm{h}(X)\}$ of the power set $\mathcal{P}(X \times X)$;

(2) $2^{\mathrm{h}(X)} \leq 2^{2^{\mathbf{m}^2}}$.

Proof. Since (2) trivially follows from (1), it suffices to show the validity of (1) or, in other words, it suffices to define effectively the required partition $\{\mathcal{W}_\alpha : \alpha \leq \mathrm{h}(X)\}$ of $\mathcal{P}(X \times X)$.

For this purpose, take any set $Z \subset X \times X$. Only two cases are possible.

(a) Z is the graph of some well-ordering of a subset of X.

In this case, the order type $\alpha = \alpha(Z)$ of Z is strictly less than $\mathrm{h}(X)$ and we put $Z \in \mathcal{W}_\alpha$.

(b) Z is not the graph of any well-ordering.

In this case, we put $Z \in \mathcal{W}_{\mathrm{h}(X)}$.

It is easy to verify that the above construction leads to the desired partition $\{\mathcal{W}_\alpha : \alpha \leq \mathrm{h}(X)\}$ of $\mathcal{P}(X \times X)$.

Lemma 3 has thus been proved.

The next auxiliary proposition is due to Halbeisen and Shelah (see [87], [88]).

Lemma 4. *Assume that X is a D-infinite set (i.e., X contains a subset equinumerous with ω) and let $\mathcal{FS}(X)$ $(= X^{<\omega})$ denote the family of all finite sequences of elements of X. Then the inequality*

$$\mathrm{card}(\mathcal{P}(X)) \leq \mathrm{card}(\mathcal{FS}(X))$$

is false.

Proof. Suppose to the contrary that this inequality is valid, so there exists an injection

$$\phi : \mathcal{P}(X) \to \mathcal{FS}(X).$$

According to our assumption, $\mathrm{card}(X) \geq \omega$. Let Y_0 be a subset of X equinumerous with ω. Obviously, we may equip this Y_0 with some well-ordering W_0.

Further, let $\mathcal{W}$ denote the family of all those well-orderings of subsets of X which extend W_0. Starting with the injection ϕ, we are going to define a function

$$\psi : \mathcal{W} \to X$$

satisfying the following condition:

(') for each $W \in \mathcal{W}$, the element $\psi(W)$ belongs to $X \setminus \mathrm{pr}_1(W)$.

For this purpose, take any well-ordering $W \in \mathcal{W}$ and denote $Y = \mathrm{pr}_1(W)$. Since Y is well-ordered, there exists an effective canonical bijection

$$h : Y \to \mathcal{FS}(Y).$$

Now, let us put

$$Z = \{z \in Y : \phi^{-1}(h(z)) \text{ exists } \& \ z \notin \phi^{-1}(h(z))\}.$$

We are going to check that $\phi(Z) \notin \mathcal{FS}(Y)$. Indeed, supposing for a moment that $\phi(Z) \in \mathcal{FS}(Y)$, we must have $h(y) = \phi(Z)$ for some element $y \in Y$, whence it follows that $y \in Z$ if and only if $y \notin Z$. The obtained contradiction shows that $\phi(Z) \notin \mathcal{FS}(Y)$. Therefore, we may take $\psi(W)$ to be equal to the first member of $\phi(Z)$ not belonging to Y.

Summarizing all the facts above, we see that an effective transfinite construction enables us to deduce that some well-ordering of the whole set X belongs to $\mathcal{W}$ (cf. the argument given in Exercise 9 from Appendix 1). But the latter yields at once a contradiction with the condition (').

Lemma 4 has thus been proved.

Remark 1. In our further consideration we need only very special case of Lemma 4. Actually, it suffices for us to know that the inequalities

$$2^{\mathbf{m}} \leq \mathbf{m} + \mathbf{m}, \quad 2^{\mathbf{m}} \leq \mathbf{m}^2$$

are false for any D-infinite cardinal number $\mathbf{m}$.

Lemma 5. *Let* **m** *and* **n** *be two cardinal numbers such that*

$$\mathbf{m} + \mathbf{m} = \mathbf{m}, \quad \mathbf{m} + \mathbf{n} = 2^{\mathbf{m}}.$$

Then the equality $\mathbf{n} = 2^{\mathbf{m}}$ *holds true.*

Proof. First of all, we easily infer from our assumptions on **m** and **n** that

$$2^{\mathbf{m}} \cdot 2^{\mathbf{m}} = 2^{\mathbf{m}+\mathbf{m}} = 2^{\mathbf{m}} = \mathbf{m} + \mathbf{n}.$$

Now, take any two sets X and Y such that

$$\mathrm{card}(X) = \mathbf{m}, \quad \mathrm{card}(Y) = \mathbf{n}, \quad X \cap Y = \emptyset.$$

As stated above, there exists a bijection

$$\phi : \mathcal{P}(X) \times \mathcal{P}(X) \to X \cup Y.$$

Consider the set $\phi^{-1}(X)$. By virtue of Cantor's theorem (cf. its proof given in Appendix 1), we can find a set $X_0 \subset X$ such that

$$(X_0, Z) \notin \phi^{-1}(X)$$

for all $Z \subset X$. So we come to the injection

$$\phi' : \{(X_0, Z) : Z \subset X\} \to Y$$

which is a restriction of ϕ. The existence of ϕ' immediately implies the inequality $2^{\mathbf{m}} \leq \mathbf{n}$. Since the opposite inequality is also valid, we can conclude by the Cantor–Bernstein theorem that $2^{\mathbf{m}} = \mathbf{n}$.

Now, we are able to demonstrate Sierpiński's fundamental result in a somewhat strengthened form suggested by Specker [256]. Let us underline once more that both these results are due to **ZF** set theory.

Theorem 1. *Let* **m** *be a nonzero cardinal number such that:*
(1) there exists no cardinal number strongly between **m** *and* $2^{\mathbf{m}}$;
(2) there exists no cardinal number strongly between $2^{\mathbf{m}}$ *and* $2^{2^{\mathbf{m}}}$.
Then $2^{\mathbf{m}} = \mathrm{h}(\mathbf{m})$, *so every set of cardinality* $2^{\mathbf{m}}$ *can be made well-ordered and, consequently, every set of cardinality* **m** *can also be made well-ordered.*

Proof. Condition (2) trivially implies the inequality $\mathbf{m} \geq 2$. According to Lemma 2, we may write

$$\mathbf{m} \leq \mathbf{m} + 1 < 2^{\mathbf{m}},$$

whence it follows that $\mathbf{m} + 1 = \mathbf{m}$. So **m** is an infinite cardinal number in the sense of Dedekind, i.e., $\omega \leq \mathbf{m}$. Further,

$$\mathbf{m} \leq \mathbf{m} + \mathbf{m} \leq 2^{\mathbf{m}} + 2^{\mathbf{m}} = 2^{\mathbf{m}+1} = 2^{\mathbf{m}},$$

and we get by virtue of Lemma 4 that $\mathbf{m} + \mathbf{m} = \mathbf{m}$. Similarly, we have

$$\mathbf{m} \leq \mathbf{m} \cdot \mathbf{m} \leq 2^{\mathbf{m}} \cdot 2^{\mathbf{m}} = 2^{\mathbf{m}+\mathbf{m}} = 2^{\mathbf{m}},$$

whence it follows (again by virtue of Lemma 4) that $\mathbf{m}^2 = \mathbf{m}$.

Now, applying Lemma 3, we infer

$$2^{\mathrm{h}(\mathbf{m})} \leq 2^{2^{\mathbf{m}^2}} = 2^{2^{\mathbf{m}}}.$$

Further, we have the following relation:

$$2^{\mathbf{m}} \leq 2^{\mathbf{m}} + \mathrm{h}(\mathbf{m}) < 2^{2^{\mathbf{m}}+\mathrm{h}(\mathbf{m})} = 2^{2^{\mathbf{m}}} \cdot 2^{\mathrm{h}(\mathbf{m})} \leq$$

$$2^{2^{\mathbf{m}}} \cdot 2^{2^{\mathbf{m}}} = 2^{2^{\mathbf{m}}+2^{\mathbf{m}}} = 2^{2^{\mathbf{m}+1}} = 2^{2^{\mathbf{m}}},$$

which implies the equality

$$2^{\mathbf{m}} + \mathrm{h}(\mathbf{m}) = 2^{\mathbf{m}}.$$

Observe now that

$$\mathbf{m} \leq \mathbf{m} + h(\mathbf{m}) \leq 2^{\mathbf{m}} + \mathrm{h}(\mathbf{m}) = 2^{\mathbf{m}}.$$

The relation $\mathbf{m} = \mathbf{m} + \mathrm{h}(\mathbf{m})$ is impossible in view of the definition of $\mathrm{h}(\mathbf{m})$. So we must have

$$\mathbf{m} + \mathrm{h}(\mathbf{m}) = 2^{\mathbf{m}}.$$

Finally, applying Lemma 5 to $\mathbf{m}$ and $\mathbf{n} = \mathrm{h}(\mathbf{m})$, we obtain $2^{\mathbf{m}} = \mathrm{h}(\mathbf{m})$.

Theorem 1 has thus been proved.

Remark 2. The above theorem shows that even two local forms of **GCH** imply the corresponding local forms of **AC**. A much simpler proof (within **ZF** set theory) of the implication **GCH** $\Rightarrow$ **AC** is outlined in Exercises 1 and 2 from this appendix.

Recall that Gödel's Constructible Universe **L** is the smallest transitive model of **ZFC** set theory, which contains the class of all von Neumann's ordinals.

As was mentioned several times in this book, **L** is also a model of **ZFC** set theory in which the Generalized Continuum Hypothesis holds (for extensive information about **L**, see, e.g., [10], [42], [103], [148]).

Some other remarkable phenomena may be observed in **L**. It is worth noticing two of them which seem to be very significant.

(i) The Constructibility Axiom $\mathbf{V} = \mathbf{L}$ implies that there exist no two-valued measurable cardinals.

(ii) The same axiom implies the existence of a Suslin line which is an alternative object to the standard real line **R**.

The statement (i) was first proved by Scott [227]. His argument uses the techniques of ultra-powers and does not rely on any specific inner properties of **L**; actually, it suffices to know the above-mentioned "abstract" definition of **L** as the smallest transitive model of **ZFC** which contains all ordinals in von Neumann's sense.

To briefly discuss the statement (ii), let us first recall the notion of a Suslin line.

A linearly ordered set $(S, \leq)$ is a Suslin line if it is densely ordered, Dedekind complete, without the least and greatest elements, has no countable everywhere dense subset, and satisfies the Suslin condition.

The latter condition means that every disjoint family of nonempty open intervals in S is at most countable. Hence S does not contain any uncountable strictly increasing (strictly decreasing) transfinite sequences of points.

Actually, the question of whether there exists a Suslin line was raised by Suslin in his note published in 1920, in the first volume of the famous international mathematical journal Fundamenta Mathematicae (see [255]).

Suslin's Hypothesis (abbreviation **SH**) is the assertion that there exists no Suslin line.

It turned out that this hypothesis is undecidable within **ZFC** set theory (see [10], [42], [101], [103], [105], [148], [254]). Indeed, as mentioned above, **SH** is false in the Constructible Universe **L**, i.e., under **V** = **L** there exists a Suslin line (cf. Exercise 11 below). On the other hand, **SH** is valid under Martin's Axiom with the negation of the Continuum Hypothesis (see exercises for Appendix 3).

EXERCISES

1*. Prove Tarski's theorem stating that the following two assertions are equivalent within **ZF** set theory:
(a) the Axiom of Choice;
(b) $\mathbf{m}^2 = \mathbf{m}$ for any infinite cardinal number $\mathbf{m}$.

Argue as follows. Since (a) $\Rightarrow$ (b) is well known, it suffices to establish the reverse implication (b) $\Rightarrow$ (a). Assume (b). Let $\mathbf{m}$ be any infinite cardinal number and let $h(\mathbf{m})$ denote the corresponding Hartogs ordinal (cardinal) number. First, check that $\mathbf{m} + 1 = \mathbf{m}$, i.e., $\mathbf{m}$ is infinite in the Dedekind sense. Then, by starting with the equality

$$(\mathbf{m} + h(\mathbf{m}))^2 = \mathbf{m} + h(\mathbf{m}),$$

demonstrate that

$$\mathbf{m} \cdot h(\mathbf{m}) = \mathbf{m} + h(\mathbf{m}).$$

Further, let A and B be any two sets satisfying the relations $\mathrm{card}(A) = \mathbf{m}$ and $\mathrm{card}(B) = h(\mathbf{m})$. As stated above, there exists a partition $\{X, Y\}$ of $A \times B$

such that $\text{card}(X) = \mathbf{m}$ and $\text{card}(Y) = h(\mathbf{m})$. For each element $a \in A$, verify that

$$Y(a) = \{b \in B : (a, b) \in Y\} \neq \emptyset$$

and, keeping in mind that both sets B and Y are well-orderable, infer that the set A is well-orderable, too. This trivially yields (a).

Remark 3. The assertion that $\mathbf{m} + \mathbf{m} = \mathbf{m}$ for all infinite cardinal numbers $\mathbf{m}$ does not imply **AC** in **ZF** set theory (see [102]).

2. Taking into account the previous exercise, give a simplified proof (within **ZF** theory) of the implication **GCH** $\Rightarrow$ **AC**.

3. Let E be a nonempty compact topological space without isolated points. Demonstrate that $\text{card}(E) \geq \mathbf{c}$.

For this purpose, construct by recursion a dyadic family of compact subsets of E with nonempty interiors.

4. Let $(E, \leq)$ be a Dedekind complete dense linearly ordered set containing at least two elements.

Demonstrate that $\text{card}(E) \geq \mathbf{c}$.

For checking the validity of this inequality, utilize the previous exercise.

5*. Show that the cardinality of any Suslin line $(S, \leq)$ does not exceed $\mathbf{c}$ (consequently, in view of Exercise 4, $\text{card}(S) = \mathbf{c}$).

For this purpose, apply the combinatorial theorem of Erdös and Rado (see Exercise 29 from Appendix 1).

6*. Assume the Continuum Hypothesis, and let E be a nonseparable topological space of cardinality $\mathbf{c}$.

Prove that there exists a subspace X of E such that:
(a) $\text{card}(X) = \mathbf{c}$;
(b) every uncountable subspace of X is nonseparable.

Argue as follows. First observe that the family of all closed separable subspaces of E has cardinality $\mathbf{c}$. Let $\{F_\xi : \xi < \omega_1\}$ denote the above-mentioned family. By using the method of transfinite recursion, construct an injective ω_1-sequence $\{x_\xi : \xi < \omega_1\}$ of points in E such that

$$x_\xi \notin \cup\{F_\zeta : \zeta < \xi\} \qquad (\xi < \omega_1).$$

Then put $X = \{x_\xi : \xi < \omega_1\}$ and check that X is as required.

Remark 4. It is clear that the argument outlined in the previous exercise is very similar to the two classical constructions of Luzin sets and Sierpiński sets (see Chapter 4 of this book). In general, analogous reasoning is applicable to many σ-ideals of subsets of a base (ground) set E.

7. Let $\mathcal{U}$ be an uncountable family of nonempty open subintervals of $\mathbf{R}$.

Demonstrate that there exists an uncountable subfamily $\mathcal{V}$ of $\mathcal{U}$ such that $\cap \mathcal{V} \neq \emptyset$.

Argue in the following manner. First, show that there are a real number $\delta > 0$ and an uncountable subfamily $\mathcal{U}'$ of $\mathcal{U}$ such that the length of any interval belonging to $\mathcal{U}'$ is strictly greater than δ. Then consider the countable set

$$D = \delta \cdot \mathbf{Z} = \{\delta \cdot m : m \in \mathbf{Z}\},$$

where $\mathbf{Z}$ denotes, as usual, the set of all integers. Finally, check that there is a point of D belonging to uncountably many members of $\mathcal{U}'$.

8*. Let $\mathcal{W}$ be an infinite family of closed bounded intervals in a Dedekind complete linearly ordered set $(E, \leq)$.

Verify the validity of the disjunction of the following two assertions:

(a) there exists an infinite disjoint subfamily $\mathcal{W}'$ of $\mathcal{W}$;

(b) there exists an infinite subfamily $\mathcal{W}''$ of $\mathcal{W}$ such that $\cap \mathcal{W}'' \neq \emptyset$.

Give a direct proof of this disjunction. On the other hand, deduce it from the infinite version of Ramsey's combinatorial theorem (see Exercise 27 in Appendix 1).

9. Assume the Continuum Hypothesis and let $(E, \leq)$ be a nonseparable dense linearly ordered set with $\text{card}(E) = \mathbf{c}$.

Demonstrate that there exists an uncountable family $\mathcal{W}$ of non-degenerate closed bounded intervals in E such that each point of E belongs to at most countably many members of $\mathcal{W}$.

In particular, if $(E, \leq)$ is a Suslin line, then for the family $\mathcal{W}$ indicated above, the disjunction of the following two assertions is false:

(a) there exists an uncountable disjoint subfamily $\mathcal{W}'$ of $\mathcal{W}$;

(b) there exists an uncountable subfamily $\mathcal{W}''$ of $\mathcal{W}$ such that $\cap \mathcal{W}'' \neq \emptyset$.

10*. A family $\{X_\alpha : \alpha < w_1\}$ of subsets of w_1 is called a diamond w_1-sequence if the following two conditions are satisfied:

(a) $X_\alpha \subset \alpha$ for each ordinal $\alpha < w_1$;

(b) for every set $X \subset w_1$, the set $\{\alpha < w_1 : X \cap \alpha = X_\alpha\}$ is stationary in w_1.

It is known that the existence of a diamond w_1-sequence follows from the Constructibility Axiom $\mathbf{V} = \mathbf{L}$ (see [10], [42], [103], [148]).

Demonstrate that if a diamond w_1-sequence does exist, then **CH** holds true and there exists a family $\{T_\alpha : \alpha < w_1\}$ of subsets of $w_1 \times w_1$ satisfying the following two conditions:

(a') $T_\alpha \subset \alpha \times \alpha$ for each ordinal $\alpha < w_1$;

(b') for every set $T \subset w_1 \times w_1$, the set $\{\alpha < w_1 : T \cap (\alpha \times \alpha) = T_\alpha\}$ is stationary in w_1.

To show (a') and (b'), pick an arbitrary bijection

$$f : w_1 \times w_1 \to w_1$$

and check that the set

$$D = \{\alpha < \omega_1 : f^{-1}(\alpha) = \alpha \times \alpha\}$$

is closed and unbounded in ω_1. Let $\{X_\alpha : \alpha < \omega_1\}$ be any diamond ω_1-sequence. For each ordinal $\alpha < \omega_1$, put

$$T_\alpha = f^{-1}(X_\alpha) \cap (\alpha \times \alpha).$$

Then condition (a') is fulfilled and $f(T_\alpha) = X_\alpha$ for all $\alpha \in D$.

Further, take an arbitrary set $T \subset \omega_1 \times \omega_1$. According to the definition of $\{X_\alpha : \alpha < \omega_1\}$, the set

$$S = \{\alpha < \omega_1 : f(T) \cap \alpha = X_\alpha\}$$

is stationary in ω_1 and so is $D \cap S$. Finally, if $\alpha \in D \cap S$, then

$$f^{-1}(f(T) \cap \alpha) = T \cap f^{-1}(\alpha) = T \cap (\alpha \times \alpha)$$

and, simultaneously,

$$f^{-1}(f(T) \cap \alpha) = T_\alpha,$$

which yields the desired result.

11*. Prove that the existence of a diamond ω_1-sequence implies the existence of a Suslin line.

Argue as follows. By using the method of transfinite recursion, construct a family $\{\preceq_\alpha : 1 \leq \alpha < \omega_1\}$ of binary relations so that these two conditions would be satisfied:

(a) for each nonzero $\alpha < \omega_1$, the relation $\preceq_\alpha$ is a linear ordering on the ordinal product $\omega \cdot \alpha$ and, in addition, $\preceq_\alpha$ is isomorphic to $(\mathbf{Q}, \leq)$;

(b) if $\alpha < \beta < \omega_1$, then $\preceq_\alpha = \preceq_\beta \cap ((\omega \cdot \alpha) \times (\omega \cdot \alpha))$.

First of all, take as $\preceq_1$ any linear ordering on ω isomorphic to $(\mathbf{Q}, \leq)$.

If $\alpha < \omega_1$ is a limit ordinal number, then put

$$\preceq_\alpha = \cup \{\preceq_\beta : 1 \leq \beta < \alpha\}.$$

If $\alpha = \beta + 1 < \omega_1$, then the linear ordering $\preceq_\beta$ on $\omega \cdot \beta$, isomorphic to $(\mathbf{Q}, \leq)$, has already been defined.

Call a subset X of $(\omega \cdot \beta) \times (\omega \cdot \beta)$ a β-foe if:

$$((x, y) \in X \ \& \ (u, v) \in X \ \& \ (x, y) \neq (u, v)) \ \Rightarrow \ (]x, y[\ \cap \ ]u, v[\ = \emptyset),$$

$$a \prec_\beta b \Rightarrow (\exists (x, y) \in X)(x \prec_\beta y \ \& \ ]a, b[\ \cap \ ]x, y[\ \neq \emptyset).$$

Clearly, the family of all β-foes in $(\omega \cdot \beta) \times (\omega \cdot \beta)$ is at most countable. Identify $\preceq_\beta$ with the natural linear ordering $\leq$ of $\mathbf{Q}$ and, therefore, identify the Dedekind completion of $\preceq_\beta$ with the natural linear ordering $\leq$ of $\mathbf{R}$. Notice that each β-foe X determines an open everywhere dense subset

$$X' = \cup\{]x, y[\; : \; (x, y) \in X\}$$

of the Dedekind completion of $\preceq_\beta$. So the Baire theorem can be applied in this situation and a bounded Dedekind cut (A, B) of $(\omega \cdot \beta, \preceq_\beta)$ can be found such that A has no maximal element, B has no minimal element and

$$(A, B) \in \cap\{X' \; : \; X \text{ is a } \beta - \text{foe}\}.$$

Now, define a linear ordering $\preceq_\alpha$ on the ordinal sum

$$\omega \cdot \alpha = (\omega \cdot \beta) + (\omega \cdot \{\beta\})$$

in such a way that:
(i) $(\omega \cdot \{\beta\}, \preceq_\alpha \cap ((\omega \cdot \{\beta\}) \times (\omega \cdot \{\beta\})))$ is isomorphic to $(\mathbf{Q}, \leq)$;
(ii) if $x \in A$ and $y \in \omega \cdot \{\beta\}$, then $x \prec_\alpha y$;
(iii) if $x \in B$ and $y \in \omega \cdot \{\beta\}$, then $y \prec_\alpha x$.
Verify that $\preceq_\alpha$ is isomorphic with $(\mathbf{Q}, \leq)$ and every β-foe is an α-foe as well. Finally, put

$$\preceq \; = \; \cup\{\preceq_\alpha \; : \; 1 \leq \alpha < \omega_1\}$$

and prove that the Dedekind completion of $(\omega_1, \preceq)$ is a Suslin line.

In order to check that there is no countable everywhere dense subset of $(\omega_1, \preceq)$, suppose to the contrary that D is such a set. Then, in view of the regularity of ω_1, the inclusion $D \subset \omega \cdot \beta$ holds true for some ordinal $\beta < \omega_1$. But then $D \cap (\omega \cdot \{\beta\}) = \emptyset$, which yields a contradiction with the everywhere density of D.

In order to show that $(\omega_1, \preceq)$ satisfies the countable chain condition, take any maximal (with respect to the inclusion relation) disjoint family $\{]a_i, b_i[\; : \; i \in I\}$ of nonempty open intervals in $(\omega_1, \preceq)$ and denote

$$T = \{(a_i, b_i) : i \in I\}.$$

Further, consider the set

$$U = \{\alpha < \omega_1 : \omega \cdot \alpha = \alpha \; \& \; T \cap (\alpha \times \alpha) \text{ is an } \alpha - \text{foe}\}$$

and verify that U is a closed unbounded subset of ω_1. Using the notation of Exercise 10, introduce the set

$$V = \{\alpha < \omega_1 : T_\alpha = T \cap (\alpha \times \alpha)\}$$

which is stationary and thus $U \cap V \neq \emptyset$. Pick any $\xi \in U \cap V$. Then T_ξ is a $(\xi + 1)$-foe and it can easily be verified by transfinite induction that the same T_ξ is an α-foe for each countable ordinal α strictly greater than ξ. Hence

$$\{]u, v[\; : \; u \prec v \; \& \; (u, v) \in T_\xi\} = \{]a_i, b_i[\; : \; i \in I\},$$

which yields $\mathrm{card}(I) \leq \omega$ and establishes the validity of the countable chain condition for $(\omega_1, \preceq)$.

Remark 5. If S is a Suslin line, then the topological square $S \times S$ does not satisfy the Suslin condition (see Exercise 9 for Appendix 3). On the other hand, it is consistent with **ZFC** theory that there are two Suslin lines S_1 and S_2 such that their topological product $S_1 \times S_2$ satisfies the Suslin condition. So one can conclude from the latter fact that S_1 is not isomorphic to S_2. Notice also that the existence of a Suslin line does not contradict the negation of **CH** (for more details, see [10], [42], [103], [148]).

12*. Prove within **ZFC** set theory that if $(S, \preceq)$ is a Suslin line, then:

(a) S contains an everywhere dense subset of cardinality ω_1 (consequently, $\mathrm{card}(S) = \mathbf{c}$);

(b) the cardinality of the Borel σ-algebra $\mathcal{B}(S)$ is equal to $\mathbf{c}$;

(c) the intersection of any countable family of open everywhere dense subsets of S has nonempty interior.

For proving (a), define a transfinite sequence $\{D_\xi : \xi < \omega_1\}$ of at most countable subsets of S in the following manner. First, put $D_0 = \{s\}$ where s is an arbitrary point of S. Then, supposing that for $\xi < \omega_1$ the partial family $\{D_\zeta : \zeta < \xi\}$ has already been constructed, consider the set $S \setminus \mathrm{cl}(\cup\{D_\zeta : \zeta < \xi\})$ which is the union of a countable disjoint family of open intervals in S. Define D_ξ as a selector of this family. Finally, put

$$D = \cup\{D_\xi : \xi < \omega_1\}$$

and verify that D is everywhere dense in S.

For proving (b), notice that the cardinality of the family of all open sets in S is equal to $\mathbf{c}$.

For proving (c), consider a countable family $\{U_i : i \in I\}$ of open everywhere dense subsets of S. Any set U_i is the union of a countable family of pairwise disjoint open intervals. Let D_i denote the set of all endpoints of those intervals. Then there exists a nonempty open set

$$U \subset S \setminus \cup\{D_i : i \in I\}.$$

Check the validity of the inclusion $U \subset \cap\{U_i : i \in I\}$.

13*. In **ZF** set theory (including Axiom 7), demonstrate H. Rubin's theorem stating that the following two assertions are equivalent:

(a) the Axiom of Choice;

(b) for any ordinal α, the power set $\mathcal{P}(\alpha)$ is well-orderable.

Notice that the implication (a) $\Rightarrow$ (b) is trivial, so only the converse implication (b) $\Rightarrow$ (a) must be established.

Assume (b). To prove (a), it suffices to show that, for every ordinal α, the set V_α of von Neumann's universe **V** is well-orderable. Use the method of transfinite induction and suppose that V_β is well-orderable for all ordinal numbers $\beta < \alpha$. If $\alpha = \beta + 1$, then (b) immediately yields that $V_\alpha = \mathcal{P}(V_\beta)$ can be made well-ordered.

It remains to envisage the situation where α is a limit ordinal.

Take any ordinal number θ such that

$$\operatorname{card}(\theta) > \sum \{\operatorname{card}(V_\beta) : \beta < \alpha\}.$$

According to (b), we may pick some well-ordering $\preceq$ of $\mathcal{P}(\theta)$. Further, define by transfinite recursion a function

$$\Phi : [0, \alpha[\; \to \mathcal{P}(V_\alpha \times V_\alpha)$$

so that $\Phi(\beta)$ is a well-ordering of V_β for each $\beta < \alpha$. Namely, put

$$\Phi(0) = \Phi(\emptyset) = \emptyset$$

and suppose that the values $\Phi(\beta)$ have already been defined for all $\beta < \xi$, where $\xi < \alpha$. Then consider two possible cases.

(i) ξ is of the form $\xi = \zeta + 1$.

In this case, we have $V_\xi = \mathcal{P}(V_\zeta)$ and, by the inductive assumption, the well-ordering $\Phi(\zeta)$ of V_ζ is already defined. There exists a unique isomorphism f between $\Phi(\zeta)$ and some proper initial subinterval of θ. Use the well-ordering $\preceq$ of $\mathcal{P}(\theta)$ with an inverse isomorphism f^{-1} and effectively introduce a concrete well-ordering $\Phi(\xi)$ of V_ξ.

(ii) ξ is a limit ordinal.

In this case, consider the set $\sum \{V_\beta \times \{\beta\} : \beta < \xi\}$ equipped with the well-ordering isomorphic to the ordinal sum of all $\Phi(\beta)$ $(\beta < \xi)$. Observe that there is a canonical surjection

$$g : \sum \{V_\beta \times \{\beta\} : \beta < \xi\} \to \cup \{V_\beta : \beta < \xi\}.$$

Keeping in mind the equality $V_\xi = \cup \{V_\beta : \beta < \xi\}$ and using g, effectively introduce a concrete well-ordering $\Phi(\xi)$ of V_ξ.

The above consideration establishes the existence of a function Φ. Now, repeat the argument of the case (ii) and endow the set $\sum \{V_\beta \times \{\beta\} : \beta < \alpha\}$ with the well-ordering isomorphic to the ordinal sum of all $\Phi(\beta)$ $(\beta < \alpha)$. Finally, taking into account the existence of a canonical surjection

$$h : \sum \{V_\beta \times \{\beta\} : \beta < \alpha\} \to \cup \{V_\beta : \beta < \alpha\}$$

and the equality $V_\alpha = \cup \{V_\beta : \beta < \alpha\}$, effectively introduce a concrete well-ordering of the set V_α.

Thus, for any ordinal α, the set V_α is well-orderable, which yields (a).

14. Work in **ZF** set theory (with Axiom 7) and demonstrate that the following two assertions are equivalent:

(a) the Axiom of Choice;

(b) every linearly orderable set is also well-orderable.

For this purpose, keep in mind Exercise 16 from Chapter 1 and the result of the previous exercise.

Remark 6. In **ZF** theory, the statement that every set is linearly orderable does not imply the Axiom of Choice (see [93], [102]).

15*. Let $\{m_i : i \in I\}$ and $\{n_i : i \in I\}$ be two families of cardinal numbers such that $m_i < n_i$ for each index $i \in I$.

Prove J. König's inequality $\sum\{m_i : i \in I\} < \prod\{n_i : i \in I\}$ (cf. Exercise 3 from Chapter 2).

Also, check that this inequality strengthens Cantor's inequality

$$\text{card}(X) < \text{card}(\mathcal{P}(X)),$$

but essentially relies on the Axiom of Choice.

Let m be an arbitrary infinite cardinal number. Deduce from J. König's inequality that $m < m^{\text{cf}(m)}$, where $\text{cf}(m)$ denotes the cofinality of m, i.e., $\text{cf}(m)$ is the least cardinal equinumerous with a cofinal subset of m.

Further, assume **GCH** and let m and n be any two infinite cardinal numbers.

By using the facts mentioned above, verify that:

(a) if $n < \text{cf}(m)$, then $m^n = m$;

(b) if $\text{cf}(m) \leq n < m$, then $m^n = 2^m$;

(c) if $m \leq n$, then $m^n = 2^n$.

Appendix 3: Martin's Axiom and its consequences in real analysis

If the Continuum Hypothesis $\mathbf{c} = \omega_1$ is assumed, then all subsets of the real line $\mathbf{R}$ are naturally divided into two classes: the first class contains all at most countable point sets and the second class contains all point sets of cardinality $\mathbf{c}$. In this respect, the structure of $\mathbf{R}$ becomes more or less observable. However, such an approach has a weak side and does not avoid certain pathologies. For instance, one must keep in mind that:

(*) under $\mathbf{CH}$ there exist extremely paradoxical subsets of $\mathbf{R}$ such as Luzin sets and Sierpiński sets (see Chapter 4);

(**) $\mathbf{CH}$ is consistent with the existence of a Lebesgue nonmeasurable set in $\mathbf{R}$ which is a continuous image of the complement of an analytic subset of $\mathbf{R}$.

Moreover, the Continuum Hypothesis maximally bounds from above the size of the continuum $\mathbf{c}$, but there is no sufficiently reasonable motivation to impose on $\mathbf{c}$ restrictions of this kind. Martin's Axiom (the abbreviation: $\mathbf{MA}$) was first introduced in the article [181]. The aim of $\mathbf{MA}$ is to withdraw such non-motivated restrictions on $\mathbf{c}$ and, simultaneously, to preserve valuable statements of real analysis, possibly in a slightly changed form.

From the viewpoint developed in the present book, one of significant facts implied by $\mathbf{MA}$ is the existence of functions acting from $\mathbf{R}$ into $\mathbf{R}$, which are absolutely nonmeasurable with respect to the class of all nonzero σ-finite diffused measures on $\mathbf{R}$ (see, e.g., Chapters 5, 12, and 13).

In this appendix we would like to recall some basic notions and statements connected with Martin's Axiom. As has been mentioned in the Preface, we are focused here on the two important consequences of $\mathbf{MA}$. Namely, in what follows we are going to demonstrate that, under $\mathbf{MA}$, the σ-ideal of all Lebesgue measure zero subsets of $\mathbf{R}$ is $\mathbf{c}$-complete ($\mathbf{c}$-additive) and so is the σ-ideal of all first category subsets of $\mathbf{R}$.

Let us begin with several auxiliary definitions and concepts from the general theory of partially ordered sets.

Let $(P, \preceq)$ be an arbitrary partially ordered set and let D be a subset of P.

We say that D is coinitial in P if for every element $p \in P$, there exists an element $q \in D$ satisfying the relation $q \preceq p$. In other words, $D \subset P$ is coinitial in P if and only if D is cofinal in the partially ordered set $(P, \succeq)$.

A nonempty set $G \subset P$ is called a filter in $(P, \preceq)$ if:

$$(\forall p \in G)(\forall q \in P)(p \preceq q \Rightarrow q \in G),$$

$$(\forall p \in G)(\forall q \in G)(\exists r \in G)(r \preceq p \ \& \ r \preceq q).$$

Notice that this definition resembles the definition of a filter in any Boolean algebra (see, for instance, [10], [29], [103], [148], [154]).

A set $Q \subset P$ is called consistent (compatible) if, for any finite subset Q' of Q, there exists an element $p \in P$ such that $p \leq q$ whenever $q \in Q'$.

Clearly, every filter in $(P, \preceq)$ is consistent (compatible).

Two elements p and q of P are called inconsistent (incompatible) if there is no $r \in P$ such that $r \preceq p$ and $r \preceq q$.

We say that a set $B \subset P$ is totally inconsistent (totally incompatible) if any two distinct elements of B are inconsistent (incompatible).

Finally, we say that $(P, \preceq)$ satisfies the countable chain condition if every totally inconsistent subset of P is at most countable.

Sometimes (especially, in topological applications), the countable chain condition is called the Suslin condition.

Example 1. Let E be a base set and let $\mathcal{T}$ be a topology on E, so we have the topological space $(E, \mathcal{T})$. Moreover, we also have the partially ordered set

$$(\mathcal{T} \setminus \{\emptyset\}, \subset)$$

canonically associated with $(E, \mathcal{T})$. It can readily be checked that the following two relations are equivalent:

(a) $(\mathcal{T} \setminus \{\emptyset\}, \subset)$ satisfies the countable chain condition;

(b) there exists no uncountable disjoint family of nonempty open sets in $(E, \mathcal{T})$.

The relation (b) is precisely the topological Suslin condition for $(E, \mathcal{T})$.

The standard formulation of Martin's Axiom (**MA**) looks as follows:

If $(P, \preceq)$ is a partially ordered set satisfying the countable chain condition and $\mathcal{D}$ is a family of coinitial subsets of P with card$(\mathcal{D}) < \mathbf{c}$, then there exists a filter $G \subset P$ which intersects every element of $\mathcal{D}$; in other words,

$$(\forall D \in \mathcal{D})(D \cap G \neq \emptyset).$$

The next statement (similar to the Baire classical theorem on category) is a purely topological equivalent of **MA** (for more details, see e.g. [148]).

If E is an arbitrary nonempty compact topological space satisfying the Suslin condition, then E cannot be covered by a family of nowhere dense subsets, whose cardinality is strictly less than $\mathbf{c}$.

Remark 1. The Continuum Hypothesis implies Martin's Axiom. Indeed, let $(P, \preceq)$ be an arbitrary partially ordered set and let $\{D_n : n < \omega\}$ be any countable family of coinitial subsets of P. Then we can easily construct (by ordinary recursion) a decreasing sequence $\{p_n : n < \omega\}$ of elements of P, such that $p_n \in D_n$ for each $n < \omega$. Indeed, if a partial finite sequence

$$p_0 \geq p_1 \geq \cdots \geq p_{n-1}$$

of elements of P has already been constructed, then we may take

$$p_n \in D_n \cap \{p \in P : p \leq p_{n-1}\}.$$

Afterwards, we put

$$G = \{p \in P : (\exists n < \omega)(p_n \preceq p)\}.$$

Clearly, G is a filter in P which intersects every D_n $(n \in \omega)$. In particular, we see that **MA** follows from **CH**.

Solovay and Tennenbaum proved in [254] that the statement

$$\textbf{MA \& } (\neg\textbf{CH})$$

is consistent with **ZFC** set theory. Furthermore, it was shown that the size of $\mathbf{c} = 2^\omega$ is not precisely determined by **MA**. For example, the statements

$$\textbf{MA \& } (2^\omega = \omega_2),$$

$$\textbf{MA \& } (2^\omega = \omega_3)$$

are consistent with **ZFC** (of course, separately), as well as many other analogous statements. For more detailed information, we refer the reader to [103] or [148].

Remark 2. In the formulation of Martin's Axiom the restriction to a family $\mathcal{D}$ of coinitial subsets with $\mathrm{card}(\mathcal{D}) < \mathbf{c}$ is not accidental. To see this, let us consider the partially ordered set $(P, \preceq)$, where P is the family of all those finite sequences whose terms belong to $\{0, 1\}$, and $p \preceq q$ means that p is an extension of q. Since P is countably infinite, $(P, \preceq)$ trivially satisfies the countable chain condition. Let $\mathcal{D}$ denote the family consisting of all sets of the form

$$A_n = \{p \in P : n \in \mathrm{dom}(p)\},$$

$$D_f = \{p \in P : \neg(p \subset f)\},$$

where $n < \omega$ and $f \in \{0, 1\}^\omega$. Observe that

$$\mathrm{card}(\mathcal{D}) = \mathbf{c}$$

and each member from $\mathcal{D}$ is coinitial in P. Suppose for a moment that G is a filter in P which intersects every set $D \in \mathcal{D}$. Then we may define

$$g = \cup \{p : p \in G\},$$

and it can easily be verified that g is a partial function acting from ω into $\{0, 1\}$. Since $G \cap A_n \neq \emptyset$ for each $n < \omega$, we see that

$$\mathrm{dom}(g) = \omega.$$

Thus, we have the function

$$g : \omega \to \{0, 1\}.$$

But we also have

$$G \cap D_f \neq \emptyset$$

for any $f \in \{0, 1\}^\omega$. So we get $g \neq f$ for every $f \in \{0, 1\}^\omega$, and this is an obvious contradiction.

Remark 3. The restriction to a partial ordering satisfying the countable chain condition is also matured in the formulation of Martin's Axiom. In fact, the following two sentences are equivalent (within **ZFC** set theory):

(i) the Continuum Hypothesis;

(ii) for any partially ordered set $(P, \preceq)$ and for every family $\mathcal{D}$ of coinitial subsets of P with $\mathrm{card}(\mathcal{D}) < \mathbf{c}$, there exists a filter $G \subset P$ which intersects each set from $\mathcal{D}$.

The proof of the equivalence of (i) and (ii) is left to the reader.

We thus conclude that the countable chain condition is substantial in the formulation of Martin's Axiom for the purpose of having an additional set theoretical statement strongly weaker than the Continuum Hypothesis.

As was mentioned earlier, in real analysis and classical theory of Lebesgue measure, Martin's Axiom quite often yields effects very similar to those which can be obtained by using the much more stronger Continuum Hypothesis. A lot of examples of such effects may be found in [10], [62], [103], [133], [148], [223].

Now, let us assume **MA** and prove in details the **c**-additivity (or, according to another terminology, **c**-completeness) of the two standard σ-ideals on **R**. The **c**-completeness of these σ-ideals plays an essential role in various topics of real analysis (see, for example, Chapter 12).

Theorem 1. *Let* **MA** *be satisfied, let* μ *be the completion of a σ-finite Borel measure on* **R** *and let* $\{X_i : i \in I\}$ *be a family of μ-measure zero subsets of* **R** *such that* $\mathrm{card}(I) < \mathbf{c}$. *Then the set* $\cup \{X_i : i \in I\}$ *is also of μ-measure zero.*

In particular, if μ *is diffused and* X *is a subset of* **R** *with* $\mathrm{card}(X) < \mathbf{c}$, *then* $\mu(X) = 0$.

Proof. Actually, it is required to demonstrate that if Martin's Axiom holds, then the σ-ideal $\mathcal{I}(\mu)$ of all μ-measure zero sets is $\mathbf{c}$-additive. In the special case $\mu = \lambda$, we obtain the $\mathbf{c}$-additivity of $\mathcal{I}(\lambda)$.

From the beginning we may assume, without loss of generality, that μ is a probability diffused measure. Take an arbitrary infinite cardinal $\kappa < \mathbf{c}$ and a family $\{X_\alpha : \alpha < \kappa\}$ of μ-measure zero subsets of $\mathbf{R}$. Fix a real $\varepsilon > 0$. It suffices to show that there exists an open set $U \subset \mathbf{R}$ such that

$$\mu(U) \leq \varepsilon, \qquad \cup \{X_\alpha : \alpha < \kappa\} \subset U.$$

For this purpose, put

$$P = \{V \subset \mathbf{R} \ : \ V \text{ is open in } \mathbf{R} \text{ and } \mu(V) < \varepsilon\}$$

and consider a partial ordering $\preceq$ on P defined by the formula:

$$U \preceq V \Leftrightarrow U \supset V.$$

First, let us establish that $(P, \preceq)$ satisfies the countable chain condition. To demonstrate this fact, suppose that $\{V_\alpha : \alpha < \omega_1\}$ is an uncountable subfamily of P. Then there exist a strictly positive real number $\varepsilon_1 < \varepsilon$ and an uncountable subset Ξ of ω_1 such that, for each $\xi \in \Xi$, the inequality $\mu(V_\xi) < \varepsilon_1$ is valid. Now, for every $\xi \in \Xi$, let J_ξ be a finite union of intervals in $\mathbf{R}$ with rational endpoints, such that

$$J_\xi \subset V_\xi, \qquad \mu(V_\xi \setminus J_\xi) < \varepsilon - \varepsilon_1.$$

Then there are two distinct ξ and ζ in Ξ for which $J_\xi = J_\zeta$. Obviously,

$$V_\xi \cup V_\zeta = V_\xi \cup (V_\zeta \setminus J_\zeta).$$

Therefore,

$$\mu(V_\xi \cup V_\zeta) \leq \mu(V_\xi) + \mu(V_\zeta \setminus J_\zeta) < \varepsilon,$$

$$V_\xi \cup V_\zeta \in P,$$

$$V_\xi \cup V_\zeta \preceq V_\xi, \qquad V_\xi \cup V_\zeta \preceq V_\zeta.$$

Thus, $(P, \preceq)$ satisfies the countable chain condition and Martin's Axiom can be applied to this partially ordered set.

Observe now that, for every ordinal $\alpha < \kappa$, the set

$$D_\alpha = \{V \in P \ : \ X_\alpha \subset V\}$$

is a coinitial subset of $(P, \preceq)$ and

$$\mathrm{card}(\{D_\alpha : \alpha < \kappa\}) \leq \kappa < \mathbf{c}.$$

Consequently, there exists a filter G in $(P, \preceq)$ such that

$$(\forall \alpha < \kappa)(D_\alpha \cap G \neq \emptyset).$$

Consider the open set

$$U = \cup G.$$

It is easy to see that

$$(\forall \alpha < \kappa)(X_\alpha \subset U).$$

Since each element of G is an open set in $\mathbf{R}$ and since $\mathbf{R}$ has a countable base, one can find a countable family $\{U_n : n < \omega\}$ of members from G such that

$$U = \cup G = \cup\{U_n : n < \omega\}.$$

Moreover, G is a directed family of sets (with respect to the inclusion relation). Therefore,

$$\mu(U_1 \cup U_2 \cup ... \cup U_n) < \varepsilon$$

for every natural number n. This fact immediately implies that

$$\mu(U) = \mu(\cup\{U_n : n < \omega\}) \leq \varepsilon.$$

Remembering that ε is an arbitrary strictly positive real number, we may conclude that the set $\cup\{X_\alpha : \alpha < \kappa\}$ is of μ-measure zero. This yields the desired result.

Remark 4. Using the same method as in the proof of Theorem 1, a more general statement can be established. Namely, if E is a metric space with a countable base and μ is the completion of a σ-finite Borel measure on E, then, under Martin's Axiom, the σ-ideal $\mathcal{I}(\mu)$ is $\mathbf{c}$-additive.

As a straightforward consequence of Theorem 1, we get the following statement.

Theorem 2. *If Martin's Axiom holds, then:*
(1) any set $X \subset \mathbf{R}$ with $\mathrm{card}(X) < \mathbf{c}$ is universal measure zero;
(2) $\mathbf{c}$ is a regular cardinal.

Proof. Notice that (1) is trivially implied by Theorem 1. In order to verify the validity of (2), suppose to the contrary that $\mathbf{c}$ is a singular cardinal, i.e.,

$$\mathbf{R} = \cup\{X_i : i \in I\},$$

where the family of sets $\{X_i : i \in I\}$ is such that

$$\mathrm{card}(I) < \mathbf{c},$$

$$(\forall i \in I)(\mathrm{card}(X_i) < \mathbf{c}).$$

Then, according to Theorem 1, all sets X_i $(i \in I)$ are of Lebesgue measure zero and the set $\cup\{X_i : i \in I\}$ must be of Lebesgue measure zero, too, which yields an obvious contradiction.

Another direct corollary of Theorem 1 looks as follows.

Theorem 3. *Let* **MA** *be satisfied, let μ be the completion of a σ-finite Borel measure on* **R**, *and let $\{X_i : i \in I\}$ be a family of μ-measurable subsets of* **R**, *such that* card$(I) <$ **c**. *Then:*
(a) the set $\cup\{X_i : i \in I\}$ is also μ-measurable;
(b) if the members of $\{X_i : i \in I\}$ are pairwise disjoint, then

$$\mu(\cup\{X_i : i \in I\}) = \sum\{\mu(X_i) : i \in I\},$$

where, by definition,

$$\sum\{\mu(X_i) : i \in I\} = \sup\{\sum\{\mu(X_j) : j \in J\} : J \subset I, \operatorname{card}(J) < \omega\}.$$

The proof of Theorem 3 is left to the reader.
Now, we would like to present a topological analogue of Theorem 1.

Theorem 4. *Assume Martin's Axiom and let $\{Y_i : i \in I\}$ be a family of first category subsets of* **R** *such that* card$(I) <$ **c**. *Then the set $\cup\{Y_i : i \in I\}$ is also of first category in* **R**.
In particular, if Y is a subset of **R** *with* card$(Y) <$ **c**, *then Y is of first category in* **R**.

Proof. In other words, it is required to demonstrate that if Martin's Axiom holds, then the σ-ideal $\mathcal{K}(\mathbf{R})$ of all first category subsets of **R** is **c**-additive.

Take any infinite cardinal $\kappa <$ **c** and a family $\{Y_\alpha : \alpha < \kappa\}$ of nowhere dense subsets of **R**. It suffices to show that

$$\cup\{Y_\alpha : \alpha < \kappa\} \in \mathcal{K}(\mathbf{R}).$$

Let $\mathcal{J}$ denote the family of all finite sequences of nonempty open intervals in **R** with rational endpoints. We introduce the set

$$P = \{(f, U) \ : \ f \in \mathcal{J} \ \& \ U \text{ is an everywhere dense open subset of } \mathbf{R}\}$$

and define a partial ordering $\preceq$ on this P. Namely, we put

$$(f, U) \preceq (g, V)$$

if and only if the relation

$$(U \subset V) \ \& \ (f \supset g) \ \& \ (\forall i \in \operatorname{dom}(f) \setminus \operatorname{dom}(g))(f(i) \subset V)$$

is valid. Further, we check that the partially ordered set $(P, \preceq)$ satisfies the countable chain condition. For this purpose, take an uncountable family

$$\{(f_\alpha, V_\alpha) : \alpha < \omega_1\}$$

of elements from P. Since the family $\{f_\alpha : \alpha < \omega_1\}$ is uncountable, too, and $\mathcal{J}$ is countable, there are two distinct ordinals $\alpha < \omega_1$ and $\beta < \omega_1$ such that $f_\alpha = f_\beta$. Now, we put

$$V = V_\alpha \cap V_\beta,$$

$$f = f_\alpha = f_\beta.$$

Then V is an everywhere dense open subset of $\mathbf{R}$ and

$$(f, V) \preceq (f_\alpha, V_\alpha), \quad (f, V) \preceq (f_\beta, V_\beta).$$

The last two relations show that $(P, \preceq)$ satisfies the countable chain condition. So Martin's Axiom is applicable to this partially ordered set.

For each $\alpha < \kappa$, each $n \in \mathbf{N}$ and for any two numbers $p \in \mathbf{Q}$ and $q \in \mathbf{Q}$ such that $p < q$, denote

$$D_\alpha = \{(f, U) \in P : Y_\alpha \cap U = \emptyset\},$$

$$E_{p,q}^n = \{(f, U) \in P : (\exists m > n)(m \in \mathrm{dom}(f) \ \& \ f(m) \cap \,]p, q[\neq \emptyset)\}.$$

It is not difficult to check that:

(a) for each $\alpha < \kappa$, the set D_α is coinitial in $(P, \preceq)$;

(b) for each $n \in \mathbf{N}$ and for all p and q from $\mathbf{Q}$ such that $p < q$, the set $E_{p,q}^n$ is coinitial in $(P, \preceq)$.

Now, we define

$$\mathcal{S} = \{D_\alpha : \alpha < \kappa\} \cup \{E_{p,q}^n : n \in \mathbf{N}, \ p \in \mathbf{Q}, \ q \in \mathbf{Q}, \ p < q\}.$$

Clearly, we may write

$$\mathrm{card}(\mathcal{S}) \leq \kappa + \omega = \kappa < \mathbf{c}.$$

Consequently, there exists a filter F in $(P, \preceq)$ which intersects all sets from the family $\mathcal{S}$. We now put

$$h = \cup\{f : (\exists U)((f, U) \in F)\}.$$

Since F is a filter, h is a function. Moreover, since the relation

$$F \cap E_{p,q}^n \neq \emptyset$$

holds for all natural numbers n and for any two rational numbers p and q such that $p < q$, the relation

$$\mathrm{dom}(h) = \mathbf{N} = \omega$$

is valid. Further, for any natural number n, we define

$$U_n = \cup\{h(m) : n < m < \omega\},$$

$$H = \cap\{U_n : n < \omega\}.$$

It is obvious that all sets U_n $(n \in \mathbf{N})$ are open in $\mathbf{R}$.

Let $n \in \mathbf{N}$. If p and q are rational numbers and $p < q$, then we have $F \cap E_{p,q}^n \neq \emptyset$. Consequently, there exists a natural number $m > n$ such that

$$h(m) \cap]p, q[\neq \emptyset.$$

Thus, for each $n \in \mathbf{N}$, the set U_n is everywhere dense and open in $\mathbf{R}$, so H is an everywhere dense G_δ-subset of $\mathbf{R}$.

Finally, notice that if $\alpha < \kappa$, then $F \cap D_\alpha \neq \emptyset$, so there exists an element (f, U) of F for which $Y_\alpha \cap U = \emptyset$. Since F is a filter, it is not difficult to verify that $H \subset U$, so $Y_\alpha \cap H = \emptyset$, too. Therefore,

$$\cup\{Y_\alpha : \alpha < \kappa\} \subset \mathbf{R} \setminus H.$$

In particular,

$$\cup\{Y_\alpha : \alpha < \kappa\} \in \mathcal{K}(\mathbf{R}),$$

which yields the desired result.

Remark 5. Evidently, the argument given in the proof of Theorem 4 works for establishing a more general theorem which states, under Martin's Axiom, that if a topological space E has a countable base, then the σ-ideal of all first category subsets of E is $\mathbf{c}$-additive.

As a consequence of Theorem 4, we get the following statement.

Theorem 5. *Assume Martin's Axiom and let $\{Y_i : i \in I\}$ be a family of subsets of $\mathbf{R}$ such that $\mathrm{card}(I) < \mathbf{c}$ and all Y_i $(i \in I)$ possess the Baire property. Then the set $\cup\{Y_i : i \in I\}$ also possesses the Baire property.*

The proof of Theorem 5 is left to the reader.

Remark 6. There are many important applications of Martin's Axiom in general topology, group theory, functional analysis, etc. We do not touch those applications in this book. In this connection, we would like to refer the reader to Fremlin's widely known monograph [62]. Also, [10], [38], [106], [148], and [223] contain very useful information about Martin's Axiom.

EXERCISES

1. Work in **ZFC** set theory and prove the equivalence of the assertions (i) and (ii) formulated in Remark 3.

2. Give a detailed proof of Theorem 3.

For this purpose, take into account that the countable chain condition is fulfilled for the Solovay algebra $\mathrm{dom}(\mu)/\mathcal{I}(\mu)$.

3. Give a detailed proof of Theorem 5.

For this purpose, take into account that the countable chain condition is fulfilled for the Cohen algebra $\mathcal{B}a(\mathbf{R})/\mathcal{K}(\mathbf{R})$, where $\mathcal{B}a(\mathbf{R})$ denotes the σ-algebra of all those subsets of $\mathbf{R}$ which possess the Baire property.

4*. Let $(P, \preceq)$ be a partially ordered set satisfying the countable chain condition and let $\{p_\alpha : \alpha < \omega_1\}$ be an uncountable family of elements of P.

Show that there exists an element $p \in \{p_\alpha : \alpha < \omega_1\}$ for which the following relation holds true:

(*) if $q \preceq p$, then q is compatible with uncountably many elements from the same family $\{p_\alpha : \alpha < \omega_1\}$.

To establish this fact, suppose to the contrary that (*) is false and construct by transfinite recursion a strictly increasing ω_1-sequence of ordinals

$$\{\alpha_\xi : \xi < \omega_1\} \subset [0, \omega_1[$$

and an injective ω_1-sequence $\{q_{\alpha_\xi} : \xi < \omega_1\}$ of elements of P such that:

(a) $q_{\alpha_\xi} \preceq p_{\alpha_\xi}$ for each ordinal $\xi < \omega_1$;

(b) if $\xi < \omega_1$, then the element q_{α_ξ} is incompatible with all elements p_α where $\alpha \geq \alpha_{\xi+1}$.

Proceeding in this manner, obtain the uncountable family $\{q_{\alpha_\xi} : \xi < \omega_1\}$ of pairwise incompatible elements of P, which yields a contradiction with the countable chain condition for $(P, \preceq)$.

5*. By definition, a partially ordered set $(P, \leq)$ satisfies the strong countable chain condition if, for any family $\{p_\xi : \xi < \omega_1\}$ of elements of P, there exists an uncountable set

$$\Xi \subset [0, \omega_1[$$

such that the partial family $\{p_\xi : \xi \in \Xi\}$ is consistent.

Assuming Martin's Axiom with the negation of the Continuum Hypothesis, prove that the following two assertions are equivalent:

(a) $(P, \leq)$ satisfies the strong countable chain condition;

(b) $(P, \leq)$ satisfies the countable chain condition.

Argue as follows. The implication (a) $\Rightarrow$ (b) is trivial, so concentrate attention on the converse implication (b) $\Rightarrow$ (a).

Suppose (b) and take a family $\{p_\alpha : \alpha < \omega_1\}$ of elements of P. By virtue of Exercise 4, it may be assumed that p_0 has the property that all elements $q \leq p_0$ are compatible with uncountably many members of $\{p_\alpha : \alpha < \omega_1\}$.

Let Q denote the family of all those finite consistent subsets of P which contain p_0 as one of their elements. Equip Q with a partial ordering $\preceq$ defined

by the formula:
$$X \preceq Y \Leftrightarrow Y \subset X \quad (X \in Q, \ Y \in Q).$$

Check that the partially ordered set $(Q, \preceq)$ satisfies the countable chain condition.

Further, for each ordinal $\alpha < \omega_1$, denote by D_α the family of all those sets $X \in Q$ which contain some element p_β, where $\alpha < \beta < \omega_1$.

Verify that D_α is coinitial in $(Q, \preceq)$.

By the assumption $\omega_1 < \mathbf{c}$, there exists a filter G in $(Q, \preceq)$ meeting all sets D_α $(\alpha < \omega_1)$. Put
$$R = (\cup G) \cap \{p_\alpha : \alpha < \omega_1\}$$

and check the validity of the following two relations:

(i) $\mathrm{card}(R) = \omega_1$;

(ii) R is consistent.

Conclude from (i) and (ii) that $(P, \leq)$ satisfies the strong countable chain condition.

6. Let $(P, \leq)$ be a partially ordered set satisfying the strong countable chain condition and let $(Q, \leq)$ be a partially ordered set satisfying the countable chain condition.

Show that the product partially ordered set

$$(P, \leq) \times (Q, \leq)$$

satisfies the countable chain condition.

Conclude that, under $\mathbf{MA}$ & $(\neg\mathbf{CH})$, the following two assertions are valid:

(a) the product of any finite family of partially ordered sets, all of which satisfy the countable chain condition, also satisfies this condition;

(b) the product of any finite family of topological spaces, all of which satisfy the Suslin condition, also satisfies this condition.

For this purpose, apply the result of Exercise 5.

7*. Work in $\mathbf{ZF}$ & $\mathbf{CC}$ theory and prove the so-called Δ-system lemma which is formulated as follows.

If $\{X_\xi : \xi < \omega_1\}$ is an arbitrary ω_1-sequence of finite sets, then there exist an uncountable set $\Xi \subset \omega_1$ and a set Y such that

$$X_\xi \cap X_\zeta = Y \quad (\xi \in \Xi, \ \zeta \in \Xi, \ \xi \neq \zeta).$$

In order to show this useful fact, first reduce the argument to the case when

$$\mathrm{card}(X_\xi) = n \quad (\xi < \omega_1)$$

for some fixed natural number n. Then argue by induction on n. Suppose that the assertion holds true for all natural numbers strictly less than n and consider the following two possibilities.

(a) There exist an element $y \in \cup\{X_\xi : \xi < \omega_1\}$ and an uncountable subset Ξ' of ω_1 such that y belongs to all members of the family $\{X_\xi : \xi \in \Xi'\}$.

In this case, use the inductive assumption to the family

$$\{X_\xi \setminus \{y\} : \xi \in \Xi'\}.$$

(b) There exists no element $y \in \cup\{X_\xi : \xi < \omega_1\}$ belonging to uncountably many members of the family $\{X_\xi : \xi < \omega_1\}$.

In this case, the set $\cup\{X_\xi : \xi < \omega_1\}$ is necessarily uncountable. Keeping in mind this circumstance, define by transfinite recursion a subset Ξ of ω_1 such that

$$X_\xi \cap X_\zeta = \emptyset \quad (\xi \in \Xi, \; \zeta \in \Xi, \; \xi \neq \zeta).$$

So in both cases (a) and (b) the required result is obtained.

8. Assume Martin's Axiom with the negation of the Continuum Hypothesis and let $\{E_i : i \in I\}$ be an arbitrary family of topological spaces satisfying the Suslin condition.

Demonstrate that the topological product $\prod\{E_i : i \in I\}$ also satisfies the Suslin condition.

For this purpose, take into account the results of Exercises 6 and 7.

9*. Let $(X, \leq)$ be a linearly ordered set satisfying the Suslin condition and suppose that X is nonseparable in its standard order topology.

Work in **ZFC** set theory and prove Kurepa's theorem stating that the topological product $X \times X$ does not satisfy the Suslin condition.

Argue as follows. First, denote by D the set of all isolated points in X. Clearly, $\mathrm{card}(D) \leq \omega$. Then, by using the method of transfinite recursion, construct three ω_1-sequences

$$\{x_\xi : \xi < \omega_1\}, \quad \{y_\xi : \xi < \omega_1\}, \quad \{z_\xi : \xi < \omega_1\}$$

of points in X such that:

$$x_\xi < y_\xi < z_\xi \quad (\xi < \omega_1),$$

$$]x_\xi, y_\xi[\; \neq \emptyset, \quad]y_\xi, z_\xi[\; \neq \emptyset \quad (\xi < \omega_1),$$

$$]x_\xi, z_\xi[\; \cap \; \{y_\zeta : \zeta < \xi\} = \emptyset \quad (\xi < \omega_1).$$

Suppose that, for an ordinal $\xi < \omega_1$, the three partial ξ-sequences

$$\{x_\zeta : \zeta < \xi\}, \quad \{y_\zeta : \zeta < \xi\}, \quad \{z_\zeta : \zeta < \xi\}$$

of points of X have already been defined. Since X is nonseparable, there exists a nonempty open interval

$$]x, z[\; \subset X \setminus \mathrm{cl}(D \cup \{y_\zeta : \zeta < \xi\}).$$

Put $x_\xi = x$ and $z_\xi = z$. Since the interval $]x_\xi, z_\xi[$ contains no isolated points, there exists $y \in]x_\xi, z_\xi[$ such that

$$]x_\xi, y[\neq \emptyset, \quad]y, z_\xi[\neq \emptyset.$$

Put $y_\xi = y$. Proceeding in this manner, get the required three ω_1-sequences of points of X and verify that the uncountable family

$$\{]x_\xi, y_\xi[\times]y_\xi, z_\xi[\ : \ \xi < \omega_1\}$$

of nonempty open sets in $X \times X$ is disjoint.

Remark 7. As we already know, in the Constructible Universe **L** of Gödel, there exists a Suslin line (see Appendix 2). Thus, in **L** there is a topological space (even compact space) X satisfying the Suslin condition, whose topological square $X \times X$ does not satisfy this condition. Respectively, in the same **L** there is a partially ordered set $(P, \leq)$ satisfying the countable chain condition, whose square $(P, \leq) \times (P, \leq)$ does not satisfy this condition. Also, it makes sense to mention here that, assuming **CH**, it is possible to establish the existence of two partially ordered sets $(P_1, \leq_1)$ and $(P_2, \leq_2)$ each of which satisfies the countable chain condition, but the product set $P_1 \times P_2$ endowed with the product partial ordering $\leq_1 \times \leq_2$ does not satisfy this condition (see [42], [103] or [148]).

10. Supposing Martin's Axiom with the negation of the Continuum Hypothesis, prove that there is no Suslin line.

For this purpose, take into account the results of Exercises 8 and 9.

11. Let E be a topological space and let $\{E_i : i \in I\}$ be a countable family of subspaces of E such that the set $\cup\{E_i : i \in I\}$ is everywhere dense in E.

Check that if all E_i $(i \in I)$ satisfy the Suslin condition, then E satisfies this condition, too.

12. Let X and Y be two topological spaces such that Y is a continuous image of X.

Show that if X satisfies the Suslin condition, then Y satisfies this condition, too.

13*. Let $\{E_i : i \in I\}$ be an arbitrary family of separable topological spaces (i.e., each space E_i $(i \in I)$ contains a countable everywhere dense subset).

Prove Marczewski's theorem stating that the topological product

$$E = \prod\{E_i : i \in I\}$$

satisfies the Suslin condition.

Argue as follows. First, for any nonzero natural number n, consider a cyclic group Z_n with $\mathrm{card}(Z_n) = n$ and equip Z_n with the discrete topology and canonical probability measure μ_n such that

$$\mu_n(\{z\}) = 1/n \quad (z \in Z_n).$$

The compact product group $Z_n^{\mathrm{card}(I)}$ carries the associated product probability measure μ which is invariant under all translations of $Z_n^{\mathrm{card}(I)}$. Since the values of μ on all basic open subsets of $Z_n^{\mathrm{card}(I)}$ are strictly positive, $Z_n^{\mathrm{card}(I)}$ trivially satisfies the Suslin condition (notice that this argument does not rely on the existence of a Haar probability measure on $Z_n^{\mathrm{card}(I)}$).

Further, equip the set $\mathbf{N}$ of all natural numbers with the discrete topology and check that $\mathbf{N}^{\mathrm{card}(I)}$ satisfies the Suslin condition.

For this purpose, keep in mind the fact stated above and utilize Exercise 11

Finally, check that there is a continuous surjection of $\mathbf{N}^{\mathrm{card}(I)}$ onto an ev erywhere dense subset of E and utilize both Exercises 11 and 12.

Remark 8. Applying an analogous method, it can be demonstrated that i $\{E_i : i \in I\}$ is any family of topological spaces whose densities do not exceed a given infinite cardinal $\mathbf{a}$, then the Suslin number of $E = \prod\{E_i : i \in I\}$ also does not exceed $\mathbf{a}$ (see, e.g., [49]).

14*. Assume Martin's Axiom and let $\mathbf{m}$ be an infinite cardinal number strictly less than $\mathbf{c}$.

Prove that the equality $2^{\mathbf{m}} = \mathbf{c}$ holds true.

Argue as follows. First, identify $\mathbf{c}$ with the least ordinal number of cardinality $\mathbf{c}$ and, analogously, identify $\mathbf{m}$ with the least ordinal number of cardinality $\mathbf{m}$. According to Exercise 3 from Chapter 13, there exists a family $\{X_\alpha : \alpha < \mathbf{c}\}$ of infinite almost disjoint subsets of ω. Consequently, the partial family $\{X_\alpha : \alpha < \mathbf{m}\}$ is almost disjoint, too.

In further consideration identify the subsets of ω with their characteristic functions.

It suffices to show the existence of an injective mapping acting from $\mathcal{P}(\mathbf{m})$ into $\mathcal{P}(\omega)$. For this purpose, it suffices to establish that, for every set $A \subset \mathbf{m}$ there is a set $X_A \subset \omega$ satisfying these two relations:

(a) if $\alpha \in A$, then the set $X_A \cap X_\alpha$ is finite;

(b) if $\alpha \in \mathbf{m} \setminus A$, then the set $X_A \cap X_\alpha$ is infinite.

So, fix a set $A \subset \mathbf{m}$ and define a partially ordered set $(P, \leq)$ (depending on A). The elements of P are pairs of the form $p = (s, F)$, where s is a finite sequence whose terms belong to $\{0, 1\}$ and F is a finite subfamily of $\{X_\alpha : \alpha \in A\}$.

For any two elements

$$p = (s, F) \in P, \quad p' = (s', F') \in P,$$

put $p' \leq p$ if and only if the following relation is fulfilled:

s' extends s, the family F is contained in F', and for each natural number $n \in \mathrm{dom}(s') \setminus \mathrm{dom}(s)$ if $s'(n) = 1$, then $n \notin \cup F$.

Verify that $(P, \leq)$ satisfies the countable chain condition.

Further, for any $\alpha \in A$, introduce the set

$$D_\alpha = \{(s, F) \in P \; : \; X_\alpha \in F\}.$$

Also, for any $\alpha \in \mathbf{m} \setminus A$ and $k < \omega$, define the set

$$D_{\alpha,k} = \{(s, F) \in P \; : \; (\exists n \geq k)(n \in X_\alpha \; \& \; s(n) = 1)\}.$$

Check that all D_α ($\alpha \in A$) and all $D_{\alpha,k}$ ($\alpha \in \mathbf{m} \setminus A$, $k < \omega$) are coinitial subsets of $(P, \leq)$.

Indeed, if $\alpha < \mathbf{m}$ and $(s, F) \in P$, then

$$(s, F \cup \{X_\alpha\}) \leq (s, P),$$

which shows that D_α is coinitial in $(P, \leq)$.

Likewise, fix $\alpha \in \mathbf{m} \setminus A$ and $k < \omega$, and take any $(s, F) \in P$. Since the family $\{X_\alpha : \alpha < \mathbf{m}\}$ is almost disjoint and consists of infinite subsets of ω, there exists a natural number n such that

$$n > k, \quad n \in X_\alpha \setminus \cup F, \quad n \notin \mathrm{dom}(s).$$

Extend s to $s' \in \{0, 1\}^{n+1}$ by putting $s'(n) = 1$ and $s'(l) = 0$ for all other natural numbers l from $\mathrm{dom}(s') \setminus \mathrm{dom}(s)$. Obviously, one has

$$(s', F) \leq (s, F),$$

which shows that $D_{\alpha,k}$ is coinitial in $(P, \leq)$.

Let G be a filter in $(P, \leq)$ meeting all the above-mentioned coinitial subsets of P, and let

$$S = \cup \{s : (\exists F)((s, F) \in G)\}.$$

Observe that S is a function acting from ω into $\{0, 1\}$, so S coincides with the characteristic function of some set $X_A \subset \omega$.

If $\alpha \in A$, then there exists $(s, F) \in G \cap D_\alpha$. Let (s', F') be any pair from G. Since G is a filter in $(P, \leq)$, there exists $(s'', F'') \in G$ such that

$$(s'', F'') \leq (s, F), \quad (s'', F'') \leq (s', F').$$

For each $n \in \mathrm{dom}(s'') \setminus \mathrm{dom}(s)$, the implication

$$s''(n) = 1 \Rightarrow n \notin X_\alpha$$

holds true, whence it follows that

$$X_A \cap X_\alpha \subset \{n \; : \; n \in \mathrm{dom}(s) \; \& \; s(n) = 1\}$$

and, consequently, the set $X_A \cap X_\alpha$ is finite.

If $\alpha \in \mathbf{m} \setminus A$ and $k < \omega$, then there exists a pair

$$(s, F) \in G \cap D_{\alpha,k}.$$

Infer that there is a natural number $n \geq k$ such that $n \in X_A \cap X_\alpha$. This shows that the set $X \cap X_\alpha$ is infinite.

Summarizing all the facts stated above, obtain the required result.

Remark 9. The partially ordered set $(P, \leq)$ described in Exercise 14 is usually called Solovay's forcing of almost disjoint sets.

Appendix 4: ω_1-dense subsets of the real line

It needless to recall here that an extensive study of structural properties of point sets on the real line $\mathbf{R}$ (and also in Euclidean space $\mathbf{R}^n$) was initiated by Cantor in his pioneer works (see [28]).

In particular, the following two classical theorems were proved by him and are widely known.

Theorem 1. *Let the set $\mathbf{Q}$ of all rational numbers be equipped with the standard linear ordering $\leq$ and let $(E, \preceq)$ be an arbitrary linearly ordered set whose cardinality does not exceed ω. Then there exists a strictly increasing mapping*

$$h : (E, \preceq) \to (\mathbf{Q}, \leq),$$

which, therefore, is an isomorphism between the linearly ordered sets $(E, \preceq)$ and $(h(E), \leq)$.

In other words, Theorem 1 states that the structure $(\mathbf{Q}, \leq)$ is universal in the class of all countable linearly ordered sets. This universal structure can be completely described in an abstract manner.

Recall that a linear ordering $\preceq$ on a set E is dense if, for any two elements $x \in E$ and $y \in E$ satisfying the relation $x \prec y$, there exists an element $z \in E$ such that $x \prec z \prec y$.

Theorem 2. *Let $(E, \preceq)$ be a countably infinite, dense linearly ordered set without the least and greatest elements. Then there exists an isomorphism*

$$g : (E, \preceq) \to (\mathbf{Q}, \leq).$$

Consequently, all countable everywhere dense subsets of $\mathbf{R}$ are mutually isomorphic.

We omit the standard proofs, within **ZF** set theory, of Theorems 1 and 2 and suggest the reader to carry out the corresponding argument (see Exercises 1 and 2 in this appendix).

Keeping in mind Theorem 2, we may introduce the following definition.

A subset X of $\mathbf{R}$ is said to be ω-dense in $\mathbf{R}$ if $\text{card}(X \cap U) = \omega$ for every nonempty open set $U \subset \mathbf{R}$.

Example 1. Clearly, the set $\mathbf{Q}$ of all rational numbers is a canonical ω-dense subset of $\mathbf{R}$. Besides, $\mathbf{Q}$ is an effective object, i.e., it exists within the framework of $\mathbf{ZF}$ set theory.

As a trivial consequence of the introduced definition, we get that if a set $X \subset \mathbf{R}$ is ω-dense in $\mathbf{R}$, then $\text{card}(X) = \omega$ (it suffices to take $U = \mathbf{R}$).

Theorem 2 readily implies another consequence which looks as follows:

if $X \subset \mathbf{R}$ is an ω-dense set, then the structures $(X, \leq)$ and $(\mathbf{Q}, \leq)$ are isomorphic.

We thus see that all ω-dense sets in $\mathbf{R}$ are isomorphic to each other.

Here we are interested in possible analogues of ω-dense sets for uncountable subsets of $\mathbf{R}$, namely, for those subsets of $\mathbf{R}$ which have cardinality ω_1.

The almost trivial considerations presented above motivate to introduce the following notion (see, e.g., [11], [225]).

A subset X of $\mathbf{R}$ is called ω_1-dense in $\mathbf{R}$ if $\text{card}(X \cap U) = \omega_1$ for every nonempty open set $U \subset \mathbf{R}$.

This definition directly implies that if a set $X \subset \mathbf{R}$ is ω_1-dense in $\mathbf{R}$, then $\text{card}(X) = \omega_1$ (again, it suffices to take $U = \mathbf{R}$).

Example 2. Let $\{\Delta_n : n < \omega\}$ denote the family of all nonempty open intervals in $\mathbf{R}$ with rational endpoints. For each natural index n, take a set $X_n \subset \Delta_n$ with $\text{card}(X_n) = \omega_1$. Then put

$$X = \cup\{X_n : n < \omega\}.$$

A straightforward verification shows that X is an ω_1-dense subset of $\mathbf{R}$.

Remark 1. The statement that there exists an ω_1-dense set in $\mathbf{R}$ cannot be proved without the aid of an uncountable form of the Axiom of Choice. Indeed, as is well known, even the existence of a subset of $\mathbf{R}$ with cardinality ω_1 implies the existence of a point set nonmeasurable in the Lebesgue sense (see Chapters 4 and 20).

Similarly to Theorem 2, one may pose the question whether any two ω_1-dense subsets of $\mathbf{R}$ are isomorphic to each other (as linearly ordered sets).

Example 3. Assume the Continuum Hypothesis $\mathbf{c} = \omega_1$ and consider any set $X \subset \mathbf{R}$ such that $\text{card}(X \cap U) = \mathbf{c}$ for every nonempty open subset U of $\mathbf{R}$ (e.g., X may be a Bernstein set in $\mathbf{R}$). Clearly, X is an ω_1-dense subset of $\mathbf{R}$. In particular, $\mathbf{R}$ itself is ω_1-dense in $\mathbf{R}$. Let t be an arbitrary point of $\mathbf{R}$. Then the set $\mathbf{R} \setminus \{t\}$ is also ω_1-dense in $\mathbf{R}$. However, there exists no isomorphism between the two linearly ordered sets $(\mathbf{R}, \leq)$ and $(\mathbf{R} \setminus \{t\}, \leq)$, because $(\mathbf{R}, \leq)$ is complete in the Dedekind sense while $(\mathbf{R} \setminus \{t\}, \leq)$ is not complete in the Dedekind sense.

In the above-mentioned work [11] Baumgartner introduced the following axiom:

(*) all ω_1-dense sets are mutually isomorphic.

In the same work Baumgartner demonstrated that (*) does not contradict the theory **ZFC** & $(\mathbf{c} = \omega_2)$.

As Example 3 shows, the implication

$$(*) \Rightarrow (\neg \mathbf{CH})$$

holds true within **ZFC** set theory.

The main goal of this appendix is to infer several other consequences from (*). Some of them are closely connected with Luzin sets and Sierpiński sets (see Chapter 4) and with universal measure zero sets (see Chapter 5).

In our presentation of this material we primarily follow [225] with some inessential simplification and modification.

First of all, we need three auxiliary propositions.

Lemma 1. *Let X be an infinite set and let $\{Y_j : j \in J\}$ be a family of sets satisfying the following two conditions:*

(1) $\mathrm{card}(J) \leq \mathrm{card}(X)$;

(2) $\mathrm{card}(X \cap Y_j) = \mathrm{card}(X)$ for each index $j \in J$.

Then there exists a partition $\{X_i : i \in I\}$ of X such that:

(a) $\mathrm{card}(I) = \mathrm{card}(X)$;

(b) $\mathrm{card}(X_i \cap Y_j) = \mathrm{card}(X)$ for all indices $i \in I$ and $j \in J$.

Proof. To obtain the required result, it suffices to apply Sierpiński's lemma on disjoint subsets (see Exercise 4 from Chapter 7 and Exercise 1 from Chapter 8).

Lemma 2. *If X is an ω_1-dense subset of $\mathbf{R}$, then there exists a partition $\{X_i : i \in I\}$ of X such that:*

(a) $\mathrm{card}(I) = \omega_1$;

(b) all the sets X_i $(i \in I)$ are ω_1-dense.

Proof. Let $\{\Delta_n : n < \omega\}$ denote the family of all nonempty open intervals in $\mathbf{R}$ whose endpoints are rational numbers. By virtue of the definition, we have

$$\mathrm{card}(X \cap \Delta_n) = \omega_1.$$

Now, we may apply Lemma 1 to the set X and to the family

$$\{Y_j : j \in J\} = \{\Delta_n : n < \omega\},$$

where $\mathrm{card}(J) = \omega < \omega_1$. In this manner, we obtain the desired partition $\{X_i : i \in I\}$ of the set X.

Lemma 2 has thus been proved (within the framework of **ZFC** set theory).

Lemma 3. *Under Baumgartner's axiom (*), the cardinality of the family $\mathcal{D}$ of all ω_1-dense sets in $\mathbf{R}$ is equal to 2^{ω_1}.*

Proof. Since the cardinality of every ω_1-dense set in $\mathbf{R}$ is ω_1, we have the trivial inequality

$$\mathrm{card}(\mathcal{D}) \leq \mathbf{c}^{\omega_1} = 2^{\omega \cdot \omega_1} = 2^{\omega_1}.$$

Let now X be an arbitrary ω_1-dense set in $\mathbf{R}$ and let $\{X_i : i \in I\}$ be a partition of X as in Lemma 2. For any nonempty subset I' of I, the set

$$X(I') = \cup\{X_i : i \in I'\}$$

is also ω_1-dense in $\mathbf{R}$. Moreover, if $I_1 \subset I$ and $I_2 \subset I$ are two distinct nonempty sets, then $X(I_1) \neq X(I_2)$. The latter fact leads at once to the inequality

$$\mathrm{card}(\mathcal{D}) \geq 2^{\omega_1}$$

and, consequently, to the required equality $\mathrm{card}(\mathcal{D}) = 2^{\omega_1}$.

Theorem 3. *Baumgartner's axiom (*) implies the Second Continuum Hypothesis of Luzin, namely, $2^\omega = 2^{\omega_1}$.*

Proof. Fix an arbitrary ω_1-dense set X in $\mathbf{R}$. By the definition, for each ω_1-dense set Y, there exists an isomorphism

$$g_{X,Y} : (X, \leq) \to (Y, \leq).$$

This $g_{X,Y}$ can be uniquely extended to an automorphism

$$g_{X,Y}^* : (\mathbf{R}, \leq) \to (\mathbf{R}, \leq).$$

Moreover, if Y and Y' are two distinct ω_1-dense sets, then

$$g_{X,Y} \neq g_{X,Y'}, \quad g_{X,Y}^* \neq g_{X,Y'}^*.$$

This circumstance and Lemma 3 imply that the cardinality of the family of all mappings $g_{X,Y}^*$ is greater than or equal to 2^{ω_1}.

Notice now that every $g_{X,Y}^*$ is a homeomorphism of $\mathbf{R}$ onto itself, so the cardinality of the family of all mappings $g_{X,Y}^*$ does not exceed 2^ω.

Comparing the obtained two inequalities, we come to the desired equality $2^\omega = 2^{\omega_1}$, which completes the proof.

Theorem 4. *It follows from Baumgartner's axiom (*) that all subsets of $\mathbf{R}$ whose cardinalities do not exceed ω_1 are universal measure zero.*

Proof. As we know, there exists a universal measure zero set $X \subset \mathbf{R}$ whose cardinality is equal to ω_1 (see, e.g., Chapter 5 or Appendix 5). It is not hard to verify that, putting

$$X' = X + \mathbf{Q} = \cup\{X + q : q \in \mathbf{Q}\},$$

we come to the ω_1-dense set X'. Of course, X' has universal measure zero as well.

Let now $Y \subset \mathbf{R}$ be any set with $\mathrm{card}(Y) = \omega_1$. As before, putting

$$Y' = Y + \mathbf{Q} = \cup\{Y + q : q \in \mathbf{Q}\},$$

we come to the ω_1-dense set Y'. According to (*), there exists an isomorphism

$$g : (X', \leq) \to (Y', \leq)$$

and this isomorphism can be uniquely extended to an automorphism

$$g^* : (\mathbf{R}, \leq) \to (\mathbf{R}, \leq)$$

which simultaneously is a homeomorphism of $\mathbf{R}$ onto itself. Consequently, its restriction

$$g = g^* | X'$$

turns out to be a Borel isomorphism between X' and Y'. Since the class of universal measure zero sets is invariant under Borel isomorphisms, we deduce that Y' has universal measure zero. Finally, remembering that $Y \subset Y'$, we conclude that Y is also universal measure zero.

Theorem 5. *Under Baumgartner's axiom (*), there are neither Luzin sets nor Sierpiński sets on* $\mathbf{R}$.

Proof. Assume (*) and suppose to the contrary that S is a Sierpiński set on $\mathbf{R}$. Since $\mathrm{card}(S) \geq \omega_1$ and every uncountable subset of S is a Sierpiński set, we may assume without loss of generality that $\mathrm{card}(S) = \omega_1$. According to Theorem 4, S must be universal measure zero. But we know that $\lambda^*(S) > 0$, where λ denotes, as usual, the standard Lebesgue measure on $\mathbf{R}$. The latter circumstance implies that S carries a nonzero σ-finite diffused Borel measure (e.g., the measure induced by λ). The obtained contradiction shows that no Sierpiński set can exist in $\mathbf{R}$.

An analogous argument works for Luzin sets. Indeed, assume again (*) and suppose to the contrary that L is a Luzin subset of $\mathbf{R}$. Since $\mathrm{card}(L) \geq \omega_1$ and every uncountable subset of L is a Luzin set, we may assume without loss of generality that $\mathrm{card}(L) = \omega_1$. Putting

$$L' = L + \mathbf{Q} = \cup\{L + q : q \in \mathbf{Q}\},$$

we come to the ω_1-dense set L' which is a Luzin set as well. Now, take an arbitrary subset Z of the Cantor discontinuum with $\mathrm{card}(Z) = \omega_1$ and define

$$Z' = Z + \mathbf{Q} = \cup\{Z + q : q \in \mathbf{Q}\}.$$

Obviously, Z' is an ω_1-dense set of first category in $\mathbf{R}$. According to (*), there exists an isomorphism

$$g : (Z', \leq) \to (L', \leq)$$

and this g can be uniquely extended to an automorphism

$$g^* : (\mathbf{R}, \leq) \to (\mathbf{R}, \leq).$$

Further, as we know, g^* is a homeomorphism of $\mathbf{R}$ onto itself, so the set

$$g^*(Z') = g(Z') = L'$$

being a g^*-image of a first category subset of $\mathbf{R}$ must be also of first category in $\mathbf{R}$. But the Luzin set L' even does not possess the Baire property.

The obtained contradiction finishes the proof of Theorem 5.

Remark 2. According to the results presented in Appendix 3, it follows from Martin's Axiom that any set $X \subset \mathbf{R}$ with $\mathrm{card}(X) < \mathbf{c}$ is of first category and has universal measure zero. Consequently, if Martin's Axiom with the negation of the Continuum Hypothesis holds, then each subset Y of $\mathbf{R}$ whose cardinality does not exceed ω_1 also is of first category and has universal measure zero, i.e., the situation is completely analogous to that when Baumgartner's axiom (*) is assumed. So the natural question arises whether $\mathbf{MA}$ & $(\neg\mathbf{CH})$ implies Baumgartner's axiom (*). In this connection, it was demonstrated by Avraham and Shelah [3] that $\mathbf{MA}$ & $(\neg\mathbf{CH})$ does not imply (*). In the same work [3] it is also proved that the following statement concerning monotone restrictions of real-valued functions does not contradict $\mathbf{ZFC}$ set theory:

Every function $f : \mathbf{R} \to \mathbf{R}$ is monotone on some uncountable subset of $\mathbf{R}$.

Consequently, the next statement is consistent with $\mathbf{ZFC}$ theory:

For every function $f : \mathbf{R} \to \mathbf{R}$, there exists an uncountable set $X \subset \mathbf{R}$ such that the restriction $f|X$ is continuous.

Therefore, if Sierpiński–Zygmund type real-valued functions are defined so that their restrictions to all uncountable subsets of $\mathbf{R}$ are discontinuous, then the result of [3] shows the consistency of the nonexistence of Sierpiński–Zygmund type functions with $\mathbf{ZFC}$ set theory.

EXERCISES

1. Give a proof of Theorem 1.

For this purpose, define the required strictly increasing mapping h by recursion. More precisely, let $(E, \preceq)$ be a countable linearly ordered set and let

$$E = \{e_1, e_2, \ldots, e_n, \ldots\}$$

be a bijective enumeration of E. Likewise, let

$$\mathbf{Q} = \{q_1, q_2, \ldots, q_n, \ldots\}$$

be a bijective enumeration of $\mathbf{Q}$. For each natural number n, define a strictly increasing mapping

$$h_n : \{e_1, e_2, ..., e_n\} \to \mathbf{Q}$$

in such a manner that h_{n+1} would be an extension of h_n.

Suppose that h_n has already been defined and let $(i_1, i_2, ..., i_n)$ denote the permutation of $\{1, 2, ..., n\}$ such that

$$e_{i_1} < e_{i_2} < ... < e_{i_n}.$$

According to the definition of h_n, one may write

$$h(e_{i_1}) < h(e_{i_2}) < ... < h(e_{i_n}).$$

Now, only three cases are possible.

(a) $e_{n+1} < e_{i_1}$. In this case, denote by m the least natural number for which $q_m < h(e_{i_1})$ and put $h_{n+1}(e_{n+1}) = q_m$.

(b) $e_{i_n} < e_{n+1}$. In this case, denote by m be the least natural number for which $h(e_{i_n}) < q_m$ and put $h_{n+1}(e_{n+1}) = q_m$.

(c) $e_{i_k} < e_{n+1} < e_{i_{k+1}}$ for some index $k \in \{1, 2, ..., n-1\}$. In this case, denote by m the least natural number for which $h(e_{i_k}) < q_m < h(e_{i_{k+1}})$ and put $h_{n+1}(e_{n+1}) = q_m$.

Proceeding in this manner, obtain the sequence $\{h_n : n < \omega\}$ of partial mappings acting from E into $\mathbf{Q}$ and verify that the common extension h of all these h_n $(n < \omega)$ is as required.

2. Starting with the result of Exercise 1 (i.e., Theorem 1) and using the fact that any surjective image of a countable set is at most countable, give an effective proof of the uncountability of $\mathbf{R}$.

Argue as follows. A straightforward consequence of Theorem 1 is that, for an arbitrary countable ordinal α, the linearly ordered set $(\mathbf{Q}, \leq)$ contains a subset isomorphic to α. This circumstance allows one to construct effectively a partition $\{Y_\alpha : \alpha < \omega_1\}$ of $\mathbf{R}$ (cf. Theorem 2 from Chapter 1). Now, for each point $x \in \mathbf{R}$, define

$$f(x) = \alpha(x) = \alpha,$$

where α is a unique ordinal such that $x \in Y_\alpha$. Finally, take into account the fact that the obtained mapping $f : \mathbf{R} \to \omega_1$ is a surjection and ω_1 is an uncountable set.

Remark 3. It makes sense to compare Exercise 2 with the Hartogs theorem stating that, for every set E, the Hartogs number $h(E)$ cannot be injectively mapped into E.

3. Give a proof of Theorem 2.

For this purpose, apply the so-called zigzag argument and, similarly to the method described in Exercise 1, define effectively the required isomorphism g by ordinary recursion.

4. Equip ω with its natural ordering $\leq$ and let $(E, \preceq)$ be an arbitrary infinite linearly ordered set.

Demonstrate that either E contains a subset isomorphic to $(\omega, \leq)$ or E contains a subset isomorphic to $(\omega, \geq)$.

On the other hand, show that there exists no uncountable linearly ordered set $(X, \leq)$ satisfying the following condition:

For any uncountable linearly ordered set $(E, \preceq)$, either E contains a subset isomorphic to $(X, \leq)$ or E contains a subset isomorphic to $(X, \geq)$.

Now, let $(F, \preceq)$ be an arbitrary infinite partially ordered set. Check that the disjunction of the following two assertions is fulfilled:

(a) F contains an infinite linearly ordered subset;

(b) F contains an infinite free subset (i.e., no two distinct elements of which are comparable with respect to $\preceq$).

For this purpose, use the infinite version of Ramsey's combinatorial theorem (see Exercise 27 from Appendix 1).

Finally, consider the real line $\mathbf{R}$ endowed with its standard ordering $\leq$ and let $\leq'$ be any well-ordering of the same $\mathbf{R}$. Define a partial ordering $\preceq$ on $\mathbf{R}$ by putting:

$$(x \preceq y) \Leftrightarrow (x \leq y \,\&\, x \leq' y) \qquad ((x, y) \in \mathbf{R} \times \mathbf{R}).$$

Verify that no uncountable subset of $(\mathbf{R}, \preceq)$ is linearly ordered and no uncountable subset of $(\mathbf{R}, \preceq)$ is free.

Remark 4. It follows from the existence of a Suslin line that there is a tree $(T, \preceq)$ with $\mathrm{card}(T) = \omega_1$ such that no uncountable subset of T is linearly ordered and no uncountable subset of T is free. For more details about such Suslin trees, see [10], [42], [103], [148].

5. Let $f : \mathbf{R} \to \mathbf{R}$ be an increasing partial mapping.

Demonstrate that the following two assertions are equivalent:

(a) f is locally bounded, i.e., for each point $x \in \mathbf{R}$, there exists a neighborhood $V = V(x)$ of x such that the restriction $f|V$ is bounded;

(b) there is an increasing function $f^* : \mathbf{R} \to \mathbf{R}$ which extends f.

Deduce from the equivalence of (a) and (b) that if $\mathrm{dom}(f)$ is everywhere dense in $\mathbf{R}$, then there always exists a unique increasing extension $f^* : \mathbf{R} \to \mathbf{R}$.

Moreover, if $\mathrm{dom}(f)$ is everywhere dense in $\mathbf{R}$ and f is strictly increasing, then there always exists a unique strictly increasing extension $f^* : \mathbf{R} \to \mathbf{R}$.

In addition, if both sets $\mathrm{dom}(f)$ and $\mathrm{ran}(f)$ are everywhere dense in $\mathbf{R}$ and f is strictly increasing, then the unique increasing extension f^* of f with $\mathrm{dom}(f^*) = \mathbf{R}$ turns out to be an automorphism of the linearly ordered set $(\mathbf{R}, \leq)$, so f^* is simultaneously a homeomorphism of $\mathbf{R}$ onto itself.

6*. Let E be an infinite set and let card$(E) = \alpha$, where α is an initial ordinal number, i.e., card$(\xi) <$ card(α) for each ordinal $\xi < \alpha$.

Demonstrate that there are exactly 2^α pairwise nonisomorphic linear orderings of E (in particular, there are exactly $2^{\mathbf{c}}$ pairwise nonisomorphic linear orderings of $\mathbf{R}$).

Argue as follows. For the sake of brevity, denote by ϕ the order type of $(\mathbf{Z}, \leq)$ where $\mathbf{Z}$ stands, as usual, for the set of all integers equipped with its standard order.

Further, if $\{i_\zeta : \zeta < \alpha\}$ is any α-sequence belonging to $\{0,1\}^\alpha$, then associate to it the order type

$$o(\{i_\zeta : \zeta < \alpha\}) = \sum \{\phi + i_\zeta + 1 : \zeta < \alpha\}.$$

Check that if $\{i_\zeta : \zeta < \alpha\}$ and $\{j_\zeta : \zeta < \alpha\}$ are two distinct α-sequences from $\{0,1\}^\alpha$, then

$$o(\{i_\zeta : \zeta < \alpha\}) \neq o(\{j_\zeta : \zeta < \alpha\}).$$

This leads to the required result.

Remark 5. As is known, the number of all pairwise nonisomorphic well-orderings of $\mathbf{R}$ is equal to $\mathbf{c}^+ = \mathbf{h}(\mathbf{c})$, so in certain models of **ZFC** set theory this number may be strictly less than the number of all pairwise nonisomorphic linear orderings of $\mathbf{R}$.

7*. Let $(E, \preceq)$ be an infinite linearly ordered set with card$(E) = \alpha$.

Show that there exists a monomorphism acting from $(E, \preceq)$ into the set $\{0,1\}^\alpha$ equipped with its lexicographical ordering $\leq$.

For this purpose, argue in the following manner. First, represent E in the form of an injective α-sequence of its elements:

$$E = \{e_\zeta : \zeta < \alpha\}.$$

Then define a mapping $f : E \to \{0,1\}^\alpha$ by putting

$$f(e_\zeta) = \{i(\zeta, \xi) : \xi < \alpha\},$$

where $i(\zeta, \xi) = 1$ if $e_\xi \preceq e_\zeta$, and $i(\zeta, \xi) = 0$ if $e_\zeta \prec e_\xi$.

Finally, verify that if $e_\zeta \prec e_\eta$, then $f(e_\zeta) < f(e_\eta)$, which yields the required result.

8*. For any ordinal number α, take the set $E = \{0,1\}^{\omega_{\alpha+1}}$ endowed with its lexicographical ordering and consider its subset $H(\alpha)$ defined as follows:

$t \in H(\alpha)$ if and only if there exists an ordinal $\xi < \omega_{\alpha+1}$ such that

$$t_\xi = 1, \quad (\forall \zeta)(\xi < \zeta < \omega_{\alpha+1} \Rightarrow t_\zeta = 0).$$

Observe that $H(\alpha)$ has neither least nor greatest elements and establish the following three properties of $H(\alpha)$:

(a) $\text{card}(H(\alpha)) = 2^{\omega_\alpha}$;

(b) if $X \subset H(\alpha)$ and $\text{card}(X) \leq \omega_\alpha$, then X is bounded from above and from below in H_α;

(c) if Y and Z are subsets of $H(\alpha)$ such that

$$\text{card}(Y) \leq \omega_\alpha, \quad \text{card}(Z) \leq \omega_\alpha, \quad (\forall y)(\forall z)((y \in Y \ \& \ z \in Z) \Rightarrow (y < z)),$$

then there exists an element $t \in H(\alpha)$ satisfying the relation

$$(\forall y)(\forall z)((y \in Y \ \& \ z \in Z) \Rightarrow (y < t < z)).$$

9*. Prove that every linearly ordered set $(E, \preceq)$ whose cardinality is equal to $\omega_{\alpha+1}$ can be isomorphically embedded in the set $H(\alpha)$ described in the previous exercise.

For this purpose, keep in mind the properties (a), (b), (c) of $H(\alpha)$ and construct the desired embedding by using the method of transfinite recursion (cf. Exercise 1).

Conclude from the facts above that, under the Generalized Continuum Hypothesis, for every infinite cardinal number of the form $\omega_{\alpha+1}$, there exists a linearly ordered set of cardinality $\omega_{\alpha+1}$ which is universal in the class of all linearly ordered sets whose cardinalities do not exceed $\omega_{\alpha+1}$.

In particular, if **CH** holds, then the set $H(0)$, equipped with its lexicographical ordering, has cardinality ω_1 and is universal in the class of all linearly ordered sets whose cardinalities do not exceed ω_1.

Remark 6. The last result substantially exploits **CH**, because there is a model of **ZFC** theory (first constructed by Shelah) in which there exists no linearly ordered set $(W, \preceq)$ with $\text{card}(W) = \omega_1$ such that $(W, \preceq)$ is universal in the class of all linearly ordered sets whose cardinalities are less than or equal to ω_1. For some other deep and interesting results in this direction, see e.g. [145].

Appendix 5: The beginnings of descriptive set theory

General set theory is concerned with abstract sets and various relationships between them. Descriptive set theory primarily deals with those subsets of the real line $\mathbf{R}$ or, more generally, of a Polish topological space E, which have rather good structure and can be described more or less effectively, e.g., without the aid of uncountable forms of the Axiom of Choice. Here we would like to recall the beginnings of this beautiful and important branch of mathematics.

From the viewpoint of real analysis and general topology, the most interesting properties of a subset of $\mathbf{R}$ are its measurability in the Lebesgue sense and the so-called Baire property which may be regarded as a certain topological analogue of measurability (see [25], [33], [115], [147], [152], [190], and [203]).

A set X in a topological space E has (possesses) the Baire property if X admits a representation in the form $X = (U \setminus Y) \cup Z$, where U is an open subset of E, and Y and Z are some first category subsets of E.

The family of all sets in E having the Baire property is denoted by $\mathcal{B}a(E)$. In fact, $\mathcal{B}a(E)$ coincides with the σ-algebra generated by the Borel σ-algebra $\mathcal{B}(E)$ and the family of all first category subsets of E.

Recall that if E is of second category, then the family of all first category subsets of E forms a proper σ-ideal in the power set $\mathcal{P}(E)$.

Let E' be a topological space and let $g : E \to E'$ be a mapping.

We say that g has the Baire property if, for each open set $V \subset E'$, the pre-image $g^{-1}(V)$ has the Baire property in E.

Observe that if there exists a first category set $X \subset E$ such that $g|(E \setminus X)$ is continuous, then g necessarily has the Baire property. The converse assertion is also true under some assumption on E'. Namely, suppose that the topology of E' is countably generated and a mapping $g : E \to E'$ has the Baire property. Let $\{V_n : n < \omega\}$ denote a countable base of E'. For each $n < \omega$, we may write

$$g^{-1}(V_n) = (U_n \setminus Y_n) \cup Z_n,$$

where U_n is an open subset of E and Y_n and Z_n are some first category sets in E. Let us put

$$X = \cup\{Y_n \cup Z_n : n < \omega\}.$$

Obviously, X is a first category subset of E and it can readily be verified that the restriction $g|(E \setminus X)$ is continuous.

Example 1. Let B be any Bernstein set in $\mathbf{R}$, let χ_B denote the characteristic function of B, and let P be a nonempty perfect subset of $\mathbf{R}$. It can easily be checked that the restriction $\chi_B|P$ is discontinuous, whence it immediately follows that B does not have the Baire property in $\mathbf{R}$.

For extensive information about the Baire property, see [25], [33], [115], [147], [152], [190], [191], [203].

Recall that the Borel σ-algebra $\mathcal{B}(E)$ of a topological space E is generated by the family of all open (equivalently, by the family of all closed) subsets of E. The following two simple auxiliary propositions turn out to be useful in many questions of descriptive set theory.

Lemma 1. *Let E be a topological space such that any open set in E is of type F_σ (or, equivalently, any closed set in E is of type G_δ).*

Then $\mathcal{B}(E)$ coincides with the monotone class generated by the family of all open sets in E (equivalently, by the family of all closed sets in E).

Lemma 2. *Let E be a topological space such that any open set in E is of type F_σ and let $\mathcal{L}$ be a class of subsets of E such that:*

(1) $\mathcal{L}$ contains the family of all open sets in E;

(2) $\mathcal{L}$ is closed under the unions of all countable disjoint families of its members;

(3) $\mathcal{L}$ is closed under the intersections of all countable families of its members.

Then $\mathcal{L}$ contains the Borel σ-algebra $\mathcal{B}(E)$.

Easy proofs of these lemmas are left to the reader (cf. Exercise 1).

Let E be a topological space and let $f : E \to \mathbf{R}$ be a function.

We say (see Chapter 1) that f is of Baire zero class if f is continuous at all points of E, i.e., f is continuous on E.

The family of all continuous functions acting from E into $\mathbf{R}$ is usually denoted by the symbol $\mathcal{C}(E, \mathbf{R})$ (or, quite frequently, by $\mathcal{C}(E)$). In accordance with the definition above, we use the notation $\mathcal{B}a_0(E, \mathbf{R})$ for the same family of functions. Thus,

$$\mathcal{B}a_0(E, \mathbf{R}) = \mathcal{C}(E, \mathbf{R}).$$

Suppose now that for an ordinal $\xi < \omega_1$, the Baire classes $\mathcal{B}a_\zeta(E, \mathbf{R})$ ($\zeta < \xi$) have already been determined.

We say that a function $f : E \to \mathbf{R}$ belongs to the class $\mathcal{B}a_\xi(E, \mathbf{R})$ if there exists a sequence of functions

$$\{f_n : n < \omega\} \subset \cup\{\mathcal{B}a_\zeta(E, \mathbf{R}) : \zeta < \xi\}$$

which pointwise converges to f, i.e.,

$$\lim_{n \to +\infty} f_n(x) = f(x) \quad (x \in E).$$

By proceeding in this way, it becomes possible to define the classes $\mathcal{B}a_\xi(E, \mathbf{R})$ for all ordinals $\xi < \omega_1$. Clearly, these classes increase by the inclusion relation. Further, putting

$$\mathcal{B}a(E, \mathbf{R}) = \cup\{\mathcal{B}a_\xi(E, \mathbf{R}) : \xi < \omega_1\},$$

we obtain the class of all Baire functions acting from E into $\mathbf{R}$ (cf. [4], [25], [103], [115], [152], [167], [197]).

In view of the regularity of ω_1, the class $\mathcal{B}a(E, \mathbf{R})$ is closed with respect to the pointwise limits of sequences of functions belonging to $\mathcal{B}a(E, \mathbf{R})$.

We say that a function $f \in \mathcal{B}a(E, \mathbf{R})$ is of Baire order $\xi < \omega_1$ if the relation

$$f \in \mathcal{B}a_\xi(E, \mathbf{R}) \setminus \cup\{\mathcal{B}a_\zeta(E, \mathbf{R}) : \zeta < \xi\}$$

holds true.

Example 2. In Chapter 1 we have mentioned some important properties of the class $\mathcal{B}a_1(\mathbf{R}, \mathbf{R})$. In addition, let us indicate that all monotone functions on $\mathbf{R}$, all semi-continuous functions on $\mathbf{R}$, and all derivatives on $\mathbf{R}$ belong to the class $\mathcal{B}a_1(\mathbf{R}, \mathbf{R})$.

The following three simple relations are valid for all Baire classes.

(1) $\mathcal{B}a_\xi(E, \mathbf{R})$ is a linear algebra over the field $\mathbf{R}$; in other words, if we have $f \in \mathcal{B}a_\xi(E, \mathbf{R})$, $g \in \mathcal{B}a_\xi(E, \mathbf{R})$, $a \in \mathbf{R}$ and $b \in \mathbf{R}$, then

$$af + bg \in \mathcal{B}a_\xi(E, \mathbf{R}), \quad f \cdot g \in \mathcal{B}a_\xi(E, \mathbf{R}).$$

(2) If $f \in \mathcal{B}a_\xi(E, \mathbf{R})$, $g \in \mathcal{B}a_\xi(E, \mathbf{R})$ and $g(x) \neq 0$ for all $x \in E$, then

$$f/g \in \mathcal{B}a_\xi(E, \mathbf{R}).$$

(3) If $f \in \mathcal{B}a_\xi(E, \mathbf{R})$ and $\phi \in \mathcal{B}a_\eta(\mathbf{R}, \mathbf{R})$, then $\phi \circ f \in \mathcal{B}a_{\xi+\eta}(E, \mathbf{R})$.

The above-mentioned relations (1), (2), and (3) can readily be checked by the method of transfinite induction.

Let us also formulate a less trivial property of any class $\mathcal{B}a_\xi(E, \mathbf{R})$. Namely, if $\{f_n : n < \omega\}$ is a sequence of functions from $\mathcal{B}a_\xi(E, \mathbf{R})$, which uniformly converges to a function $f : E \to \mathbf{R}$, then f also belongs to $\mathcal{B}a_\xi(E, \mathbf{R})$.

Recall that the symbol $\mathcal{B}(E, \mathbf{R})$ denotes the family of all Borel functions acting from E into $\mathbf{R}$. The following statement shows a close connection between Baire and Borel functions.

Theorem 1. *Suppose that E is a perfectly normal topological space, i.e., E is normal and every open set in E is of type F_σ. Then the equality*

$$\mathcal{B}a(E, \mathbf{R}) = \mathcal{B}(E, \mathbf{R})$$

holds true. In particular, this equality is fulfilled for every metric space E.

Keeping in mind Lemma 1, the Tietze-Urysohn theorem on extensions of real-valued continuous functions (see, e.g., [49], [152]), and the relation

$$\mathcal{B}a(E, \mathbf{R}) = \cup \{\mathcal{B}a_\xi(E, \mathbf{R}) : \xi < \omega_1\},$$

one can prove Theorem 1 by using the method of transfinite induction on $\xi < \omega_1$. We leave the corresponding technical details to the reader (see Exercise 3).

Recall that ω^ω denotes, as usual, the canonical Baire space of topological weight ω. We also recall that the symbol $\omega^{<\omega}$ denotes the family of all finite sequences of natural numbers. In what follows, we do not assume that this family is equipped with any topology. For each $s \in \omega^{<\omega}$, we put

$$\mathrm{lh}(s) = \mathrm{card}(\mathrm{dom}(s)).$$

Actually, $\mathrm{lh}(s)$ coincides with the length of s. Further, if

$$s = (s_1, s_2, ..., s_k) \in \omega^{<\omega}$$

and $n < \omega$, then the symbol $s * n$ denotes the extended sequence

$$(s_1, s_2, ..., s_k, s_{k+1}) \in \omega^{<\omega},$$

where $s_{k+1} = n$. In our further considerations, it will be convenient to identify each natural number k with the set $\{0, 1, ..., k-1\}$. In fact, this assumption is redundant since in modern set theory the natural numbers and, more generally, the ordinal numbers are usually introduced by the von Neumann scheme (see [10], [103], [148], [164] or Exercise 5 from Appendix 1).

Let us briefly touch upon the important notion of A-operation over appropriately indexed countable families of subsets of a base (ground) set E.

If a countable family of sets $\{X_s : s \in \omega^{<\omega}\} \subset \mathcal{P}(E)$ is given, then the set

$$A(\{X_s : s \in \omega^{<\omega}\}) = \bigcup_{\phi \in \omega^\omega} (\cap \{X_{\phi|k} : k < \omega\})$$

is called the result of A-operation applied to this family (cf. [25], [33], [103], [115], [152], [154], [167], [191]).

In many cases we may assume, without loss of generality, that an initial family $\{X_s : s \in \omega^{<\omega}\}$ is regular, i.e.,

$$(\forall n < \omega)(\forall s \in \omega^{<\omega})(X_{s*n} \subset X_s).$$

Indeed, if necessary, we may replace this family by $\{Y_s : s \in \omega^{<\omega}\}$, where

$$Y_\emptyset = E, \quad Y_{n_0 n_1 ... n_k} = X_{n_0} \cap X_{n_0 n_1} \cap ... \cap X_{n_0 n_1 ... n_k}$$

for any nonempty finite sequence $(n_0, n_1, ..., n_k)$ of natural numbers. Clearly,

$$A(\{X_s : s \in \omega^{<\omega}\}) = A(\{Y_s : s \in \omega^{<\omega}\}).$$

Also, suppose for a while that an initial regular family $\{X_s : s \in \omega^{<\omega}\}$ satisfies the following condition:

$$(\forall s \in \omega^{<\omega})(\forall t \in \omega^{<\omega})((s \neq t \ \& \ \mathrm{lh}(s) = \mathrm{lh}(t)) \Rightarrow X_s \cap X_t = \emptyset).$$

Then we have the equality

$$A(\{X_s : s \in \omega^{<\omega}\}) = \bigcap_{k<\omega} (\bigcup_{s \in \omega^{<\omega}, \ \mathrm{lh}(s)=k} X_s),$$

which is easily verified. Consequently, in this special case, the A-operation is reduced to the operations of countable unions and countable intersections.

If a class $\mathcal{L}$ of subsets of E is given, then the symbol $\mathcal{A}(\mathcal{L})$ stands for the class of all those sets which can be represented in the form $A(\{X_s : s \in \omega^{<\omega}\})$, where $\{X_s : s \in \omega^{<\omega}\} \subset \mathcal{L}$. The members from $\mathcal{A}(\mathcal{L})$ are usually called analytic sets over the original class $\mathcal{L}$.

It readily follows from the definition of the class $\mathcal{A}(\mathcal{L})$ that if $\mathcal{L}$ is closed under finite unions and finite intersections, then the same property holds true for $\mathcal{A}(\mathcal{L})$. Moreover, under the assumption of the closeness of $\mathcal{L}$ with respect to finite unions and finite intersections, one can assert that $\mathcal{A}(\mathcal{L})$ is closed under countable unions and countable intersections. Indeed, consider an arbitrary countable family of sets $\{W_i : i < \omega\} \subset \mathcal{A}(\mathcal{L})$. Then, for each index $i < \omega$, we may write

$$W_i = A(\{X_s^i : s \in \omega^{<\omega}\}),$$

where $\{X_s^i : s \in \omega^{<\omega}\} \subset \mathcal{L}$ is a regular family of sets. Let us define

$$Y_\emptyset = E, \quad Y_{i n_0 n_1 ... n_k} = X_{n_0 n_1 ... n_k}^i$$

for any $i < \omega$ and for any $(n_0, n_1, ..., n_k) \in \omega^{k+1}$. Applying the A-operation to the obtained family $\{Y_s : s \in \omega^{<\omega}\} \subset \mathcal{L}$, one can easily infer that

$$\cup\{W_i : i < \omega\} \in \mathcal{A}(\mathcal{L}).$$

Further, let us define

$$Z_{n_0 n_1 ... n_k} = \bigcap_{i \leq k} (\cup\{X_{m_0 m_1 ... m_k}^i : m_0 + m_1 + ... + m_k \leq n_0 + n_1 + ... + n_k\})$$

for any $(n_0, n_1, ..., n_k) \in \omega^{k+1}$. Applying the A-operation to the obtained family $\{Z_s : s \in \omega^{<\omega}\} \subset \mathcal{L}$ and taking into account König's lemma on ω-trees with finite levels (see Exercise 20 from Appendix 1), one can derive that

$$\cap\{W_i : i < \omega\} \in \mathcal{A}(\mathcal{L}).$$

From the point of view of numerous applications, the most important case is when the role of a ground set E is played by an uncountable Polish topological space and A-operation is applied to appropriately indexed countable families of closed subsets of E, i.e., the role of $\mathcal{L}$ is played by the class of all closed subsets of E. In this way we come to the analytic (Suslin) subsets of E. The next statement shows that they admit a rather nice topological description (see [115], [152], [167], [191]).

Theorem 2. *Let E be a Polish space and let X be a nonempty analytic subset of E. Then there exists a continuous surjection of ω^ω onto X. Conversely, any continuous image (in E) of the space ω^ω is a nonempty analytic subset of the space E.*

Proof. For every $s \in \omega^{<\omega}$, denote

$$U(s) = \{z \in \omega^\omega : (\forall i)(i < \mathrm{lh}(s) \Rightarrow z(i) = s(i))\}.$$

It is obvious that the family of sets $\{U(s) : s \in \omega^{<\omega}\}$ forms a base of ω^ω consisting of closed-open sets. Since the given Polish space E is nonempty, it is not difficult to construct a regular system $\{F_s : s \in \omega^{<\omega}\}$ of nonempty closed subsets of E such that:

(1) $\lim_{\mathrm{lh}(s) \to +\infty} \mathrm{diam}(F_s) = 0$, where the symbol $\mathrm{diam}(F_s)$ stands for the diameter of F_s;

(2) $E = A(\{F_s : s \in \omega^{<\omega}\})$.

Actually, relation (2) shows that E is a continuous image of the Baire space ω^ω. Indeed, for any $z \in \omega^\omega$, we may put

$$\{\phi(z)\} = F_{z_0} \cap F_{z_0 z_1} \cap \ldots \cap F_{z_0 z_1 \ldots z_k} \cap \ldots \, .$$

It is easy to check that this formula determines the continuous surjection

$$\phi : \omega^\omega \to E.$$

Since the given nonempty set $X \subset E$ is analytic, we can write

$$X = A(\{X_s : s \in \omega^{<\omega}\}),$$

where $\{X_s : s \in \omega^{<\omega}\}$ is a regular system of closed subsets of E. Taking into account the above relations (1) and (2), we may assume without loss of generality that

$$\lim_{\mathrm{lh}(s) \to +\infty} \mathrm{diam}(X_s) = 0.$$

Now, let us put

$$Z = \{z \in \omega^\omega : \cap\{X_{z|k} : k < \omega\} \neq \emptyset\}.$$

It can readily be checked that Z is a nonempty closed subset of ω^ω. Consequently, Z is a nonempty Polish space as well. Let us define a mapping

$$g : Z \to E$$

in the following manner: for each point $z \in Z$, let the corresponding value $g(z)$ be equal to a unique point of the set $\cap\{X_{z|k} : k < \omega\}$.

An easy verification shows that g is continuous and $g(Z) = X$.

Keeping in mind the same relations (1) and (2) and applying their analogues to Z, we infer that there exists a continuous surjection

$$\psi : \omega^\omega \to Z.$$

So we come to the required continuous surjection $g \circ \psi$ of the space ω^ω onto the set X.

Conversely, let Y be a subset of E such that there exists a continuous surjection

$$h : \omega^\omega \to Y.$$

Then, for each point $z \in \omega^\omega$, we may write

$$\{h(z)\} \subset \cap\{h(U(z|k)) : k < \omega\} \subset \cap\{\mathrm{cl}(h(U(z|k))) : k < \omega\}.$$

By virtue of the continuity of h, we get

$$\lim_{k \to +\infty} \mathrm{diam}(\mathrm{cl}(h(U(z|k)))) = 0,$$

whence it immediately follows that

$$\{h(z)\} = \cap\{\mathrm{cl}(h(U(z|k))) : k < \omega\}.$$

Finally, denoting

$$P_s = \mathrm{cl}(h(U(s))) \qquad (s \in \omega^{<\omega}),$$

we conclude that, for the given set Y, the equality

$$Y = A(\{P_s : s \in \omega^{<\omega}\})$$

is fulfilled, which completes the proof of Theorem 2.

By using the above theorem, it can readily be shown that a metrizable continuous image of a Suslin space is also Suslin, and the topological product of a countable family of Suslin spaces is Suslin, too.

If E is an arbitrary topological space in which every open set is of type F_σ, then we obviously have the inclusion $\mathcal{B}(E) \subset \mathcal{A}(E)$. Therefore, Theorem 2 directly implies that any nonempty Borel subset of a Polish space E can be regarded as a continuous image of the Baire space ω^ω. In this context, it should be noticed that, for Borel subsets of Polish spaces, a much stronger result can be established. Namely, we have the next important statement essentially due to Luzin (see [115], [152], [167], [191], [264]).

Theorem 3. *Let E be a Polish space. Every Borel subset of E may be regarded as a bijective continuous image of some Polish space.*

Proof. Denote by $\mathcal{L}$ the class of all those subsets of E which are bijective continuous images of Polish spaces.

Since any open subset U of E is a Polish space (see Exercise 5), we obtain at once that U belongs to $\mathcal{L}$.

Let $\{X_i : i \in I\}$ be an arbitrary disjoint countable family of subsets of E belonging to the class $\mathcal{L}$. For each index $i \in I$, denote by P_i a Polish space such that there exists a continuous bijection $f_i : P_i \to X_i$. Let P be the topological sum of the family of spaces $\{P_i : i \in I\}$. Clearly, P is a Polish space, too. Without loss of generality, we may treat each P_i $(i \in I)$ as a subset of P. Let

$$f : P \to \cup\{X_i : i \in I\}$$

denote the common extension of all mappings f_i $(i \in I)$. It can easily be seen that f is a continuous bijection, which immediately yields

$$\cup\{X_i : i \in I\} \in \mathcal{L}.$$

Let now $\{Y_i : i \in I\}$ be an arbitrary countable family of subsets of E belonging to the class $\mathcal{L}$. Again, for each index $i \in I$, denote by Q_i a Polish space such that there exists a continuous bijection $h_i : Q_i \to Y_i$. Both topological products $\prod\{Q_i : i \in I\}$ and E^I are Polish spaces. Define a continuous mapping

$$h : \prod\{Q_i : i \in I\} \to E^I$$

by putting

$$h(q) = \{h_i(q_i) : i \in I\} \qquad (q = \{q_i : i \in I\} \in \prod\{Q_i : i \in I\}).$$

Obviously, h is injective and continuous. Further, denote

$$\Delta = \{z \in E^I : (\forall i \in I)(\forall j \in I)(\mathrm{pr}_i(z) = \mathrm{pr}_j(z))\}$$

and observe that Δ is a closed set in E^I. Now, put

$$Q = h^{-1}(\Delta).$$

Since Q is a closed subset of $\prod\{Q_i : i \in I\}$, we see that Q is a Polish space as well. Additionally, the restricted mapping

$$h|Q : Q \to (\prod\{Y_i : i \in I\}) \cap \Delta$$

is a continuous bijection. Taking into account the simple fact that the set $\prod\{Y_i : i \in I\} \cap \Delta$ is homeomorphic to the set $\cap\{Y_i : i \in I\}$, we infer that $\cap\{Y_i : i \in I\}$ belongs to the class $\mathcal{L}$.

Finally, Lemma 2 and all the facts above enable us to conclude that $\mathcal{L}$ contains the Borel σ-algebra $\mathcal{B}(E)$, which completes the proof.

In connection with Theorem 3, see also Exercise 9 of this appendix.

The following statement (due to Alexandrov and Hausdorff) is an important consequence of Theorem 3.

Theorem 4. *Any uncountable Borel subset of a Polish space contains a topological copy of Cantor's discontinuum and, therefore, has cardinality* **c**.

Proof. Let E be a Polish space and let B be an uncountable Borel set in E. By virtue of Theorem 3, there exist a Polish space P and a bijective continuous mapping

$$f : P \to B.$$

The space P contains a topological copy C of Cantor's space $\{0,1\}^\omega$ (cf. Exercise 6 from this appendix). Since C is compact, the restriction $f|C$ is a homeomorphism between C and $f(C)$, which yields the required result.

It is a remarkable fact that the analogue of Theorem 4 is valid for all uncountable Suslin subsets of Polish spaces. The proof of this analogue needs a slightly different argument (see Exercise 5 from Chapter 20).

We also have the following profound, important and useful for various applications statement (see again [115], [152], [167], [191], [264]).

Theorem 5. *Let E_1 and E_2 be two Polish spaces, X be a Borel subset of E_1, and let a mapping $g : X \to E_2$ be injective and continuous. Then $g(X)$ is a Borel subset of E_2.*

Proof. In view of Exercises 8 and 9, it suffices to show that for every Polish space E and for any injective continuous mapping $g : \omega^\omega \to E$, the set $\mathrm{ran}(g)$ is Borel in E. In order to establish this fact, fix a natural number k and consider the corresponding disjoint countable family of Suslin sets

$$g(U(s)) \quad (s \in \omega^k).$$

It follows from the generalized separation principle (see Exercise 16 in Chapter 20) that there are pairwise disjoint Borel sets Y_s ($s \in \omega^k$) in E for which the relations

$$g(U(s)) \subset Y_s \quad (s \in \omega^k)$$

are satisfied. Now, we define by recursion the following Borel sets:

$$Y_n^* = Y_n \cap \mathrm{cl}(g(U(n))) \quad (n < \omega),$$

$$Y_{s*n}^* = Y_{s*n} \cap \mathrm{cl}(g(U(s*n))) \cap Y_s^* \quad (s \in \omega^{<\omega}, \ n < \omega).$$

Using the method of induction on $\mathrm{lh}(s)$, it is not difficult to check the inclusions

$$Y_s^* \subset Y_s, \quad g(U(s)) \subset Y_s^* \subset \mathrm{cl}(g(U(s))) \quad (s \in \omega^{<\omega}).$$

Consequently, for any $z \in \omega^\omega$, we get the equality

$$\{g(z)\} = \cap \{Y^*_{z|k} : k < \omega\},$$

which directly implies that

$$g(\omega^\omega) = \bigcup_{z \in \omega^\omega} \left(\bigcap_{k < \omega} Y^*_{z|k} \right).$$

Keeping in mind the circumstance that for each $k < \omega$, the family $\{Y^*_s : s \in \omega^k\}$ of Borel sets in E is disjoint, we come to the relation

$$g(\omega^\omega) = \bigcap_{k < \omega} \left(\bigcup_{s \in \omega^{<\omega}, \; \mathrm{lh}(s)=k} Y^*_s \right),$$

from which it follows that $g(\omega^\omega)$ is a Borel subset of E.

Theorem 5 has thus been proved.

Theorems 3 and 5 together allow to deduce that the Borel subsets of Polish topological spaces coincide with the bijective continuous images of G_δ-subsets of the canonical Baire space ω^ω (cf. Exercise 9). In addition to this result, it directly follows from Theorem 5 that if X and Y are two Borel subsets of Polish spaces and

$$f : X \to Y$$

is a continuous bijection, then $f^{-1} : Y \to X$ is a Borel bijection and, consequently, f turns out to be a Borel isomorphism between X and Y.

The assertion of Theorem 5 remains valid for an arbitrary injective Borel mapping $g : X \to E_2$ (see Exercise 10). In this context, it should also be mentioned that a much stronger result holds true. Namely, let E and E' be two Polish spaces and let $f : E \to E'$ be a partial Borel mapping defined on a Borel subset of E and satisfying the inequality

$$\mathrm{card}(f^{-1}(y)) \leq \omega$$

for every point $y \in E'$. Then, for any Borel set $B \subset E$, the image $f(B)$ is a Borel subset of E'. Moreover, the set $\mathrm{dom}(f)$ admits a representation

$$\mathrm{dom}(f) = \cup\{X_n : n < \omega\},$$

where all X_n $(n < \omega)$ are pairwise disjoint Borel subsets of E and for each index $n < \omega$, the restriction $f|X_n$ is a Borel isomorphism between X_n and $f(X_n)$ (for more details, see, e.g., [115], [152], [167], [191]).

In descriptive set theory, general topology, the theory of stochastic processes, and other fields of modern mathematics an important role is played by various statements on the existence of selectors with nice structural properties. Here we would like to touch upon only two statements of this type.

First, let us recall that the pair $(X, \mathcal{S})$ is an abstract measurable space if X is a nonempty base (ground) set and $\mathcal{S}$ is some σ-algebra of subsets of X.

For instance, if E is an arbitrary nonempty topological space, then the pair $(E, \mathcal{B}(E))$ is a canonical example of a measurable space.

Let $(X, \mathcal{S})$ be a measurable space, Y be a topological space, and let

$$F : X \rightarrow \mathcal{P}(Y)$$

be a set-valued mapping such that $F(x)$ is a nonempty closed subset of Y for every point $x \in X$. In the sequel only those set-valued functions F will be considered, which satisfy this condition.

We say that F is a weakly measurable set-valued mapping if, for any open subset U of Y, we have $\{x \in X : F(x) \cap U \neq \emptyset\} \in \mathcal{S}$.

It is not difficult to check the validity of the following auxiliary proposition.

Lemma 3. *Let (Y, d) be a separable metric space. For a set-valued mapping $F : X \rightarrow \mathcal{P}(Y)$, these two assertions are valid:*

(1) F is weakly measurable if and only if, for each point y from some everywhere dense subset of Y, the function

$$g_y : X \rightarrow \mathbf{R}$$

defined by the formula

$$g_y(x) = d(F(x), y) \qquad (x \in X)$$

is measurable with respect to the σ-algebras $\mathcal{B}(Y)$ and $\mathcal{S}$;

(2) if F is weakly measurable, then the graph of F, i.e., the set

$$\{(x, y) \in X \times Y : y \in F(x)\},$$

is a measurable subset of the product space $(X \times Y, \mathcal{S} \otimes \mathcal{B}(Y))$.

The proof of Lemma 3 is left to the reader.

Let f be a mapping acting from X into Y. We recall that f is a selector of a given set-valued mapping $F : X \rightarrow \mathcal{P}(Y)$ if for any element $x \in X$, we have $f(x) \in F(x)$. In view of the relation $(\forall x \in X)(F(x) \neq \emptyset)$, the Axiom of Choice immediately yields the existence of a selector of F. But if we want to get a measurable selector, then we must use a more delicate additional argument. For our further purposes, we also need the next simple auxiliary proposition.

Lemma 4. *Let (X, d) be an arbitrary metric space. Then there exist a Banach space $(E, || \cdot ||)$ and an isometric embedding*

$$\phi : (X, d) \rightarrow (E, || \cdot ||).$$

In particular, if (X,d) is a complete metric space, then the image $\phi(X)$ is a closed subset of the space E. Moreover, the space E can be chosen in such a way that

$$\mathrm{w}(E) \leq \mathrm{w}(X) + \omega,$$

where $\mathrm{w}(E)$ (respectively, $\mathrm{w}(X)$) denotes the topological weight of E (respectively, of X).

Proof. Without loss of generality, we may assume that X is a nonempty space. Let t be a fixed element of X. Further, for each element $y \in X$, let us consider a mapping $f_y : X \to \mathbf{R}$ defined by the formula

$$f_y(x) = d(x,y) - d(x,t) \quad (x \in X).$$

It is easy to see that

$$|f_y(x)| \leq d(y,t) \quad (x \in X).$$

Hence the function f_y is bounded. Now, we can consider the family E of all bounded functions f acting from X into $\mathbf{R}$. Equip E with the standard norm

$$||f|| = \sup_{x \in X} |f(x)|.$$

Obviously, E is a Banach space with respect to this norm. Moreover, it is not difficult to check that $||f_y - f_z|| = d(y,z)$ for any two elements y and z from X. Consequently, a mapping $\phi : X \to E$ defined by the formula

$$\phi(y) = f_y \quad (y \in X)$$

is an isometric embedding of X into E. A priori, the topological weight of E may be strictly greater than the topological weight of X. But if we take the closed vector subspace of E generated by the set $\phi(X)$, then the weight of this subspace does not exceed $\mathrm{w}(X) + \omega$. This finishes the proof of Lemma 4.

Of course, various important statements are known in the literature, which are concerned with embeddings of metric spaces and are much stronger than Lemma 4. However, for our further considerations, this lemma is completely sufficient.

Now, we can formulate and prove a famous theorem concerning the existence of measurable selectors. This important theorem was established by Kuratowski and Ryll-Nardzewski in [155].

Theorem 6. *Let $(X,\mathcal{S})$ be a measurable space, let Y be a Polish space and let $F : X \to \mathcal{P}(Y)$ be a set-valued mapping satisfying the following two conditions:*

(i) for each element $x \in X$, the set $F(x)$ is nonempty and closed in Y;
(ii) F is weakly measurable.

Then there exists a mapping $f : X \to Y$ such that:
(a) f is a selector of F;
(b) f is a measurable mapping acting from $(X, \mathcal{S})$ into $(Y, \mathcal{B}(Y))$.

Proof. Taking into account Lemma 4, we may assume without loss of generality that Y is a separable Banach space. Let d be a complete metric on Y agreed with the norm of Y; in other words, for any two points y and z from Y, we put

$$d(y, z) = ||y - z||.$$

Furthermore, denote by $\{y_k : 1 \leq k < \omega\}$ a countable family of points of Y which is everywhere dense in Y. Using ordinary recursion, we shall construct a sequence $\{f_n : 1 \leq n < \omega\}$ of mappings acting from X into Y and satisfying the following relations:

(1) $\mathrm{ran}(f_n) \subset \{y_k : 1 \leq k < \omega\}$ for any natural number $n \geq 1$;
(2) $d(f_n(x), f_{n-1}(x)) < 1/2^{n-1}$ for each natural number $n \geq 2$ and for each element $x \in X$;
(3) $d(F(x), f_n(x)) < 1/2^{n-1}$ for any natural number $n \geq 1$ and for any element $x \in X$;
(4) f_n is a measurable mapping acting from $(X, \mathcal{S})$ into $(Y, \mathcal{B}(Y))$ for any natural number $n \geq 1$.

Let x be an arbitrary point of the space X. Then we put $f_1(x) = y_k$ where k is the smallest natural number for which $d(F(x), y_k) < 1$. Evidently, we obtain a certain mapping

$$f_1 : X \to Y.$$

Since F is a weakly measurable set-valued mapping, it is easy to check that f_1 is a measurable mapping acting from $(X, \mathcal{S})$ into $(Y, \mathcal{B}(Y))$ (cf. Lemma 3).
Suppose now that a partial finite sequence of functions

$$\{f_1, f_2, \ldots, f_n\}$$

satisfying the above relations (1)–(4) has already been defined. Fix again an arbitrary element $x \in X$ and consider the following two sets:

$$V = \{y \in Y : d(y, f_n(x)) < 1/2^n\},$$

$$W = \{y \in Y : d(y, F(x)) < 1/2^n\}.$$

Clearly, V and W are open subsets of the space Y. Moreover, since

$$d(F(x), f_n(x)) < 1/2^{n-1},$$

there exists a point $z \in F(x)$ such that $d(z, f_n(x)) < 1/2^{n-1}$. Now, it can easily be seen that the point

$$y = (z + f_n(x))/2$$

belongs to the set $V \cap W$. Hence $V \cap W$ is a nonempty open subset of Y. Let k be the smallest natural number for which $y_k \in V \cap W$. Let us put

$$f_{n+1}(x) = y_k.$$

Taking into account the fact that x is an arbitrary element of X, we have certain mapping

$$f_{n+1} : X \to Y.$$

Since our set-valued mapping F is weakly measurable, it is not difficult to check that f_{n+1} is a measurable mapping acting from $(X, \mathcal{S})$ into $(Y, \mathcal{B}(Y))$. Also, it is not hard to show that all relations (1)–(4) remain true for the partial sequence

$$\{f_1, \ f_2, \ \dots, \ f_n, \ f_{n+1}\}.$$

Proceeding in such a manner, we are able to construct the required infinite sequence of functions

$$\{f_1, \ f_2, \ \dots, \ f_n, \ \dots \}.$$

Finally, we put

$$f(x) = \lim_{n \to +\infty} f_n(x) \qquad (x \in X).$$

Notice that the function f is well-defined because, in view of relation (2), the sequence of functions $\{f_n : 1 \leq n < \omega\}$ uniformly converges on X.

Thus, we have the mapping f acting from X into Y. Now, using relation (3), we can readily conclude that f is a measurable selector of the original set-valued mapping F.

The proof of Theorem 6 is completed.

Remark 1. The assumption that Y is a Polish topological space is essential in the formulation of the theorem of Kuratowski and Ryll–Nardzewski. However, in some particular cases, the assertion of this theorem remains true if we replace Y by a nonseparable complete metric space Y'. Moreover, sometimes we do not need even the completeness of the space Y', for example, in those cases where all sets $F(x)$ are compact and nonempty (cf. [31]).

Remark 2. Various situations closely connected with the theorem of Kuratowski and Ryll-Nardzewski are discussed in [63] where it is shown that certain additional set-theoretical axioms are necessary for appropriate generalizations of this theorem.

Some applications of Theorem 6 may be found in [115], [153], and [155] (in this connection, see also [31] and [63]). Here we would like to apply this theorem for deducing one classical result concerning the uniformization of an analytic set lying in the product of two Polish topological spaces. The above mentioned result is due to Luzin, Jankov, and von Neumann (cf. [115], [167], [264]).

Theorem 7. *Let E_1 and E_2 be two Polish spaces, let A be an analytic subset of the product space $E_1 \times E_2$ and let S be the σ-algebra of subsets of $\mathrm{pr}_1(A)$, generated by the family of all analytic sets in $\mathrm{pr}_1(A)$. Then there exists a mapping*

$$f : \mathrm{pr}_1(A) \to E_2$$

satisfying the following relations:
(1) the graph $\mathrm{Gr}(f)$ of f is contained in the set A;
(2) f is a measurable mapping acting from $(\mathrm{pr}_1(A), S)$ into $(E_2, \mathcal{B}(E_2))$.
In particular, f is measurable with respect to the completion of an arbitrary σ-finite Borel measure given on the Suslin space $\mathrm{pr}_1(A)$ (or, equivalently, on the space E_1).

Proof. We may assume that $A \neq \emptyset$. Since A is an analytic set, there exists a continuous function

$$g : \omega^\omega \to E_1 \times E_2$$

such that $g(\omega^\omega) = A$ (see Theorem 2 of this appendix).
Denote by G the graph of g, i.e., put

$$G = \{(t, x, y) \in \omega^\omega \times E_1 \times E_2 : g(t) = (x, y)\}.$$

Obviously, G is a closed subset of the product space $\omega^\omega \times E_1 \times E_2$. Now, we define a set-valued mapping

$$\Phi : \mathrm{pr}_1(A) \to \mathcal{P}(\omega^\omega \times E_1 \times E_2)$$

by the formula

$$\Phi(x) = G \cap (\omega^\omega \times \{x\} \times E_2) \qquad (x \in \mathrm{pr}_1(A)).$$

It is not difficult to check that Φ satisfies the assumptions of Theorem 6. According to that theorem, there exists a selector

$$\phi : \mathrm{pr}_1(A) \to \omega^\omega \times E_1 \times E_2$$

of Φ which is measurable with respect to the σ-algebras $\mathcal{B}(\omega^\omega \times E_1 \times E_2)$ and S. It remains to put

$$f = \mathrm{pr}_3 \circ \phi$$

and to verify that f is as required. The corresponding easy details are left to the reader.

So far, we were concerned with basic properties of Borel and analytic subsets of a Polish space. As was mentioned earlier, these classes of sets are important from the point of view of numerous applications. However, descriptive set theory also deals with more general classes of sets, which are of paramount interest from the viewpoint of logical foundations of contemporary mathematics. For example, one such class is formed by the so-called projective subsets of the real line, which were introduced and thoroughly studied by Luzin and Sierpiński beginning from 1925. Here we do not intend to discuss very interesting and deep properties of projective sets. Notice that the most natural structural properties of these sets (e.g., the Lebesgue measurability or the Baire property) turn out to be independent of **ZFC** theory. However, they become provable under the assumption of the existence of large cardinals of certain types.

Below, we only present the classical definition of projective sets and a few results about them.

Let E be an arbitrary Polish space. We define the classes of sets

$$\mathcal{P}r_0(E), \ \mathcal{P}r_1(E), \ \ldots, \ \mathcal{P}r_n(E), \ \ldots$$

by ordinary recursion (see, e.g., [152], [154]). First, let us put

$$\mathcal{P}r_0(E) = \mathcal{B}(E).$$

Suppose now that for a natural number $n \geq 1$, the classes $\mathcal{P}r_k(E)$, where $k < n$, have already been determined.

If n is odd, then we define $\mathcal{P}r_n(E)$ as the class of all continuous images (in E) of the sets from $\mathcal{P}r_{n-1}(E)$.

If n is even, then we define $\mathcal{P}r_n(E)$ as the class of all complements of the sets from $\mathcal{P}r_{n-1}(E)$. Finally, we put

$$\mathcal{P}r(E) = \cup\{\mathcal{P}r_n(E) : n < \omega\}.$$

By definition, any member of $\mathcal{P}r(E)$ is called a projective subset of E. We thus see that

$$\mathcal{P}r_0(E) = \mathcal{B}(E), \quad \mathcal{P}r_1(E) = \mathcal{A}(E),$$

i.e., the families of Borel and analytic sets are only the first two steps in the Luzin-Sierpiński projective hierarchy.

It should be mentioned that nowadays a different notation for the classes of projective sets is usually exploited. Namely, for $n \geq 1$, the classes $\mathcal{P}r_{2n}(E)$ are denoted by $\Pi_n^1(E)$ while the classes $\mathcal{P}r_{2n-1}(E)$ are denoted by $\Sigma_n^1(E)$. Notice also that all the above classes are preserved under Borel isomorphisms (consequently, under homeomorphisms) between Polish topological spaces (see [115], [152]).

ZFC theory is not strong enough to establish fundamental properties of projective sets, such as measurability in the Lebesgue sense, Baire property, and perfect subset property. The progress in this direction can be achieved by assuming some additional axioms. For instance, according to Solovay's result, the existence of a two-valued measurable cardinal implies the Lebesgue measurability of any Σ_2^1-subset of the real line (for more details, see [103]).

Another approach to various properties of projective sets is based on a certain variant of the Axiom of Determinacy ($\mathbf{AD}$). In order to introduce $\mathbf{AD}$, we need several auxiliary definitions.

An infinite game of two persons I (he) and II (she) is a pair of the form

$$G_Z = (\omega^\omega, Z),$$

where $Z \subset \omega^\omega$. Any sequence

$$(z_0, z_1, ..., z_k, ...) = (x_0, y_0, x_1, y_1, ..., x_n, y_n, ...) \in \omega^\omega$$

may be considered as a concrete realization of G_Z (called also a play corresponding to G_Z).

The partial sequence $(x_0, x_1, ..., x_n, ...)$ is regarded as the moves of I and the partial sequence $(y_0, y_1, ..., y_n, ...)$ is treated as the moves of II.

We say that a partial function $\sigma : \omega^{<\omega} \to \omega$ is a strategy for I (in the game G_Z) if $\mathrm{dom}(\sigma)$ consists of all those sequences from $\omega^{<\omega}$ whose lengths are even natural numbers.

Analogously, we say that a partial function $\tau : \omega^{<\omega} \to \omega$ is a strategy for II (in the same game) if $\mathrm{dom}(\tau)$ consists of all those sequences from $\omega^{<\omega}$ whose lengths are odd natural numbers.

The player I follows strategy σ in $(x_0, y_0, x_1, y_1, ..., x_n, y_n, ...)$ if

$$\sigma(\emptyset) = x_0, \quad \sigma(x_0, y_0) = x_1, \quad \sigma(x_0, y_0, x_1, y_1) = x_2, \quad \ldots .$$

The player II follows strategy τ in $(x_0, y_0, x_1, y_1, ..., x_n, y_n, ...)$ if

$$\tau(x_0) = y_0, \quad \tau(x_0, y_0, x_1) = y_1, \quad \tau(x_0, y_0, x_1, y_1, x_2) = y_2, \quad \ldots .$$

If $(x_0, y_0, x_1, y_1, ..., x_n, y_n, ...) \in Z$, then I is declared a winner in the game G_Z, otherwise II is declared a winner in the same game.

A strategy σ is called a winning strategy for I (in the game G_Z) if I turns out to be a winner whenever he follows σ.

Analogously, a strategy τ is called a winning strategy for II if II turns out to be a winner whenever he follows τ.

We say that a game G_Z is determined if either I has a winning strategy or II has a winning strategy.

Now, we are able to formulate the Axiom of Determinacy which was first considered by Mycielski and Steinhaus in [196]:

($\mathbf{AD}$) For each set $Z \subset \omega^\omega$, the game G_Z is determined.

It can readily be shown that, within the framework of $\mathbf{ZF}$ theory, $\mathbf{AD}$ implies $\mathbf{CC(R)}$ (see Exercise 19 from this appendix). Also, it was proved that $\mathbf{AD}$ implies the nonexistence of cardinal numbers strictly between ω and $\mathbf{c}$, so the Continuum Hypothesis is trivially solved by $\mathbf{AD}$.

Besides, by assuming the existence of some large cardinals, it was demon strated that the theory

ZF & DC & AD

is consistent (cf. also [114]). However, **AD** contradicts the Axiom of Choice, so **AD** cannot be added to **ZFC** set theory as a new axiom. Moreover, even the existence of a well-ordering of **R** implies the negation of **AD** (see Exercise 20).

The Axiom of Projective Determinacy is a restricted form of **AD** and is formulated as follows:

(**PD**) For each projective set $Z \subset \omega^\omega$, the game G_Z is determined.

This axiom is acceptable in the sense that, by assuming again the existence of certain large cardinals, it can be shown that the theory

ZFC & PD

is consistent (see [182]). Another essential advantage of **PD** is that it has nice consequences in classical descriptive set theory. Indeed, it was deduced from **PD** that any projective set is Lebesgue measurable (moreover, is absolutely measurable with respect to the class of completions of all σ-finite Borel measures on ω^ω), possesses the Baire property and either is countable or contains a nonempty perfect subset. Also, under **PD**, every projective subset of the product space $\omega^\omega \times \omega^\omega$ admits a uniformization by a projective subset (for more details, see [10], [103], [115], [191]).

Various important connections between the theory of infinite games and theory of large cardinals are discussed in [110], [115], [191]. Some interesting facts are concerned with the so-called topological games (see, e.g., [260]), the prototype of which is the famous Banach–Mazur game (see [203]) whose determinacy is closely connected with the Baire property of subsets of **R**.

EXERCISES

1. Give a detailed proof of Lemmas 1 and 2.

Actually, deduce Lemma 1 from the more general fact stating that if $\mathcal{R}$ is an algebra of subsets of a base set E, then the σ-algebra generated by $\mathcal{R}$ coincides with the monotone class generated by $\mathcal{R}$.

Further, let $\mathcal{L}$ be a family of subsets of E such that:

(a) $\mathcal{R} \subset \mathcal{L}$;

(b) $\mathcal{L}$ is closed under the unions of all countable disjoint families of its members;

(c) $\mathcal{L}$ is closed under the intersections of all countable families of its members.

Verify that $\mathcal{L}$ contains the σ-algebra generated by $\mathcal{R}$, so Lemma 2 is valid, too.

2. Let ξ be an arbitrary at most countable ordinal number.

Prove that the Baire class $Ba_\xi(E, \mathbf{R})$ is closed with respect to the uniform convergence of sequences of its elements.

3*. Let E be an arbitrary topological space.
By using the method of transfinite induction, prove that:
(a) $Ba(E, \mathbf{R}) \subset B(E, \mathbf{R})$;
(b) if E is perfectly normal, then $Ba(E, \mathbf{R}) = B(E, \mathbf{R})$.
To demonstrate the validity of (b), argue in the following manner. First, reduce the argument to the case of the characteristic function of any Borel subset X of E, i.e., prove that the function χ_X belongs to $Ba(E, \mathbf{R})$. For this purpose, use the Tietze-Urysohn theorem, Lemma 1 and transfinite induction over the order of X.

4. Let E and E' be two Polish topological spaces and let

$$f : E \to E'$$

be a Borel mapping.
Show that, for any Suslin set $X \subset E$, its image $f(X)$ is a Suslin subset of E'.

5*. Prove that any open subset of a complete metric space is metrizable by a complete metric (in particular, any open subset of a Polish space is also Polish).
Argue as follows. Let (E, d) be a complete metric space and let U be an open subset of E such that $U \neq E$. Consider a mapping

$$g : U \to E \times \mathbf{R}$$

defined by the formula

$$g(u) = (u, 1/d(u, E \setminus U)) \quad (u \in U)$$

and check that g is a homeomorphism between U and the closed subset $g(U)$ of the product space $E \times \mathbf{R}$.
Infer from the above result Alexandrov's classical theorem stating that any G_δ-set in a complete metric space is metrizable by a complete metric (in particular, any G_δ-subset of a Polish space is also Polish).
For this purpose, argue similarly to the proof of Theorem 3.

6. Let E be a nonempty complete metric space without isolated points.
Show that there exists a subset X of E which is homeomorphic to the Cantor space $\{0, 1\}^\omega$.
For this purpose, construct a dyadic system of closed balls in E which produces the required X.
Conclude from the above result that any uncountable Polish space contains a topological copy of the Cantor space $\{0, 1\}^\omega$.

7. Let E be a zero-dimensional Polish topological space and let Z be a subset of E such that:

(a) Z is of type G_δ;

(b) Z is everywhere dense in E;

(c) $E \setminus Z$ is everywhere dense in E.

Prove that Z is homeomorphic to the canonical Baire space ω^ω.

For this purpose, show that Z can be expressed as the result of A-operation applied to some regular system $\{F_s : s \in \omega^{<\omega}\}$ of nonempty closed-open sets in E satisfying the relations:

(i) $\lim_{\mathrm{lh}(s)\to+\infty}\mathrm{diam}(F_s) = 0$;

(ii) $F_s \cap F_t = \emptyset$ for any two distinct sequences $s \in \omega^{<\omega}$ and $t \in \omega^{<\omega}$ such that $\mathrm{lh}(s) = \mathrm{lh}(t)$.

8. Prove that an arbitrary uncountable zero-dimensional Polish topological space E can be represented in the form

$$E = Z \cup D \quad (Z \cap D = \emptyset),$$

where Z is homeomorphic to ω^ω and D is at most countable.

In order to show this fact, denote by P the set of all condensation points of E and observe that P is an uncountable perfect subset of E whose complement is countable. Let D_0 be a countable everywhere dense subset of P. Put

$$Z = P \setminus D_0, \quad D = (E \setminus P) \cup D_0$$

and, keeping in mind the result of Exercise 7, check that Z and D yield the required decomposition of E.

9. Demonstrate that any Borel subset of a Polish space can be regarded as a bijective continuous image of some Polish subspace of the Baire space ω^ω.

For this purpose, take into account the existence of a canonical continuous bijection acting from the space ω^ω onto the interval $]0, 1]$, which implies the existence of a continuous bijection of $\omega^\omega = \omega^{\omega \times \omega}$ onto the product space $]0, 1]^\omega$. Further, use the fact that $]0, 1]^\omega$ contains topological copies of all Polish spaces and apply Theorem 3.

10. Let E_1 and E_2 be two Polish spaces, X be a Borel subset of E_1 and let

$$g : X \to E_2$$

be an injective Borel mapping.

Prove that $g(X)$ is a Borel subset of E_2.

Argue in the following manner. In the product set $X \times E_2$ consider the graph of g, i.e., consider the set

$$G = \{(x, y) \in X \times E_2 : g(x) = y\}.$$

First, show that G is a Borel subset of $X \times E_2$. For this purpose, denote by d any metric on E_2 compatible with the topology of E_2 and introduce a function

$$\phi : X \times E_2 \to [0, +\infty[$$

defined by the formula

$$\phi(x, y) = d(g(x), y) \qquad (x \in X, y \in E_2).$$

Check that ϕ is a Borel mapping and $G = \phi^{-1}(0)$, which yields that G is Borel. Finally, consider the injective continuous mapping

$$\mathrm{pr}_2|G : G \to E_2$$

and apply to it Theorem 5 of this appendix.

11*. Let X and X' be any two uncountable Borel subsets of Polish spaces E and E' respectively.

Demonstrate that there exists a Borel isomorphism

$$f : X \to X'.$$

For this purpose, first show that the Cantor space $\{0, 1\}^\omega$ is Borel isomorphic to the Hilbert cube $[0, 1]^\omega$. Then, applying the topological universality of $[0, 1]^\omega$ in the class of all separable metrizable spaces, Theorem 4 from this appendix and the Cantor–Bernstein theorem from Appendix 1, obtain the required result.

12*. In this exercise, the extended real line

$$[-\infty, +\infty] = \{-\infty\} \cup \mathbf{R} \cup \{+\infty\}$$

is assumed to be equipped with the standard linear ordering and topology. So $[-\infty, +\infty]$ becomes isomorphic to the closed interval $[-1, 1]$.

Let E be an uncountable Polish topological space.

Prove the classical Lebesgue theorem stating that, for any ordinal $\xi < \omega_1$, there exists a mapping

$$\Phi_\xi : E \times [0, 1] \to [-\infty, +\infty]$$

satisfying the following two conditions:

(i) Φ_ξ is Borel;

(ii) for each function $f \in Ba_\xi(E, \mathbf{R})$, there is a point t from $[0, 1]$ such that $\Phi_\xi(\cdot, t) = f$.

In order to demonstrate this important statement, first observe that in view of Exercise 11, it suffices to reduce the argument to the case $E = [0, 1]$.

Consider any Peano type mapping

$$\phi = (\phi_k)_{k<\omega} : [0,1] \to [0,1]^\omega.$$

This phrase simply means that ϕ is a continuous surjection. Let $\{Q_k : k < \omega\}$ denote the sequence of all polynomials on $[0,1]$ with rational coefficients. Show that one can put

$$\Phi_0(x,t) = \limsup_{k\to+\infty} Q_k(x)\phi_k(t) \quad (x \in [0,1], t \in [0,1]).$$

If $\xi < \omega_1$ and the mapping Φ_ξ is already defined, then one can put

$$\Phi_{\xi+1}(x,t) = \limsup_{k\to+\infty} \Phi_\xi(x,\phi_k(t)) \quad (x \in [0,1], t \in [0,1]).$$

Finally, if $\xi < \omega_1$ is a limit ordinal, then one can take a strictly increasing sequence $\{\xi_k : k < \omega\}$ of ordinals such that $\lim_{k\to+\infty}\xi_k = \xi$ and put

$$\Phi_\xi(x,t) = \limsup_{k\to+\infty} \Phi_{\xi_k}(x,\phi_k(t)) \quad (x \in [0,1], t \in [0,1]).$$

This procedure allows one to define the required mapping Φ_ξ for any ordinal number $\xi < \omega_1$. By starting with the existence of Φ_ξ for each $\xi < \omega_1$, derive that

$$Ba_\xi([0,1], \mathbf{R}) \setminus \cup\{Ba_\zeta([0,1], \mathbf{R}) : \zeta < \xi\} \neq \emptyset$$

and that the same holds true for an arbitrary uncountable Polish space E instead of $[0,1]$.

Argue in the following manner. Suppose to the contrary that

$$Ba_\xi([0,1], \mathbf{R}) \setminus \cup\{Ba_\zeta([0,1], \mathbf{R}) : \zeta < \xi\} = \emptyset$$

and deduce from this assumption that

$$Ba([0,1], \mathbf{R}) = Ba_\xi([0,1], \mathbf{R}).$$

Then, by using the mapping Φ_ξ, construct a Borel mapping

$$\Phi : [0,1] \times [0,1] \to [0,1]$$

such that for any Borel function $f : [0,1] \to [0,1]$, there exists a point $t \in [0,1]$ for which $\Phi(\cdot,t) = f$.

Further, define a Borel mapping Ψ by the formula

$$\Psi(x,t) = \lim_{n\to+\infty} \frac{n\Phi(x,t)}{1 + n\Phi(x,t)} \quad (x \in [0,1], t \in [0,1])$$

and check that:
(a) $\mathrm{ran}(\Psi) = \{0,1\}$;

(b) for any Borel function $g : [0, 1] \rightarrow \{0, 1\}$, there exists a point $t \in [0, 1]$ such that $\Psi(\cdot, t) = g$.

Finally, put

$$h(x) = 1 - \Psi(x, x) \quad (x \in [0, 1])$$

and verify that:

(c) the function h is Borel;

(d) $\mathrm{ran}(h) \subset \{0, 1\}$.

Now, keeping in mind relation (b), pick a point $t \in [0, 1]$ for which $h = \Psi(\cdot, t)$ and infer the equalities

$$h(t) = 1 - \Psi(t, t) = \Psi(t, t), \quad \Psi(t, t) = 1/2,$$

contradicting relation (a).

Thus, the obtained contradiction yields the existence of Baire functions of any order $\xi < \omega_1$ and, consequently, the existence of Borel sets in $[0, 1]$ of the same order ξ.

13*. Let E be a base set, $\mathcal{L}$ be a class of subsets of E and let $\{X_s : s \in \omega^{<\omega}\}$ be a countable system of sets from $\mathcal{L}$.

By using the method of transfinite recursion on $\xi < \omega_1$, define the following sets:

(a) $X_s^0 = X_s$ for all $s \in \omega^{<\omega}$;

(b) $X_s^{\xi+1} = X_s^\xi \cap (\cup\{X_{s*n}^\xi : n < \omega\})$ for any $\xi < \omega_1$ and for all $s \in \omega^{<\omega}$;

(c) $X_s^\xi = \cap\{X_s^\zeta : \zeta < \xi\}$ for a limit ordinal $\xi < \omega$ and for each $s \in \omega^{<\omega}$.

Check (by using transfinite induction) that if $\zeta \leq \xi < \omega_1$, then

$$X_s^\xi \subset X_s^\zeta \quad (s \in \omega^{<\omega}).$$

Further, for any $\xi < \omega_1$, put:

$$Y_\xi = \cup\{X_n^\xi : n < \omega\};$$

$$T_\xi = \cup\{(X_s^\xi \setminus X_s^{\xi+1}) : s \in \omega^{<\omega}\};$$

$$Z_\xi = Y_\xi \setminus T_\xi.$$

Verify that all the sets

$$X_s^\xi, \ Y_\xi, \ T_\xi, \ Z_\xi \quad (\xi < \omega_1, s \in \omega^{<\omega})$$

belong to the σ-ring generated by the given class $\mathcal{L}$.

Prove the following important equalities due to Sierpiński:

$$A(\{X_s : s \in \omega^{<\omega}\}) = \cap\{Y_\xi : \xi < \omega_1\} = \cup\{Z_\xi : \xi < \omega_1\}.$$

In order to establish these equalities, argue in the following manner. First, show the validity of the inclusion

(i) $\cup\{Z_\xi : \xi < \omega_1\} \subset A(\{X_s : s \in \omega < \omega\})$.

Take an arbitrary $z \in \cup\{Z_\xi : \xi < \omega_1\}$. Then for some ordinal $\xi < \omega_1$, one must have $z \in Y_\xi$ and $z \notin T_\xi$. Consequently, for some natural number n_0, the relations

$$z \in X_{n_0}^\xi, \quad z \notin X_{n_0}^\xi \setminus X_{n_0}^{\xi+1}$$

are satisfied, whence it follows that $z \in X_{n_0}^{\xi+1}$. Keeping in mind the inclusion

$$X_{n_0}^{\xi+1} \subset \cup\{X_{n_0 n}^\xi : n < \omega\},$$

obtain that $z \in X_{n_0 n_1}^\xi$ for some natural number n_1. Continuing this process by ordinary recursion, construct an infinite sequence

$$\phi = (n_0, n_1, n_2, ...) \in \omega^\omega$$

such that

$$z \in \cap\{X_{\phi|k}^\xi : k < \omega\} \subset \cap\{X_{\phi|k} : k < \omega\}$$

and, therefore, $z \in A(\{X_s : s \in \omega^{<\omega}\})$. This result obviously yields (i).

Now, check the validity of the inclusion

(ii) $A(\{X_s : s \in \omega^{<\omega}\}) \subset \cap\{Y_\xi : \xi < \omega_1\}$.

For this purpose, show that if $\xi < \omega_1$, $r < \omega$ and $\phi \in \omega^\omega$, then the relation

$$\cap\{X_{\phi|k} : k < \omega\} \subset X_{\phi|r}^\xi$$

holds true (use the method of transfinite induction on $\xi < \omega_1$). Consequently,

$$\cap\{X_{\phi|k} : k < \omega\} \subset X_{\phi(0)}^\xi \subset \cup\{X_n^\xi : n < \omega\} = Y_\xi,$$

which immediately yields (ii).

Finally, verify the validity of the relation

(iii) $\cap\{T_\xi : \xi < \omega_1\} = \emptyset$.

To show this fact, suppose otherwise, i.e., suppose that there exists an element $x \in \cap\{T_\xi : \xi < \omega_1\}$. Then for any ordinal $\xi < \omega_1$, one can find a sequence $s = s(\xi) \in \omega^{<\omega}$ such that

$$x \in X_s^\xi \setminus X_s^{\xi+1}.$$

Since $\omega^{<\omega}$ is countable and ω_1 is uncountable, there are two distinct ordinals $\xi < \omega_1$ and $\zeta < \omega_1$ for which $s = s(\xi) = s(\zeta)$. One may assume, without loss of generality, that $\xi < \zeta$. But in this case the relation $x \in X_s^\zeta \subset X_s^{\xi+1}$ must be true, thus a contradiction follows.

Taking into account the relations (i), (ii) and (iii) established above, derive the validity of Sierpiński's equalities.

Infer from these equalities that any analytic set A over the class $\mathcal{L}$ can be represented as the intersection of an ω_1-sequence of members from the σ-ring generated by $\mathcal{L}$, and A can also be represented as the union of an ω_1-sequence of members from the same σ-ring.

Conclude that if D is a co-Suslin subset of a Polish topological space, then

$$\mathrm{card}(D) \in \{0, 1, ..., n, ..., \omega, \omega_1, \mathbf{c}\}.$$

The same relation remains valid if D is any Borel image (contained in a Polish space) of a co-Suslin set.

14*. We preserve the notation of Exercise 13. Suppose that a base set E is equipped with a topology satisfying the Suslin condition.

Show that if all sets from $\{X_s : s \in \omega^{<\omega}\}$ have the Baire property, then the set

$$X = A(\{X_s : s \in \omega^{<\omega}\})$$

has the Baire property, too.

For this purpose, take $s \in \omega^{<\omega}$ and consider the sets

$$X_s^\xi \setminus X_s^{\xi+1} \quad (\xi < \omega_1),$$

which are pairwise disjoint. Assuming without loss of generality that E is a Baire space, derive that there exists an ordinal $\alpha(s) < \omega_1$ such that all sets $X_s^\xi \setminus X_s^{\xi+1}$ are of first category in E whenever $\alpha(s) \leq \xi < \omega_1$. Further, put

$$\alpha = \sup\{\alpha(s) : s \in \omega^{<\omega}\}$$

and conclude that the set T_α is of first category in E. Taking into account the inclusions

$$X \setminus Z_\alpha \subset Y_\alpha \setminus Z_\alpha \subset T_\alpha,$$

obtain the desired result.

Suppose now that E is equipped with a σ-finite complete measure μ. Show that if all sets from a family $\{X_s : s \in \omega^{<\omega}\}$ are μ-measurable, then the set $X = A(\{X_s : s \in \omega^{<\omega}\})$ is μ-measurable, too.

For this purpose, apply a method analogous to that described above.

Remark 3. In fact, the second part of Exercise 14 directly follows from its first part. In order to establish this implication, it suffices to consider a von Neumann topology associated with a given σ-finite complete measure μ (see [203]). In addition, the invariance of the Baire property under A-operation remains valid for an arbitrary topological space E (see, e.g., [152]).

15. Prove that there exists a nonempty perfect subset P of $\mathbf{R}$ which is linearly independent over the field $\mathbf{Q}$ of all rational numbers.

By using the Kuratowski–Zorn lemma, extend P to a Hamel basis of $\mathbf{R}$ and conclude that there are Hamel bases in $\mathbf{R}$ which contain nonempty perfect subsets.

Remark 4. In connection with the result of Exercise 15, it should be noticed that von Neumann was able to construct a nonempty perfect set of algebraically independent real numbers (see, for instance, [268]).

16. Let E and E' be two Polish topological spaces and let

$$f : E \to E'$$

be a Borel mapping such that

$$(\forall y \in E')(\mathrm{card}(f^{-1}(y)) \leq \omega).$$

Check that if $X \subset E$ is a universal measure zero set, then its image $f(X)$ is universal measure zero, too.

Taking into account the above-mentioned fact, prove that there exists an uncountable universal measure zero subset of $\mathbf{R}$ which is simultaneously a vector space over the field $\mathbf{Q}$.

For this purpose, start with the existence of a nonempty perfect set $P \subset \mathbf{R}$ linearly independent over $\mathbf{Q}$ and pick an uncountable universal measure zero set $X \subset P$. Denote by Y the linear hull of X (over the same $\mathbf{Q}$) and check that Y is a universal measure zero subset of $\mathbf{R}$ (cf. Remark 1 from Chapter 13).

17. Show that there exists a Borel isomorphism $f : [0,1] \to [0,1]$ which does not preserve the Baire property of subsets of $[0,1]$.

On the other hand, let E_1 and E_2 be two uncountable Polish spaces without isolated points. Prove that there exists a Borel isomorphism $g : E_1 \to E_2$ satisfying the following condition: a set $X \subset E_1$ is of first category in E_1 if and only if its image $g(X)$ is of first category in E_2.

Derive from this condition that both mappings g and g^{-1} preserve the Baire property.

In order to establish the existence of the required Borel isomorphism g, use Exercise 8 of this appendix.

18*. Let $X \subset \mathbf{R}$ be an analytic non-Borel set (such a set exists in view of Suslin's theorem). According to one of Sierpiński's equalities,

$$X = \cup\{Z_\xi : \xi < \omega_1\},$$

where all Z_ξ ($\xi < \omega_1$) are some Borel subsets of $\mathbf{R}$ (see Exercise 13). Verify that

$$Z_\xi \setminus \cup\{Z_\zeta : \zeta < \xi\} \neq \emptyset$$

for uncountably many ordinals $\xi < \omega_1$. By starting with this fact and taking into account the argument presented in Exercise 14, give a construction of an uncountable universal measure zero set in $\mathbf{R}$.

19. Prove, within $\mathbf{ZF}$ set theory, that $\mathbf{AD}$ implies $\mathbf{CC}(\mathbf{R})$.

Argue as follows. Assume **AD** and take any countable family $\{C_k : k < \omega\}$ of nonempty subsets of ω^ω. Associate with this family the game G_Z defined in the following manner:

II wins in a play $(x_0, y_0, x_1, y_1, ..., x_n, y_n, ...) \Leftrightarrow (y_0, y_1, ..., y_n, ...) \in C_{x_0}$.

Describe the set Z of this game and check that:

(a) if all sets C_k $(k < \omega)$ are Borel in the space ω^ω, then Z is also Borel in ω^ω;

(b) I does not have a winning strategy in G_Z.

Consequently, II must have a winning strategy τ in G_Z. Finally, for each $k < \omega$, apply the strategy τ to

$$(x_0, x_1, x_2, ..., x_n, ...) = (k, 0, 0, ..., 0, ...) \in \omega^\omega$$

and obtain the required result.

Remark 5. According to Martin's fundamental theorem [180], all games corresponding to Borel sets $Z \subset \omega^\omega$ are determined. If there exists a two-valued measurable cardinal, then all games corresponding to analytic sets $Z \subset \omega^\omega$ are also determined. Extensive information about infinite games and various forms of determinacy may be found in [10], [103], [115], [191], [260].

20*. Work in **ZF** set theory and, supposing that **R** (equivalently, ω^ω) is well-orderable, prove that **AD** is false.

For this purpose, consider two transfinite **c**-sequences

$$\{\sigma_\alpha : \alpha < \mathbf{c}\}, \quad \{\tau_\alpha : \alpha < \mathbf{c}\},$$

the first of which consists of all strategies for I and the second consists of all strategies for II. Define by transfinite recursion two **c**-sequences

$$\{z_\alpha : \alpha < \mathbf{c}\}, \quad \{z'_\alpha : \alpha < \mathbf{c}\}$$

of points in ω^ω as follows. Assume that, for $\alpha < \mathbf{c}$, the partial α-sequences

$$\{z_\beta : \beta < \alpha\}, \quad \{z'_\beta : \beta < \alpha\}$$

of points in ω^ω have already been defined and

$$\{z_\beta : \beta < \alpha\} \cap \{z'_\beta : \beta < \alpha\} = \emptyset.$$

Take σ_α and choose a point

$$y = (y_0, y_1, ..., y_n, ...) \in \omega^\omega$$

so that the relation

$$(x_0, y_0, x_1, y_1, ..., x_n, y_n, ...) \in \omega^\omega \setminus \{z_\beta : \beta < \alpha\}$$

would be satisfied, where

$$x_0 = \sigma_\alpha(\emptyset), \quad x_1 = \sigma_\alpha(x_0, y_0), \quad x_2 = \sigma_\alpha(x_0, y_0, x_1, y_1), \ldots .$$

Then put $z'_\alpha = (x_0, y_0, x_1, y_1, \ldots, x_n, y_n, \ldots)$. Further, take τ_α and choose a point

$$x' = (x'_0, x'_1, \ldots, x'_n, \ldots) \in \omega^\omega$$

so that the relation

$$(x'_0, y'_0, x'_1, y'_1, \ldots, x'_n, y'_n, \ldots) \in \omega^\omega \setminus \{z'_\beta : \beta \leq \alpha\}$$

would be satisfied, where

$$y'_0 = \tau_\alpha(x'_0), \quad y'_1 = \tau_\alpha(x'_0, y'_0, x'_1), \quad y'_2 = \tau_\alpha(x'_0, y'_0, x'_1, y'_1, x'_2), \ldots .$$

Then put $z_\alpha = (x'_0, y'_0, x'_1, y'_1, \ldots, x'_n, y'_n, \ldots)$.

Proceeding in this manner, the desired two **c**-sequences of points in ω^ω will be constructed and it directly follows from the described construction that

$$\{z_\alpha : \alpha < \mathbf{c}\} \cap \{z'_\alpha : \alpha < \mathbf{c}\} = \emptyset.$$

Finally, denote $Z = \{z_\alpha : \alpha < \mathbf{c}\}$ and verify that the game G_Z is not determined.

Remark 6. There is a famous theorem of Zermelo [277], according to which any finite game of two persons is determined. For more information about this theorem, we refer the reader to [226].

21. Let L be a Luzin set in $\mathbf{R}$ and let X be an analytic set in L (obtained by applying the A-operation to some countable family of closed subsets of L).

Show that X is a Borel set in L.

For this purpose, utilize the result of Exercise 14.

Let S be a Sierpiński set in $\mathbf{R}$ and let Y be an analytic set in S (obtained by applying the A-operation to some countable family of closed subsets of S).

Show that Y is a Borel set in S.

For this purpose, keep in mind the fact that S is a Luzin set in the topological space $(\mathbf{R}, \mathcal{T}_d)$, where $\mathcal{T}_d$ denotes the density topology of $\mathbf{R}$ (see Exercise 6 from Chapter 4), and utilize again the result of Exercise 14.

22*. Let H be an arbitrary Hamel basis of $\mathbf{R}$.

Prove Sierpiński's theorem stating that H is not an analytic subset of $\mathbf{R}$.

For this purpose, take into account the facts that any analytic set in $\mathbf{R}$ is measurable with respect to the Lebesgue measure λ and that λ possesses the Steinhaus property (cf. Exercise 1 from Chapter 3).

Remark 7. As was established by Miller, in Gödel's Constructible Universe **L** there exists a Hamel basis of $\mathbf{R}$ which is a co-analytic subset of $\mathbf{R}$.

23. Supposing that there exists a well-ordering of $\mathbf{R}$ whose graph is a projective subset of $\mathbf{R}^2$, demonstrate that any projective set in $\mathbf{R}^2$ admits a uniformization by some projective set.

Argue as follows. Let $\preceq$ be a projective well-ordering of $\mathbf{R}$ and let Z be an arbitrary projective set in $\mathbf{R}^2$. For each point $x \in \mathrm{pr}_1(Z)$, put

$$f(x) = \inf_{\preceq}\{y : (x, y) \in Z\}$$

and verify that the graph of the obtained function $f : \mathrm{pr}_1(Z) \to \mathbf{R}$ is a projective subset of $\mathbf{R}^2$. This yields at once the desired result.

24. Let $n \geq 1$ be a natural number. Assuming that any Π_n^1-subset of the plane $\mathbf{R}^2$ admits a uniformization by a Π_n^1-subset of $\mathbf{R}^2$, show that any Σ_{n+1}^1-subset of $\mathbf{R}^2$ admits a uniformization by a Σ_{n+1}^1-subset of $\mathbf{R}^2$.

For this purpose, use an argument similar to the proof of Theorem 7.

Remark 8. Luzin and Novikov presented an example of some Borel set $B \subset \mathbf{R}^2$ with $\mathrm{pr}_1(B) = \mathbf{R}$ which does not admit a uniformization by a Borel set (see [167]). According to Kondo's theorem proved within **ZFC** theory, every Π_1^1-subset of the plane $\mathbf{R}^2$ can be uniformized by a Π_1^1-subset of $\mathbf{R}^2$. As a straightforward consequence of this theorem, one obtains within the same theory that every Σ_2^1-subset of $\mathbf{R}^2$ can be uniformized by a Σ_2^1-subset of $\mathbf{R}^2$. In general, uniformization of projective sets in $\mathbf{R}^2$ of higher levels needs additional set-theoretical assumptions (for more details, see e.g. [25], [103], [115], [191]).

25. Deduce from Kondo's theorem that any Σ_2^1-set can be considered as a bijective continuous image of a Π_1^1-set.

26*. Suppose that there exists a well-ordering $\preceq$ of $[0, 1]$ for which the following two conditions are fulfilled:

(') $\preceq$ is isomorphic to the natural well-ordering of ω_1;
(") the graph of $\preceq$ is a projective subset of $[0, 1]^2$.
Demonstrate that there exists a function

$$\phi : [0, 1] \to [0, 1]$$

whose graph is a projective subset of $[0, 1]^2$ and which is absolutely nonmeasurable with respect to the class of all nonzero σ-finite diffused measures on $[0, 1]$ (see Chapter 5).

For this purpose, argue step by step as follows.

(a) First, denote by E the compact metric space consisting of all nonempty closed subsets of $[0, 1]$, and consider its subspace E' consisting of all nonempty nowhere dense closed subsets of $[0, 1]$.

Check that E' is of type G_δ in E, so E' can be treated as a Polish topological space.

(b) Further, identify the Baire canonical space ω^ω with the set I of all irrational numbers in $[0, 1]$ and introduce a continuous surjection

$$\Phi : I \to E'.$$

Then, for each point $x \in [0, 1]$, consider the set

$$Z(x) = \{y : x \preceq y \ \& \ y \notin \cup\{\Phi(i) : i \preceq x\}\}$$

and define the subset Z of the plane $\mathbf{R}^2$ by putting

$$Z = \cup\{\{x\} \times Z(x) : x \in [0, 1]\}.$$

Verify that Z is a projective subset of the square $[0, 1]^2$ and $\mathrm{pr}_1(Z) = [0, 1]$.
(c) Taking into account the result of Exercise 23, establish the existence of a function

$$\phi : [0, 1] \to [0, 1]$$

whose graph is a projective subset of $[0, 1]^2$ and is contained in Z.
Check that the range $\mathrm{ran}(\phi)$ of ϕ is a Luzin set in $[0, 1]$ which simultaneously is a projective subset of $[0, 1]$ and

$$\mathrm{card}(\phi^{-1}(t)) \leq \omega$$

for every point $t \in [0, 1]$. Keeping in mind Theorem 2 of Chapter 5, conclude that ϕ is absolutely nonmeasurable with respect to the class of all nonzero σ-finite diffused measures on $[0, 1]$.
Deduce from the above result that there is a countable family $\{T_j : j \in J\}$ of projective subsets of $[0, 1]$ such that no nonzero σ-finite diffused measure μ on $[0, 1]$ satisfies the relation

$$\{T_j : j \in J\} \subset \mathrm{dom}(\mu).$$

For this purpose, take some countable base $\{U_n : n < \omega\}$ of open sets in $[0, 1]$ and some countable base $\{V_n : n < \omega\}$ of open sets in the space $\mathrm{ran}(\phi) \subset [0, 1]$. Denote

$$\{T_j : j \in J\} = \{U_n : n < \omega\} \cup \{\phi^{-1}(V_n) : n < \omega\}$$

and show that the family $\{T_j : j \in J\}$ is as required.

27. Assuming that there exists a well-ordering of $\mathbf{R}$ whose graph is a Σ_2^1 subset of the plane $\mathbf{R}^2$, prove that there exists a Vitali set in $\mathbf{R}$ which is a Σ_2^1-subset of $\mathbf{R}$.

Remark 9. The situations described in Exercises 26 and 27 are realizable in certain models of **ZFC** theory, e.g., in Gödel's Constructible Universe **L**.

Bibliography

[1] R.M. Aron, F.J. Garcia-Pacheco, D. Perez-Garcia, J.B. Seoane-Sepulveda, *On dense lineability of sets of functions on* **R**, Topology, v. 48, 2009, pp. 149–156.

[2] A. Ascherl, J. Lehn, *Two principles of extending probability measures*, Manuscr. Math., v. 21, 1977, pp. 43–50,

[3] U. Avraham, S. Shelah, *Martin's Axiom does not imply that every two ω_1-dense sets of reals are isomorphic*, Israel Journal of Mathematics, v. 38, n. 1–2, 1981, pp. 161–176.

[4] R. Baire, *Lecons sur les fonctions discontinues*, Paris, 1905.

[5] M. Balcerzak, K. Ciesielski, T. Natkaniec, *Sierpiński-Zygmund functions that are Darboux, almost continuous, or have a perfect road*, Arch. Math. Logic, v. 37, n. 1, 1997, pp. 29–35.

[6] S. Banach, *Sur le probleme de la mesure*, Fund. Math., v. 4, 1923, pp. 7–33.

[7] S. Banach, *Über die Baire'sche Kategorie gewisser Funktionenmengen*, Studia Math., v. 3, 1931, pp. 174–179.

[8] S. Banach, C. Kuratowski, *Sur une généralisation du probléme de la mesure*, Fund. Math., v. 14, 1929, pp. 127–131.

[9] A. Bartoszewicz, S. Gläb, D. Pellegrino, J.B. Seoane-Sepulveda, *Algebrability, non-linear properties, and special functions*, Proc. Amer. Math. Soc., v. 141, 2013, pp. 3391–3402.

[10] J. Barwise (editor), *Handbook of Mathematical Logic*, North-Holland Publ. Co., Amsterdam, 1977.

[11] J. E. Baumgartner, *All ω_1-dense sets of reals can be isomorphic*, Fund. Math., v. 79, 1973, pp. 101–106.

[12] J. L. Bell, D. H. Fremlin, *A geometric form of the Axiom of Choice*, Fund. Math., v. 77, 1972, pp. 167–170.

[13] L. Bernal-González, D. Pellegrino, J. B. Seoane-Sepúlveda, *Linear subsets of nonlinear sets in topological vector spaces*, Bull. Amer. Math. Soc., v. 51, issue 1, 2014, pp. 71–130.

[14] F. Bernstein, *Zur Theorie der trigonometrischen Reihen*, Sitzungsbe* Sächs. Akad. Wiss. Leipzig. Math.-Natur., Kl. 60, 1908, pp. 325–338.

[15] C. E. Blair, *The Baire category theorem implies the principle of deper* *dent choices*, Bull. Acad. Polon. Sci., Ser. Math., v. 25, 1977, pp. 933–934

[16] A. Blass, *Existence of bases implies the Axiom of Choice*, Contemporar* Mathematics, v. 31, 1984, pp. 31–33.

[17] V. Bogachev, *Measure Theory*, Springer-Verlag, Berlin-Heidelberg, 2007

[18] N. Bourbaki, *Theory of Sets*, Hermann et Cie, Paris, 1968.

[19] N. Bourbaki, *General Topology*, Springer-Verlag, Berlin, 1989.

[20] N. Bourbaki, *Integration*, v. I, II, Springer-Verlag, Berlin, 2004.

[21] J. B. Brown, *Restriction theorems in real analysis*, Real Anal. Exchang* v. 20, n. 2, 1994–95, pp. 510–526.

[22] J. B. Brown, K. Prikry, *Variations on Lusin's theorem*, Trans. Amer* Math. Soc. v. 302, n. 1, 1987, pp. 77–86.

[23] A. Bruckner, *Differentiation of Real Functions*, Springer-Verlag, Berlin* 1978.

[24] J. Brzuchowski, J. Cichon, E. Grzegorek, C. Ryll-Nardzewski, *On th* *existence of nonmeasurable unions*, Bull. Acad. Pol. Sci., Ser. Sci. Math., v* 27, 1979, pp. 447–448.

[25] L. Bukovský, *The Structure of the Real Line*, Birkhäuser, Basel, 2011.

[26] V. V. Buldygin, A. B. Kharazishvili, *Geometric Aspects of Probabilit* *Theory and Mathematical Statistics*, Kluwer Academic Publishers, Dordrecht* 2000.

[27] C. Burstin, Die Spaltung des Kontinuums in c im L. Sinne nicht messbar* Mengen, Sitzungsb. der Akad. d. Wiss. in Wien, Math.-nat. Kl., II a, v. 125 1916, pp. 209–217.

[28] G. Cantor, *Gesammelte Abhandlungen*, Springer-Verlag, Berlin, 1932.

[29] C. C. Chang, H. J. Keisler, *Model Theory*, North-Holland Publishin* Co., Amsterdam, 1990.

[30] J. Cichon, A. Jasinski, *A note on complex unions of subsets of the rea* *line*, Acta Universitatis Carolinae, Math. et Phys., v. 42, n. 2, 2001, pp. 11–15

[31] J. Cichon, A. Kharazishvili, *On ideals with projective bases*, Georgian Mathematical Journal, v. 9, n. 3, 2002, pp. 461–472.

[32] J. Cichon, A. B. Kharazishvili, B. Weglorz, *On sets of Vitali's type*, Proc. Amer. Math. Soc., v. 118, 1993, pp. 1243–1250.

[33] J. Cichon, A. B. Kharazishvili, B. Weglorz, *Subsets of the Real Line*, Lodz University Press, Lodz, 1995.

[34] J. Cichon, M. Morayne, R. Ralowski, Cz. Ryll-Nardzewski, Sz. Zeberski, *On nonmeasurable unions*, Topology and its Applications, v. 154, issue 4, 2007, pp. 884–893.

[35] J. Cichon, P. Szczepaniak, *Hamel-isomorphic images of the unit ball*, Mathematical Logic Quarterly, v. 56, issue 6, 2010, pp. 625–630.

[36] K. Ciesielski, *How good is Lebesgue measure?* Math. Intelligencer, v. 11, n. 2, 1989, pp. 54–58.

[37] K. Ciesielski, *Set-theoretic real analysis*, Journal of Applied Analysis, v. 3, n. 2, 1997, pp. 143–190.

[38] K. Ciesielski, *Set Theory for the Working Mathematician*, Cambridge University Press, Cambridge, 1997.

[39] P. J. Cohen, *The independence of the continuum hypothesis*, Proc. Nat. Acad. Sci. U.S.A., v. 50, 1963, pp. 1143–1148; v. 51, 1964, pp. 105–110.

[40] P. J. Cohen, *Set Theory and the Continuum Hypothesis*, Benjamin, New York, 1966.

[41] W. W. Comfort, *Topological groups*, in: Handbook of Set-Theoretic Topology, edited by K. Kunen and J.E. Vaughan, North-Holland Publ. Co., Amsterdam, 1984, pp. 1143–1263.

[42] K. J. Devlin, *Constructibility*, Springer-Verlag, Berlin, 1984.

[43] R. Diaconescu, *Axiom of Choice and complementation*, Proc. Amer. Math. Soc., v. 51, 1975, pp. 176–178.

[44] F. G. Dorais, R. Filipow, *Algebraic sums of sets in Marczewski-Burstin algebras*, Real Analysis Exchange, v. 31, n. 1, 2005–2006, pp. 133–142.

[45] F. G. Dorais, R. Filipow, T. Natkaniec, *On some properties of Hamel bases and their applications to Marczewski measurable functions*, Central European Journal of Mathematics, v. 11, issue 3, 2013, pp. 487–502.

[46] G. Drago, P. D. Lamberti, P. Toni, *A "bouquet" of discontinuous functions for beginners in mathematical analysis*, Amer. Math. Monthly, v. 118, n. 9, 2011, pp. 799–811.

[47] F. R. Drake, *Set Theory: An Introduction to Large Cardinals*, North-Holland Publ. Co., Amsterdam, 1974.

[48] E. Engeler, *Metamathematik der Elementarmathematik*, Springer-Verlag, Berlin, 1983.

[49] R. Engelking, *General Topology*, PWN, Warszawa, 1985.

[50] P. Erdös, A. Hajnal, A. Máté, R. Rado, *Combinatorial Set Theory: Partition Relations for Cardinals*, North-Holland Publ. Co., Amsterdam, 1984.

[51] P. Erdös, S. Kakutani, *On nondenumerable graphs*, Bull. Amer. Math. Soc., v. 49, 1943, pp. 457–461.

[52] P. Erdös, K. Kunen, R. D. Mauldin, *Some additive properties of sets of real numbers*, Fund. Math., v. 113, n. 3, 1981, pp. 187–199.

[53] P. Erdös, R.D. Mauldin, *The nonexistence of certain invariant measures*, Proc. Amer. Math. Soc., v. 59, 1976, pp. 321–322.

[54] P. Erdös, R. Rado, *A partition calculus in set theory*, Bull. Amer. Math. Soc., v. 62, 1956, pp. 427–489.

[55] V. V. Fedorchuk, *A compact space having the cardinality of the continuum with no convergent sequences*, Math. Proc. Camb. Phil. Society, v. 81, 1977, pp. 177–181.

[56] S. Feferman, *Independence of the axiom of choice from the axiom of dependent choices*, The Journal of Symbolic Logic, v. 29, 1964, p. 226.

[57] S. Feferman, A. Levy, *Independence results in set theory by Cohen's method*, Notices Amer. Math. Soc., v. 10, 1963, p. 593.

[58] U. Felgner, *Comparison of the axioms of local and universal choice*, Fund. Math., v. 71, 1971, pp. 43–62.

[59] U. Felgner, T. J. Jech, *Variants of the axiom of choice in set theory with atoms*, Fund. Math., v. 79, 1973, pp. 79–85.

[60] A. A. Fraenkel, Y. Bar-Hillel, A. Levy, *Foundations of Set Theory*, North-Holland Publishing Co., Amsterdam, 1973.

[61] C. Freiling, *Axioms of symmetry: throwing darts at the real number line*, The Journal of Symbolic Logic, v. 51, 1986, pp. 190–200.

[62] D. H. Fremlin, *Consequences of Martin's Axiom*, Cambridge University Press, Cambridge, 1984.

[63] D. H. Fremlin, *Measure-additive coverings and measurable selectors*, Dissertationes Math., v. 260, 1987.

[64] D. H. Fremlin, *Measure Theory, v. 4: Topological Measure Spaces*, University of Essex, Torres Fremlin, 2006.

[65] D. H. Fremlin, *Measure Theory, v. 5: Set-theoretic Measure Theory*, University of Essex, Torres Fremlin, 2008.

[66] D. H. Fremlin, *Measure Theory, v. 2: Broad Foundations*, University of Essex, Torres Fremlin, 2010.

[67] D. H. Fremlin, *Measure Theory, v. 1: The Irreducible Minimum*, University of Essex, Torres Fremlin, 2011.

[68] D. H. Fremlin, *Measure Theory, v. 3: Measure Algebras*, University of Essex, Torres Fremlin, 2012.

[69] H. Friedman, *A definable nonseparable invariant extension of Lebesgue measure*, Illinois Journal of Mathematics, v. 21, issue 1, 1977, pp. 140–147.

[70] H. Friedman, *A consistent Fubini-Tonelli theorem for non-measurable functions*, Illinois Journal of Mathematics, v. 24, 1980, pp. 390–395.

[71] H. Friedman, *On definability of nonmeasurable sets*, Canadian Journal of Mathematics, v. 32, n. 3, 1980, pp. 653–656.

[72] L. Fuchs, *Infinite Abelian Groups*, vol. 1, Academic Press, New York-London, 1970.

[73] J. L. Gamez-Merino, G. A. Munoz-Fernandez, V. M. Sanchez, J. B. Seoane-Sepulveda, *Sierpiński–Zygmund functions and other problems of lineability*, Proc. Amer. Math. Soc., v. 138, n. 11, 2010, pp. 3863–3876.

[74] J. L. Gamez-Merino, G. A. Munoz-Fernandez, J. B. Seoane-Sepulveda, *Lineability and additivity in $\mathbf{R}^{\mathbf{R}}$*, Journal of Mathematical Analysis and Applications, v. 369, issue 1, 2010, pp. 265–272.

[75] D. Garciá, B. C. Grecu, M. Maestre, J. B. Seoane-Sepúlveda, *Infinite-dimensional Banach spaces of functions with nonlinear properties*, Math. Nachr., v. 283, n. 5, 2010, pp. 712–720.

[76] R. J. Gardner, W. F. Pfeffer, *Borel measures*, in: Handbook of Set-Theoretic Topology, edited by K. Kunen and J. E. Vaughan, North-Holland Publ. Co., Amsterdam, 1984, pp. 961–1043.

[77] B. R. Gelbaum and J. M. H. Olmsted, *Counterexamples in Analysis*, Holden-Day, San Francisco, 1964.

[78] B. V. Gnedenko and A. N. Kolmogorov, *Limit Distributions for Sums of Independent Random Variables*, Gosudarstv. Izdat. Tehn.-Teor. Lit., Moscow-Leningrad, 1949 (in Russian).

[79] K. Gödel, *The consistency of the axiom of choice and of the generalized continuum hypothesis*, Proc. Nat. Acad. Sci. U.S.A., v. 24, 1938, pp. 556–557

[80] C. Goffman, C. J. Neugebauer, T. Nishiura, *Density topology and approximate continuity*, Duke Math. Journal, v. 28, 1961, pp. 497–505.

[81] N. Goodman, J. Myhill, *Choice implies excluded middle*, Zeitschr. math. Logik Grundlagen Math., v. 24, 1978, p. 461.

[82] T. Gowers (editor), *The Princeton Companion to Mathematics*, Princeton University Press, Princeton, 2008.

[83] E. Grzegorek, *Solution of a problem of Banach on σ-fields without continuous measures*, Bull. Acad. Polon. Sci., Ser. Sci. Math., v. 28, 1980, pp. 7–10.

[84] V. I. Gurarii, *Subspaces and bases in spaces of continuous functions*, Dokl. Akad. Nauk SSSR, v. 167, 1966, pp. 971–973 (in Russian).

[85] V. I. Gurarii, *Linear spaces composed of everywhere nondifferentiable functions*, C. R. Acad. Bulgare Sci., vol. 44, 1991, pp. 13–16 (in Russian).

[86] A. Hajnal, I. Juhász, *Discrete subspaces of topological spaces*, Indag. Math., v. 29, 1967, pp. 343–356.

[87] L. Halbeisen, *Combinatorial Set Theory*, Springer-Verlag, Berlin, 2011

[88] L. Halbeisen, S. Shelah, *Consequences of arithmetic for set theory*, The Journal of Symbolic Logic, v. 59, 1994, pp. 30–40.

[89] P. R. Halmos, *Measure Theory*, D. Van Nostrand, New York, 1950.

[90] G. Hamel, *Eine Basis aller Zahlen und die unstetigen Lösungen der Funktionalgleichung $f(x + y) = f(x) + f(y)$*, Math. Ann., v. 60, 1905, pp. 459–462.

[91] L. A. Harrington, M. D. Morley, A. Scedrov, S. G. Simpson, *Harvey Friedman's Research on the Foundations of Mathematics*, Elsevier, Amsterdam, 1985.

[92] F. Hartogs, *Über das Problem der Wohlordnung*, Mathematische Annalen, v. 76, 1915, pp. 436–443.

[93] H. Herrlich, *Axiom of Choice*, Springer-Verlag, Berlin, 2006.

[94] G. Hesse, *Zur Topologisierbarkeit von Gruppen*, Dissertation, University of Hannover, Hannover, 1979.

[95] E. Hewitt, K. A. Ross, *Abstract Harmonic Analysis*, vol. 1, Springer-Verlag, Berlin, 1963.

[96] E. Hewitt and K. Stromberg, *Real and Abstract Analysis*, Springer-Verlag, New York, 1965.

[97] D. Hilbert, *Mathematische Probleme*, Nachrichten von der Königlichen Gesellschaft der Wissenschaften zu Göttingen, 1900, pp. 253–297.

[98] A. E. Holroyd and T. Soo, *A nonmeasurable set from coin flips*, Amer. Math. Monthly, **116**, 2009, pp. 926–928.

[99] A. Hulanicki, *Invariant extensions of the Lebesgue measure*, Fund. Math., v. 51, 1962, pp. 111–115.

[100] V. Jarnik, *Sur la dérivabilité des fonctions continues*, Spisy Privodov, Fak. Univ. Karlovy, v. 129, 1934, pp. 3–9.

[101] T. J. Jech, *Non-provability of Souslin's Hypothesis*, Commentationes Mathematicae Universitatis Carolinae, v. 8, 1967, pp. 291–305.

[102] T. J. Jech, *The Axiom of Choice*, North-Holland Publishing Company, Amsterdam, 1973.

[103] T. J. Jech, *Set Theory*, Springer-Verlag, Berlin, 2003.

[104] R. B. Jensen, *Independence of the axiom of dependent choices from the countable axiom of choice*, The Journal of Symbolic Logic, v. 31, 1966, p. 294.

[105] R. B. Jensen, *Souslin's hypothesis is incompatible with* $\mathbf{V} = \mathbf{L}$, Notices Amer. Math. Soc., v. 15, 1968, p. 935.

[106] I. Juhász, *Cardinal Functions in Topology*, Math. Centre, Amsterdam, 1971.

[107] S. Kakutani, J. C. Oxtoby, *Construction of a nonseparable invariant extension of the Lebesgue measure space*, Annals of Mathematics, v. 52, 1950, pp. 580–590.

[108] A. Kanamori, *The emergence of descriptive set theory*, Synthese, v. 251, 1995, pp. 241–262

[109] A. Kanamori, *The mathematical development of set theory from Cantor to Cohen*, The Bulletin of Symbolic Logic, v. 2, n. 1, 1996, pp. 1–71.

[110] A. Kanamori, *The Higher Infinite*, Springer-Verlag, Berlin-Heidelberg, 2003.

[111] A. Kanamori, *Zermelo and set theory*, The Bulletin of Symbolic Logic, v. 10, n. 4, 2004, pp. 487–553.

[112] A. Kanamori, D. Pincus, *Does GCH imply AC locally?* in: Paul Erdös and his Mathematics, II, Budapest, 2002, pp. 413–426.

[113] A. S. Kechris, *The perfect set theorem and definable well-orderings c the continuum*, The Journal of Symbolic Logic, v. 43, n. 4, 1978, pp. 630–634.

[114] A. S. Kechris, *The axiom of determinacy implies dependent choices i $L(\mathbf{R})$*, The Journal of Symbolic Logic, v. 49, 1984, pp. 161–173.

[115] A. S. Kechris, *Classical Descriptive Set Theory*, Springer-Verlag, Nev York-Berlin, 1995.

[116] A. B. Kharazishvili, *An example of an invariant measure*, Dokl. Akad Nauk SSSR, v. 220, n. 1, 1975, pp. 44–46 (in Russian).

[117] A. B. Kharazishvili, *On certain types of invariant measures*, Dokl Akad. Nauk SSSR, v. 222, n. 3, 1975, pp. 538–540 (in Russian).

[118] A. B. Kharazishvili, *Some Questions of Set Theory and Measure The ory*, Tbilisi University Press, Tbilisi, 1978 (in Russian).

[119] A. B. Kharazishvili, *The axiom of choice in the general theory of sys tems*, Bull. Acad. Sci. GSSR, v. 94, n. 3, 1979, pp. 533–536 (in Russian)

[120] A. B. Kharazishvili, *Nonmeasurable Hamel bases*, Dokl. Akad. Nau SSSR, v. 253, n. 5, 1980, pp. 1068–1070 (in Russian).

[121] A. B. Kharazishvili, *Elements of the Combinatorial Theory of Infinit Sets*, Izd. Tbil. Gos. Univ., Tbilisi, 1981 (in Russian).

[122] A. B. Kharazishvili, *The absolute nonmeasurability of the unit ball i an infinite-dimensional separable Hilbert space*, Bull. Acad. Sci. GSSR, v. 107 n. 1, 1982, pp. 17–20 (in Russian).

[123] A. B. Kharazishvili, *Invariant Extensions of the Lebesgue measure* Tbilisi University Press, Tbilisi, 1983 (in Russian).

[124] A. B. Kharazishvili, *Some properties of isodyne topological spaces*, Bull Acad. Sci. GSSR, v. 127, n. 2, 1987, pp. 261–264 (in Russian).

[125] A. B. Kharazishvili, *On separable supports of Borel measures*, Georgia Mathematical Journal, v. 2, n. 1, 1995, pp. 45–53.

[126] A. B. Kharazishvili, *Transformation Groups and Invariant Measures* World Scientific, London-Singapore, 1998.

[127] A. B. Kharazishvili, *Applications of Point Set Theory in Real Analysis* Kluwer Academic Publishers, Dordrecht, 1998.

[128] A. B. Kharazishvili, *Nonmeasurable Sets and Functions*, Elsevier, Am sterdam, 2004.

[129] A. B. Kharazishvili, *On absolutely nonmeasurable additive functions* Georgian Mathematical Journal, v. 11, n. 2, 2004, pp. 301–306.

[130] A. B. Kharazishvili, *The algebraic sum of two absolutely negligible sets can be an absolutely nonmeasurable set*, Georgian Mathematical Journal, v. 12, n. 3, 2005, pp. 455–460.

[131] A. B. Kharazishvili, *On additive absolutely nonmeasurable Sierpiński–Zygmund functions*, Real Analysis Exchange, v. 31, n. 2, 2005–2006, pp. 553–560.

[132] A. B. Kharazishvili, *On measurable Sierpiński–Zygmund functions*, Journal of Applied Analysis, v. 12, n. 2, 2006, pp. 283–292.

[133] A. B. Kharazishvili, *Strange Functions in Real Analysis*, 2nd edition, Chapman and Hall, New York, 2006.

[134] A. B. Kharazishvili, *A nonseparable extension of the Lebesgue measure without new null sets*, Real Analysis Exchange, vol. 33, no. 1, 2007–2008, pp. 259–268.

[135] A. B. Kharazishvili, *On nonmeasurable functions of two variables and iterated integrals*, Georgian Mathematical Journal, v. 16, n. 4, 2009, pp. 705–710.

[136] A. B. Kharazishvili, *On sets with homogeneous linear sections*, Proc. A. Razmadze Mathematical Institute, vol. 151, 2009, pp. 124–128.

[137] A. B. Kharazishvili, *Topics in Measure Theory and Real Analysis*, Atlantis Press and World Scientific, Amsterdam–Paris, 2009.

[138] A. B. Kharazishvili, *On a relationship between the measurability and continuity of real-valued functions*, Georgian Mathematical Journal, v. 17, n. 4, 2010, pp. 649–661.

[139] A. B. Kharazishvili, *Measurability properties of Vitali sets*, Amer. Math. Monthly, v. 118, n. 10, 2011, pp. 693–703.

[140] A. B. Kharazishvili, *A large group of absolutely nonmeasurable additive functions*, Real Analysis Exchange, v. 37, n. 2, 2011–2012, pp. 467–476.

[141] A. B. Kharazishvili, *Sums of absolutely nonmeasurable functions*, Georgian Mathematical Journal, v. 20, n. 2, 2013, pp. 271–282.

[142] A. B. Kharazishvili, A. Kirtadze, *On the measurability of functions with respect to certain classes of measures*, Georgian Mathematical Journal, v. 11, n. 3, 2004, pp. 489–494.

[143] A. B. Kharazishvili, A. Kirtadze, *On weakly metrically transitive measures and nonmeasurable sets*, Real Analysis Exchange, vol. 32, no. 2, 2007, pp. 553–562.

[144] K. Kodaira, S. Kakutani, *A nonseparable translation invariant exten-sion of the Lebesgue measure space*, Annals of Mathematics, v. 52, 1950, pp. 574–579.

[145] M. Kojman, S. Shelah, *Nonexistence of universal orders in many car-dinals*, The Journal of Symbolic Logic, v. 57, n. 3, 1992, pp. 875–891.

[146] P. Komjáth and V. Totik, *Ultrafilters*, Amer. Math. Monthly, v. 115, n. 1, 2008, pp. 33–44.

[147] M. Kuczma, *An Introduction to the Theory of Functional Equations and Inequalities: Cauchy's Equation and Jensen's Inequality*, PWN, Katowice, 1985.

[148] K. Kunen, *Set Theory*, North-Holland Publishing Company, Amster-dam, 1980.

[149] K. Kunen, *A compact L-space under CH*, Topology and its Applica-tions, v. 12, 1981, pp. 283-287.

[150] K. Kunen, *Random and Cohen Reals*, in: Handbook of Set-Theoretic Topology, edited by K. Kunen and J. E. Vaughan, North-Holland Publ. Co., Amsterdam, 1984, pp. 887–911.

[151] K. Kuratowski, *Une méthode d'élimination des nombres transfinis des raisonnements mathématiques*, Fund. Math., v. 3, 1922, pp. 76–108.

[152] K. Kuratowski, *Topology*, vol. 1, Academic Press, New York-London, 1966.

[153] K. Kuratowski, *Topology*, v. 2, Academic Press, New York-London, 1969.

[154] K. Kuratowski, A. Mostowski, *Set Theory*, North-Holland Publ. Co., Amsterdam, 1967.

[155] K. Kuratowski, Cz. Ryll-Nardzewski, *A general theorem on selectors*, Bull. Acad. Polon. Sci., Ser. Math., v. 13, n. 6, 1965, pp. 397–402.

[156] K. Kuratowski, W. Sierpiński, *Le théorème de Borel–Lebesgue dans la théorie des ensembles abstraits*, Fund. Math., v. 2, 1921, pp. 172–178.

[157] D. Kurepa, *La condition de Souslin et une propriété caractéristique des nombres réels*, C. R. Acad. Sci. Paris, v. 231, 1950, pp. 1113–1114.

[158] D. Kurepa, *The Cartesian multiplication and the cellularity numbers*, Publ. Inst. Math. Beograd, v. 2, 1962, pp. 121–139.

[159] A. G. Kurosh, *The Theory of Groups*, Izd. Nauka, Moscow, 1967 (in Russian).

[160] Y. Kuznetsova, *On continuity of measurable group representations and homomorphisms*, Studia Math., v. 210, 2012, pp. 197–208.

[161] M. Kysiak, *Nonmeasurable algebraic sums of sets of reals*, Colloq. Math., v. 102, n. 1, 2005, pp. 113–122.

[162] M. Laczkovich, *Paradoxes in measure theory*, in: Handbook of Measure Theory, edited by E. Pap, vol. 1, Elsevier, Amsterdam, 2002, pp. 83–123.

[163] H. Lebesgue, *Sur les fonctions représentables analytiquement*, J. Math. Pures et Appl., ser. 6, t. 1, fasc. 2, 1905, pp. 139–216.

[164] A. Levy, *Basic Set Theory*, Dover Publications, Inc., Mineola, New York, 2002.

[165] A. Lindenbaum, *Sur quelques propriétés des fonctions de variable réelle*, Ann. Soc. Pol. Math., v. 6, 1927, pp. 129–130.

[166] T. P. Lukashenko, *Boks integrable nonmeasurable functions*, Mat. Zametki, v. 17, n. 1, 1975, pp. 49–56 (in Russian).

[167] N. N. Luzin, *Collected Works*, volumes 1-3, Izd. Acad. Nauk SSSR, Moscow, 1953–1959 (in Russian).

[168] N. Luzin, W. Sierpiński, *Sur une décomposition d'un intervalle en une infinité non dénombrable d'ensembles non mesurables*, Comptes Rendus de l'Academie des Sciences (Paris), t. 165, 1917, pp. 422–424.

[169] R. D. Mabry, *Sets which are well-distributed and invariant relative to all isometry invariant total extensions of Lebesgue measure*, Real Analysis Exchange, v. 16, 1990–1991, pp. 425–459.

[170] R. D. Mabry, *Some remarks concerning the uniformly gray sets of G. Jacopini*, Rend. Accad. Naz. Sci., Mem. Mat. Appl., v. 22, n. 5, 1998, pp. 43–49.

[171] R. D. Mabry, *Stretched shadings and a Banach measure that is not scale-invariant*, Fund. Math., v. 209, 2010, pp. 95–113.

[172] P. Maddy, *Believing the axioms. I*, The Journal of Symbolic Logic, v. 53, n. 2, 1988, pp. 481–511.

[173] P. Maddy, *Believing the axioms. II*, The Journal of Symbolic Logic, v. 53, n. 3, 1988, pp. 736–764.

[174] E. Marczewski (E. Szpilrajn), *Sur l'extension de la mesure lebesguienne*, Fund. Math., v. 25, 1935, pp. 551–558.

[175] E. Marczewski (E. Szpilrajn), *Remarque sur les produits cartésiens d'espaces topologiques*, Dokl. Akad. Nauk SSSR, v. 31, 1941, pp. 525–527.

[176] E. Marczewski (E. Szpilrajn), *On problems of the theory of measure*, Uspekhi Mat. Nauk, v. 1, n. 2 (12), 1946, pp. 179–188 (in Russian).

[177] E. Marczewski (E. Szpilrajn), *Séparabilité et multiplication Cartésienne des espaces topologiques*, Fund. Math., v. 34, 1947, pp. 127–143.

[178] E. Marczewski, *On compact measures*, Fund. Math., v. 40, 1953, pp. 113–124.

[179] E. Marczewski, *Collected Mathematical Papers*, Polish Academy of Sciences, Warszawa, 1996.

[180] D. A. Martin, *Borel determinacy*, Annals of Mathematics, v. 102, n 2, 1975, pp. 363–371.

[181] D. Martin, R. Solovay, *Internal Cohen extensions*, Ann. Math. Logic v. 2, n. 2, 1970, pp. 143–178.

[182] D. A. Martin, J. R. Steel, *A proof of projective determinacy*, Journa of the American Mathematical Society, v. 2, n. 1, 1989, pp. 71–125.

[183] G. Matusik, *On the lattice generated by Hamel functions*, Real Analysis Exchange, v. 36, n. 1, 2010, pp. 65–78.

[184] R. D. Mauldin, *The existence of nonmeasurable sets*, Amer. Math Monthly, v. 86, 1979, pp. 45–46.

[185] S. Mazurkiewicz, *Sur un ensemble plan qui a avec chaque droite deux et seulement deux points communs*, C. R. Varsovie, v. 7, 1914, pp. 382–384.

[186] S. Mazurkiewicz, *Sur les fonctions non-derivables*, Studia Math., v. 3 1931, pp. 92–94.

[187] E. Mendelson, *Introduction to Mathematical Logic*, D. Van Nostranc Company, Inc., Princeton, 1964.

[188] A. W. Miller, *Special subsets of the real line*, in: Handbook of Set-Theoretic Topology, edited by K. Kunen and J. E. Vaughan, North-Holland Publ. Co., Amsterdam, 1984, pp. 201–233.

[189] G. H. Moore, *Zermelo's Axiom of Choice*, Springer-Verlag, New York 1982

[190] J. C. Morgan, II, *Point Set Theory*, Marcel Dekker, Inc., New York 1990.

[191] Y. N. Moschovakis, *Descriptive Set Theory*, North-Holland Publ. Co. Amsterdam, 1980.

[192] J. Mycielski, *On the axiom of determinateness*, Fund. Math., v. 54 1964, pp. 205–224.

[193] J. Mycielski, *Independent sets in topological algebras*, Fund. Math., v. 55, 1964, pp. 139–147.

[194] J. Mycielski, *On the axiom of determinateness*, Fund. Math., v. 59, 1966, pp. 203–212.

[195] J. Mycielski, *Finitely additive invariant measures I*, Colloq. Math., v. 42, 1979, pp. 309–318.

[196] J. Mycielski, H. Steinhaus, *A mathematical axiom contradicting the axiom of choice*, Bull. Acad. Polon. Sci., v. 10, 1962, pp. 1–3.

[197] I. P. Natanson, *Theory of Functions of a Real Variable*, 3rd edition, Izdat. Nauka, Moscow, 1974 (in Russian).

[198] T. Natkaniec, H. Rosen, *An example of an additive almost continuous Sierpiński–Zygmund function*, Real Analysis Exchange, v. 30, 2004–2005, pp. 261–266.

[199] J. Neveu, *Bases Mathématiques du Calcul des Probabilités*, Masson et Cie, Paris, 1964.

[200] H. Okamoto, M. Wunsch, *A geometric construction of continuous strictly increasing singular functions*, Proc. Japan Acad., Ser. A Math. Sci., v. 83, n. 7, 2007, pp. 114–118.

[201] A. Yu. Ol'shanskii, *A remark on a countable nontopologized group*, Vestnik Moscow University, Ser I, Mat. Mekh., v. 3, 1980, p. 103 (in Russian).

[202] A. J. Ostaszewski, *Beyond Lebesgue and Baire III: Steinhaus' theorem and its descendants*, Topology and its Applications, v. 160, issue 10, 2013, pp. 1144–1154.

[203] J. C. Oxtoby, *Measure and Category*, Springer-Verlag, Berlin, 1971.

[204] J. C. Oxtoby, S. Ulam, *Measure-preserving homeomorphisms and metrical transitivity*, Annals of Mathematics, v. 42, n. 2, 1941, pp. 874–920.

[205] G. Pantsulaia, *An application of independent families of sets to the measure extension problem*, Georgian Mathematical Journal, v. 11, n. 2, 2004, pp. 379–390.

[206] E. Pap (editor), *Handbook of Measure Theory*, v. 1–2, Elsevier, Amsterdam, 2002.

[207] G. Peano, *Démonstration de l'intégrabilité des équations différentielles ordinaires*, Mathematische Annalen, v. 37, 1890, pp. 182–228.

[208] W. F. Pfeffer, K. Prikry, *Small spaces*, Proc. London Math. Soc., v. 58, n. 3, 1989, pp. 417–438.

[209] Sh. S. Pkhakadze, *On the iterated integrals*, Proc. A. Razmadze Math. Institute, v. 20, 1954, pp. 167–209 (in Russian).

[210] Sh. S. Pkhakadze, *The Theory of Lebesgue measure*, Proc. A. Razmadze Math. Institute, v. 25, 1958, pp. 3–272 (in Russian).

[211] Sz. Plewik, *Towers are always universally measure zero and always of first category*, Proc. Amer. Math. Soc., v. 119, 1993, pp. 865–868.

[212] K. Plotka, *Sum of Sierpiński–Zygmund and Darboux like functions*, Topology and its Applications, v. 122, n. 3, 2002, pp. 547–564.

[213] Yu. V. Prokhorov, *Convergence of random processes and limit theorem in probability theory*, Teor. Veroyatnost. i Primenen. v. 1, 1956, pp. 177–238 (in Russian).

[214] J. Raisonnier, *A mathematical proof of S. Shelah's theorem on the measure problem and related results*, Israel Journal of Mathematics, v. 48, n. 1 1984, pp. 48–56.

[215] R. Ralowski, S. Zeberski, *On nonmeasurable images*, Czechoslovak Mathematical Journal, v. 60, n. 2, 2010, pp. 423–434.

[216] I. Reclaw, *On nonmeasurable unions of sections of a Borel set*, Tatra Mt. Math. Publ., v. 28, 2004, pp. 71–73.

[217] A. Robinson, *Non-Standard Analysis*, North-Holland Publ. Co., Amsterdam, 1965.

[218] L. Rodriguez-Piazza, *Every separable Banach space is isometric to a space of continuous nowhere differentiable functions*, Proc. Amer. Math. Soc. v. 123, n. 12, 1995, pp. 3649–3654.

[219] C. A. Rogers, *A linear Borel set whose difference set is not a Borel set*, Bull. London Math. Soc., v. 2, n. 1, 1970, pp. 41–42.

[220] A. Roslanowski, S. Shelah, *Measured creatures*, Israel Journal of Mathematics, v. 151, 2006, pp. 61–110.

[221] F. Rothberger, *Eine Äquivalenz zwischen der Kontinuumhypothese und der Existenz der Lusinschen und Sierpińskischen Mengen*, Fund. Math., v. 30, 1938, pp. 215–217.

[222] H. Rubin, J. E. Rubin, *Equivalents of the Axiom of Choice*, II, North Holland Publishing Co., Amsterdam, 1985.

[223] M. E. Rudin, *Lectures on Set-Theoretic Topology*, American Mathematical Society, Providence, Rhode Island, 1975.

[224] W. Rudin, *Real and Complex Analysis*, McGraw-Hill, New York, 1987

[225] D. I. Saveljev, *A note on connection of linear orderings with measure and category*, Usp. Mat. Nauk, v. 53, issue 6 (324), 1998, pp. 261–262 (in Russian).

[226] U. Schwalbe, P. Walker, *Zermelo and the early history of game theory*, Games and Economic Behavior, v. 34, issue 1, 2001, pp. 123–137.

[227] D. Scott, *Measurable cardinals and constructible sets*, Bull. Acad. Polon. Sci., v. 9, 1961, pp. 521–524.

[228] S. Shelah, *On a problem of Kurosh, Jónsson groups, and applications*, in: Word Problems II, edited by S. I. Adian, W. W. Boone, and G. Higman, North-Holland Publ. Co., Amsterdam, 1980, pp. 373–394.

[229] S. Shelah, *Can you take Solovay's inaccessible away?* Israel Journal of Mathematics, v. 48, n. 1, 1984, pp. 1–47.

[230] S. Shelah, *Cardinal Arithmetic*, Oxford University Press, Oxford, 1994.

[231] J. R. Shoenfield, *Mathematical Logic*, Addison-Wesley, Reading, MA, 1967.

[232] W. Sierpiński, *L'axiome de M. Zermelo et son role dans la Théorie des Ensembles et l'Analyse*, Bull. Intern. Acad. Sci. Cracovie, Ser. A, 1918, pp. 97–152.

[233] W. Sierpiński, *Sur un théoréme équivalent á l'hypothése du continu*, Bull. Intern. Acad. Sci. Cracovie, Ser. A, 1919, pp. 1–3.

[234] W. Sierpiński, *Sur la question de la mesurabilité de la base de M. Hamel*, Fund. Math., vol. 1, 1920, pp. 105–111.

[235] W. Sierpiński, *Sur un probléme concernant les ensembles mesurables superficiellement*, Fund. Math., v. 1, 1920, pp. 112–115.

[236] W. Sierpiński, *Les exemples effectifs et l'axiome du choix*, Fund. Math., v. 2, 1921, pp. 112–118.

[237] W. Sierpiński, *Sur l'équivalence de trois propriétés des ensembles abstraits*, Fund. Math., v. 2, 1921, pp. 179–188.

[238] W. Sierpiński, *Zermelo's axiom and its role in set theory and analysis*, Mat. Sbornik, v. 31, issue 1, 1922, pp. 94–128 (Russian).

[239] W. Sierpiński, *Fonctions additives non complètement additives et fonctions nonmesurables*, Fund. Math., v. 30, 1938, pp. 96–99.

[240] W. Sierpiński, *L'hypothèse généralisée du continu et l'axiome du choix*, Fund. Math., v. 34, 1947, pp. 1–5.

[241] W. Sierpiński, *General Topology*, Toronto, 1952.

[242] W. Sierpiński, *On the congruence of sets and their equivalence by finite decomposition*, Lucknow University, Lucknow, 1954.

[243] W. Sierpiński, *Cardinal and Ordinal Numbers*, PWN, Warszawa, 1958.

[244] W. Sierpiński, *Oeuvres Choisies*, v. II, PWN, Warszawa, 1975.

[245] W. Sierpiński, *Oeuvres Choisies*, v. III, PWN, Warszawa, 1976.

[246] W. Sierpiński, A. Zygmund, *Sur une fonction qui est discontinue sur tout ensemble de puissance du continu*, Fund. Math., v. 4, 1923, pp. 316–318.

[247] J. C. Simms, *Sierpiński's theorem*, Simon Stevin, v. 65, n. 1–2, 1991, pp. 69–163.

[248] S. G. Simpson, *Subsystems of Second Order Arithmetic*, Springer-Verlag, Berlin, 1999.

[249] S. G. Simpson (editor), *Reverse Mathematics 2001*, ASL Publisher, San Diego, 2005.

[250] A. V. Skorokhod, *Integration in Hilbert space*, Springer-Verlag, Berlin, 1974.

[251] R. M. Smullyan, M. Fitting, *Set Theory and the Continuum Problem*, Dover Publications, New York, 2010.

[252] S. Solecki, *Measurability properties of sets of Vitali's type*, Proc. Amer. Math. Soc., v. 119, n. 3, 1993, pp. 897–902.

[253] R. M. Solovay, *A model of set theory in which every set of reals is Lebesgue measurable*, Annals of Mathematics, Ser. 2, v. 92, 1970, pp. 1–56.

[254] R. M. Solovay, S. Tennenbaum, *Iterated Cohen extensions and Souslin's problem*, Annals of Mathematics, v. 94, n. 2, 1971, pp. 201–245.

[255] M. Souslin, *Problème 3*, Fund. Math., v. 1, 1920, p. 223.

[256] E. Specker, *Verallgemeinerte Kontinuumshypothese und Auswahlaxiom*, Archiv der Mathematik, v. 5, 1954, pp. 332–337.

[257] H. Steinhaus, *Sur les distances des points dans les ensembles de mesure positive*, Fund. Math., v. 1, 1920, pp. 93–104.

[258] F. Tall, *The density topology*, Pacific Journal of Mathematics, v. 62, 1976, pp. 275–284.

[259] A. Tarski, *On well-ordered subsets of any set*, Fund. Math., v. 32, 1939, pp. 176–183.

[260] R. Telgársky, *Topological games: The 50th anniversary of the Banach–Mazur game*, Rocky Mountain Journal of Mathematics, v. 17, n. 2, 1987, pp. 227–276.

[261] S. Tennenbaum, *Souslin's problem*, Proc. Nat. Acad. Sci. USA, v. 59, n. 1, 1968, pp. 60–63.

[262] R. Thomas, *A combinatorial construction of a nonmeasurable set*, Amer. Math. Monthly, v. 92, 1985, pp. 421–422.

[263] S. Ulam, *Zur Masstheorie in der allgemeinen Mengenlehre*, Fund. Math., v. 16, 1930, pp. 140–150.

[264] V. A. Uspenskii, *N. N. Luzin's contribution to the descriptive theory of sets and functions: concepts, problems, predictions*, Usp. Mat. Nauk, v. 40, issue 3 (243), 1985, pp. 85–116 (in Russian).

[265] N. N. Vakhanyia, V. I. Tarieladze, and S. A. Chobanyan, *Probability Distributions in Banach Spaces*, Nauka, Moscow, 1985 (in Russian).

[266] G. Vitali, *Sul problema della misura dei gruppi di punti di una retta*, Tip. Gamberini e Parmeggiani, Bologna, 1905.

[267] M. L. Wage, *Products of Radon spaces*, Uspekhi Mat. Nauk, v. 35, n. 3, 1980, pp. 151–153 (in Russian).

[268] S. Wagon, *The Banach–Tarski Paradox*, Cambridge University Press, Cambridge, 1985.

[269] H. Weber, E. Zoli, *Finite partitions of topological groups into congruent thick subsets*, Journal of Applied Analysis, v. 14, n. 1, 2008, pp. 1–12.

[270] W. Wilczyński, A. Kharazishvili, *On translations of measurable sets and of sets with the Baire property*, Bull. Acad. Sci. Georgia, v. 145, n. 1, 1992, pp. 43-46 (in Russian).

[271] W. H. Woodin, *The Continuum Hypothesis*, Notices of the American Mathematical Society; part 1: v. 48, n. 6, 2001, pp. 567-576; part 2: v. 48, n. 7, 2001, pp. 681-690.

[272] P. Zakrzewski, *Measures on algebraic-topological structures*, in: Handbook of Measure Theory, edited by E. Pap, North-Holland Publ. Co., Amsterdam, 2002, pp. 1091–1130.

[273] P. Zakrzewski, *On a construction of universally small sets*, Real Analysis Exchange, v. 28, n. 1, 2002–2003, pp. 215–221.

[274] Sz. Zeberski, *On completely nonmeasurable unions*, Mathematical Logic Quarterly, v. 53, issue 1, 2007, pp. 38–42.

[275] E. Zermelo, *Beweis dass jede Menge wohlgeordnet werden kann*, Math. Ann., v. 59, 1904, pp. 514–516.

[276] E. Zermelo, *Untersuchungen über die Grundlagen der Mengenlehre*, Math. Ann., v. 65, 1908, pp. 261–281.

[277] E. Zermelo, *Über eine Anwendung der Mengenlehre auf die Theorie des Schachspiels*, in: Proceedings of Fifth Congress of Mathematicians (Cambridge 1912), Cambridge University Press, 1913, pp. 501–504.

[278] E. Zermelo, Über Grenzzahlen und Mengenbereiche. Neue Untersuchungen über die Grundlagen der Mengenlehre, Fund. Math., v. 16, 1930, pp. 29–47.

[279] E. Zoli, *On the equivalence between CH and the existence of certain I-Luzin subsets of* **R**, Georgian Mathematical Journal, v. 11, n. 1, 2004, pp. 195–202.

Subject Index

Absolutely measurable
  function, 67
  set, 67
Absolutely negligible set in a group,
  76
Absolutely nonmeasurable
  function, 68
  set, 68
  set in a group, 76
Absolute null space, 73
Accumulation point, 4
Admissible
  class of subsets of the plane, 294,
    308
  group of transformations, 271,
  measure, 78
Alexandrov, A.D.'s theorem, 98
Alexandrov's one-point compacti-
  fication, 80
Alexandrov's theorem, 33
  on G_δ-sets, 399
Almost disjoint family of sets, 124,
  131
Almost invariant family of sets, 271
Almost symmetric set, 294
Almost translation invariant set, 133
Analytic set, 88
A-operation, 309, 384
Arcwise connected topological space,
  223
Aronszajn tree, 207, 334
Axiom of Choice (**AC**), 1

Axiom of countable choice (**CC**),
  21
Axiom of dependent choices (**DC**),
  22
Axiom of Determinacy (**AD**), 31,
  397
Axiom of Projective Determinacy
  (**PD**), 398

Baire canonical space, 116, 384
Baire class
  zero, 8
  one, 8
  two, 8
Baire order of a function, 383
Baire property, 381
Baire theorem, 23
Baire topological space, 19
Banach limit, 146
Banach theorem, 139
Baumgartner's axiom, 372, 373
Bernstein set, 37
Bichromatic graph, 45
Borel σ-algebra of a topological
  space, 382
Borel selector, 52
Branch of a tree, 334
Burali-Forti theorem, 330

Cantor–Bernstein theorem, 326
Cantor's inequality, 325, 326
Cantor's theorem, 325

Cantor set, 16
Cartesian product, 2
    of two sets, 322
Cauchy's functional equation, 36
Cavalieri's principle, 239
Character of a commutative group,
    225
Choice function, 9
Circle group, 178
Co-analytic set, 92
Cofinal subset of a partially ordered
    set, 108
Cohen algebra, 134
Coinitial subset of a partially or-
    dered set, 108
Compact topological space, 17
Comprehension Principle, 321
Condensation point, 4
Conditionally complete linearly or-
    dered set, 17
Consistent set, 356
Constructibility Axiom, 92
Constructible Universe, 332
Continuous function
    in the Cauchy sense, 15
    in the Heine sense, 15
Continuous measure, 152
Continuous nowhere approximately
    differentiable function, 222
Continuous nowhere differentiable
    function, 209
Continuum Hypothesis (**CH**), 53
Countable chain condition, 356
Countably compact topological space,
    49
Countably generated σ-algebra, 79
Countably quasi-compact topolog-
    ical space, 49
C-property of Luzin, 49

Dedekind complete linearly ordered
    set, 17
Density topology, 59
D-finite set, 323
Diamond ω_1-sequence, 349
Dieudonné measure, 102
D-infinite set, 323
Difference set, 118
Diffused measure, 152
Direct summand in a commutative
    group, 281
Dirichlet function, 15
Divisible group, 234

Element of a set, 317
Empty set, 318
Erdös–Rado combinatorial theorem,
    338

Filter in a partially ordered set,
    356
Finite version
    of **AC**, 10
    of Ramsey's theorem, 338
Foundation Axiom, 320
Freiling's axiom of symmetry, 64
Fubini's theorem, 239
Function approximately differenti-
    able at a point, 222
Function measurable in the Luzin
    sense, 92
Function totally discontinuous with
    respect to a class of topolo-
    gies, 154

Generalized Continuum Hypothe-
    sis (**GCH**), 53
Generalized Luzin set, 111
Generalized Sierpiński set, 166, 179

Hall's combinatorial theorem, 335
Hamel basis, 36
Hartogs ordinal number of a set, 331, 342
Hausdorff–Tarski theorem, 335
Height of a tree, 334
Henry's theorem, 102
Hereditarily Lindelöf topological space, 17
Hilbert's first problem, 305

Inconsistent elements in a partially ordered set, 356
Independent family of sets, 335
Inductive set, 323
Infinite game of two persons, 397
Infinite set, 319
Infinite version of Ramsey's theorem, 337
Infinity Axiom, 319
Injective commutative group, 280
Inner regular measure, 97
Intersection of sets, 321

Jordan measure, 149

König, J.'s inequality, 354
König's lemma, 334
Kulikov's theorem, 232
Kuratowski's lemma on closed projection, 220
Kuratowski–Ryll-Nardzewski theorem, 392, 393
Kurepa's theorem, 366

Lavrentiev's theorem on extensions of continuous functions, 165
of homeomorphisms, 165
Level of a tree, 334

Lindelöf topological space, 17
Luzin–Jankov–von Neumann theorem, 287, 395
Luzin set, 62
Luzin's separation principle, 314

Marczewski's method of extending σ-finite measures, 82
Marczewski's theorem, 367
Martin's Axiom (**MA**), 63, 356
Measurable selector, 52, 287
Measure space, 50
Member of a set, 317
Membership relation, 317
Method of transfinite
chains of half-open intervals, 33
recursion, 324
Metrical transitivity, 40
Mid-point convex function, 135

Negligible set in a group, 82
Nonseparable measure, 79
Nonstationary set, 176, 337
Normal extension of a measure, 299

One-dimensional unit torus, 178
Ordered pair, 322
Orientation preserving isometric transformation, 271
Oscillation of a function at a point, 152
Outer regular measure, 97

Partial homomorphism of commutative groups, 234
Partially ordered set, 108
Peano type curve, 16

Perfect
    mapping, 221
    probability space, 94
    subset property, 311
Perfectly meager set, 59
Polish topological space, 20
Power set of a set, 318
Principle of transfinite induction, 324
Probability space, 93
Projective hierarchy, 309
Projective set, 309

Quasi-compact topological space, 17
Quasi-cyclic group, 279

Radon measure, 87, 88
Radon–Nikodym operation, 142
Radon topological space, 88
Random variable, 94
Rank of a set, 332
Real character of a commutative group, 226
Real-valued measurable cardinal, 327
Regularity Axiom, 320
Regular probability measure, 160
Relatively measurable
    function, 68
    set, 68
Replacement Axiom, 319
Root of a tree, 334
Rubin, H.'s theorem, 352

Scattered topological space, 31
Second Continuum Hypothesis, 316
Selector of a family of sets, 2
Separable measure, 79
Separation Principle, 321

Set of type G_δ, 14
Set-theoretical equation, 333
Set-valued mapping, 391
Shelah's inequality, 66
Sierpiński's equalities, 403
Sierpiński's lemma on disjoint subsets, 117
Sierpiński's problem, 268
Sierpiński set, 54
Sierpiński–Zygmund function, 38
    in the strong sense, 212
Sierpiński–Zygmund type function, 169
    in the measure-theoretical sense, 219
Solovay algebra, 134
Solovay's forcing of almost disjoint sets, 370
Sorgenfrey line, 83
Sorgenfrey's topology, 83
Stationary set, 337
Steinhaus property, 36
Step function, 107
Strategy of a player, 397
Strong countable chain condition, 364
Strongly inaccessible cardinal, 327
Strong measure zero set, 193
Strong Steinhaus property, 278
Subset of a set, 318
Suslin condition for a measure, 50
Suslin line, 60, 347
Suslin number of a Boolean algebra, 206
Suslin set, 88
Suslin's theorem, 313

Tarski's theorem, 347
Topological equivalent of **MA**, 356
Torsion free commutative group, 230

Torsion subgroup of a commutative group, 230
Totally imperfect set, 47
Totally inconsistent set, 356
Transitive closure of a set, 329
Transitive set, 323
Translation invariant measure, 76
Translation quasi-invariant measure, 76
Tree, 334
Two-valued measurable cardinal, 31, 327

Ulam's theorem, 101
Ulam's transfinite matrix, 336
Union of all members of a set, 319
Universally Baire set, 65

Universal measure zero space, 73

Vitali set, 36, 174
Vitali's theorem, 110
Vitali type function, 149
von Neumann ordinal, 330
von Neumann universe, 320, 331

Weakly admissible group of transformations, 272
Weakly inaccessible cardinal, 327
Weakly measurable set-valued mapping, 391
Winning strategy, 397

Zermelo–Fraenkel set theory, 2